SYNTHETIC METALS
FOR NON-LINEAR OPTICS AND ELECTRONICS

EUROPEAN MATERIALS RESEARCH SOCIETY SYMPOSIA PROCEEDINGS

Volume 1: Ceramic Materials Research (ed. R.J. Brook)
Volume 2: Photon, Beam and Plasma Assisted Processing (eds. I.W. Boyd and E.F. Krimmel)
Volume 3: Deep Implants (eds. G.G. Bentini, A. Golanski and S. Kalbitzer)
Volume 4: Metastable Alloys: Preparation and Properties (eds. K. Samwer, M. von Allmen, J. Bøttiger and B. Stritzker)
Volume 5: Superconducting and Low-Temperature Particle Detectors (eds. G. Waysand and G. Chardin)
Volumes 6A, 6B: High T_c Superconductors (eds. P.F. Bongers, C. Schlenker and B. Stritzker)
Volume 7: Solid State Ionics (eds. M. Balkanski and C. Julien)
Volume 8: Rare-Earth Permanent Magnets (ed. I.R. Harris)
Volume 9: Defects in Silicon (eds. C.A.J. Ammerlaan, A. Chantre and P. Wagner)
Volumes 10A, 10B: Silicon Molecular Beam Epitaxy (eds. E. Kasper and E.H.C. Parker)
Volume 11: Acoustic, Thermal Wave and Optical Characterization of Materials (eds. G.M. Crean, M. Locatelli and J. McGilp)
Volume 12: Beam Processing and Laser Chemistry (eds. I.W. Boyd and E. Rimini)
Volume 13: Interfaces in Biomaterials Sciences (eds. D. Muster and G. Hastings)
Volumes 14A, 14B: High T_c Superconductor Materials (eds. H.-U. Habermeier, E. Kaldis and J. Schoenes)
Volume 15: Metal Matrix Composites (eds. G. Chadwick and L. Froyen)
Volume 16: Magnetic Thin Films, Multilayers and Superlattices (eds. A. Fert, G. Güntherodt, B. Heinrich, E.E. Marinero and M. Maurer)
Volume 17: Metallurgical Coatings and Materials Surface Modifications (eds. H. Hintermann and J. Spitz)
Volume 18: Surface Processing and Laser Assisted Chemistry (eds. I.W. Boyd, E. Fogarassy and M. Stuke)
Volume 19: Materials for Optoelectronic Devices, OEICs and Photonics (eds. H. Schlötterer, M. Quillec, P.D. Greene and M. Bertolotti)
Volume 20: SiC, Natural and Synthetic Diamond and Related Materials (eds. A.A. Gippius, R. Helbig and J.P.F. Sellschop)
Volume 21: Analytical Techniques for the Characterization of Compound Semiconductors (eds. G. Bastard and H. Oppolzer)
Volume 22: Micronic Integrated Sensors (ed. J.L. Robert)
Volume 23: High Energy and High Dose Ion Implantation (eds. S.U. Campisano, J. Gyulai, P.L.F. Hemment and J.A. Kilner)
Volume 24: Laser Surface Processing and Characterization (ed. I.W. Boyd)
Volume 25: Nuclear Methods in Semiconductor Physics (eds. G. Langouche, J.C. Soares and J.P. Stoquert)
Volume 26: Clays and Hydrosilicate Gels in Nuclear Fields (ed. A. Meunier)
Volume 27: Chemistry of Cements for Nuclear Applications (eds. P. Barret and F.P. Glasser)
Volume 28: Nuclear Materials for Fission Reactors (eds. Hj. Matzke and G. Schumacher)
Volume 29: Modifications Induced by Irradiation in Glasses (ed. P. Mazzoldi)
Volume 30: Polyconjugated Materials (ed. G. Zerbi)
Volume 31: SiGe Based Technologies (eds. E. Kasper, Y. Shiraki and T.P. Pearsall)
Volume 32: Materials Surface Processing (eds. M. Stuke, E.E. Marinero and I. Nishiyama)
Volume 33: Chemistry for Electronic Materials (eds. Y. Pauleau, G. Wahl, K.F. Jensen and T. Hirai)
Volume 34: Semiconductor Materials Analysis and Fabrication Process Control (eds. G. Crean, R. Stuck and J. Woollam)
Volume 35: Synthetic Materials for Non-Linear Optics and Electronics (eds. C. Taliani, Z.V. Vardeny and Y. Maruyama)
Volume 36: CdTe and Related Cd Rich Alloys (ed. R. Triboulet, W.R. Wilcox and O. Oda)
Volume 37: Single Chamber Processing (eds. Y.I. Nissim and A. Katz)

SYNTHETIC METALS FOR NON-LINEAR OPTICS AND ELECTRONICS

PROCEEDINGS OF SYMPOSIUM E ON
SYNTHETIC METALS FOR NON-LINEAR OPTICS AND ELECTRONICS
OF THE 1992 E-MRS SPRING CONFERENCE

STRASBOURG, FRANCE, JUNE 2–4, 1992

Edited by:

C. TALIANI
CNR, Bologna, Italy

Z. V. VARDENY
University of Utah, Salt Lake City, UT, USA

Y. MARUYAMA
Institute of Molecular Science, Okazaki, Japan

1993
NORTH-HOLLAND
AMSTERDAM - LONDON - NEW YORK - TOKYO

Printed on acid-free paper

ISBN: 0 444 89909 X

Published by:
North-Holland
Elsevier Science Publishers B.V.
Sara Burgerhartstraat 25
P.O. Box 211
1000 AE Amsterdam
The Netherlands

Reprinted from
SYNTHETIC METALS vol. 54

The manuscripts for the Proceedings were received by the Publisher: mid-August – mid-October 1992

Printed in the Netherlands

Contents

Elsevier Sequoia

4. Poster Session: Synthetic Materials for Molecular Electronics

5. Ferroelectricity and Ferromagnetism in Organics

6. Poster Session – Nonlinear Optics: Materials and Devices

7. Processing and Nanotechnologies

8. Non-Carbon Based Polymers

9. Devices

Preface

This volume presents the Proceedings of the Symposium on 'Synthetic Electroactive Materials' (SEM) in the context of the International Conference on Electronic Materials (ICAM) and the Spring European Material Research Society (E-MRS) Meeting held in Strasbourg, France, June 2–5, 1992.

The presence of this Symposium within the framework of ICAM indicates, I believe, that this field has grown from the level of a scientific curiosity to a mature field of materials science bringing important technological perspectives to electronic applications.

Two areas are particularly promising for the development of a new electronics based on SEM materials: (i) the integration of organic materials into planar silicon technology like, for instance, the development of 'all organic' field-effect transistors (FET) and the new organic light emitting diodes (LED); and (ii) the development of a totally new electronics in which photons, rather than electrons, carry the information and SEM materials act as switching devices. Both aspects and more are covered by this Symposium.

Particular emphasis has been given to the study of oligomers as model compounds and promising SEM materials. In this field we observe the merging of two scientific cultures: on the one hand, the consolidated molecular crystal science which grew in the late fifties from the early works of A. S. Davydov, D. P. Craig and M. Pope, and, on the other hand, the vast area of science which expanded tremendously from the seventies, stemming from the discovery of conductivity in doped polyacetylene. Scientific achievements from both areas are essential to understand the properties of oligomers and this Symposium has been, in this respect, a fruitful meeting.

The recent discovery of high temperature superconductivity (HTSC) in doped fullerene C_{60} opens a vast new field of research and gives evidence of the great potential that doped molecular crystals may have, both in the understanding of the HTSC mechanism and in the development of new HTSC materials.

More than one hundred scientists have taken part in the Symposium giving 39 oral presentations, of which 12 were invited, and about 50 posters. This volume contains 52 papers covering areas like generalities of conjugated polymers and oligomers, photoexcitations and fast phenomena in conjugated systems, nonlinear optical properties of organics, MX chain systems, processing and nanotechnology, and devices.

I wish to express appreciation to Y. Maruyama and Z. V. Vardeny, who acted as co-chairmen of this Symposium and contributed to make it truly international, and to members of the International Scientific Committee for the fruitful cooperation and advice in planning the programme. Special acknowledgement is given to Mrs M. Adloff, Mrs M. Cobut and the staff of

E-MRS for organizational support, and to Mrs B. Gualandi of ISM-CNR for her tremendous effort in handling the secretarial work.

December 1992 Carlo Taliani

Sponsors

This Conference was held under the auspices of:

The Council of Europe
The Commission of the European Communities

It is our pleasure to acknowledge with gratitude the financial assistance provided by

Banque Populaire	(France)
Centre de Recherches Nucléaires	(France)
Centre National de la Recherche Scientifique	(France)
Elsevier Science Publishers B.V.	(The Netherlands)
Office du Tourisme de la Ville de Strasbourg	(France)
The Commission of the European Communities	
The Council of Europe	
The European Parliament	
Ville de Strasbourg	(France)

and with respect to Symposium E:

Xerox Corporation	(USA)
Perkin Elmer S.A.	(France)

1. Generalities on Synthetic Materials for Electronics and Photonics

Molecular crystals: past and present

A. Bree
Chemistry Department, University of British Columbia, Vancouver, BC, V6T 1Z1 (Canada)

Abstract

Five different types of binding in crystals are generally recognized. These are (with common examples chosen from organic materials given, in some cases, in parentheses): molecular (anthracene, C_{60}), ionic (K_3C_{60}, perhaps some charge-transfer complexes), covalent (diamond?), metallic (the choice of organic here depends on the definition of a metal) and amorphous (polymers, for example, polyacetylene or polythiophene might be of current interest). This paper will focus on molecular crystals, in particular describing the developments over the past 40 years in linear and nonlinear optical properties and in electrical behaviour. Studies in two areas of molecular crystals were given a great impetus just before 1950. Davydov (*Zh. Eksp. Theor. Fiz.*, *18* (1948) 210) extended the earlier work of Frenkel and Peierls, presenting a clear and quantitative model for exciton behaviour, while Akamatu and Inokuchi (*J. Chem. Phys.*, *18* (1950) 810) as well as Vartanian (*Zh. Fiz. Khim.*, *22* (1948) 769) began an investigation of the electrical conductance and photoconductance in these samples. Where these early ideas have led will be the substance of this paper.

Introduction

This brief review of the early work on molecular crystals is directed towards materials scientists. The focus will be on organic crystals and the emphasis will be to relate the earlier work to current materials. As the title of the symposium suggests, the prime concern will be with nonlinear optics and electrical conductivity. There have been a number of books published which review some of the properties of molecular crystals in more depth than is possible here [1–4]. In what is necessarily a short paper, one can only provide a brief account of the considerable achievements that have been made; the topics selected reflect no more than a personal choice.

Strictly when speaking of nonlinear processes, one should specify a particular process, the power level and the wavelength region to be used. However, all susceptibilities (and molecular hyperpolarizabilities) ultimately depend on the energies of the states of the system and the strengths of the transitions between them [5, 6]. For this reason, before we discuss any nonlinear optical property, it is useful to consider what interactions in a molecular crystal lead to the existence of a Frenkel exciton band [7].

The usual model for a molecular crystal is based on the notion that the forces which act between atoms in one molecule are much greater

0379-6779/93/$6.00

than those which act between molecules. In this spirit, we can write the crystal Hamiltonian as

$$\mathcal{H}_{\text{crystal}} = \sum_{k=1}^{N} \mathcal{H}_k + \sum_{k=1}^{N} \sum_{l>k}^{N} V_{kl} \tag{1}$$

in which $V_{kl} \ll \mathcal{H}_k$. In this expression, $\mathcal{H}_k$ is the Hamiltonian for molecule k in isolation and V_{kl} describes the interaction between molecules k and l in the crystal. Perturbation theory is an appropriate approximation for this problem and we see that the crystal properties should (nearly) be the sum of the molecular properties. Product functions of the form

$$\phi^{\mathrm{g}} = \prod_j \varphi_j \tag{2}$$

and

$$\phi_i^r = \varphi_i^r \prod_{j \neq i} \varphi_j \tag{3}$$

being eigenfunctions of the zeroth-order problem are chosen as basis functions. Here, φ_i^r refers to the wave function of an isolated molecule in its rth excited state occupying site i in the lattice; ϕ^{g} is the wave function for the non-degenerate ground state in the crystal.

It is the term V_{kl} that lifts the N-fold degeneracy of the zero-order problem for an excited state of the crystal and gives rise to the exciton band. It becomes important to examine the nature of the interaction term $\langle \varphi_k \varphi_l^r | V_{kl} | \varphi_k^r \varphi_l \rangle$ in one of several different interesting circumstances:

(i) When the transition from the ground to the excited electronic state is allowed, the interaction can be approximated to that between two transition dipole moments centred on the two interacting molecules and leads to exciton bandwidths that fall typically in the range 100–10 000 cm^{-1}.

(ii) For a symmetry-forbidden, spin-allowed transition, an extension to the above approximation is needed. The expansion of the term $1/r$ from the interaction V_{kl} must now extend beyond the simple dipole–dipole limit. When the transition dipole is small, the addition of higher multipole interactions is necessary to bring the calculations in closer agreement with the observed splittings and bandwidths [2] of about 10–100 cm^{-1}.

(iii) In the treatment of triplet excitons, intermolecular electron exchange terms become dominant. For organic molecules, spin–orbit coupling is weak. In particular, for planar aromatic hydrocarbons, the coupling is especially weak since all one- and two-centre integrals that appear in the spin–orbit coupling matrix vanish; in fact, the radiative lifetime increases from about 10^{-9} s for an allowed transition to about 30 s for one that is spin-forbidden. For heterocyclic and halogen-containing compounds (and, perhaps, for a non-planar molecule like C_{60}), the effect of spin–orbit coupling in Coulomb integrals may need to be included (as well as in the exchange terms, which do not vanish even when spin–orbit coupling is zero). The bandwidth in this case is about 10 cm^{-1}.

(iv) The change in a vibrational transition in going from the vapour to the solid is described in terms of the dependence of the sum of atom–atom interactions on adjacent molecules as the molecules undergo their vibrational motions. The assumed form of the interaction models dispersion attraction at large r and exchange repulsion at small r. Observed bandwidths are of the order of 1 cm^{-1}.

All the types of interactions listed above give a finite contribution to the bandwidth, but it is assumed for the convenience of calculation that one type of interaction will dominate. All modelling of observed phenomena follows the dictum that some part of the complete Hamiltonian may be neglected to make the modelling tractable.

Nonlinear optics

For a solid, the *induced polarization density is*

$$P - P_0 = \chi^{(1)} \cdot E + \chi^{(2)} \cdots EE + \chi^{(3)} \cdots EEE + \cdots \tag{4}$$

where E is the electric field (usually) associated with a radiation field. The $\chi^{(i)}$ represent material properties that we would like to understand and control. On the other hand, for a molecule we use a similar expression for the *induced dipole moment* as

$$\mu - \mu_0 = \alpha^{(1)} \cdot F + \alpha^{(2)} \cdots FF + \alpha^{(3)} \cdots FFF + \cdots \tag{5}$$

where F is a local electric field; for example, if the molecule is in an environment where the refractive index differs from unity, a local field correction should be applied. $\alpha^{(2)}$, $\alpha^{(3)}$, etc., are molecular hyperpolarizabilities.

The hyperpolarizabilities (and the susceptibilities of the solid) involve a sum-over-states, e.g., for a two-photon absorption in a molecule the appropriate tensor has the form

$$\alpha_{xy}(\text{TPA}) = \frac{1}{2}\sum_i \left[\frac{\langle f|x|i\rangle\langle i|y|g\rangle}{E_i - h\nu} + \frac{\langle f|y|i\rangle\langle i|x|g\rangle}{E_i - h\nu} \right] \tag{6}$$

where the sum is over a complete set of intermediate states $|i\rangle$. It is this assumed knowledge of all states that makes the calculation of the various $\alpha^{(i)}$ values quite difficult. The situation is made more complicated by the fact that there is little experimental data for the absolute cross sections for the various nonlinear processes and so it is difficult to judge how accurate any calculated values may be. The calculation of the $\chi^{(i)}$ values is even more of a problem since different molecular electronic states mix in the crystal through the intermolecular interaction operator V_{kl}.

One of the big hopes in the use of organics for nonlinear optical purposes was to have available the chance of 'designing' new materials, i.e., if it were possible to devise a new molecular solid using some a priori basic principles, then one could set about its synthesis by means available to the organic chemist. If the use of calculation is to be regarded as a suspect design tool (or, at least, one that requires some sort of confirmation), how else can we set about devising a suitable molecule for our purposes? This question was recently addressed by Meredith [8] and Sherwood [9].

To enhance the $\alpha^{(i)}$ values, inspection of eqn. (6) (chosen as an example) indicates that we should choose chromophores with intense transitions at low energy. This is really a restatement of Miller's rule [10] that says that a material will have a strong nonlinear response if its linear response is strong. Meredith [8] has listed other useful strategies. For example, it might be helpful to extend the range of resonance within a molecule; this would have the effect of lowering the energies of the molecular states. If there were more than one chromophore associated with the molecule, then the hyperpolarizability for the collection could be approximated by the tensor sum of the individual tensors of each chromophore. Finally, the two-level system approximation was considered [8, 9]. This approximation is best fitted to charge-transfer compounds where donor and acceptor substituents are separated by some conjugated framework. A prototype is 4-nitroaniline. The belief here is that the intense, low-energy, charge-transfer transition will be dominant in the sum-over-states.

An extension of the 'design' strategy is that once a useful organic material has been produced, it may be refined by making systematic small changes in the molecular structure. One flaw in this idea is that small changes in molecular structure can give rise to large changes in the crystal space group. Where crystal symmetry is important, this is a serious impediment. However, it should be noted at this point that two molecules with exceptionally large hyperpolarizabilities were specifically designed to produce just this effect: DCNP (3-(1,1-dicyanoethenyl)-phenyl-1,5-dihydro-1H-pyrazole) was designed from theoretical considerations by Morley and Pugh [11] and NPP (*N*-(4-nitrophenyl)-(L)-prolinol) was devised by Zyss *et al.* [12].

There are several problems commonly associated with the use of organics as nonlinear elements, especially as harmonic generators. Melt-grown samples, a common technique for obtaining single crystals of large dimensions, have a low optical quality because of inherent crystal strains and faults caused by their growth through a large temperature gradient; solution-grown crystals (grown slowly at a constant temperature) generally have a higher optical quality [9]. In some applications they have been found to have a low resistance to damage from high power lasers; this is important in frequency doubling where the efficiency of the process increases with pump power. On the other hand, Sherwood [9] reports a few organic materials that can withstand very high powers indeed (up to 500 MW/cm^2); this point has also been made by Chemla and Zyss [13]. The following experimental result may expose one potential problem.

We had found [14] that a crystal of the heterocyclic compound, benzimidazole, acted as an efficient harmonic generator for visible light.

Benzimidazole has four polar molecules packed in an orthorhombic unit cell. The crystals cleave with remarkable facility along the (100) plane and it was normal to this plane that doubling of light with wavelength near 650 nm was observed. However, the dispersion of the refractive index showed that an (010) section of benzimidazole should be near the phase-matching condition at 1064 nm and so should allow doubling of the output from the Nd/YAG laser. We tried the experiment last year. A melt-grown crystal, about 9 mm thick, was carefully aligned in the low-power beam and then the power was turned up to maximum (700 mJ in a 8 ns pulse covering a doughnut-shaped area of about 0.5 cm^2). In the minute that it took to decide that no doubled green light could be seen, the previously transparent benzimidazole crystal had darkened uniformly throughout its length and numerous cracks had developed, presumably due to overheating. The power in the tightly focused visible light (from a Nd/YAG pumped dye laser) was about the same (or, perhaps, even more powerful — it was difficult to determine the beamwaist accurately) as for the unfocused infrared beam at 1064 nm. A rough calculation showed that the power in either beam was at least one order of magnitude less than that required for breakdown (i.e., to break bonds or to cause ionization). Yet the infrared beam damaged the crystal while the visible light did not. The infrared beam must have been partially absorbed by the sample. What caused this absorption? The only reasonable explanation is to suppose that the excitation frequency had the bad luck to catch the upper edge of the vibrational transition $v = 0 \rightarrow v = 3$ of the CH stretching mode. CH stretches are highly anharmonic and, at this energy, a change to a local mode is probably underway [15]; in this case, the absorption profile is probably irregular and may involve a quantum of the NH stretch as well. Of course, local mode absorption into states with many quanta of the CH stretching mode are possible and these transitions run into the visible. However, the absorption strength falls off logarithmically with the vibrational quantum number [16] so that absorption in the visible, even if it fell near the $v = 5$ or the $v = 6$ level, would be very weak.

Other weak absorption can occur in the visible. Absorption directly into the lowest energy triplet electronic state has been observed in organic crystals about 1 cm thick but, in this case, the absorption is so weak that it could be observed only by carefully cooling the crystal to 4.2 K to sharpen the bands [17]. At room temperature, this is unlikely to be a cause for concern. The point to be made here, though, is that states normally considered to be 'dark' states may give rise to very weak absorption and be a cause of optical damage at high power. This type of damage may be avoided by a careful selection of the wavelength range to be used.

Electrical conductivity

Molecular crystals are, generally speaking, fine insulators. Anthracene, taken as a typical example, has an intrinsic conductivity of

$\approx 10^{-22}\ \mathrm{ohm^{-1}\ cm^{-1}}$, about the same as fused silica. However, anthracene is a semiconductor, its conductivity increasing exponentially with temperature.

A brief historical survey of the studies of the electrical conductivity in anthracene is helpful in that it illustrates how our understanding of the underlying mechanisms involved have developed. The story begins in 1948 with the study of some phthalocyanines by Vartanian in Russia [18] and Eley in the UK [19] and was followed soon after by a paper published by Akamatu and Inokuchi [20] dealing with the conductivity of some dye materials. The 1950s focused on the measurement of photocurrents in anthracene, which gave a linear response to the incident light intensity over many decades. The photocurrent response curve mimicked the shape of the absorption spectrum, which led to the idea that a surface current was being observed and an estimate of the 'effective surface depth' of about 50 layers was made [21]. Since the surface was implicated, Bree and Lyons [22] studied the effect of removing the air from around the sample. The photocurrent was reduced almost to zero if the oxygen was removed; nitrogen was not effective in producing carriers. Indeed, the photocurrent was shown to vary with oxygen pressure in a way that reproduced a Langmuir adsorption isotherm. By using guard-ring electrodes connected to ground, it became possible to remove the surface current and meaure a residual bulk effect [23]. By the end of the decade attention was turning to the bulk (intrinsic) photocurrent.

In the early 1960s, a variety of novel electrode materials were tried, some acting as blocking electrodes and some providing for charge injection; a full account of this is given by Pope and Swenberg [3]. With at least one ohmic contact the current through the crystal is space-charge-limited and the theory to describe this was applied to organics for the first time by Mark and Helfrich [24]. At the same time, experiments were underway to measure drift mobilities [25] calculated from the transit time through the crystal by a plane of charge carriers generated at one surface by a fast pulse. Drift mobilities were found to be low ($\mu \approx 1\ \mathrm{cm^2\ s^{-1}\ V^{-1}}$ although they were anisotropic, see ref. 26) compared with the more standard inorganic materials like Ge and Si. These low mobilities led to a discussion of whether this conductance should be described by a band model or a hopping model. The band model is appropriate for wide-band materials where $\mu \gg 1\ \mathrm{cm^2\ s^{-1}\ V^{-1}}$ and the mean free path covers many lattice spacings. On the other hand, the hopping model applies to a narrow-band substance with $\mu \ll 1\ \mathrm{cm^2\ s^{-1}\ V^{-1}}$ and the carrier is scattered at every lattice site. If the carrier spends a time at a lattice site that is long compared with the lattice relaxation time, then the lattice may deform about the charged site to form a polaron. In most molecular crystals only nearest neighbours are involved in the deformation which is known as a small polaron. Typical organic molecular crystals generally fall in an intermediate category so that, in fact, the band model was applied at low temperature (below about 100 K) while the hopping model seemed more appropriate at higher temperatures.

Intrinsic carrier photogeneration is associated with charge-transfer (CT) states, where a promoted electron is transferred to the nearest or next-nearest neighbouring molecular site but remains correlated with its parent hole. Because of the anisotropy of these crystals, CT states are distorted, tightly bound Wannier excitons. The non-conducting ion-pair states were first discussed in organic crystals by Lyons [27] who predicted their energies to lie within the more intense Frenkel exciton bands. Thermal, or field-assisted, dissociation of these germinate pairs produces charge carriers according to an earlier theory of Onsager [28]. The dielectric constant of an organic crystal is smaller than for inorganic materials and the attraction between an electron and hole is accordingly strong. The quantum efficiency for carrier photogeneration is small, geminate recombination being the dominant process following light absorption. An excellent review of this work, to 1980, has been given by Silinsh [29], one of its principal participants.

Electro-reflection [30] and electro-absorption [31] experiments, that detect a modulated response to an applied alternating electric field, are important techniques that allow experimental access to CT states since it permits the weak transitions to CT states to be more clearly observed even in the presence of much stronger molecular absorption. The interpretation of these results has proven to be something of an on-going experience and the exact mechanism for this dissociation is still under active investigation [32].

The situation can be even more complicated because real crystals possess more states than described so far, due to the presence of chemical impurities, lattice defects and surface states, all of which can provide a local modification of the true bulk states. While optical absorption will not usually be so sensitive to these extra states, electrical conductance probably will since they may act to trap carriers or to scatter them. A more complete discussion of many of these processes is given by Pope and Swenberg [3].

Finally, it is worth pointing out that a knowledge of the CT states in a molecular crystal is important also because they influence the energies in the Frenkel exciton manifold (through off-diagonal coupling) especially for very weak molecular transitions [33].

References

1 F. Gutmann and L. E. Lyons, *Organic Semiconductors*, Wiley, New York, 1967.
2 D. P. Craig and S. H. Walmsley, *Excitons in Molecular Crystals*, Benjamin, New York, 1968.
3 M. Pope and C. E. Swenberg, *Electronic Processes in Organic Crystals*, Oxford University Press, London, 1982.

4 J. D. Wright, *Molecular Crystals*, Cambridge University Press, Cambridge, 1989.
5 N. Bloembergen, *Nonlinear Optics*, Benjamin, New York, 1965.
6 D. P. Craig and T. Thirunamachandran, *Molecular Quantum Electrodynamics*, Academic Press, Orlando, FL, 1984.
7 J. Frenkel, *Phys. Rev.*, *37* (1931) 17 and 1276.
8 G. R. Meredith, in C. Flytzanis and J. L. Oudar (eds.), *Nonlinear Optics: Materials and Devices*, Springer, Berlin, 1986.
9 J. N. Sherwood, *Philos. Trans. R. Soc. London, Ser. A*, *330* (1990) 127.
10 R. C. Miller, *Appl. Phys. Lett.*, *5* (1964) 17.
11 J. O. Morley and D. Pugh, in R. A. Hahn and D. Bloor (eds.), *Organic Materials for Nonlinear Optics*, Royal Society of Chemistry, London, 1989.
12 J. Zyss, J. Nicoud and M. Coquillary, *J. Chim. Phys.*, *81* (1984) 4160.
13 D. S. Chemla and J. Zyss, in D. S. Chemla and J. Zyss (eds.), *Nonlinear Optical Properties of Organic Molecules and Crystals*, Vol. 1, Academic Press, Orlando, FL, 1987.
14 A. Bree and R. Zwarich, *J. Chem. Phys.*, *90* (1989) 5901.
15 B. R. Henry and W. Siebrand, *J. Chem. Phys.*, *49* (1968) 5369.
16 M. S. Burberry and A. C. Albrecht, *J. Chem. Phys.*, *70* (1979) 147.
17 D. M. Hanson and G. W. Robinson, *J. Chem. Phys.*, *43* (1965) 4174; R. H. Clarke and R. M. Hochstrasser, *J. Chem. Phys.*, *47* (1967) 1915.
18 A. T. Vartanian, *Zh. Fiz. Khim.*, *22* (1948) 769.
19 D. D. Eley, *Nature (London)*, *162* (1948) 819.
20 H. Akamatu and H. Inokuchi, *J. Chem. Phys.*, *18* (1950) 810.
21 L. E. Lyons, *J. Chem. Phys.*, *23* (1955) 220.
22 A. Bree and L. E. Lyons, *J. Chem. Phys.*, *25* (1956) 384 and 1284.
23 D. M. J. Compton, W. G. Schneider and T. C. Waddington, *J. Chem. Phys.*, *27* (1957) 160.
24 P. Mark and W. Helfrich, *J. Chem. Phys.*, *33* (1962) 205.
25 R. G. Kepler, *Phys. Rev.*, *119* (1960) 1226; O. H. Le Blanc, *J. Chem. Phys.*, *33* (1960) 626.
26 L. B. Schein, *Phys. Rev. B*, *15* (1977) 1024.
27 L. E. Lyons, *J. Chem. Soc.*, (1957) 5001; *Aust. J. Chem.*, *10* (1957) 365.
28 L. Onsager, *Phys. Rev.*, *54* (1938) 554.
29 E. A. Silinsh, *Organic Molecular Crystals: Their Electronic States*, Springer, Berlin, 1980.
30 S. C. Abbi and D. M. Hanson, *J. Chem. Phys.*, *60* (1974) 319.
31 L. Sebastian, G. Weiser and H. Bassler, *Chem. Phys.*, *61* (1981) 125; L. Sebastian, G. Weiser, G. Peter and H. Bassler, *Chem. Phys.*, *75* (1983) 103.
32 P. Petelenz and D. Mucha, *Chem. Phys.*, *154* (1991) 145.
33 R. Silbey, J. Jortner, M. T. Vala and S. A. Rice, *J. Chem. Phys.*, *42* (1965) 2948; P. Petelenz and A. Eilmes, *Chem. Phys.*, *155* (1991) 247.

Photoconduction and photorefraction in molecularly doped polymers

J. Campbell Scott*, L. Th. Pautmeier and W. E. Moerner
IBM Research Division, IBM Almaden Research Center, 650 Harry Road, San Jose, CA 95120-6099 (USA)

Abstract

During the past 20 years organic photoconductors have found increasing application in electrophotographic copying and printing. Charge carrier mobility in these materials is achieved by doping a polymer matrix with up to 50% by weight of a molecular donor (acceptor) which provides sites for the hopping of holes (electrons). Sensitization to the desired wavelength of light is obtained via a suitable charge generation dye or pigment, frequently in the form of a second layer distinct from the transport layer. More recently, optically nonlinear polymers have been synthesized by the incorporation of conjugated donor–acceptor species into macromolecular chains and/or networks. When the donor–acceptor chromophores are aligned, for example, by poling in an electric field, second-order optical nonlinearity, including an electro-optic effect, is observed. Photoconductivity and nonlinearity can be combined, for example, by doping a nonlinear polymer with a molecular charge transport agent, to yield a potentially photorefractive material. The photorefractive effect in several such formulations has been confirmed and characterized by a combination of absorption, photoconductivity, mobility, electro-optic, four-wave mixing (holography) and grating phase measurements. The results are discussed in terms of a mechanism involving (i) charge generation on the nonlinear chromophore, (ii) transport of holes by hopping among molecular donor sites, (iii) charge trapping to develop a space-charge electric field and (iv) modulation of the refractive index via the electro-optic effect.

1. Introduction

Molecularly doped polymers (MDPs) have been commercialized as *electronic materials* for more than 20 years [1]. They are used as the photosensitive element in electrophotographic copiers and printers, and since their introduction in 1970 have come to dominate this billion dollar market, largely replacing amorphous selenium and its alloys. Their success depends on the exploitation of several unique electrical and optical properties, and on the development of the physics, chemistry and materials science necessary to optimize those properties. In this paper, we summarize the phenomena which make MDPs useful as photoreceptors and describe current efforts to extend their utility to applications, such as

*Author to whom correspondence should be addressed.

0379-6779/93/$6.00

holographic storage and optical switching, which require a photorefractive response [2].

The photoconductor in electrophotography is coated onto a conducting ground plane, usually an aluminum drum or a belt of aluminized Mylar™. The top surface must accept and retain, in the dark, charge from a corona such that its voltage approaches 1 kV in magnitude. When it is illuminated by imaging the document to be copied or by scanning with a laser it must efficiently discharge to near zero volts. The discharge process involves, first, the generation of electron–hole pairs. In multilayered photoconductors, this is accomplished in a layer of organic pigment, the charge generation layer, location adjacent to the ground plane. (There may be a thin intervening polymer layer to assist adhesion and/or to limit dark charge injection from the metal into the organic materials.) Next, the appropriate sign of charge carrier travels through the overlying layer, the transport layer, to neutralize the surface charge. Efficient charge transport through the transport layer requires sufficiently high mobility and a low density of traps. In Section 2, we outline relevant features of charge generation, transport and trapping as they are currently understood for organic materials.

The photorefractive effect [3] also requires photoconduction in order to generate and redistribute charges, leading to an inhomogeneous space-charge field in the bulk of the sample. In addition, the material must exhibit an electro-optic effect so that the space-charge field modulates the index of refraction. In the MDPs introduced in Section 3, the electro-optic response is achieved by including in the polymer chromophores with high second-order non-linearities (β coefficient) [4]. These chromophores are aligned (or poled) by applying a sufficiently high external electric field. The photoconductive properties of the photorefractive formulations are summarized in Section 4 and the optical properties, including characterization of holographic grating formation, are given in Section 5.

Since the majority of the work reported here is already published or submitted for publication, this paper is in the form of a review and tutorial.

2. Organic photoconductors for electrophotography

In general the transport layer in an electrophotographic photoconductor conducts only one sign of charge, usually holes. The formulation which achieves this property consists of a polymer binder (e.g. polycarbonate or polyester) molecularly doped with a compound, the transport molecule, which has a low oxidation potential and a stable cation. Charge transport then occurs by 'hopping' of the hole from molecule to molecule.

Charge carrier mobility in molecularly doped polymers has been thoroughly studied experimentally for two decades [5, 6]. Even so, some details of the mechanisms are not yet entirely understood [6, 7]. The method of choice to measure the mobility is the transient time-of-flight technique [8]. The sample is prepared as a layer of order 10 μm thick between two electrodes, one of which is (semi-)transparent. A voltage applied between the electrodes produces a uniform electric field across the sample. Electron–hole pairs are generated by a short light pulse, incident through one electrode and strongly absorbed, say within 0.1 μm. Carriers of one sign, determined by the orientation of the applied field, propagate across the sample and generate a current which is measured in the external circuit. As the carriers reach the second electrode they cease to contribute to the current which therefore falls to zero. The mean mobility of the carriers is then given by the ratio of their mean velocity and the applied field:

$$\mu = \langle v \rangle / E = (\mathrm{L}/E) \langle t_{\mathrm{a}}^{-1} \rangle$$

where E is the applied field and t_{a} is the arrival time of a given carrier. Note that the mean *inverse* arrival time enters the expression for mobility [9].

Although many details of the behavior of mobilities in MDPs are still a matter of debate, it is generally agreed that the magnitude of the mobilities (from below 10^{-8} to as high as 10^{-3} $\mathrm{cm}^2/(\mathrm{V\,s})$ at room temperature), their thermal activation and strong concentration dependence indicate hopping transport; i.e., the carriers reside on each molecular site for times long compared to the time required to transfer from one site to another. Particularly puzzling is the electric field dependence of the mobility which in most cases appropriates the form $\ln \mu \sim E^{1/2}$ [9–11], reminiscent of Poole–Frenkel release from charged traps [12]. However, there are considerable problems with a simple Poole–Frenkel interpretation of the data, especially the fact that the slope of $\ln \mu$ versus $E^{1/2}$ reverse sign at high temperature in some materials [13].

There is a growing consensus that mobility in molecularly doped polymers is governed to a considerable extent by the random electric potential resulting from molecular dipoles in the solid solution [14–16]. The dipoles reside both on the polymer binder, for example, the carbonate group in polycarbonate, and on neighboring charge transport molecules.

In organic photoconductors, traps are extremely scarce and carrier ranges regularly exceed the thickness of the charge transport layer which is typically 20 μm. This large value frequently surprises those who are familiar with crystalline inorganic semiconductors where impurity levels are pushed to part-per-billion range in order to reduce the trap limitation on mobility. The purity of commercial polymers is more typically in the percent range. The apparent discrepancy is easily resolved by recalling that hole transport molecules are selected to have an extremely low oxidation potential. Only impurities with a yet lower oxidation potential can act as traps and, among all possible impurities, a vanishingly small number are so readily ionized as the transport molecule. Furthermore, molecules which are easily ionized may not be stable as cations and therefore react or decompose to form benign impurities.

Charge generation in multilayered electrophotographic photoconductors is achieved in an organic pigment coated as a thin layer (typically 0.1 μm thick) under the transport layer [1]. Charge generation efficiency is conveniently measured by the xerographic discharge technique [17, 18]. In this method the sample is coated onto a conducting ground plane and the top surface charged with a corona. The surface potential is measured using a non-contacting electrometer. The change in potential (ΔV) resulting from a pulse of light exposure is measured. If all carriers generated reach the surface of the sample without trapping, then the quantum efficiency for generation is simply given by the ratio of carriers generated to photons absorbed:

$$\phi = \frac{C\Delta V}{(It_p/h\nu)(1 - \exp(-\alpha L))}$$

where C is the sample capacitance per unit area, I is the light intensity at photon energy $h\nu$, t_p is the temporal length of the light pulse, α is the absorption coefficient and L is the thickness of the sample. If the carrier range is not long compared to the sample thickness this formula must be modified by a factor involving the ratio $(L/\mu\tau E)$ and it is necessary to measure the thickness dependence of the voltage change in order to determine the efficiency.

The generation process is understood in terms of the dissociation of a molecular excitation [18, 19]. In a single-particle picture, the absorbed light quantum excites an electron from the highest occupied molecular orbital (HOMO) to the lowest unoccupied molecular orbital (LUMO). Then, either the electron in the LUMO or the hole left in the HOMO is transferred to a neighboring molecule. In the case of hole transfer, an energetically favorable pathway is established when the HOMO of the transport molecule lies at a higher energy than the HOMO of the generator. The transfer process is also energetically enhanced by application of an electric field, a mechanism which can be quantified in terms of a modified Onsager process [20]. In contrast to homogeneous organic semiconductors [21], the species which ultimately accepts the charge is the transport molecule in the polymer matrix. Therefore, the mechanism in at least some aggregate or multilayer photoconductors involves migration of the excitation to the particle surface before dissociation [22]. Hence a solid-state structure which permits exciton diffusion favors efficient generation. Quantum efficiencies approaching unity have been achieved in a multilayer organic photoconductor [18].

3. Formulation of photorefractive polymers

The properties of organic photoconductors make them ideal candidates as the basis of polymeric photorefractives. The generation, transport and trapping processes are reasonably well characterized, if not fully understood. What is missing is an electro-optic (EO) response. In the formulations described in this section, the EO response is achieved by incorporating suitable nonlinear chromophores into the structure of the binder polymer itself.

High second-order nonlinearity is found in asymmetrically substituted conjugated molecules, e.g., aromatics or olefins with an electron donor (D) at one end and an acceptor (A) at the other [4]. *para*-Nitroaniline is prototypical of such a D-π-A molecule [23].

To date we have examined polymers which incorporate the nonlinear optical (NLO) chromophore into two distinctly different polymer chemistries: epoxies and acrylics. In the epoxies, multifunctional amines such as 4-nitro-1,2-phenylenediamine (NPDA) [24, 25] or *N*-(2-aminophenyl)-4-nitroaniline (NAN) [26] are employed. The amine groups serve double duty: as the donor of the D-π-A structure and as the crosslinker for epoxy polymerization. Epoxides which we have examined include bisphenol-A-diglycidylether (bisA) and *N*,*N*-diglycidyl-4-nitroaniline (NNDN) [26] which also contains a nonlinear chromophore. In the acrylics, the NLO chromophore, e.g., nitroaminotolane (NAT), or *para*-nitroaniline (PNA) [27–29] is attached as a pendant side group to the polyacrylate. The pendant side-group approach to polymer synthesis has also been taken by groups at Kodak [30], Lowell [31] and Arizona [32], all of whom report photoconductivity and electro-optic response.

These polymers then serve as binders for the transport molecule. As discussed below, the NLO chromophore can act as a charge generation agent and, therefore, it is not absolutely necessary to include another sensitizer in the formulation. However, the NLO species which we have characterized to date are rather inefficient generators, with values of ϕ ranging from less than 10^{-3} for fields of order 10^5 V/cm, to about 10^{-2} at 6×10^5 V/cm [33]. Therefore, enhancement may be obtained by the addition of a sensitizer, such as an organic pigment [30] or electron acceptor which forms a charge transfer complex with the transport donor [34].

Several ancillary considerations enter the choice of compounds to use in the formulation. The diffraction efficiency for holographic gratings is proportional to the square of the sample thickness which should therefore be maximized (we typically use several hundred microns). Then the absorption coefficient should be adjusted (or the wavelength of measurement selected) such that the optical density of the sample is of order unity, a value which optimizes the trade-off between photogeneration and the absorption of the diffracted beam. In practice, if a sensitizer is used, this condition limits its concentration to less than 1%.

Charge carrier mobility is a strong function of transport agent concentration. Therefore, it is important to have the maximum level consistent with polymer mechanical properties and with the competing requirement of high NLO chromophore concentration. Good mobility can be obtained with 30% or more transport agent. At concentrations above about 40 or 50% crystallization may occur, even for relatively compatible polymer dopant systems. In the formulations which we have found to give the fastest and greatest photorefractive response, we have used 30% of various aminobenzaldehyde-hydrazone derivatives, for example, diethylaminobenzaldehyde diphenylhydrazone (DEH) [23, 35, 36].

To date, we have not deliberately included trapping species in the formulation, but have relied on naturally occurring concentrations. Photoconductivity and photorefraction data (see below) reveal that this may not be the ideal situation, since the concentration is lower than the optimum by about two orders of magnitude. It would be beneficial to add a lower oxidation potential compound at the level of 10–100 ppm.

It must be emphasized that photoconductivity and electro-optical response are merely necessary conditions. Their presence does not guarantee that a given formulation will be photorefractive. Experimental proof requires the observation of phase-shifted gratings, as discussed below.

4. Photoconductivity of photorefractive polymers

The results of an extensive study of the photoconductivity of the epoxy polymer bisA-NPDA doped with 30% by weight of DEH have recently been submitted for publication [33]. Results for PMMA-PNA:DEH were included in a second paper [29]. It was shown that bisA-NPDA without dopant has a photoconductivity at least 10^5 times less than when doped. The DEH dopant imparts a photoconductivity ranging from 10 pS cm/W at 650 nm to 10 nS cm/W at 400 nm. The action spectrum follows the absorption of NPDA indicating that the generation process commences with excitation of the NLO chromophore. In PMMA-PNA:DEH there is evidence for absorption due to the formation of a charge transfer complex which enhances the generation efficiency in the red [27, 28]. Holes are the dominant carrier, with a mobility of 10^{-8} $cm^2/(V\ s)$ at 10^5 V/cm in bisA-NPDA:DEH.

The carrier range in these materials is rather difficult to estimate accurately. Time-of-flight transients [33] show a shoulder which indicates that a substantial fraction of the carriers reach the collecting electrode without deep trapping. On the other hand, the transient current decreases significantly at short times, implying that the carrier population is equilibrating with a distribution of traps. The electric field dependence of the mobility deviates markedly from $\ln \mu \sim E^{1/2}$ behavior, suggesting the presence of shallow traps (i.e. states at about kT below the transport levels) which are easily depopulated in the presence of a field. The observation also of 'persistent' conductivity, a long-lived enhancement in dark conductivity after exposure to light, implies that many carriers can be stored temporarily in shallow traps, later to be thermally excited into conducting states. The picture which emerges is of a wide energy distribution of sites available for carrier hopping, virtually all of which may be DEH molecules, but in strongly perturbed environments due to the presence of the large dipole moments of the NLO species. The distinction between transport sites and trapping sites is therefore blurred and the effective trap density for long-term space-charge storage may merely reflect the tail of this wide distribution.

These observations have major implications for the creation and lifetime of holographic gratings. Shallow traps are detrimental because not only do they reduce the carrier mobility and slow the hologram formation but, once occupied, they also provide a source of carriers which can be thermally excited in the dark, leading to a persistent conductivity which accelerates grating decay. The concentration of effective (i.e. long-lived) trapping centers in bisA-NPDA:DEH has been estimated from holographic grating efficiency measurements to be of order 10^{15} cm^{-3} [24]. This is two orders of magnitude lower than the optimum value which would make the carrier range comparable to the holographic grating period. Therefore reliance on 'accidental' traps in the extreme tail of the energy distribution induced by disorder is not the best approach to materials capable of sustaining long-term holographic storage. Rather, specific trapping species with energies many times kT below the conduction states should be introduced to the formulation. Suitable compounds are those with oxidation potentials several tenths of an electronvolt less than the transport agent. The difficult feature of this approach will be to find a compound with a cation absorption profile which permits grating erasure by optical excitation of the trapped hole back into the conduction levels.

5. Holographic gratings in photorefractive polymers

The necessary requirements for observation of the photorefractive effect in a given material are (i) photoconductivity and (ii) an electro-optic response. Photoconductivity in MDPs has been characterized above. The electro-optic coefficient can be determined by using a Mach–Zehnder interferometer, in one arm of which the sample, sandwiched between elctrodes, is subject to an electric field [37].

However, these two properties are not sufficient to guarantee that the material will exhibit a useful photorefractive response. It is further necessary to demonstrate grating formation and, moreover, to show that the gratings are indeed photorefractive in origin and not due to some other photophysical or photochemical phenomenon such as photochromism. Gratings can be conveniently written by the interference of two coherent laser beams within the sample. The distinguishing feature of photorefraction is that the resulting index grating has a spatial phase shift relative to the intensity pattern. This phase shift is $\pi/2$ at zero applied electric field, when the carrier motion is purely diffusive, goes through a minimum at intermediate fields and returns to $\pi/2$ at high field where carrier drift moves the mobile charges to the minima of the intensity pattern [38].

Grating formation can be detected in a four-wave mixing experiment [24]. Two writing beams are obliquely incident on the sample in such a way that the normal to the planes of maximum intensity has a component along the applied electric field, thus ensuring that drift of the mobile holes spatially separates them from the immobile electrons. The reading beam counter-propagates along the direction of one of the writing beams and, therefore, any grating present must satisfy the Bragg condition for diffraction into the direction counter to the second write beam. The diffracted beam is reflected from a beam splitter into the detector.

The materials which we have examined to date do not support permanent poling. Hence, the NLO chromophores must be aligned by the application of an external electric field which couples to their dipole moment. The same applied field also enhances the charge generation efficiency. Such an arrangement, although perhaps not technologically ideal, permits some additional important tests of photorefraction. For example, if after the grating is established the electric field is turned off, then the EO effect and the photorefractive response vanish as the NLO dipoles relax to random orientations. However, the stored space charge remains in place and diffraction can be recovered by turning the field back on to realign the NLO chromophores. This test already distinguishes photorefraction from photochromism and any other effect which does not depend on an applied electric field.

The spatial phase shift of the grating can be measured with a two-wave mixing technique [39, 40]. The non-zero phase shift couples energy from one of the interfering writing beams to the other. (This phenomenon is the basis of the application of photorefractive materials for optical amplification, beam fanning, etc. [2, 3].) The experiment consists of first writing the grating, then translating the sample along the grating normal. As the grating translates the direction of coupling oscillates back and forth, and can be detected in the modulation of the intensities of the two transmitted beams. No other known mechanism for grating formation yields $\pi/2$ phase-shifted gratings. The technique not only permits the measurement of the phase and amplitude of the index grating, but also of any absorption grating which may be present. The results [40] show an index grating phase shift which tends to $\pi/2$ at high electric fields.

It should be remarked that some materials which we have examined exhibit photochromism. These form amplitude gratings with zero phase shift at zero applied electric field. Moreover, they are not erasable. In all of the materials which we have reported as photorefractive, a phase-shifted index grating dominates by an order of magnitude the photochromic response at fields of order 10^4 V/cm or higher.

The competition with photochromism suggests a clear criterion for the choice of NLO chromophore: it should *not* be susceptible to photochemical change of any kind, including photoinduced isomerization which is inherent in stilbenes and in many azo-dyes [4].

6. Conclusions

There are four steps in the mechanism leading to photorefraction: charge generation, transport, trapping and electro-optic response.

Many organic chromophores exhibit charge generation, including those which also show optical nonlinearity. Charge generation efficiency can be considerably enhanced by addition of an organic sensitizer such as the pigments used in electrophotography or electron acceptors which form charge transfer complexes with hole transport agents. The optimum concentration of sensitizer is dictated by considerations of sample thickness and optical density at the wavelength of interest.

Suitable transport molecules are well known to the electrophotographic industry. Monopolar transport is readily achieved by dissolving such compounds into a host polymer at the level of 25–50% resulting in mobilities which can exceed 10^{-4} $cm^2/(V\ s)$. In the photorefractive formulations which we have discussed, the mobilities are four orders of magnitude lower. This may reflect the choice of transport agent (DEH) which was most readily available to us and may also be limited by the high degree of energetic disorder introduced by the dipolar NLO chromophores. Charge transporting polymers, such as polysilanes [41] and poly(*p*-phenylenevinylenes) [42], may provide suitable alternatives as high mobility polymer hosts to be doped with NLO molecules.

Hole traps in organic photoconductors are donor molecules of exceptionally low oxidation potential. Very few of these are present as naturally occurring impurities and the optimum concentration must be achieved by deliberate addition.

Several classes of D-π-A molecules have emerged as candidates for incorporation into optical nonlinear polymers. The same requirements apply for photorefractive materials as for electro-optic modulators, second harmonic generation, etc., namely the highest possible hyperpolarizability, the ability to pole and the stability of the poled polymer. In addition the material must give no photochromic response, and must have chemical and physical compatibility with the dopant species to be added to it.

The field of organic photorefractivity is still in its infancy. There are many directions to follow in the synthesis and formulation of materials in order to optimize them for applications in optical communication and data storage.

References

1 P. M. Borsenberger and D. S. Weiss, in A. S. Diamond (ed.), *Handbook of Imaging Materials*, Marcel Dekker, New York, 1991, Ch. 9.
2 P. Günter and J.-P. Huignard (eds.), *Photorefractive Materials and Their Applications*, Vol. 1, Springer, Berlin, 1988; P. Günter and J.-P. Huignard (eds.), *Photorefractive Materials and Their Applications*, Vol. 2, Springer, Berlin, 1989.
3 T. J. Hall, R. Jaura, L. M. Connors and P. D. Foote, *Prog. Quantum Electron.*, *10* (1985) 77.
4 D. S. Chemla and J. Zyss (eds.), *Nonlinear Optical Properties of Organic Molecules and Crystals*. Academic Press, Orlando, FL, 1987, Vols. 1 and 2.
5 A. V. Vannikov, *Fiz. Tverd. Tela (Leningrad)*, *9* (1966) 1367 (*Sov. Phys. Solid State*, *9* (1967) 1068).
6 L. B. Schein, *Philos. Mag. B*, *65* (1992) 795.
7 B. Movaghar, *J. Mol. Electron.*, *3* (1987) 183; *4* (1988) 79.
8 F. K. Dolezalek, in J. Mort and P. M. Pai (eds.), *Photoconductivity and Related Phenomena*. Elsevier, Amsterdam, 1976, p. 27.
9 J. C. Scott, L. Th. Pautmeier and L. B. Schein, *Phys. Rev. B*, in press.
10 W. D. Gill, *J. Appl. Phys.*, *43* (1972) 5033.
11 L. B. Schein, A. Peled and D. Glatz, *J. Appl. Phys.*, *66* (1989) 686.
12 J. Frenkel, *Phys. Rev.*, *54* (1938) 647.
13 A. Peled, L. B. Schein and D. Glatz, *Phys. Rev. B*, *41* (1990) 10 835.
14 A. V. Vannikov, A. Yu, Kryukov, A. G. Tyurin and T. S. Zhuravleva, *Phys. Status Solidi (a)*, *115* (1989) K47.
15 H.-J. Yuh and D. M. Pai, *Philos. Mag. Lett.*, *62* (1990) 61.
16 P. M. Borsenberger and H. Bässler, *J. Chem. Phys.*, *95* (1991) 5327.
17 H. T. Li and P. J. Regensburger, *J. Appl. Phys.*, *34* (1963) 1730.
18 P. J. Melz, R. B. Champ, L. S. Chang, C. Chiou, G. S. Keller, L. C. Liclican, R. R. Neiman, M. D. Shattuck and W. J. Weiche, *Photogr. Sci. Eng.*, *21* (1977) 73.
19 P. Nielsen, *Photogr. Sci. Eng.*, *18* (1974) 186.
20 C. L. Braun, *J. Chem. Phys.*, *80* (1984) 4147; T. E. Goliber and J. H. Perlstein, *J. Chem. Phys.*, *80* (1984) 4162.
21 M. Pope and C. E. Swenberg, *Electronic Processes in Organic Crystals*, Clarendon Press, Oxford, 1982.
22 Z. D. Popovic, A. M. Hor and R. O. Loutfy, *Chem. Phys.*, *127* (1988) 451.
23 S. R. Marder, D. N. Beratan and L. T. Cheng, *Science*, *252* (1991) 451.
24 S. Ducharme, J. C. Scott, R. J. Twieg and W. E. Moerner, *Phys. Rev. Lett.*, *66* (1991) 1846.
25 M. Eich, B. Reck, D. Y. Yoon, C. G. Willson and G. C. Bjorklund, *J. Appl. Phys.*, *66* (1989) 3241.
26 D. Jungbauer, B. Reck, R. J. Twieg, D. Y. Yoon, C. G. Willson and J. D. Swalen, *Appl. Phys. Lett.*, *56* (1990) 2610.
27 S. M. Silence, C. A. Walsh, J. C. Scott, T. J. Matray, R. J. Twieg, F. Hache, G. C. Bjorklund and W. E. Moerner, *Opt. Lett.*, submitted for publication.
28 W. E. Moerner, C. A. Walsh, S. M. Silence, R. J. Twieg, T. J. Matray, J. C. Scott, V. Y. Lee, F. Hache and G. C. Bjorklund, *Mater. Res. Soc. Conf. Proc.*, (1992) in press.
29 J. C. Scott, L. Th. Pautmeier, W. E. Moerner, C. A. Walsh, S. M. Silence, T. J. Matray and R. J. Twieg, *Mater. Res. Soc. Conf. Proc.*, (1992) in press.
30 J. S. Schildkraut, *Appl. Phys. Lett.*, *58* (1991) 340.
31 L. Li, J. Y. Lee, Y. Yang, J. Kumar and S. K. Tripathy, *Appl. Phys. B*, *53* (1991) 279.
32 K. Tamura, A. B. Padias, H. K. Hall and N. Peyghambarian, *Appl. Phys. Lett.*, *60* (1992) 1803.
33 J. C. Scott, L. Th. Pautmeier and W. E. Moerner, *J. Opt. Soc. Am.*, *Part B*, submitted for publication.
34 P. J. Melz, *J. Chem. Phys.*, *57* (1972) 1694.

35 W. E. Moerner, C. Walsh, J. C. Scott, S. Ducharme, D. M. Burland G. C. Bjorklund and R. J. Twieg, *Non-Linear Optical Properties of Organic Materials IV, Proc. SPIE, 1560* (1991) 278.
36 J. C. Scott, S. Ducharme, R. J. Twieg and W. E. Moerner, *ACS Polym. Prepr., 32* (1991) 107.
37 S. Ducharme, J. Feinberg and R. Neurgaonkar, *IEEE J. Quantum Electron., 23* (1987) 2116.
38 G. C. Valley and M. B. Klein, *Opt. Eng., 22* (1983) 704.
39 K. Sutter and P. Günter, *J. Opt. Soc. Am., Part B, 7* (1990) 2274.
40 C. A. Walsh and W. E. Moerner, *J. Opt. Soc. Am., Part B*, in press.
41 M. Abkowitz, H. Bässler and M. Stolka, *Philos. Mag. B, 63* (1991) 201.
42 M. Gailberger and H. Bässler, *Phys. Rev. B, 44* (1991) 8643.

Synthetic Metals, *54* (1993) 21–32

Nonlinear optical properties of sublimed C_{60} thin films

F. Kajzar
Commissariat à l'Energie Atomique, Direction des Technologies Avancées LETI, DEIN/LPEM, CEN Saclay, 91191 Gif-Sur-Yvette Cédex (France)

C. Taliani, R. Zamboni, S. Rossini and R. Danieli
Istituto di Spettroscopia Molecolare del CNR, via de' Castagnoli, 1-40126 Bologna (Italy)

Abstract

Results of optical second-harmonic generation and wave-dispersed third-harmonic generation measurements in C_{60} thin films are reported and discussed. In third-harmonic generation measurements two strong resonances in cubic susceptibility $\chi^{(3)}(-3\omega; \omega, \omega, \omega)$ are observed. The first one occurring at 1.3 μm fundamental wavelength is interpreted in terms of a three-photon resonance with vibronic components of gerade states made partially allowed by vibronic mixing with the higher T_{1u} state. The second resonance at 1.064 μm is interpreted as a three-photon resonance with the T_{1u} state. The measured resonant values of $\chi^{(3)}(-3\omega; \omega, \omega, \omega)$ are 6.1×10^{-11} and 8.2×10^{-11} e.s.u. at 1.3 and 1.064 μm, respectively. The second-harmonic generation measurements performed at 1.064 μm on native thin films lead to a significant second-order susceptibility $d_{pp}(-2\omega; \omega, \omega) = 3.8 \times 10^{-9}$ e.s.u. (1.6 pm/V).

1. Introduction

Interest in organic materials for applications in photonics is based mainly on their fast, electronic origin and large nonlinear optical response due to the large hyperpolarizability of delocalized π-electrons. On the other hand, organic materials are characterized by interesting physicochemical properties, like processability, versatility and good optical quality. They are also expected to integrate well with semiconductor technology.

One-dimensional conjugated polymers, like polydiacetylenes, polyacetylenes and polythiophenes [1], are indeed characterized by a large nonlinear optical response and fast response time, but it is difficult to obtain good optical quality, stable thin films of these materials.

Polydiacetylenes are of repute for their excellent crystallinity which leads to unwanted light scattering in thin films. Polyacetylenes are known to be unstable in air. However, these materials are characterized by a residual absorption in the near-IR due to overtones of C–H and O–H stretching vibrations. Consequently it limits their application in the interesting (telecommunication window) operation wavelength range. For these reasons the fullerenes in general, and C_{60} in particular, are

0379-6779/93/$6.00

interesting materials for applications in photonics because they do not have C–H stretching vibrations. The electronic bands in C_{60} are narrower [2] than in conjugated polymers, where they are broadened due to conformations and conjugation length distribution. Consequently one expects resonances to be also narrower.

Large nonlinear optical response in C_{60} has been reported by third-harmonic generation on thin films [3] and in solution [4], as well as by degenerate four-wave mixing [5] in solution differing by two orders of magnitude from the previous two methods. All these measurements have been performed at 1.064 μm fundamental wavelength.

Also, a significant second-order nonlinear optical response from C_{60} has been reported very recently by Wang *et al.* [6] by optical second-harmonic generation in contrast with what one could expect for, presumably, centrosymmetric evaporated thin films. In fact no bulk quadratic susceptibility should be observed, at least at dipolar approximation. A second-order activity of C_{60} thin films has also been observed by Hoshi *et al.* [3].

In this paper we report the results of a systematic study of nonlinear optical properties of C_{60} thin films by optical third-harmonic generation (THG) and second-harmonic generation (SHG). The THG measurements were taken as a function of fundamental beam wavelength. This technique allows us to determine precisely electronic cubic susceptibility $\chi^{(3)}(-3\omega; \omega, \omega, \omega)$ and to learn more about the electronic structure of these materials. In fact, this technique gives not only the magnitude of the nonlinear susceptibility, but when done as a function of fundamental wavelength allows us also to locate the allowed and forbidden excited states. The SHG measurements were performed at 1.064 μm fundamental wavelength leading to the determination of quadratic susceptibility $\chi^{(2)}(-2\omega; \omega, \omega)$.

2. Experimental

2.1. Thin film preparation

C_{60} was prepared by the usual arc method. C_{60} powder was extracted with benzene and then chromatographically separated from higher fullerenes. C_{60} thin films were obtained by high vacuum sublimation of pure C_{60} powder on high purity and calibrated fused silica substrates. Three thin films with different thicknesses ($a = 0.054$ μm, $b = 0.071$ μm and $c = 0.088$ μm) were used in THG experiments. A direct thickness measurement of sample c gives 0.080 μm. This value has been correlated with the ellipsometric extinction coefficient determination by Ren *et al.* [7] giving 0.088 μm in a good agreement with our determination. As the ellipsometric technique gives much more precise data we have consequently chosen the last value as the film thickness. Thicknesses of other films were consequently determined from absorption spectra.

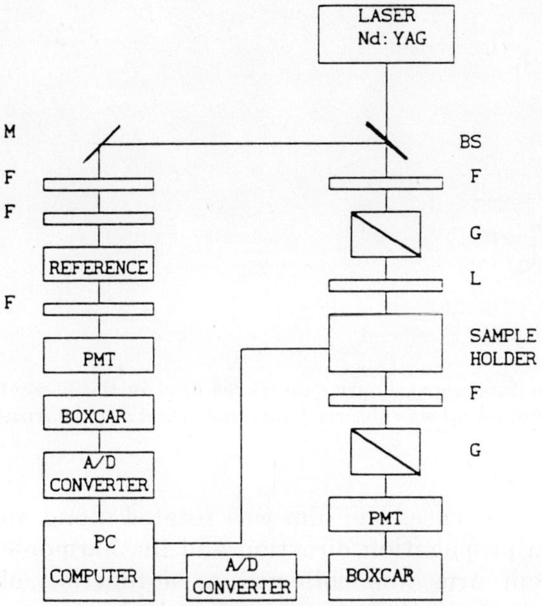

Fig. 1. Schematic representation of experimental setup: F, filters; L, converging lens; P–B, Pellin–Brocca prism; BS, beam splitter; M, mirror; G, Glan polarizer; PMT, photomultiplier.

2.2. Harmonic generation measurements

The second- and third-harmonic generation measurements were performed using the apparatus shown schematically in Fig. 1. The light source is a Q-switched Nd:YAG laser operating at 1.062 μm fundamental wavelength with 13 ns pulse duration. After frequency doubling with a phase-matched KDP single crystal, the SHG was used to pump the dye laser. Tuneable visible light was subsequently shifted in a high pressure hydrogen cell (56 bars) by stimulated Raman scattering to near-infrared. This system allows us to obtain a tuneable laser source between 0.532 and 1.907 μm; the last wavelength is directly obtained by Raman shifting of the fundamental. Since the fundamental beam used is not a monomode, the THG and SHG reference intensities were taken as the third-harmonic generation on a thin film (polydiacetylene or a doped polymer) or second-harmonic generation on a poled polymer film, respectively, allowing us to correct precisely the fundamental beam fluctuations.

The film deposited on a high grade silica substrate was mounted on a goniometer. In the case of THG measurements it was placed in a vacuum chamber in order to avoid environmental effects [8]. The third-harmonic generation signal has been referenced to a graded silica thin plate with a known nonlinear optical response, the measurements being done alternatively on the studied film and silica substrates. In the case of SHG measurements, a small second-harmonic signal, at the limit of dielectric breakdown, was detected only in the p–p fundamental-harmonic polariza-

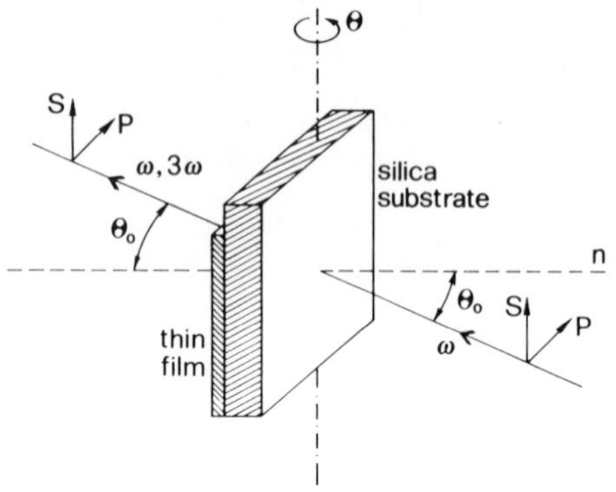

Fig. 2. Polarization configurations for fundamental-harmonic waves used in the present experiment. A third-harmonic generation signal was observed only in s–s and p–p polarization configurations.

tion configuration (Fig. 2). In both cases the film was rotated along an axis perpendicular to the beam propagation direction and the harmonic intensity from the measurement arm was collected as a function of incidence angle. The experiment was fully computer controlled.

From the third-harmonic generation measurements between two independently orientable polarizers (Fig. 2) we have found no third-harmonic generation for s–p and p–s polarization configurations showing that the measured film is isotropic.

2.3. Third-harmonic intensities

In general, the resultant harmonic field at the output of a nonlinear medium (or media) is a superposition of harmonic fields generated in separate media. For the case considered here (thin film deposited on a transparent substrate (graded silica) placed in a vacuum chamber) only harmonic fields generated in the substrate and thin film contribute to the measured signal. In the case when both fields are comparable and the substrate characteristics are well known (thickness and refractive indices), it is possible to determine not only the module but also the phase of $\chi^{(3)}(-3\omega;\omega,\omega,\omega)$ ($\chi^{(3)}=|\chi^{(3)}|\exp(i\Phi)$ susceptibility [9]. Otherwise one can determine only the module of $\chi^{(3)}$. In the present case it was possible to determine the phase of $\chi^{(3)}$ only at 1.907 μm. Examples of incidence angle dependence of harmonic intensities at two different wavelengths from the thin film plus substrate assembly and from the substrate alone are shown in Fig. 3. One observes a decrease of substrate coherence length and relative increase of harmonic intensity from the substrate–C_{60} film assembly as compared to the substrate itself with decreasing wavelength. This is due to two factors: one is an increase of thin film cubic susceptibility and the second is connected with the decrease of substrate coherence length; thus a decrease of effective interaction length compared to a constant thin film thickness being below its coherence length

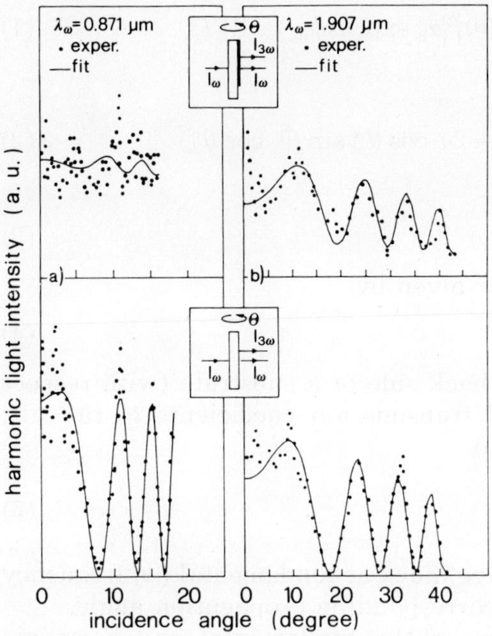

Fig. 3. Incidence angle dependence of harmonic intensity at two different wavelengths: (a) 0.871 μm and (b) 1.907 μm. Points represent experimental values, whereas solid lines the fitted ones. The lower Figures show harmonic intensities from substrate alone.

($l < l_c$). As the nonlinear optical response of thin film (Fig. 3) was significantly larger than that of substrate it was impossible to determine precisely the phase of the cubic susceptibility $\chi^{(3)}(-3\omega;\ \omega,\ \omega,\ \omega)$ except at 1.907 μm fundamental wavelength. This was found to be small ($\Phi = 0.1$ rad) corresponding to a cubic susceptibility with both real and imaginary parts positive. The solid lines in Fig. 3 indicate the calculated harmonic intensities using formulae derived in ref. 9.

The homogeneity of the thin film was checked by translating the film perpendicularly to the beam propagation direction. Constant THG intensity proves good film homogeneity. The sensitivity is proportional to the square of the variation of thin film thickness.

2.4. Second-harmonic generation

For second-harmonic generation the situation is much simpler than that for THG because neither air nor substrate contribute to the resultant harmonic coefficients; the former being considered as a linear medium.

As we have already mentioned, a small signal was detected only at p–p fundamental-harmonic polarization configuration. The measured second-harmonic intensity as a function of incidence angle was fitted with the formulae for C_v symmetry [10] (symmetry of orientation of crystallites on silica substrate) which read:

$$I_{2\omega}^{\mathrm{pp}}(\theta)=\frac{32\pi^3}{c\alpha_{\mathrm{h}}^2}\left|\frac{\chi_{\mathrm{pp}}^{(2)}}{\Delta\varepsilon}\right|^2|P_{\mathrm{pp}}(\theta)T_{\mathrm{pp}}(\theta)A_{\mathrm{pp}}(\theta)|^2|\alpha_{\mathrm{h}}\exp(\mathrm{i}\Delta\phi_{\mathrm{pp}})-1|^2I_{\omega}^2 \tag{1}$$

where

$$P_{\mathrm{pp}}(\theta)=(\sin^2\theta_{\omega}^{\mathrm{p}}+a\cos^2\theta_{\omega}^{\mathrm{p}})\sin\theta_{2\omega}^{\mathrm{p}}+2a\cos\theta_{\omega}^{\mathrm{p}}\sin\theta_{\omega}^{\mathrm{p}}\cos\theta_{2\omega}^{\mathrm{p}} \tag{2}$$

is the projection factor with

$$a=\chi_{113}^{(2)}(-2\omega;\omega,\omega)/\chi_{333}^{(2)}(-2\omega;\omega,\omega) \tag{3}$$

T_{pp} is an overall tranmission factor given by

$$T_{\mathrm{pp}}(\theta)=(t_{\mathrm{ag}}^{\mathrm{p}}t_{\mathrm{gf}}^{\mathrm{p}})^2 \tag{4}$$

for the thin film deposited on the back side of a substrate (with respect to the light source) with Fresnel transmission coefficients t_{ij} (a = air, g = substrate, f = thin film) equal to

$$t_{ij}^{\mathrm{pp}}=\frac{2n_i^{\mathrm{p}}\cos\theta_i^{\mathrm{p}}}{n_i^{\mathrm{p}}\cos\theta_j^{\mathrm{p}}+n_j^{\mathrm{p}}\cos\theta_i^{\mathrm{p}}} \tag{5}$$

where n_i is the ith medium refractive index at fundamental (ω) frequency for p polarized wave and θ is the corresponding propagation angle.

The phase mismatch $\Delta\phi$ between the fundamental and harmonic waves is given by

$$\Delta\phi=\frac{4\pi}{\lambda_{\omega}}(n_{\omega}^{\mathrm{p}}\cos\theta_{\omega}^{\mathrm{p}}-n_{2\omega}^{\mathrm{p}}\cos\theta_{2\omega}^{\mathrm{p}})\times 1 \tag{6}$$

where n_{ω} and $n_{2\omega}$ are the nonlinear medium refractive indices at ω and 2ω frequencies, respectively, $\theta_{\omega}^{\mathrm{p}}$ and $\theta_{2\omega}^{\mathrm{p}}$ are the corresponding propagation angles and 1 is the thin film thickness.

The factor α_{h} in eqn. (1) takes account of thin film absorption at harmonic frequency and is given by [10]

$$\alpha_{\mathrm{h}}=\exp(4\pi\kappa_{2\omega}^{\mathrm{p}}\times 1/\lambda_{\omega}\cos\theta_{2\omega}^{\mathrm{p}}) \tag{7}$$

where $\kappa_{2\omega}^{\mathrm{p}}$ is the extinction coefficient at harmonic frequency for p polarized wave.

The factor A in eqn. (1) arises from boundary conditions [8].

The harmonic intensities were fitted as a function of incidence angle using eqn. (1) and the refractive indices determined by Ren *et al.* [7]. An example of such a fit is shown in Fig. 4 with $a=0.53$. The value of quadratic susceptibilty was determined by comparison with SHG from an α-quartz single crystal plate under the same experimental conditions. Using for α-quartz the value reported by Meredith [11] ($d_{11}=\chi^{(2)}(-2\omega;\omega,\omega)/2=2.3\times10^{-9}$ e.s.u.) we obtained a comparable value of 3.8×10^{-9} e.s.u. for the studied film. This is about two times larger than the value reported by Wang *et al.* [6] ($d_{\mathrm{pp}}=2.1\times10^{-9}$ e.s.u.). It is worth noticing that it is a resonant value, the harmonic wavelength falling within the material absorption range.

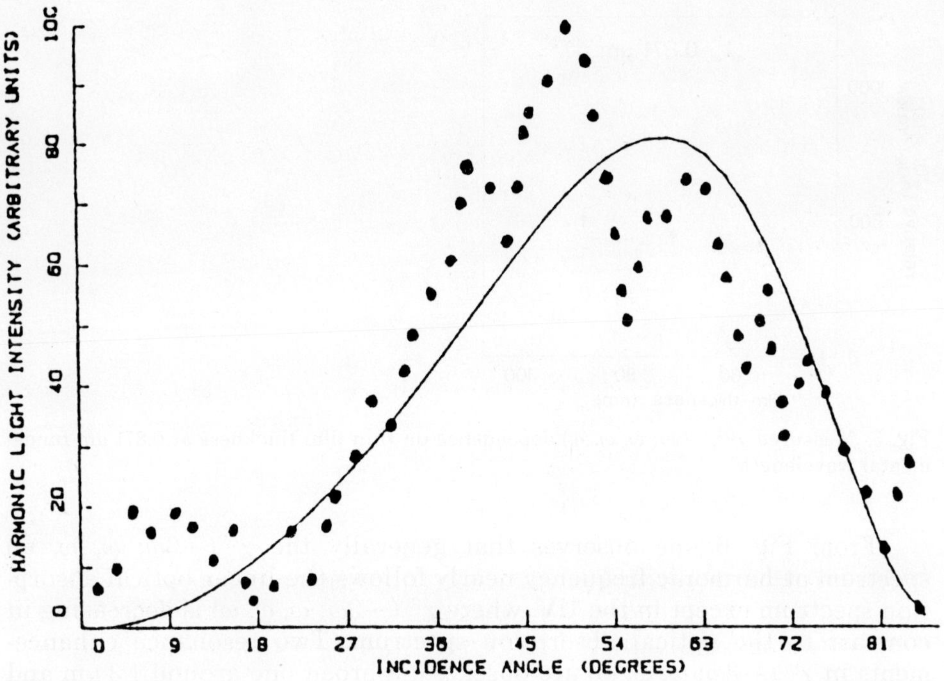

Fig. 4. Incidence angle dependence of second-harmonic intensity from a 0.071 μm thick film at 1.064 μm fundamental wavelength. Points represent experimental values, whereas the solid line shows the fitted ones using eqn. (1).

3. Results

3.1. Third-harmonic generation

First of all we have verified for the three films studied that the measured $\chi^{(3)}(-3\omega;\, \omega, \omega, \omega)$ does not depend on thickness. It proves the correctness of the technique used and shows that we observe a bulk and not a surface effect. This is shown in Fig. 5.

The measured cubic susceptibility $\chi^{(3)}(-3\omega;\, \omega, \omega, \omega)$ is plotted in Fig. 6 as a function of fundamental wavelength together with the thin film optical absorption spectrum. The relative precision in $\chi^{(3)}(-3\omega;\, \omega, \omega, \omega)$ values is very high (2–3%) when taken within the same dye emission range. The absolute precision depends on the thin film homogeneity and the reliability of the standard value used. As already mentioned we have used a fused silica plate as standard with $\chi^{(3)}(-3\omega;\, \omega, \omega, \omega) = 2.8 \times 10^{-14}$ e.s.u. as determined by Buchalter and Meredith [12] at 1.907 μm fundamental wavelength. The values for smaller wavelengths have been determined using Miller's rule (3.0, 3.1 and 3.2×10^{-14} e.s.u. at 1.3, 1.06 and 0.85 μm, respectively).

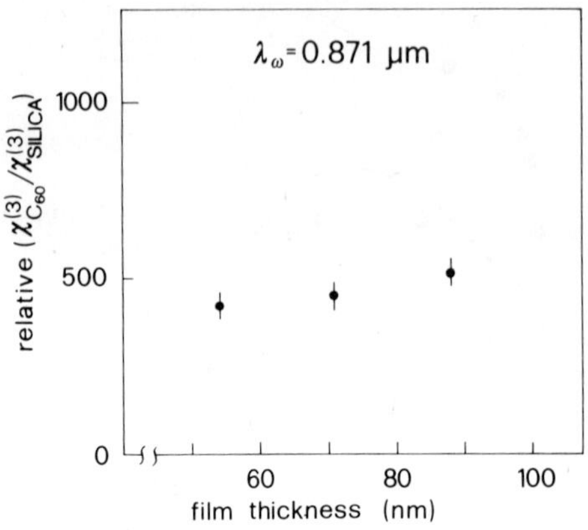

Fig. 5. Measured $\chi^{(3)}(-3\omega;\, \omega,\, \omega,\, \omega)$ dependence on thin film thickness at 0.871 μm fundamental wavelength.

From Fig. 6 one observes that generally the $\chi^{(3)}(-3\omega;\, \omega,\, \omega,\, \omega)$ spectrum at harmonic frequency nearly follows the linear optical absorption spectrum except in the UV, where $\chi^{(3)}(-3\omega;\, \omega,\, \omega,\, \omega)$ is decreasing in contrast to the optical absorption spectrum. Two resonance enhancements in $\chi^{(3)}(-3\omega;\, \omega,\, \omega,\, \omega)$ are observed: a broad one around 1.3 μm and another one with larger $\chi^{(3)}(-3\omega;\, \omega,\, \omega,\, \omega)$ values at 1.064 μm, both coinciding with peaks of the linear optical absorption spectrum. The resonance at 1.064 μm is clearly a three-photon resonance; the triple of the fundamental photon energy being equal to the T_{1u} one-photon allowed transition energy at 3.5 eV. The second broad resonance at around 1.3 μm falls in the range of weak one-photon absorption with a characteristic appearance of a forbidden transition made partially allowed by vibronic coupling. Along these lines Negri *et al.* [13] have interpreted the matrix

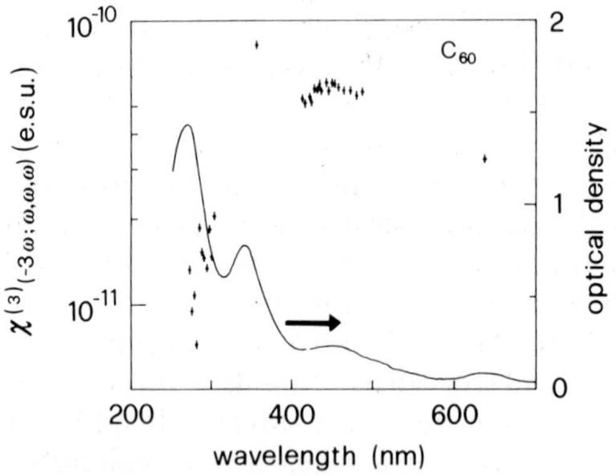

Fig. 6. Wavelength variation of cubic susceptibility measured for the 0.071 μm thick film. Solid line shows optical absorption spectrum.

isolation spectrum performed by Leach *et al.* [14]. The lowest electronic level is assigned to T_{1g}. Vibronic bands involving T_{1u} vibrations built on the T_{1g} origin have a total T_{1u} character. Intensity *borrowing* from the allowed T_{1u} state may than occur, inducing some one-photon allowed character to the vibronic levels.

Other gerade electronic states are assigned in the same spectral range [13] having T_{2g} and G_g symmetry. The broad resonance at 1.3 nm in the THG spectrum reflects therefore a three-photon resonance with vibronic components with total T_{1u} character made partially allowed by vibronic coupling.

3.2. Second-harmonic generation

Our measurements confirm second-order nonlinear optical activity of evaporated C_{60} thin films. We found a $d_{pp}(-2\omega;\ \omega,\ \omega)$ value equal to 3.8×10^{-9} e.s.u. (1.6 pm/V) about twice as large as that found by Wang *et al.* [6] (2.1×10^{-9} e.s.u.). From the incidence angle dependence of second-harmonic intensities we have found the ratio $a = \chi^{(2)}_{113}(-2\omega;\ \omega,\ \omega)/\chi^{(2)}_{333}(-2\omega;\ \omega,\ \omega) = 0.53$, a value close to that reported by Wang *et al.* [6] ($a = 0.48$). It corresponds to $d_{sp} = 2 \times 10^{-9}$ e.s.u. (0.8 pm/V).

4. Discussion

The measured values of $\chi^{(3)}(-3\omega;\ \omega,\ \omega,\ \omega)$ are indeed large and comparable with the values observed in conjugated polymers. The largest values are found in oriented polyacetylene but they are three-photon resonant values, thus not practically useful. Moreover in oriented conjugated polymers, $\chi^{(3)}(-3\omega;\ \omega,\ \omega,\ \omega)$ is strongly enhanced in the polymer chain direction with a large refractive index (of 3.2 and 5 at fundamental and harmonic frequencies, respectively [15]), whereas it is significantly lower in C_{60} [7]. Consequently the efficiency, which is inversely proportional to $n^2_{\omega}n_{2\omega}$, will also be lower.

Our results disagree with earlier determinations of Hoshi *et al.* [3], Qihuang Gong *et al.* [4] and Blau *et al.* [5], but agree within experimental accuracy with recent determinations by Wang *et al.* [6] at 1.907 μm.

Recently Kafafi *et al.* [16] reported a value of 7×10^{-12} e.s.u. from degenerate four-wave mixing measurements at 1.062 μm for Kerr susceptibility $\chi^{(3)}(-3\omega;\ \omega,\ \omega,\ \omega)$. This value is about one order of magnitude smaller than the value we obtained by THG measurements. The difference may well be accounted for by the three-photon resonance enhancement in the THG process as claimed here. The THG measurements by Hoshi *et al.* [3] were done in air with an unfocused laser beam. This procedure may lead to large errors as discussed in ref. 8 due to the air contribution to harmonic intensities.

From the thin film measurements it is possible to obtain the molecular hyperpolarizability through the following relation:

$$\chi^{(3)}(-3\omega;\omega,\omega,\omega)=\tfrac{1}{5}Nf\gamma(-3\omega;\omega,\omega,\omega) \tag{8}$$

where N is the number of molecules per unit volume and f is the local field factor given by

$$f=f_{3\omega}(f_{\omega})^{3} \tag{9}$$

For a spherical molecule and at the given frequency, the local field factor is given by

$$f_{\omega,3\omega}=\frac{n^{2}_{\omega,3\omega}+2}{3} \tag{10}$$

Taking the refractive indices determined by Ren *et al.* [7] and the solid C_{60} film density $d=1.65\ \mathrm{g/cm^3}$, one obtains a value of 1.1×10^{-32} e.s.u. (1.5×10^{-42} SI) at 1.907 μm for $\gamma(-3\omega;\omega,\omega,\omega)$ which is four orders of magnitude smaller than that reported by Blau *et al.* [5] from solution measurements.

The value at 1.907 μm cannot be still considered as an off-resonance value. Already the optical absorption spectrum (Fig. 6) shows a very weak absorption at harmonic wavelength. This may be due to impurity. Our determination gives at this wavelength a complex cubic susceptibility with a small phase $\Phi=0.1$ rad. This means that the real part of $\chi^{(3)}$ is positive. This is expected for the situation when an electronic state is close to, but still at larger energies than, the triple of the fundamental photon energy equal to 1.95 eV. Reber *et al.* [17] from C_{60} thin film low temperature luminescence measurements located an electronic transition at 1.8 eV. Thus, the value measured at 1.907 μm may contain a three-photon resonant contribution with this state the symmetry of which has been assigned to T_{1g} [13]. Our data are consistent with a T_{1g} state at 1.95 eV above the fundamental state. In that case the three-photon energy with 1.907 μm fundamental wavelength lies slightly below the T_{1g} state. Consequently, as argued above, $\chi^{(3)}(-3\omega;\omega,\omega,\omega)$ should be complex with a positive real part, as observed.

The measured low value of quadratic susceptibility $\chi^{(2)}(-2\omega;\omega,\omega)$ is in agreement with that expected for a centrosymmetric structure. This may be a surface, quadrupolar or purely impurity effect. The last possibility may explain the difference with the measurements reported by Wang *et al.* [6]. Their observed increase of $\chi^{(2)}(-2\omega;\omega,\omega)$ after corona poling may well be accounted for by impurity and/or charge injection [18].

5. Conclusions

Nonlinear optical properties of C_{60} thin films have been measured by SHG and wave-dispersed THG transmission techniques. The SHG measurements confirm earlier observations of second-order nonlinear optical response of evaporated thin films [3–6] with two nonzero tensor components $\chi^{(2)}_{113}$ and $\chi^{(2)}_{333}$ (the 3 axis being normal to the thin film surface) with $\chi^{(2)}_{113}/\chi^{(2)}_{333} = 0.53$.

The wave-dispersed THG measurements show two large resonance enhancements at 1.064 and 1.3 μm fundamental wavelengths. The first one is assigned to a three-photon resonance with the T_{1u} one-photon allowed electronic level transition and the second one can be tentatively assigned as a three-photon resonance with vibronic components of gerade states made partially allowed by vibronic mixing with the higher T_{1u} state. The observed large cubic susceptibility $\chi^{(3)}(-3\omega;\ \omega,\ \omega,\ \omega)$ values and the absence of C–H and O–H stretches suggest C_{60} and, more generally, fullerenes as very exciting materials for photonic device applications.

Acknowledgements

The authors would like to thank Messrs A. Lorin, P. Nardin, G. Tasini and P. Mei for technical assistance. Partial financial support by the CNR under the P.F. Superconductive and Cryogenics Technologies and Special Materials for Advanced Technologies programmes is also gratefully acknowledged.

References

1 See, e.g., (a) J. Messier, F. Kajzar, P. N. Prasad and D. Ulrich (eds.), *Nonlinear Optical Effects in Organic Polymers*, Kluwer, Dordrecht, 1989; (b) J. L. Brédas and R. R. Chance (eds.), *Conjugated Polymeric Materials: Opportunities in Electronics, Optoelectronics, and Molecular Electronics*, Kluwer, Dordrecht, 1990; (c) L. Y. Chiang, P. M. Chaikin and D. O. Cowan (eds.), *Advanced Organic Solid State Materials*, Materials Research Society, Pittsburgh, PA, 1990; (d) J. Messier, F. Kajzar and P. Prasad (eds.), *Organic Molecules for Nonlinear Optics and Photonics*, Kluwer, Dordrecht, 1991.

2 (a) E. Sohmen, J. Fink and W. Kratchmer, *Z. Phys. B*, *86* (1992) 87; (b) J. Fink, E. Sohmen, M. Merkel, A. Masaki, H. Romberg, M. Alexander, M. Knupfer, M. S. Golden, P. Adelmann and B. Renker, *Synth. Met.*, in press.

3 H. Hoshi, N. Nakamura, Y. Maruyama, T. Nakagawa, S. Suzuki, H. Shiromaru and Y. Achiba, *Jpn. J. Appl. Phys.*, *30* (1991) L1397.

4 Qihuang Gong, Yuxing Sun, Zongju Xia, Y. H. Zou, Zhenan Gu, Xihuang Zhou and Di Qiang, *J. Appl. Phys.*, *71* (1992) 3025.

5 W. J. Blau, H. J. Byrne, D. J. Cardin, T. J. Dennis, J. P. Hare, H. W. Kroto, R. Taylor and D. R. M. Walton, *Phys. Rev. Lett.*, *67* (1991) 1423.

6 X. K. Wang, T. G. Zhang, W. P. Lin, Shen Zhong Liu, G. K. Wong, M. M. Kappes, R. P. H. Chang and J. B. Ketterson, *Appl. Phys. Lett.*, *60* (1992) 810.

7 S. L. Ren, Y. Wang, A. M. Raon, E. McRae, J. M. Holden, T. Hager, Kai An Wang, Wan-Tse Lee, H. F. Ni, J. Selegue and P. C. Eklund, *Appl. Phys. Lett.*, *59* (1992) 2678.

8 F. Kajzar and J. Messier, *Phys. Rev. A*, *32* (1985) 2352.

9 F. Kajzar, J. Messier and C. Rosilio, *J. Appl. Phys.*, *60* (1986) 3040.

10 F. Kajzar, P. A. Chollet, I. Ledoux, J. Le Moigne, A. Loris and G. Gadret, in J. Messier, F. Kajzar and P. Prasad (eds.), *Organic Molecules for Nonlinear Optics and Photonics*, Kluwer, Dordrecht, 1991, p. 403.

11 G. R. Meredith, *Phys. Rev. B*, *24* (1981) 5522.

12 B. Buchalter and G. Meredith, *Appl. Opt.*, *21* (1982) 3221.
13 F. Negri, G. Orlandi and F. Zerbetto, in C. Taliani, G. Ruani and R. Zamboni (eds.), *Fullerenes: Status and Perspectives*, World Scientific Publishing, Singapore, 1992, p. 87.
14 S. Leach, M. Vervloet, A. Desprès, E. Breheret, J. P. Hare, T. J. Dennis, H. W. Kroto, R. Taylor and D. R. M. Walton, *Chem. Phys.*, *160* (1992) 451.
15 G. Leising, *Phys. Rev. B*, *38* (1988) 10 313.
16 Z. H. Kafafi, J. R. Lindle, R. G. S. Pong, F. J. Bartoli, L. J. Lingg and J. Milliken, *Chem. Phys. Lett.*, *188* (1992) 492.
17 C. Reber, L. Yee, J. McKiernan, J. I. Zink, R. S. Williams, W. M. Tong, D. A. A. Ohlberg, R. L. Whetten and F. Diederich, *J. Phys. Chem.*, *95* (1991) 2127.
18 S. Yitzchaik, G. Berkovic and V. Krongauz, *Nonlinear Opt.*, in press.

2. Conjugated Polymers and Oligomers

Electronic excitations and alternation of conjugated polymers

Z. G. Soos[a], S. Ramasesha[a,b], D. S. Galvão[c,d], R. G. Kepler[e] and S. Etemad[c]

[a]*Department of Chemistry, Princeton University, Princeton, NJ 08544 (USA)*

[b]*Solid State and Structural Chemistry Unit, Indian Institute of Science, Bangalore 560012 (India)*

[c]*Bell Communications Research, Red Bank, NJ 07701 (USA)*

[d]*Universidade Estadual de Campinas, Departmento de Fisica Aplicada, 13081 Campinas, SP (Brazil)*

[e]*Sandia National Laboratory, Albuquerque, NM 87185 (USA)*

Abstract

The electronic excitations and fluorescence of conjugated polymers are related to large or small alternation δ of the transfer integrals $t(1 \pm \delta)$ along the backbone. The fluorescence of polysilanes (PSs) and poly(*para*-phenylenevinylene (PPV) is linked to large δ, which places the one-photon gap E_g below the lowest two-photon gap E_a and reduces distortions due to electron–phonon (e–p) coupling. In contrast to small $\delta \sim 0.1$ in π-conjugated polymers, such as polyacetylene (PA), *para*-conjugated phenyls lead to an extended π-system with increased alternation, to states localized on each ring and to charge-transfer excitations between them. Surprisingly good agreement is found between semiempirical parametric method 3 (PM3) bond lengths and exact Pariser–Parr–Pople (PPP) π-bond orders for *trans*-stilbene, where the PPV bipolarons are confined to two phenyls. Stilbene spectra are consistent with increased alternation and small e–p distortions.

1. Introduction

The Su–Schrieffer–Heeger (SSH) model [1] for polyacetylene (PA) focuses on electron–phonon (e–p) coupling in polymers with a degenerate ground state. The Raman and IR signatures of charged solitons in *trans*-PA have been extensively probed in chemically doped or photoexcited samples with variable disorder or isotopic labeling [2, 3]. Neutral solitons have been more elusive. Similar e–p contributions due to bipolarons have been demonstrated in π-conjugated polymers, such as polythiophenes (PTs) or polydiacetylenes (PDAs) with nondegenerate ground states [3]. However, chemical doping leads to conducting polymers with broadly similar properties [4] and blurs the sharp theoretical distinction between degenerate and nondegenerate ground states.

The Peierls or Peierls–Hubbard instability of a half-filled π-band in *trans*-PA leads to alternating transfer integrals $t(1 \pm \delta)$ along the conju-

0379-6779/93/$6.00

gated backbone, with $\delta \sim 0.1$. The alternation δ in PDAs or PTs is slightly larger and fixed or locked in by the nondegenerate ground state. Although both PDAs and PTs have four carbons per unit cell, their conjugated backbones are still one dimensional and well represented by alternating t integrals. More complicated topologies occur in conjugated polymers that incorporate phenyl rings. The ground states of poly(*para*-phenylenevinylene) (PPV) or poly(*para*-phenylene) (PP) are clearly nondegenerate. The effective alternation [5] is considerably larger than in PA, PDA or PT. We develop in this paper the contrasting physical properties of polymers with small and large alternation.

In a parallel development, σ-conjugation has been studied [6, 7] in polysilanes (PSs). The one-dimensional $(Si)_n$ backbone consists of a half-filled band of sp^3 orbitals, with each Si contributing two orbitals and two electrons. The resulting Hückel model due to Sandorfy [8] has large alternation, since t integrals between sp^3 orbitals on the same and adjacent Si are quite different; $\delta \sim 1/3$ is needed [9] for various PS excitations within the Pariser–Parr–Pople (PPP) model. The strong fluorescence and negligible Stokes shifts characteristic of PSs are quite different from the weak emission and significant e–p coupling of PA or PDAs. Such experimental observations are particularly intriguing for closely related theoretical models.

The excitations [10] of poly(di-n-hexylsilane (PDHS) in Fig. 1. resemble those of organic molecular solids like anthracene. Electron–electron (e–e) correlations are now crucial. The singlet exciton at $E_g \sim 3.4$ eV is

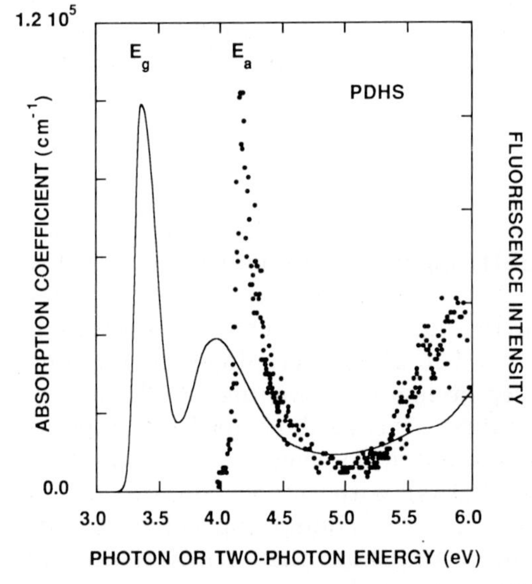

Fig. 1. One- and two-photon spectra of poly(di-n-hexylsilane) films at 295 K. The linear absorption shows the one-photon gap E_g, while the fluorescence intensity gives the two-photon gap E_a.

over an electronvolt below the threshold for direct generation of charge carriers. The fluorescence is even sharper and is understood [6] in terms of exciton migration to long segments with narrow absorption and emission. Site-selective spectroscopy [11] in the red edge is one of several ways of demonstrating small Stokes' shifts. Thus e–p coupling is weak for these neutral excitations. The two-photon spectrum in Fig. 1 is monitored through the fluorescence intensity and has prominent features around $E_a \sim 4.2$ eV and above 5 eV. The PDHS phosphorescence [12] below 20 K shows the singlet exciton to be only 0.12 eV below E_g. PDHS is highly crystalline, while poly(methylphenylsilane) (PMPS) is amorphous. Its one- and two-photon spectra are broadened to higher energy relative to Fig. 1 and the gap E_b for direct generations [13] of carriers is $\sim$4.6 eV.

In this paper, we link large and small alternation to the different photophysics of PS or PPV and PA or PDAs. When the chemical structure produces a large gap between the valence and conduction bands, we have reduced e–p or vibronic coupling in a semiconductor or molecule. Concomitantly, large δ shifts the lowest two-photon excitation above the one-photon absorption in conjugated polymers with intermediate e–e correlations and reduces the single–triplet splitting. Thus, the one- and two-photon spectra of finite polyenes [14, 15] are reversed compared to Fig. 1.

We sketch in Section 2 the interplay between alternation and e–e correlations in Hubbard and PPP models of conjugated polymers, with particular attention to the δ-dependence of the crossover from single-particle to correlated states. A dimer or molecular limit is developed for strongly alternating chains with intermediate e–e correlations. In Section 3 we summarize the canonical transformation leading to localized and extended states with large alternation in PPV and in related molecules such as styrene and stilbene. Several excited-state geometries of stilbene are examined in Section 4 by widely different theoretical methods.

2. Triplet and singlet energy gaps of alternating chains

Theoretical models of conjugated polymers are highly idealized: the restriction to π-electrons with e–p and e–e interactions provides initial insights into systems with many conformational degrees of freedom, complex morphologies and extensive solid-state redox chemistry. Current models are consequently phenomenological and aimed at organizing and rationalizing observations about neutral or charged excitations. The SSH and related e–p models [1, 2] for polymers with small alternation $t(1 \pm \delta)$ along the backbone focus on the instability of half-filled bands with $\delta = 0$. PPP and molecular exciton models focus instead on e–e correlations. Semiempirical parameters are chosen separately for every polymer in SSH theory, while molecular parameters have been transferred [15] to polymers in PPP theory. We summarize below the interplay of δ and e–e correlations in Hubbard and PPP models, with special emphasis on the spin and charge excitations of systems with electron–hole (e–h) symmetry [16], in order to understand the qualitatively different photophysics of PPV or PSs and PA or PDAs.

The half-filled Hückel model with $\delta = 0$ is a one-dimensional metal that is unstable to both spin and charge-density waves. On-site repulsion $U > 0$ in the regular Hubbard chain opens a one-photon gap, $E_g(U, 0)$, for dipole-allowed excitations, but there is no gap in the spin wave spectrum; both $E_T(U, 0)$ for the lowest triplet and $E_a(U, 0)$ for the lowest two-photon excitation vanish rigorously [17]. The separation of charge and spin excitations is a consequence of e–h symmetry that holds for extended Hubbard or PPP models with arbitrary nearest-neighbor t integrals and equal site energies [15]. But exact results are not available for E_T, E_a or E_g in infinite chains. The separation becomes particularly simple in the atomic limit, $U \gg t$, with $E_g \sim U$ and a spin-$\frac{1}{2}$ Heisenberg chain with $J = 2t^2/U$ for the spin excitations.

Alternating transfer integrals $t(1 \pm \delta)$, on the other hand, open a gap $E_g = 4t\delta$ between the filled valence band and the empty conduction band. Triplet, one-photon and two-photon excitations are all fixed at E_g in infinite Hückel chains. The single-particle or band states are stable against perturbations small compared to E_g and the universal behavior found in regular chains, where $E_a = 0$ is always below $E_g(U, 0) > 0$ is lost. Since a two-photon-allowed singlet can be constructed from two triplets, the crossover from band to correlated states occurs when $E_a(U, \delta)$ is given by $2E_T(U, \delta)$ and

$$E_a(U, \delta) = 2E_T(U, \delta) \sim E_g(U, \delta) \tag{1}$$

At larger U, the lowest two-photon singlet falls below the lowest one-photon singlet. We then regain the large U regime in which E_g eventually goes to U and the spin waves are described by an alternating spin-$\frac{1}{2}$ Heisenberg chain. There is no δ-dependence for $U \gg t$ and the charge-spin pattern is again universal.

The relevance of the crossover of $2E_T$ and E_g in eqn. (1) for conjugated polymers thus depends on both e–e correlations and alternation. Moreover, E_g is associated with singlet excitons in PDAs and PSs and Hubbard models must minimally be extended to nearest-neighbor Coulomb interactions V. The e–h symmetry of alternating Hubbard or PPP models distinguishes sharply between triplet, one-photon and two-photon excitations. Their appropriate signatures must then be sought in polymer spectra, preferably with minimal assumptions about parameters. Intermediate e–e correlations are a reasonable starting point for conjugated polymers.

To illustrate the crossover in eqn. (1), we consider the limit of strong dimerization ($\delta \sim 1$) that corresponds to weakly interacting molecules in a chain. Analytical solutions are well known for the Hubbard or PPP dimer. The lowest one-photon excitation for a Hubbard chain of dimers involves charge-transfer (CT) between dimers [9] and is denoted as $E_{CT}(U, \delta)$. The crossover (1) for noninteracting Hubbard dimers with transfer integral $t(1+\delta)$ is

$$E_{CT}(U, \delta) - 2E_T(U, \delta) = U - 2t(1+\delta) \tag{2}$$

The two-photon excitation becomes the lowest singlet when U becomes larger than the single-particle gap, $2t(1+\delta)$, of the dimer. In the $U=0$ limit, conversely, we have $E_T = E_g$ for any δ. On introducing an intradimer V, we may shift the one-photon absorption of the dimer below E_{CT}; the crossover with $2E_T$ then occurs at $U = \sqrt{2}t(1+\delta)$ for dimers. We have $E_g < E_a$ for the PDHS excitations in Fig. 1, in contrast to $E_a < E_g$ in π-conjugated oligomers [14], and variable δ in PPP models [9] with molecular e–e correlations accounts naturally for the crossover.

The dimer limit is an idealization for conjugated polymers that highlights alternation effects, in contrast to single-particle or strongly correlated limits that emphasize e–p and e–e contributions, respectively. An obvious theoretical problem is to find the crossover for intermediate δ as accurately as possible between the known limits of $U=0$ at $\delta=0$ and various dimer values; the crossover [9] is around $\delta=0.1$ for Hubbard models with $U=2t$. Since the dimer and CT excitations are mixed at intermediate δ, we again have complicated correlated states. The dimer states are then approximate and suggestive in the same way that single-determinantal functions or valence-bond diagrams are approximate for intermediate e–e correlations.

Although we have focused on the interplay of alternation and e–e correlations for the lowest excitations in various symmetries, alternation also affects e–p coupling [18]. The Peierls instability at $\delta=0$ leads in the SSH model to topological solitons in PA. Bandgap modulation in polymers with small δ accounts for a variety of dopant and photoinduced gap states [2, 3]. Such valence contributions become less important in molecules or semiconductors with an intrinsically large gap between filled and empty orbitals. The Stokes shift is a direct measure of e–p relaxation and is small [11] (<300 cm^{-1}) in both PPV and PS. While e–p coupling should no more be neglected at large δ than e–e interactions at small δ, we emphasize that alternation changes their roles.

3. Fluorescence and effective alternation of PPV

The quantum efficiency [19, 20] of ~5% of PPV and its derivatives is intermediate between the strong PS luminescence and the orders of magnitude weaker or extrinsic emission of PDAs or *trans*-PA. We recently associated [5] polymer fluorescence with a dipole-allowed lowest singlet excitation, as illustrated for PS in Fig. 1. Kasha's rule [21] for molecules is that fluorescence is from the lowest singlet of whatever symmetry.

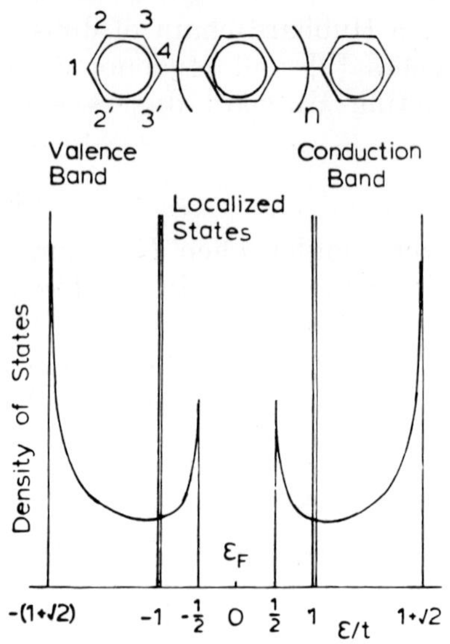

Fig. 2. Idealized geometry of planar poly(*para*-phenylene) and the resulting density of localized and extended states in Hückel theory. The extended and localized states are even and odd, respectively, on interchanging sites 2 and 2′ or 3 and 3′.

Since fluorescence is a competition among radiative and nonradiative processes, modern methods may detect some emission from excited states at short times or from dipole-forbidden states following an intense laser pulse. The association of *strong* fluorescence with a lowest *dipole-allowed* singlet takes into account the lower excitation energies and many non-radiative channels of conjugated polymers.

The phenyl rings of PPV have uniform bond lengths that resist significant changes even in excited states. To demonstrate their role in increasing the effective alternation, we consider a Hückel model for the simpler PP backbone in Fig. 2, without the ethylenic links. We assume for simplicity an idealized planar geometry appropriate for PPV and equal bond lengths. The reflection symmetry of sites 2 and 2′ or 3 and 3′ in Fig. 2 is assumed in Hückel or SSH models of PPV and is broken only on considering second-neighbor Coulomb interactions. Exact PPP solutions [5] for the actual structure of styrene, the PPV unit cell, or stilbene can readily be classified by this reflection.

The six π-orbitals of each phenyl are resolved into four even and two odd MOs under 2,2′ reflection by the transformation:

$$\phi_{2,3}^{(\pm)} = (\phi_{2,3} \pm \phi_{2',3'})/\sqrt{2} \qquad (3)$$

The odd $(-)$ orbitals produce a bonding and antibonding level at $\pm t_{23}$ that is strictly localized on each ring. The resulting spikes in the density of states in Fig. 2 correspond to a third of the PP states, or to a fourth of the PPV states for eight π-electrons per unit cell.

The even (+) orbitals in eqn. (3) form an extended one-dimensional network the alternation of which can be properly compared to PA or PDAs. Transfer integrals $\langle \phi_1 | H | \phi_2^{(+)} \rangle$ with bridgehead carbons are increased to $\sqrt{2}t_{12}$. The extended π-system for *equal* bond lengths in PP thus has *alternating* t integrals of $\sqrt{2}, 1, \sqrt{2}, 1, \ldots$ that correspond to $2\delta = (\sqrt{2} - 1)$. The Hückel energies form a valence and conduction band:

$$\epsilon_k = \pm 2\bar{t}(\cos^2 k + \delta^2 \sin^2 k)^{1/2} \quad (4)$$

with $2\bar{t} = t(\sqrt{2} + 1)$. The resulting density of states is shown in Fig. 2. The optical gap, $E_g = t$, is obtained at the band edges, $k = \pi/2$ in eqn. (4). Although the PPV gap is slightly smaller [5] due to the weaker alternation of the ethylenic part, the topological gap due to the phenyls again raises δ to higher values than expected for partial single and double bonds.

The symmetry of the valence and conduction bands in Fig. 2 follows from e–h or alternancy symmetry and requires equal site energies (Hückel α's) in the PP or PPV backbone. The valence band is completely filled in either case. The lowest-energy excitations around $E_g = t$ involve extended states that are even under 2,2′ or 3,3′ reflection. Charge-transfer (CT) excitations [5] between extended and localized states are possible above $1.5t$ in Fig. 2, or ~ 1 eV higher in the Hückel limit. Since the electron count in the localized states is changed by one, CT states are odd under 2,2′ or 3,3′ reflection.

The sulfur atoms in PTs break the e–h symmetry and the resulting site-energy differences [22] increase the bandgap in Hückel theory, thereby increasing δ relative to PA. PT spectra [23] show the S atoms to be a perturbation to a nondegenerate, *cis*-PA backbone. Larger alternation rationalizes an order-of-magnitude smaller quantum efficiency for PT fluorescence [20] compared to PPV. Moreover, the two-photon (E_a) excitation [24] in a PT oligomer containing six rings is slightly (0.11 eV) above the one-photon gap.

The interplay between the one- and two-photon gaps of conjugated molecules and polymers has been a paradigm of e–e correlations [15], since single-particle theory leads to $E_g < E_a$ even in the self-consistent-field limit. The increased alternation of t integrals due to the transformation (3) is consequently suggestive but incomplete until e–e effects are addressed. The band-edge electron–hole excitation in Fig. 2 is nondegenerate, while different k, k' values in eqn. (4) lead to doubly degenerate excitations on interchanging the electron and hole, and this degeneracy follows from alternate symmetry.

The alternancy symmetry becomes the e–h symmetry for interacting electrons [16]. We associate even (+) e–h symmetry with the covalent ground state. Dipole-allowed (one-photon) transitions are then rigorously to ionic states with odd (−) e–h symmetry. Every degenerate k, k' electron–hole excitation can be resolved into linear combinations that are even and odd under e–h symmetry. Our experience [25] with exact PPP or Hubbard states has been that e–e interactions shift ionic (odd, −) states to higher energy than covalent (even, +) states, as follows from first-order perturbation theory.

The issue for PPV, then, is whether the lowest two-photon excitation of the extended states in Fig. 2 remains above E_g when intermediate e–e correlations are included. The localized states, which are clearly buried in the band, are high-energy excitations of the polymer. Their role for discrete molecular states is discussed below. We note that the idealized planar PP backbone in Fig. 2 is quite inappropriate for that polymer due to steric repulsions that lead to noncoplanar rings. While PPV is planar, its larger unit cell leads to an extended π-system with six sites per unit cell that preclude analytical solution even in the Hückel limit.

4. Stilbene and its bipolarons

trans-Stilbene has recently acquired a new role as an oligomer of PPV, although its *cis–trans* isomerization and photophysics have fascinated researchers for decades. Neutral or charged bipolarons in PPV are delocalized over five or six phenyls, but the corresponding excitations of *trans*-stilbene are confined to two rings and thus maximize e–p coupling. We will discuss elsewhere one-photon, two-photon and triplet–triplet stilbene spectra [26]. We focus here on the relative position of E_a and E_g and on geometrical changes on forming bipolarons.

Both *ab initio* and semiempirical methods have recently been applied [27] to the ground-state geometry and rotational barriers of stilbene. We are using PM3, an efficient new semiempirical method [28] for accurate geometries, to probe rotational barriers in oligomers with up to six phenyls. The simpler stilbene results are essentially at a self-consistent-field level. The 14 π-electrons of stilbene are at our current limit for exact PPP solutions and lead to some 2.8 and 5.0×10^6 singlet and triplet VB diagrams, respectively. Thus PM3 and PPP represent quite different approximations about the relevant electrons and e–e correlations.

For PPP calculations, we adopt the stilbene geometry in Fig. 3. All bond angles are 120°, while the bond lengths are 1.40 Å in the ring and 1.45 and 1.35 Å for the ethylenic part. We retain as usual the molecular parameters [15]: $t = 2.40$ eV at $R = 1.40$ Å, $\delta = 0.07$, on-site correlations $U = 11.26$ eV, and the Ohno interpolation for Coulomb interactions from U at $R = 0$ to e^2/R at large R. These values are transferable among conjugated hydrocarbons, both molecular and polymeric. Inversion and e–h symmetry lead to a $^1A_g{}^+$ ground state, two-photon excitations to $m^1A_g{}^+$ and one-photon excitations to $n^1B_u{}^-$ states.

The two lowest excitations in each subspace are shown in Fig. 3, together with the large dipole transition moment to $1^1B_u{}^-$. The two degenerate CT excitations [5] expected for two rings in Hückel theory are unequivocally the covalent pair at 3.97 eV in Fig. 3. Their degeneracy is hardly lifted and the coefficients of selected symmetry-related VB dia-

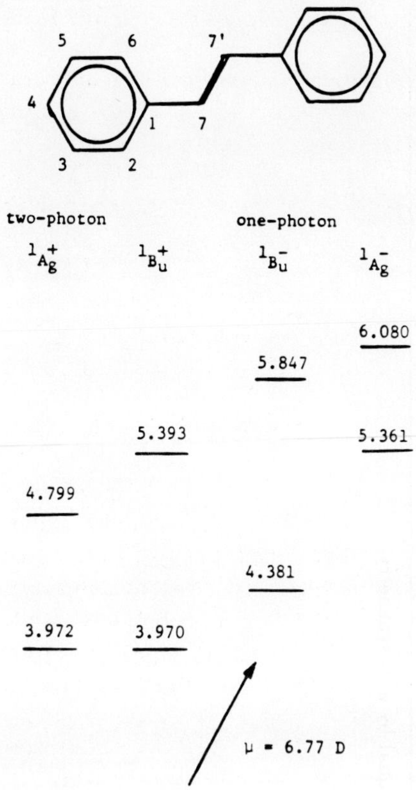

Fig. 3. Idealized planar geometry of *trans*-stilbene with 120° bond angles. The two lowest singlet excitations in each subspace are exact PPP solutions for molecular parameters, while μ is the transition dipole from the ground state to $1^1B_u^-$. The site numbering applies to Table 1.

grams have almost the same magnitudes but opposite phases. Similarly, $1^1B_u^-$ and $3^1A_g^+$ in Fig. 3 are excitations of the extended π-system, while the ionic partners of the CT states are higher-energy A^- and B^- singlets. The lowest two-photon excitation of the extended π-system remains above $E_g = E(1^1B_u^-)$ in PPP theory, as expected for an increased effective δ.

The lowest stilbene triplet at 2.518 eV is slightly higher than the solid-state value [29] of 2.2 eV. We find a weak T–T transition at 2.697 eV and a strong one at 3.753 eV that corresponds to the observed 3.2 eV transition [29]. More detailed comparisons will require relaxed excited states and site energies. The one-photon excitation of the stilbene dianion at 2.321 eV agrees with the broad absorption peak [30] at 2.5 eV. We also predict a weak absorption at 0.73 eV that involves a CT excitation of the dianion.

The PM3 bond lengths in Table 1 are labeled according to Fig. 3. A single-determinantal function is often adequate for a closed shell, here the ground state of stilbene or its charged bipolaron. The singlet and triplet

TABLE 1

Stilbene bond lengths (Å) and π-bond orders (eqn. (5)) labeled according to Fig 3; excited state values are relative to the $1^1A_g^+$ ground state

State		Ring bonds						Bridge bonds	
		12	16	23	56	34	45	17	77′
$1^1A_g^+$:	R (Å)	1.399	1.399	1.389	1.388	1.391	1.391	1.457	1.342
	p	0.607	0.610	0.655	0.646	0.636	0.639	0.309	0.855
$^1B_u^-$:	ΔR (Å)	0.019	0.017	−0.010	0.001	0.010	−0.003	−0.042	0.060
	Δp	−0.129	−0.142	0.044	0.034	−0.059	−0.036	0.197	−0.397
$1^3B_u^+$:	ΔR (Å)	0.029	0.028	−0.032	−0.011	0.006	0.005	−0.076	0.115
	Δp	−0.116	−0.118	0.022	0.031	−0.045	0.048	0.227	−0.495
BP[a]:	ΔR (Å)	0.054	0.052	−0.018	−0.022	0.015	0.024	−0.090	−0.113
	ΔR (Å)	0.046	0.047	−0.020	−0.023	0.010	0.018	−0.090	0.098
	Δp	−0.184	−0.175	0.073	0.094	−0.096	−0.129	0.285	−0.302

[a]Bipolaron; PM3 bond lengths for ±2; PPP bond orders are equal by e–h symmetry.

bipolarons are also associated with a single determinant in one-electron theory, in contrast to $m^1A_g^+$ states. The optimized ground-state geometry in Table 1 closely resembles other stilbene calculations and confirms the chemical expectation of two benzenes and partial single and double bonds. The other PM3 bond lengths in Table 1 were optimized separately and are given relative to the ground state. They clearly show the expected quinoidal distortion of the charged and triplet bipolaron, and the smaller distortion of the singlet bipolaron.

The π-electron bond orders in Table 1, on the other hand, were obtained from Coulson's formula [31]:

$$p_{ij} = -\frac{\partial}{\partial t_{ij}} \langle X|H|X\rangle/2 = -\sum_{\sigma} \langle a_{i\sigma}^{+}a_{j\sigma} + a_{j\sigma}^{+}a_{i\sigma}\rangle/2 \tag{5}$$

where $|X\rangle$ is the exact PPP eigenstate for the initial geometry. To a first approximation, we have $p \sim 0.8$, 0.3 and 0.6, respectively, for double, single and benzene bonds, and the bond order decreases with increasing bond length. The general agreement between the PM3 ground state and the PPP bond orders in Table 1 is hardly surprising. But the striking correspondence of the slight changes in the benzene moiety is quite remarkable and demonstrates how completely the changes are due to π-electrons. Moreover, Δp and ΔR have opposite signs for the charged and triplet bipolarons in Table 1, while the singlet bipolaron has two exceptions for small ΔR values. The approximate 2,6 or 3,5 reflection of stilbene is also shown clearly in Table 1. The absence of e–h symmetry in the PM3 calculation leads to slightly different bond lengths for the positive and negative bipolarons.

5. Discussion

There is remarkable agreement in Table 1 between the PM3 geometries and the PPP bond orders, even for excited states that can reasonably be approximated by single determinants. Quantum-chemical methods are at their best for the ground-state potential, where a mean-field analysis usually suffices. The PM3 results for stilbene extend to selected excited states and their close relation to π-bond orders provides surprisingly detailed support for elementary resonance ideas. Such comparisons between completely different semiempirical and correlated calculations have been unfortunately, but understandably, rare since different goals and techniques are involved. Yet critical evaluation of correlation contributions clearly requires such combined analysis.

The double-bond vibration (7,7′ in Fig. 3) in the singlet excited state of stilbene is hardly reduced [32] from the ground state. The absorption and fluorescence spectra indicate planar states with modest distortions. The triplet excited state [29] and the dianion structure [33] in a salt are also planar and bear directly on the importance of e–p coupling. The 7,7′

change in the $1^1B_u^-$ π-bond order in Table 1 is considerably smaller than the Hückel result, but may still be larger than observed. A proper assessment of stilbene spectra and excited-state geometries requires more work and is in progress [26].

The π-electron contribution to the ground-state alternation δ of a polymer is related in Hückel theory to the difference $\Delta p(\delta) = p_+(\delta) - p_-(\delta)$ between the bond orders of double and single bonds. If we take K for the harmonic σ-electron force constant and $\alpha = (\partial t/\partial R)_0$ as the linear e–p coupling constant, the dimensionless e–p coupling constant at alternation [18] δ is

$$\frac{|t|K}{\alpha^2} = \frac{\Delta p(\delta)}{\delta} = \frac{4}{\pi q}[K(q) - E(q)] \tag{6}$$

where $q^2 = 1 - \delta^2$ and K, E are complete elliptic integrals. The Peierls instability of the regular chain leads to divergence as $\delta \to 0$. Except for $\delta < 0.1$, however, $\Delta p(\delta)/\delta$ is small. In effect, Δp values do not depend strongly on alternation for the lowest one- or two-photon excitations and the decrease is due to the denominator. Coulomb contributions [18] to the dimerization involve the δ-dependence of V, which also varies as $\Delta p(\delta)/\delta$ in lowest order. The effective alternation of $\delta = (\sqrt{2} - 1)/2$ for the extended PPV states is consistent with weak e–p distortion. As noted in the Introduction, increasing δ is less effective energetically when E_g is already large at $\delta = 0$.

In summary, half-filled bands with $\delta = 0$ are unstable while $\delta \sim 1$ corresponds to molecular crystals with localized e–e and vibronic interactions. Conjugated polymers with alternating transfer integrals have intermediate δ. We have distinguished between the photophysics associated with (i) small $\delta \sim 0.1$, strong e–p coupling and a two-photon gap E_a below the one-photon gap E_g, and (ii) large $\delta \sim 1/3$ in PS, weak e–p coupling and strong fluorescence from $E_g < E_a$. We have also shown how phenyl rings lead to large effective δ in PPV and in stilbene.

Acknowledgements

We gratefully acknowledge partial support of the work at Princeton by the National Science Foundation, NSF-DMR 8921072, support of the work at Sandia Laboratories by the US Department of Energy, DE-AC04-76DP00789 and support for D.S.G. from the Brazilian Agency, FAPESP.

References

1 A. J. Heeger, S. Kivelson, J. R. Schrieffer and W. P. Su, *Rev. Mod. Phys.*, *60* (1988) 781.
2 T. A. Skotheim (ed.), *Handbook of Conducting Polymers*, Vols. 1 and 2, Marcel Dekker, New York, 1986.

3 *Synth. Met.*, *41–43* (1991); *27–29* (1988/89).
4 R. H. Baughman, J. L. Brédas, R. R. Chance, R. L. Elsenbaumer and L. W. Shacklette, *Chem. Rev.*, *82* (1982) 209.
5 Z. G. Soos, S. Etemad, D. S. Galvao and S. Ramasesha, *Chem. Phys. Lett.*, *194* (1992) 341.
6 R. D. Miller and J. Michl, *Chem. Rev.*, *89* (1989) 1359.
7 J. M. Zeigler and F. W. G. Fearon (eds.), *Silicon-Based Polymer Science: A Comprehensive Resource*, *ACS Symposium Series 381*, American Chemical Society, Washington, DC, 1990.
8 C. Sandorfy, *Can. J. Chem.*, *33* (1955) 1337.
9 Z. G. Soos and G. W. Hayden, *Chem. Phys.*, *143* (1990) 199.
10 Z. G. Soos and R. G. Kepler, *Phys. Rev. B*, *43* (1991) 11 908; R. G. Kepler, C. H. Seager, Z. G. Soos and P. C. M. McWilliams, *Synth. Met.*, *50* (1992) 387.
11 A. Elschner, R. F. Mart, L. Pautmeier, H. Bässler, M. Stolka and K. McGrane, *Chem. Phys.*, *150* (1991) 81; U. Rauscher, H. Bässler, D. D. C. Bradley and M. Hennecke, *Phys. Rev. B*, *42* (1990) 9830.
12 C. Walsh, W. P. Ambrose, D. M. Burland, R. D. Miller and D. S. Tinto, *Inorg. Organomet. Polym.*, *2* (1992) 285; O. Ito, M. Terazima and T. Azumi, *J. Am. Chem. Soc.*, *112* (1990) 444.
13 R. G. Kepler and Z. G. Soos, *Phys. Rev. B*, *43* (1991) 12 530.
14 B. S. Hudson, B. E. Kohler and K. Schulten, in E. Lim (ed.), *Excited States*, Vol. 6, Academic Press, Orlando, FL, 1982, p. 1.
15 Z. G. Soos and G. W. Hayden, in T. A. Skotheim (ed.), *Electroresponsive Molecular and Polymeric Systems*, Vol. 1, Marcel Dekker, New York, 1988, p. 197.
16 S. R. Bondeson and Z. G. Soos, *J. Chem. Phys.*, *71* (1979) 3807; *73* (1980) 598; S. R. Bondeson and Z. G. Soos, *Chem. Phys.*, *44* (1979) 403.
17 A. A. Ovchinnikov, *Sov. Phys. JETP*, *30* (1970) 1160.
18 G. W. Hayden and Z. G. Soos, *Phys. Rev. B*, *38* (1988) 6075.
19 N. L. Colaneri, D. D. C. Bradley, R. H. Friend, P. L. Burns, A. B. Holmes and C. W. Spangler, *Phys. Rev. B*, *42* (1990) 11 670; J. H. Burroughes, D. D. C. Bradley, A. R. Brown, R. N. Marks, K. MacKay, R. H. Friend, P. L. Burns and A. B. Holmes, *Nature (London)*, *347* (1990) 539.
20 L. S. Swanson, P. A Lane, J. Shinar and F. Wudl, *Phys. Rev. B*, *44* (1991) 10 617.
21 M. Kasha, *Discuss. Faraday Soc.*, *9* (1950) 14.
22 Z. G. Soos and G. W. Hayden, *Synth. Met.*, *28* (1989) D543.
23 A. O. Patil, A. J. Heeger and F. Wudl, *Chem. Rev.*, *88* (1988) 183.
24 N. Periasamy, R. Danieli, G. Ruani, R. Zamboni and C. Taliani, *Phys. Rev. Lett.*, *68* (1992) 919.
25 Z. G. Soos and S. Ramasesha, *Phys. Rev. B*, *29* (1989) 5410.
26 Z. G. Soos, S. Ramasesha and D. S. Galvão, *Phys. Rev. B*, in press.
S. Ramasesha, Z. G. Soos and D. Galvao, unpublished results.
27 O. Lhost and J. L. Brédas, *J. Chem. Phys.*, *96* (1992) 5279.
28 J. J. P. Stewart, *J. Comput. Chem.*, *10* (1989) 209 and 221; P. Scano and C. Thompson, *J. Comput. Chem.*, *12* (1991) 172.
29 W. G. Herkstroeter and D. S. McClure, *J. Am. Chem. Soc.*, *90* (1968) 4522.
30 H. Suzuki, K. Koyano and T. L. Kunii, *Bull. Chem., Jpn.*, *45* (1972) 1972.
31 C. A. Coulson, *Proc. Roy. Soc. London, Ser. A*, *169* (1939) 413.
32 R. H. Dyck and D. S. McClure, *J. Chem. Phys.*, *36* (1962) 2326.
33 M. Walczak and G. Stucky, *J. Am. Chem. Soc.*, *98* (1976) 5531.

Anion and cation absorption spectra of conjugated oligomers and polymer

M. Deussen and H. Bässler
Fachbereich Physikalische Chemie und Zentrum für Materialwissenschaften der Philipps-Universität Marburg, Hans-Meerwein-Strasse, W-3550 Marburg (Germany)

Abstract

The absorption spectra of chemically oxidized and reduced phenylenevinylene oligomers and the phenyl-substituted polymer are reported. Based upon the variation of transition energy with chain length at constant spectral shape and the agreement between photoinduced and chemically induced absorption, it is proposed that at moderate doping levels the absorption is due to a Franck–Condon allowed transition of radical anions and cations not reflecting polaron binding energies.

1. Introduction

Generating charge carriers by photoexciting the π-electron system of a conjugated polymer gives rise to new transient absorption features below the absorption edge of the neutral species [1, 2]. Their analysis is usually based upon the one-electron semiconductor band picture for systems with strong electron–phonon coupling [1, 3, 4]. A photon is considered to raise an electron for the π-type valence band of the polymer chain to the conduction band, followed by rapid self-localization due to local (structural) chain relaxation. This leads to the formation of localized polaron levels inside the bandgap. The photoinduced (PI) spectra are assigned to transitions among the localized levels and between localized levels and continuum band states, respectively. A triplet feature is considered as a signature of a singly charged polaron while a pair of charges (bipolaron) gives rise to a doublet. Collapse of a pair of oppositely charged polarons leads to fluorescence with a Stokes shift of twice the polaron binding energy. Coulombic [5] and electron correlation [6] effects known to be important in extended π-electron systems are ignored within the framework of this model.

The aim of the present paper is to disseminate the idea that PI spectra of conjugated molecules can be rationalized in much simpler terms adopting the concepts known from molecular spectroscopy. The conclusion that can be drawn from recent site-selective [7] and electromodulation [8] spectroscopy is that conjugated polymers consist of an array of segments the lengths (effective conjugation lengths) of which are less than the polymer chain length. The main absorption band has been interpreted as

0379-6779/93/$6.00

an inhomogeneously broadened singlet exciton transition subject to weak electron–phonon coupling only. We show that the absorption spectra of chemically produced anions and cations of oligo(phenylenevinylene)s (OPV) and the corresponding polymer are nothing but the Franck–Condon allowed transitions of those species after charging.

2. Experimental and results

Anions and cations of both unsubstituted OPV(n) (up to $n = 7$, where n is the total number of phenyl and phenylene rings) and poly(phenyl-phenylenevinylene) (PPPV) as well as cations of OPV bearing t-butyl substituents at the terminal ring *meta* positions (B-OPV(n) with $n = 5$–7) were prepared chemically by reaction at room temperature of a THF solution with a potassium mirror and a CH_2Cl_2 solution with $SbCl_5$, respectively. The concentrations were of the order of 10^{-6} to 10^{-5} M; for details see ref. 9. Absorption spectra were recorded at various stages of the reaction.

When OPV(3) reacts with potassium, a set of three bands (labelled A_1, A_2, A_3) is observed, which grow in intensity yet remain spectrally invariant until the absorption band of the neutral educt is completely bleached (Fig. 1). Each band shows vibronic structure. In the presence of excess potassium and after extended reaction times (about 20 min, depending on reaction conditions) the A-type band system is replaced by a new band system (DA_1, DA_2) with concomitant appearance of an isosbestic point and under fulfilment of the sum rule. The conversion is reversible as earlier work has indicated [10]. Complementary results have been reported by Schenk *et al.* [11] for the higher-membered B-OPVs. Since the species responsible for the A-type bands give rise to an ESR signal [12], it is straightforward to associate it with the absorption spectrum of the monovalent OPV radical anion. It is similarly straightforward to

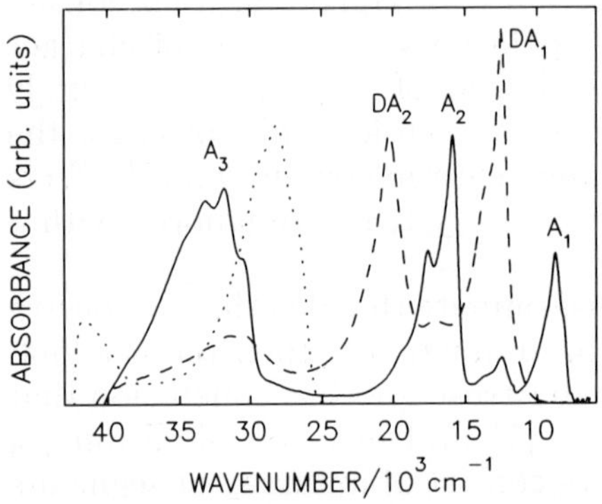

Fig. 1. Spectra of OPV(3) after reduction with potassium in THF solution at room temperature: (······), neutral OPV(3); (———), radical anion; (– – –), dianion.

assign DA-type spectra to the divalent anion. The immediate conclusion is that the divalent species is not formed until all of the neutral educt has been converted to the radical anion indicating that

$$OPV^- + OPV^- \longrightarrow OPV^{2-} + OPV$$

is an *endothermic* reaction as can be expected from the two distinct reduction steps observed in cyclic voltammetry [13].

In accord with what is known about the spectroscopy of radical anions and cations of polycyclic aromatic ring systems [14], radical cations of OPV absorb almost isoenergetically with the respective OPV^- [15], irrespective of chain length. Obviously, the radical cation is energetically favoured relative to the dication and the oxidation potential of $SbCl_5$ is insufficient for further oxidizing $OPV(3)^+$, while $B\text{-}OPV(7)^+$ is readily converted to the dication. Figure 2(a) compares the C-type spectra (C stands for radical cation) of OPV(3) and B-OPV(n) with $n = 5$–7.

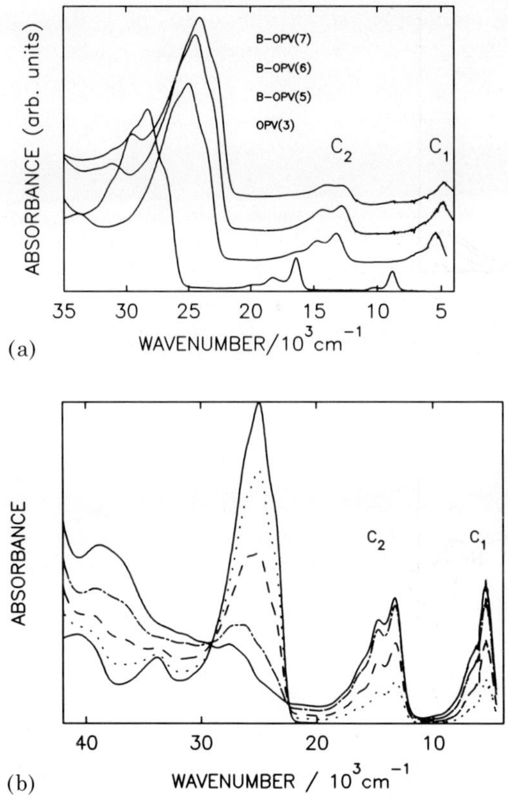

Fig. 2. (a) Comparison of spectra of OPV and B-OPV radical cations. CH_2Cl_2 solutions of the oligomers were oxidized with $SbCl_5$ at room temperature. (b) B-OPV(5) spectra after oxidation with $SbCl_5$ in CH_2Cl_2 at ambient temperature.

Upon gradual oxidation of PPPV in solution the C_1/C_2 bands grow in intensity and shift towards higher energies. At the same time the absorption band of the neutral polymer is progressively bleached starting from the low energy edge. New features (DC_1, DC_2) are not observed until the $PPPV^0$ band has been bleached completely, yet higher energy components of C_1/C_2 persist (Fig. 3).

3. Discussion

Based upon the facts (i) that the collapse of two like charges to form a doubly charged species is an endothermic process and (ii) that the photoinduced (PI) absorption spectrum of PPV [16] coincides with the C/A-type spectra (yet not the DA/DC-type ones) of PPPV at low doping (Fig. 3(a)), we conclude that the assignment of the PI spectra to a

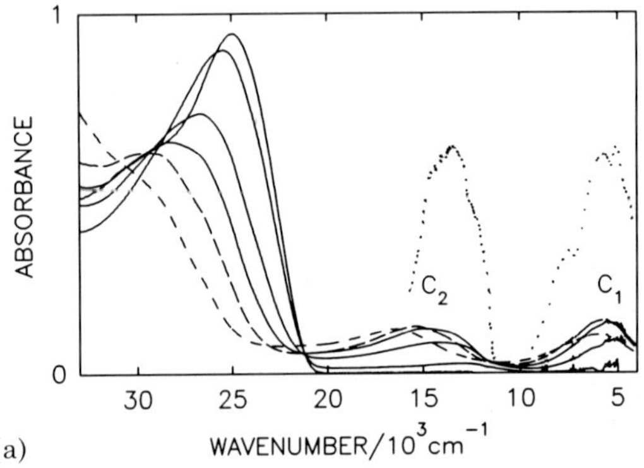

(a)

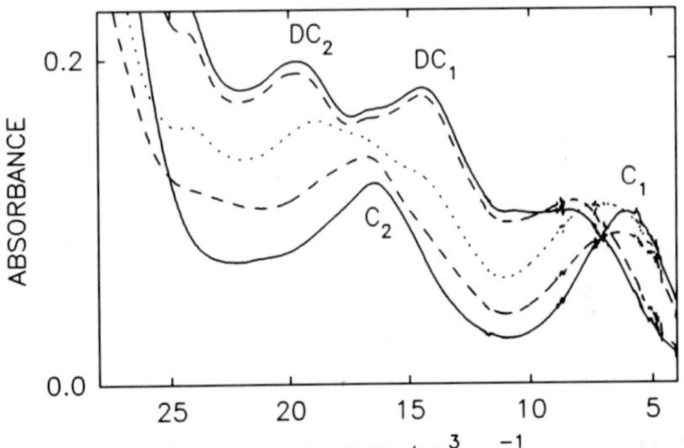

(b)

Fig. 3. (a) Evolution of the absorption spectrum of PPPV in CH_2Cl_2 during oxidation with $SbCl_5$ at room temperature. The photoinduced absorption spectrum of PPV (···) is included for comparison (taken from ref. 16). (b) Evolution of the PPPV absorption spectrum at high $SbCl_5$ concentrations. C = bands due to radical cation segments; DC = bands due to doubly charged segments.

bipolaron is erroneous. At low charge densities, as always realized in a photoexcitation experiment, radical anions and cations are the energetically preferred species, indicating that the gain in lattice relaxation energy upon double occupancy does not overcompensate the coulombic repulsion. Otherwise the case of a negative Hubbard U would be realized. We conjecture that the absence of an ESR signal associated with the PI spectrum is related to the low charge density concomitant with line broadening due to a distribution of g values.

Since a molecule like OPV(3) has no continuum states the spectra of the oligomer radical ions have undoubtedly to be interpreted as Franck–Condon allowed vertical transitions among the molecular orbitals of the charged species. This assignment is well accepted for polycyclic aromatic molecules [14] and stilbene [17]. The essential point we wish to make is that this assignment remains valid also for the polymer, maintaining that the polymer behaves like an array of oligomers with statistically varying lengths reflecting the local degree of ordering [7]. The basic arguments are the following:

(i) Upon increasing the oligomer length the ion spectra only show a bathochromic shift, their profiles remaining unchanged (Fig. 2(a)). This is not what would be expected in the case of a mechanism-switching in the sense that in the polymer the low energy band is due to a transition between a localized and a continuum state of the same electronic parentage but characterized by different configurational coordinates.

(ii) Upon plotting the energies of the A_1/C_1 and A_2/C_2 bands and the OPV^0 transition energies versus reciprocal chain length, a family of parallel straight lines is recovered, which is usually observed for electronic transitions in linear chromophores and rationalized in terms of the particle-in-a-box model [18, 19]. Within the framework of that model the slope $d(\Delta E)/d(L^{-1})$ reflects the decrease of the electronic level spacing with increasing chain length and the ordinate intercept is a measure of the amplitude of the periodic potential determined by bond alternation. The former does not to first order depend on the electronic orbital involved while the latter does.

(iii) The transition energies of the polymer and longer oligomer ions depend on the effective conjugation length of the absorbing charged chromophore. In a CH_2Cl_2 solution of PPPV a broad-range configuration is likely to exist. Upon oxidation the most extended segments will be charged first in accord with the hypsochromic shift of the C_1/C_2 bands and $PPPV^0$ bleaching (Fig. 3(a)). A similar phenomenon has been observed by Goedel *et al.* [20] upon electrooxidation of poly(3-cyclohexylthiophene). Upon photoexcitation of a PPV bulk polymer the charges will also tend to relax to segments of greater effective conjugation lengths in accord with a sharp bleaching feature at the red edge of the PPV^0 band. In any case, projecting the transition energy of a polymer/long oligomer onto the line ΔE versus L^{-1} in Fig. 4 yields an estimate of the effective conjugation length over which the charge is spread. It is noteworthy that the transi-

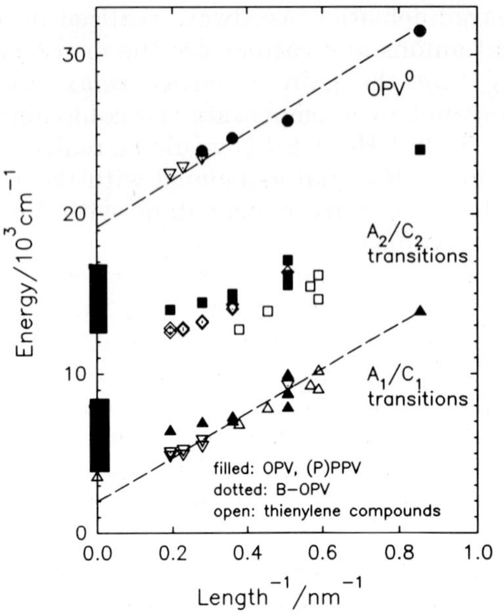

Fig. 4. Compilation of transition energies of a variety of arylenevinylene oligomers and polymers as a function of reciprocal molecular length. The bars at $L^{-1}=0$ indicate the energy range over which the C_1 and C_2 bands of PPPV are shifted with increasing degree of oxidation. The positions of the PI bands reported for PPV fall within these ranges. (Literature data taken from: OPV^0 [21], OPV^- [10], thienylene oligomers [22], poly(phenylenevinylene) PI [16], poly(thiophene) PI [23].)

tion energies of B-$OPV^{\pm}$ strictly obey this linear relationship, whereas the data for the unsubstituted OPVs show some deviation indicative of molecular distortion because of the poor solubility.

(iv) The essential parameter for analysing the observed ion transitions in terms of the conventional polaron model is $2\omega_0/2\Delta_0$ [3], equivalent to the ratio of the C_2/A_2 band energy to the transition energy of the neutral species. In the polymer → oligomer series this ratio decreases from $\simeq 0.58$ to $\simeq 0.53$ only, incompatible with the large variation of the confinement length and the small value of the electron–phonon coupling parameter inferred from the vanishing Stokes shift of the fluorescence spectra [7].

Summary

We note that both the chemically induced and photoinduced absorption spectra of charged species in phenylenevinylene- and thienylene-based compounds can consistently be interpreted in terms of Franck–Condon allowed transitions among the electronic orbitals of finite and (in the case of moderate doping) singly charged chain segments comprising several repeat units. We do wish to emphasize that this interpretation does not rule out configurational changes upon oxidation/reduction. These do occur but are not reflected in the ion absorption spectra because the latter probe the electronic energy band scheme of the chromophore *after* bond length adjustment has occurred.

Acknowledgements

We are indebted to Professor K. Müllen for providing us with B-OPV samples and to H. Martelock and M. J. Oberski for the preparation of PPPV and OPV. Financial support by the Fonds der Chemischen Industrie is gratefully acknowledged.

References

1 A. J. Heeger, S. Kivelson, J. R. Schrieffer and W. P. Su, *Rev. Mod. Phys.*, *60* (1988) 781.
2 R. H. Friend, D. D. C. Bradley and P. M. Townsend, *J. Phys. D: Appl. Phys.*, *20* (1987) 1367.
3 K. Fesser, A. R. Bishop and D. K. Campbell, *Phys. Rev. B*, *27* (1983) 4804.
4 M. J. Rice and S. R. Phillpot, *Phys. Rev. Lett.*, *58* (1987) 937.
5 L. Sebastian and G. Weiser, *Chem. Phys.*, *62* (1981) 447.
6 B. E. Kohler, *J. Chem., Phys.*, *93* (1990) 5838.
7 U. Rauscher, H. Bässler, D. D. C. Bradley and M. Hennecke, *Phys. Rev. B*, *42* (1990) 9830; S. Heun, R. F. Malirt, A. Greiner, U. Lemmer, H. Bässler, D. A. Halliday, D. D. C. Bradley, P. L. Burn and A. B. Holmes, to be published.
8 A. Horvath, G. Weiser and H. Bässler, *J. Phys.: Condens. Matter*, *173* (1992) in press.
9 M. Deussen and H. Bässler, *Chem. Phys.*, *164* (1992) 247.
10 J. Oberski, A. Greiner and H. Bässler, *Chem. Phys. Lett.*, *184* (1991) 391.
11 R. Schenk, H. Gregorius and K. Müllen, *Adv. Mater.*, *3* (1991) 492.
12 P. Brendel, A. Grupp, M. Mehring, R. Schenk, K. Müllen and W. Huber, *Synth. Met.*, *45* (1991) 49.
13 J. Heinze, J. Mortensen, K. Müllen and R. Schenk, *J. Chem. Soc., Chem. Commun.*, (1987) 701.
14 D. Distler and G. Hohlneicher, *Ber. Bunsenges. Phys. Chem.*, *74* (1970) 960.
15 C. W. Spangler and T. J. Hall, *Synth. Met.*, *44* (1991) 85.
16 N. F. Colaneri, D. D. C. Bradley, R. H. Friend, P. L. Burn and A. B. Holmes, *Phys. Rev. B*, *42* (1990) 11 670.
17 A. Heidenreich, N. Münzel and A. Schweig, *Z. Naturforsch, Teil A*, *41* (1986) 1415.
18 R. H. Baughman and R. R. Chance, *J. Polym. Sci., Polym. Phys. Ed.*, *14* (1976) 2037.
19 G. Wenz, M. A. Müller, M. Schmidt and G. Wegner, *Macromolecules*, *17* (1984) 830.
20 W. A. Goedel, N. S. Somanathan, V. Enkelmann and G. Wegner, *Macromol. Chem.*, *193* (1992) 1206.
21 R. F. Mahrt, J. Yang, A. Greiner and H. Bässler, *Makromol. Chem. Rapid Commun.*, *11* (1990) 415.
22 (a) D. Fichou, G. Horowitz and F. Garnier, *Synth. Met.*, *39* (1990) 125; (b) D. Fichou, G. Horowitz, B. Xu and F. Garnier, *Synth. Met.*, *39* (1990) 243.
23 Z. Vardeny, E. Ehrenfreund, O. Brafman, M. Nowak, H. Schaffer, A. J. Heeger and F. Wudl, *Phys. Rev. Lett.*, *56* (1986) 671.

Electronic levels ordering in α-sexithienyl

R. Zamboni, N. Periasamy*, G. Ruani and C. Taliani
Istituto di Spettroscopia Molecolare del CRN, via de' Castagnoli, 1-40126 Bologna (Italy)

Abstract

The electronic level ordering in a polycrystalline thin film of α-sexithienyl (T_6) has been determined by photoluminescence, one- and two-photon excitation, and photoconductivity measurements. The lowest 1^1B_u Frenkel exciton is located at 17 452 cm^{-1}. From site-selective photoluminescence we locate an extended inhomogeneous distribution of localized isolated T_6 levels (X-traps) below the 1^1B_u Frenkel exciton extending for 2000 cm^{-1}. The origin of the lowest 2^1A_g Frenkel exciton is located at 18 350 cm^{-1}, i.e. 898 cm^{-1} above the 1^1B_u. The photoconductivity action spectrum shows an onset at about the same energy of the 2^1A_g exciton band and reaches a maximum 1450 cm^{-1} above the 1^1B_u exciton band. T_6 is proposed as a good model compound for *real* polythiophene. The difference in energetics and photophysical properties due to different electronic correlation effects between polyenes and polythiophene is discussed.

1. Introduction

Conjugated systems constitute a well-established field of interest and there is a good possibility for these new materials to find commercial use in a large variety of applications. Flexibility, chemical tailoring, processability and low cost are among the major advantages of conjugated systems. Knowledge of the electronic structure is a crucial factor for the understanding of both basic theoretical aspects as well as the possible implications that the electronic level ordering has in the relevant charge transport and nonlinear optical properties [1, 2] of these materials. There are two classes of conjugated systems: one with degenerate ground state (DGS) and the other with nondegenerate ground state (NDGS). Polythiophene (PT) is a prototype material of the latter but defects, both of conformation and configuration with consequent distribution of chain and conjugation lengths, increase the complexity of the *real* material. As a result, there is a scattering of spectroscopic features and physical properties depending on sample preparation and processing. For instance, the photoluminescence spectrum of PT depends substantially on sample preparation as reported in the literature [3–7]. Consequently, the intrinsic electronic properties of the *idealized infinite* PT chain cannot therefore be easily derived by measuring and examining the spectra of *real* PT.

*Permanent address: Tata Institute of Fundamental Research, Bombay 400005, India.

0379-6779/93/$6.00

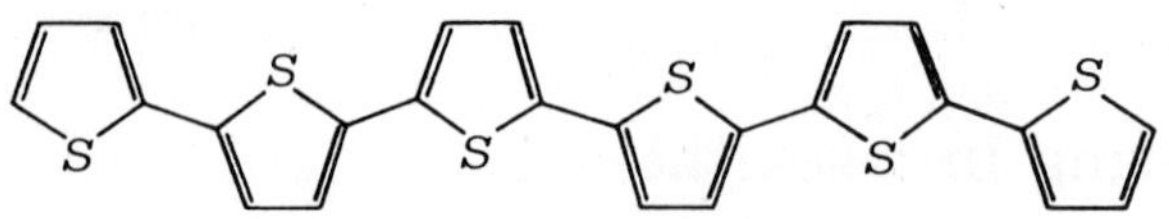

Fig. 1. Molecular structure of α-sexithienyl (T_6).

There are several reasons supporting the choice of α-sexithienyl (T_6) as a model compound for PT. First of all, T_6 is a well-defined chemical system. Figure 1 shows the molecular structure. Furthermore, the crystal structure of this 22.5 Å long thiophene oligomer has been resolved and refined [8]. This allows a precise structural definition of the condensed phase overcoming one of the major problems of *real* polymers, i.e. the impossibility of defining in a unique way the investigated compounds because of their intrinsic disorder, as mentioned above. In addition, there are experimental results supporting the appropriateness of T_6 as a model compound for PT [9–11] showing how fluorescence and photoinduced spectra are very similar for the two systems. Oligoenes extending to eight double bonds have been used as a model in order to understand the energy ordering of the low-lying A_g and B_u electronic levels in polyacetylene [12, 13]. Experimentally [13] it has been found that in short polyenes the lowest excited state is the covalent 2^1A_g. This result is in agreement with theoretical calculation and models which involve explicitly electron–electron (e–e) interaction [14]. T_6 can be regarded as a oligoene in which the carbon–carbon double bonds are locked by sulfur atoms which possess two lone electron pairs.

The aim of this paper is to contribute to the understanding of the electronic level ordering of PT by an extensive study of two-photon excitation (TPE) spectroscopy, photoluminescence and photoconductivity measurements on T_6.

2. Experimental

T_6 was chemically synthesized by catalytic coupling of Grignard salts obtained from 2,2′-dithienyl with 5,5′-dibromo 2,2′dithienyl by using [1,2-bis(diphenylphosphino)ethane] nickel chloride as a catalyst. This synthetic route is advantageous since it uses less catalyst (by two orders of magnitude) than that reported in ref. 15. Thin films of T_6 were prepared on different substrates by vacuum sublimation. The tendency of T_6 molecules to self-assembly gives rise to highly oriented polycrystalline thin films when the growing technique is vacuum sublimation. We have demonstrated by X-ray diffraction in asymmetric grazing incidence geometry that the long unit cell axis is almost perpendicular to the substrate [16]. In this arrangement the long molecular axis (L) makes an angle of 32° with respect to the substrate normal [8]. Crystallites have an average dimension of about 30 nm [16]. Photoluminescence spectra were obtained by exciting the isooriented polycrystalline film sample of T_6 by a nanosecond laser pulse at repetition rate of 20 Hz. The film was oriented at 45° with respect to the exciting and detected beams. The emission was dispersed in a Spex 1402 double monochromator and detected by a cooled GaAs photomultiplier. The output signal of the detected emission was fed to a boxcar averager with a gate width of 5 ns for processing. The excitation spectra were measured by monitoring an emission band and scanning the excitation wavelength of the dye laser using several dyes to scan the range.

The two-photon excitation experiment was performed by a tuneable nanosecond Raman shifted pulsed laser. Emission was detected by a photomultiplier tube. Pulse-to-pulse energy fluctuations were corrected by monitoring the second-harmonic generation (SHG) of a reference KDP powder. Both TPE and SHG signals were fed to a boxcar averager and for each laser shot computer analysed and controlled. The experimental setup and more details are reported in ref. 9.

Samples for photoconductivity measurements consisted of vacuum-sublimed thin films on a interdigitized two-electrode test pattern. Gold electrodes were separated by 10 μm of SiO_2; the channel width was 1.2 cm. For a detailed description of the device used see ref. 17. The photoconductivity experiment was performed at room temperature and the sample excited by a xenon lamp followed by a 1/4 m monochromator. The sample was oriented perpendicular to the incoming exciting light. The photocurrent was amplified with a low noise operational amplifier and fed to a voltage follower. The normalized photoconductivity action spectrum was obtained by means of two lock-in (sample and reference) detection channels (PAR 5101) and followed by a ratiometer (PAR 188).

3. Results and discussion

The emission spectrum of T_6 thin film at 4.2 K excited at 532 nm is shown in Fig. 2 together with the photoexcitation spectrum. An identical photoluminescence spectrum was obtained by exciting at 355 nm. The spectrum consists of three peaks at 588, 640 and 696 nm. The band separation is about 1382 cm^{-1} indicating a vibronic origin. We should note that the photoluminescence of highly crystalline chemically synthesized PT [7] is qualitatively very similar to the emission spectrum of T_6 (Fig. 2). The quantitative differences of the T_6 and PT spectra will be discussed in a forthcoming paper [18]. The excitation spectrum of the 588 nm emission band (Fig. 2) shows a prominent peak located at 573 nm (17 452 cm^{-1}). Identical excitation spectra were obtained by monitoring the other two emission bands at 640 and 696 nm. The photoluminescence decay time is less than 600 ps. This rather fast decay time indicates a spin-allowed transition supporting the assignment of the 573 nm band to

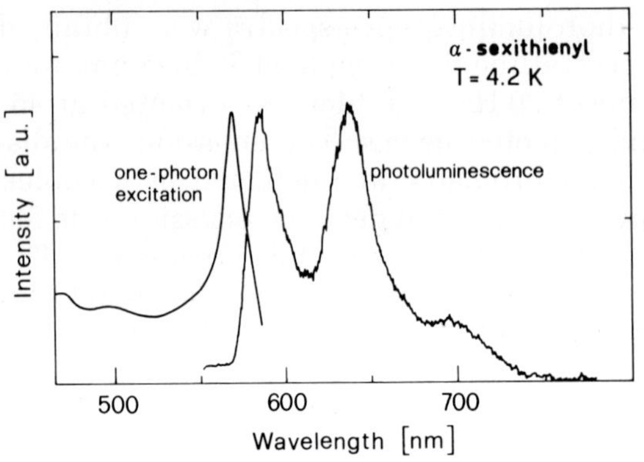

Fig. 2. One-photon excitation of the 588 nm emission band (left side) and photoluminescence (right side) spectra of α-sexithienyl (T_6) thin films at 4.2 K. The peak of the one-photon excitation spectrum is at 573 nm. Photoluminescence is excited at 532 nm.

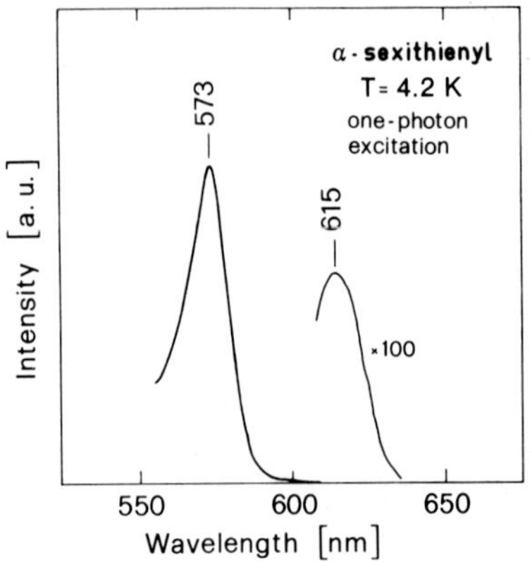

Fig. 3. One-photon excitation spectrum of the 640 nm emission band of α-sexithienyl (T_6) thin films at 4.2 K. Magnification of the low energy side shows a peak at 615 nm.

the origin of the 1^1B_u exciton band in the condensed phase of T_6. As reported above, the three emission bands share the same excitation spectrum proving their vibronic nature. The T_6 fluorescence is therefore a typical Franck–Condon (FC) emission involving intramolecular vibrations. In Fig. 3 we show the excitation spectrum of the 640 nm emission band. A quite weak band at 615 nm is clearly present. In order to investigate the nature of the weak excitation band below the lowest Frenkel singlet exciton located at 573 nm, we performed site-selective photoluminescence spectroscopy.

The emission spectra obtained by tuning the exciting wavelength as much as 2000 cm^{-1} below the 1^1B_u is shown in Fig. 4. Each spectrum consists of two prominent bands, the peak positions of which shift to lower energies as the laser excitation energy decreases. Similarity with

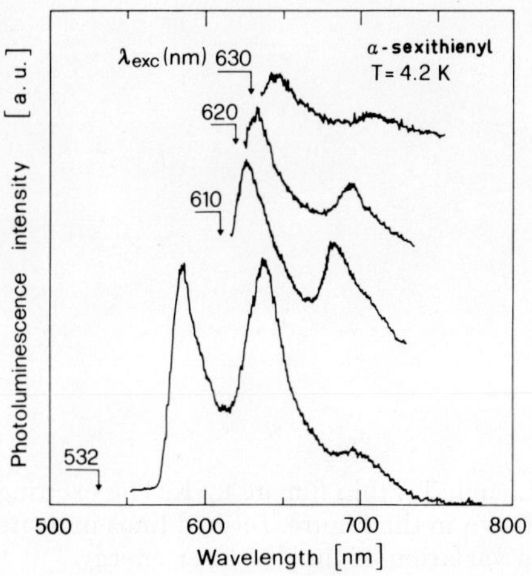

Fig. 4. Site-selective emission spectra of α-sexithienyl (T_6) thin films at 4.2 K. The lower spectrum is obtained by photoexciting the T_6 above the 1^1B_u Frenkel exciton located at 573 nm. The red-shifted emission spectra are obtained by tuning the exciting laser wavelength below the 1^1B_u Frenkel exciton.

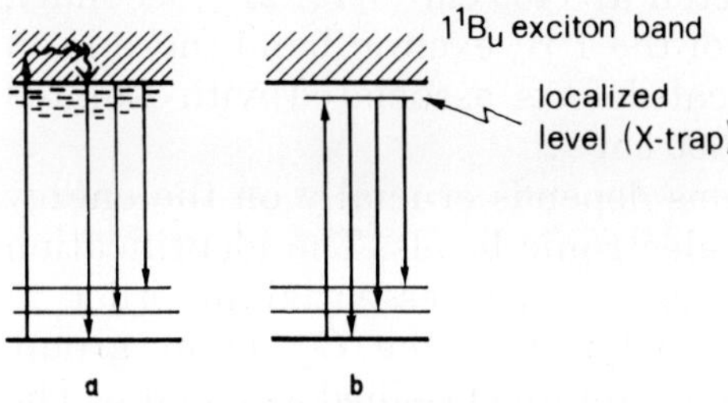

Fig. 5. Schematic diagrams of the levels of the lowest electronic transition in the T_6 condensed phase: (a) excitation relaxes in the 1^1B_u Frenkel exciton band and emits radiatively from the bottom of the band (573 nm) to the ground state; (b) below the ungerade Frenkel exciton there is a distribution of localized levels extending for 2000 cm^{-1} constituted by T_6 molecules at defect sites spatially separated which act as emitting sites (X-traps).

the emission spectrum allow us to identify the emitter as T_6 molecules. In addition, the spectrum is similar to that of the T_6 in dilute solution [19]. From these results we draw the conclusion that the states extending 2000 cm^{-1} below the exciton band are due to T_6 molecules located in physical defect sites of the crystal. These sites are isolated from each other and therefore act as emitting sites since excitation energy cannot migrate to other sites [20]. In Fig. 5 we show a scheme of these sites (X-traps). In this diagram we first emphasize that in the condensed phase the lowest 1^1B_u electronic transition is a Frenkel exciton band with the origin located at 17 452 cm^{-1}. The transition between the ground state and the lowest exciton band is accompanied by more than one vibration of the

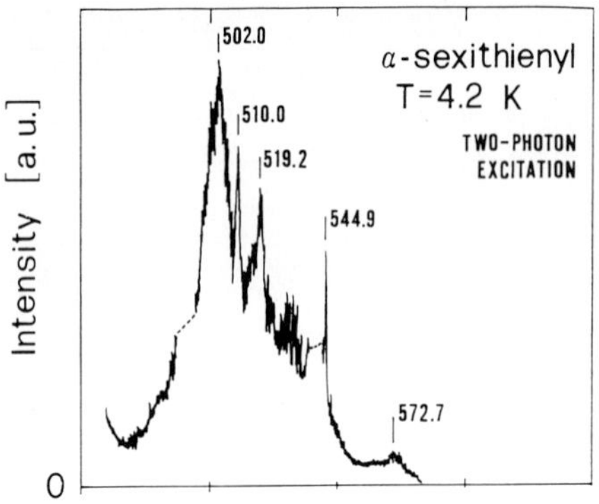

Fig. 6. Two-photon excitation spectrum of α-sexithienyl (T_6) thin film at 4.2 K. The exciting laser radiation wavelength is at twice the value shown in the Figure. Dashed lines indicate spectral regions of rejected data because of sharp variation of the dye laser energy.

fundamental state as suggested by the progressively larger half-width of the two bands at 640 and 696 nm (see Fig. 2). We were able to fit the spectrum by considering two FC vibronic progressions involving the C–C stretch at 1219 cm^{-1} and the C=C stretch at 1459 cm^{-1} [18, 21]. Secondly, we emphasize that below the bottom of the 1^1B_u exciton band there is an inhomogeneous distribution of physical defects associated with isolated T_6 molecules extending for at least 2000 cm^{-1}.

Photophysics of conjugated systems depends crucially on the energy order of gerade (g) and ungerade (u) electronic levels. The identification of the relative location of these states in T_6 is essential in order to understand the energetics of PT. T_6 belongs to the C_{2h} point group. Molecular electronic energy levels are therefore classified as g and u. The mutual exclusion between g and u states for one-photon and two-photon absorption processes should be also valid in the solid. The TPE spectrum of T_6 is shown in Fig. 6. We would like to recall that in sublimed thin films T_6 molecules stack with the long axis making a small angle with the normal to the substrate and that the film is oriented at 45° with respect to the incoming exciting laser beam and the outcoming emitted signal. The relevance of this specific sample orientation will be discussed below. The weak band at 572.7 nm coincides with the peak of the one-photon excitation spectrum and consequently is assigned to the 1^1B_u origin made partially two-photon allowed. Originally [9] we interpreted this effect in terms of reduction of molecular symmetry caused by rotation around the inter-ring single bond or by local crystal distortion which decreases the factor group symmetry. Now the proposed crystal structure [8] indicates that the molecule is not located at the inversion centre of the crystal. This causes a weak mix of g and u states making u states partially allowed in the TPE spectrum. Prominent bands are located at 544.9, 519.2, 510.0 and 502.0 nm. The sharp band (25 cm^{-1} width) at 544.9 nm (18 350 cm^{-1}) is assigned to the origin of the 2^1A_g exciton band (i.e. 898 cm^{-1} above the ungerade lowest exciton). Other intense bands located at 910, 1257 and 1570 cm^{-1} above the origin are assigned to totally symmetric vibronic levels of the 2^1A_g electronic manifold. The possible corresponding gerade vibrations in the ground state, measured by Fourier transform Raman scattering [21], are the ring stretch (863 cm^{-1}, the C–H deformation or ring breathing (1219 cm^{-1}) and the C=C stretch (1459 cm^{-1}). The remarkable frequency increase observed for the C=C stretch indicates that the equilibrium geometry in the excited state is strongly different along this normal coordinate. A similar situation has been observed in polyenes [12].

We would like to stress the relative location of the 2^1A_g state above the 1^1B_u. This implies that photoexcitation in the 2^1A_g exciton band relaxes in the 1^1B_u electronic manifold before being emitted radiatively from the bottom of the 1^1B_u exciton band to the ground state. Since photoluminescence of PT is very similar to that of T_6 [7, 18], we infer that the electronic energetics of *real* PT follows the same excited state ordering of T_6. Moreover, the location of the lowest gerade state in T_6 above the ungerade one implies that e–e correlations are weaker in T_6 than in short polyenes where the 2^1A_g is below the 1^1B_u [12–14]. Since T_6 is a model compound for PT as well as oligoenes are for polyacetylene, we can draw the conclusion that the location of 2^1A_g marks the relevant difference between the energetics and photophysics of these two systems.

We have demonstrated that neutral excitations are Frenkel excitons, that the electronic energy relaxes radiatively from the 1^1B_u to the ground state via Franck–Condon fluorescence and that the 2^1A_g exciton band is higher than the 1^1B_u exciton band. But where do charges become separated giving rise to charge transport? The nonzero intensity of the one-photon excitation (OPE) spectrum at higher energies than 573 nm (see Fig. 2), as well as the drop of quantum yield fluorescence pumping in resonance with the maximum of the one-photon absorption peak at 480 nm [21, 22], suggests that a new channel of nonradiative decay opens up at higher energies than the 1^1B_u exciton band. Since the formation of an electron–hole pair is the natural choice for nonradiative decay, we performed photoconductivity experiments in order to measure the threshold of charge carrier creation in T_6.

The sample setup has been described in the Experimental Section. Here we recall that the same sample setup was used for field effect transistor (FET) characterization and that T_6 is up to now the organic FET with the best performance [23, 17].

As in the case of TPE and OPE the sample is a highly ordered polycrystalline thin film. Contrary to the experiments, in which the film was oriented at 45°, here we have taken the action spectrum at normal incidence. This implies that, while TPE and OPE analyse the full exciton band structure of the crystal, the action spectrum is investigating the *bc* plane. Since the short in-plane molecular axis (M) and the plane normal (N) have strong projections on this plane we select charged excitation polarized parallel to these directions.

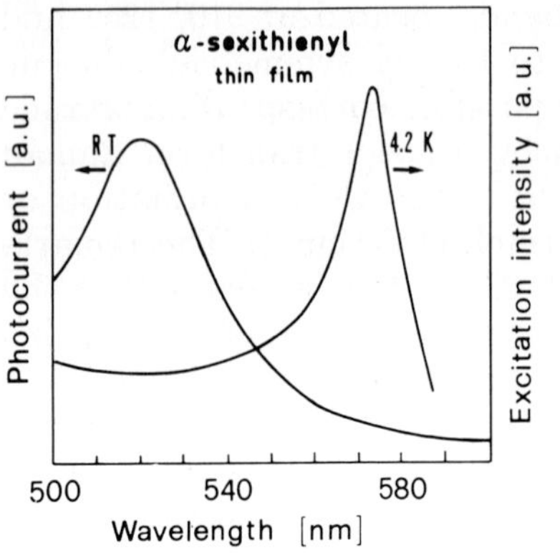

Fig. 7. Photoconductivity action spectrum (left side) and one-photon excitation spectrum (right side) of α-sexithienyl (T_6) thin films. The broad peak of photocurrent at room temperature is located at 520 nm. The peak of one one-photon excitation spectrum at 4.2 K is located at 573 nm.

Dark conductivity at room temperature for an applied bias voltage of 9 V was negligible. Photoconductivity increased linearly with light intensity. The photoconductivity action spectrum is shown together with the one-photon excitation spectrum in Fig. 7. It shows a broad band in the 400–550 nm region peaked at approximately 520 nm. The comparison between photoexcitation and photoconductivity action spectra suggests that they are qualitatively complementary. The two spectra cross each other at approximately 544 nm. Incidentally, the 2^1A_g exciton band origin is located approximately at the same energy.

The location of 2^1A_g at about the same energy of the opening of a new radiationless channel suggests two hypotheses. First, the 2^1A_g exciton band provides an efficient direct crossing to extended ionized states. Generation of extended states by crossing from neutral covalent Frenkel excitons has been suggested by Petelenz and Mucha [24] in anthracene. Intersystem crossing would most probably be very much dependent on the nature of the electronic state prepared by two-photon absorption. The second hypothesis suggests that the 2^1A_g gives rise to fission into two triplets as predicted in ref. 25.

Further experiments are needed to analyze these aspects. In addition we should investigate directly the location of intermolecular charge transfer excitations. Preliminary electromodulation experiments are under way in order to study these aspects.

4. Conclusions

The electronic level ordering of solid T_6 is summarized in Fig. 8. T_6 behaves like a typical molecular solid. Low-lying electronic excitations in

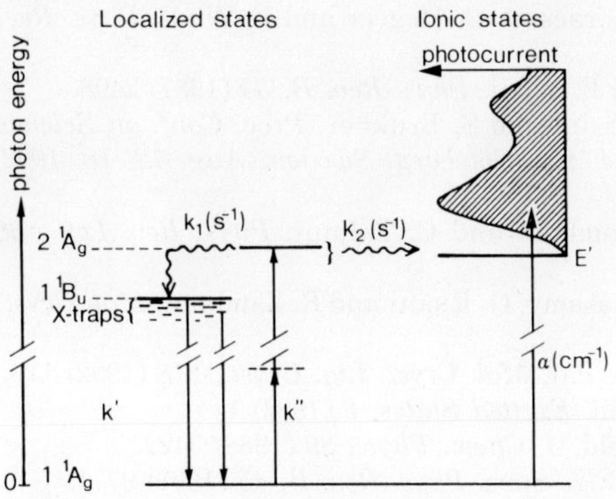

Fig. 8. Schematic diagram of electronic level ordering in T_6. Full arrows indicate radiative absorption or emission. Wavy arrows indicate nonradiative processes. The lowest 1^1B_u emitting level is located at 17 452 cm^{-1}. Below the 1^1B_u there is a 2000 cm^{-1} extended localized emitting level (X-traps) band. The origin of the lowest gerade (2^1A_g) Frenkel exciton band is located at 18 350 cm^{-1}.

the solid are neutral excitations localized at the molecule forming fairly narrow bands called Frenkel excitons. These excitations preserve the major character of the molecule itself as proved by vibronic FC emission involving intramolecular vibrations. The lowest electronic excitation is constituted by a neutral 1^1B_u Frenkel exciton. The second electronic excitation is a neutral *forbidden* 2^1A_g Frenkel exciton with a very narrow bandwidth. The charge carrier generation threshold is located at about the energy of the 2^1A_g exciton band. Since T_6 is a good model compound for *real* PT, the same energetics of T_6 may be extended to the energetics of *real* PT.

Acknowledgements

We wish to thank Dr R. Danieli and Mrs S. Rossini for chemical preparation of T_6, Messrs G. Tasini, P. Mei, A. Martiniello and L. Minghetti for technical assistance, and P. F. Materiali Speciali per Tecnologie Avanzate of CNR for financial support.

References

1 P. N. Prasad and J. K. Nigam (eds.), *Frontiers of Polymer Research*, Plenum Press, New York, 1991.
2 J. L. Brédas and R. R. Chance (eds.), Conjugated Polymeric Materials: *Opportunities in Electronics, Optoelectronics and Molecular Electronics*, Kluwer, Dordrecht, 1991.
3 S. Hayashi, K. Kaneto and K. Yoshino, *Solid State Commun., 61* (1987) 249.
4 J. L. Suvajol, D. Chenouni and S. Hasoon, *Synth. Met., 28* (1989) C293.
5 C. Taliani, R. Danieli, R. Zamboni, P. Ostoja and W. Porzio, *Synth. Met., 18* (1987) 117.

6 M. Kobayashi, J. Chen, T. C. Chung, F. Moraes, A. J. Hegeer and F. Wudl, *Phys. Rev. B*, *30* (1984) 702.
7 Z. Vardeny, E. Ehrenfreund, J. Shinar and F. Wudl, *Phys. Rev. B*, *35* (1987) 2498.
8 W. Porzio, S. Destri, M. Mascherpa, S. Rossini and S. Brukner, *Proc. Conf. on Science and Technology of Synthetic Metals (ICSM '92), Gőteborg, Sweden, Aug. 12–18, 1992; Synth. Met.*, *55–57* (1993) in press.
9 N. Periasamy, R. Danieli, G. Ruani, R. Zamboni and C. Taliani, *Phys. Rev. Lett.*, *68* (1992) 2359.
10 C. Taliani, R. Danieli, R. Lazzaroni, N. Periasamy, G. Ruani and R. Zamboni, *Mol. Cryst. Liq. Cryst.*, *217* (1992) 101.
11 R. Zamboni, G. Ruani, C. Taliani and A. J. Pal, *Mol. Cryst. Liq. Cryst.*, *218* (1992) 113.
12 B. S. Hudson, B. E. Kohler and K. Schluten, *Excited States*, *6* (1982) 1.
13 B. E. Kohler, C. Spangler and C. Westerfield, *J. Chem. Phys.*, *89* (1988) 5422.
14 P. C. M. McWilliams, G. W. Hayden and G. Z. Soos, *Phys. Rev. B*, *43* (1991) 9777.
15 J. Kagan and S. K. Arora, *Heterocycles*, *20* (1983) 1937.
16 P. Ostoja, S. Guerri, S. Rossini, M. Servidori, C. Taliani and R. Zamboni, *Synth. Met.*, in press.
17 P. Ostoja, S. Guerri, M. Impronta, P. Zabberoni, R. Danieli, S. Rossini, C. Taliani and R. Zamboni, *Adv. Mater. Opt. Electron.*, *1* (1992) 127.
18 N. Periasamy, C. Taliani, G. Ruani and R. Zamboni, to be published.
19 R. Zamboni, N. Periasamy, G. Ruani and C. Taliani, to be published.
20 (a) D. P. Craig, L. A. Dissado and S. H. Walmsley, *Chem. Phys. Lett.*, *46* (1977) 191; (b) D. P. Craig, L. A. Dissado and S. H. Walmsley, *Chem. Phys.*, *46* (1980) 87.
21 R. Lazzaroni, A. J. Pal, S. Rossini, G. Ruani, R. Zamboni and C. Taliani, *Synth. Met.*, *42* (1991) 2359.
22 C. Taliani, R. Danieli, R. Lazzaroni, N. Periasamy, G. Ruani and R. Zamboni, *Proc. Conf. on Science and Technology of Synthetic Metals (ICSM '92), Gőteborg, Sweden, Aug. 12–18, 1992; Synth. Met.*, *55–57* (1993) in press.
23 G. Horowitz, X. Z. Peng, D. Fichou and F. Garnier, *J. Appl. Phys.*, *19* (1990) 1489.
24 P. Petelenz and D. Mucha, *Chem. Phys.*, *154* (1991) 145.
25 P. Tavan and K. Schulten, *Phys. Rev. B*, *36* (1987) 4337.

Sample dependence of the magnetic behaviour of polyacetylene. A pulsed ESR comparative study

P. Petit, M. Bernard, J.-J. André, D. Reibel and C. Mathis
Institut Charles Sadron (CRM-EAHP) (CNRS-ULP), 6 rue Boussingault, 67083 Strasbourg Cédex (France)

Abstract

A recent electron spin echo (ESE) study has questioned the existence of mobile defects in *trans*-$(CH)_x$. Our results confirm the very different behaviour of the magnetic species present in Shirakawa *trans*-$(CH)_x$ and Durham *trans*-$(CH)_x$. Pure *trans*-S-$(CH)_x$ presents at room temperature the only highly mobile species. The presence of an ESE in the D-$(CH)_x$ sample confirms the inhomogeneity of the local environment of the spin which may result from the presence of chemical defects inherent of this synthetic route. These chemical defects shorten the lengths of the conjugated segments leading to trapped solitons.

Introduction

The very peculiar properties of polyacetylene $(CH)_x$, i.e., magnetic and electrical properties, have focused the attention of a great number of physicists and chemists all over the world for more than a decade and papers are still being published. Among the most important results, obtained on the *trans* isomer of $(CH)_x$, is the existence of a highly mobile topological defect called a 'soliton' carrying an electronic $\frac{1}{2}$ spin and no charge. The existence of such a defect, which was already predicted in the early 1960s [1] has been modelled by Su *et al.* [2] and extensively studied by magnetic resonance techniques [3]. Until very recently, the solitonic nature of the defects in *trans*-$(CH)_x$ was not contested. Thomann *et al.* [4], using the very powerful electron spin echo (ESE) technique, put in question all the preceding works arguing that, although the spin wave function of the solitons is extended on 14 atoms of carbon, the defect is actually localized on the chains at any temperature. However, these authors never indicated clearly which type of polyacetylene they used. In the present paper we compare two samples of *trans*-$(CH)_x$, i.e., S-$(CH)_x$ and D-$(CH)_x$, obtained, respectively, by the Shirakawa [5] and Durham [6] synthetic routes and show, by a pulsed ESR study, that the nature of the defects is totally different in both samples.

0379-6779/93/$6.00

Experimental results and discussion

Films of the *cis* isomer of S-$(CH)_x$ were prepared at low temperature according to a modified Shirakawa technique [7]. By heating these films at 180 °C for 60 min under high vacuum, *trans* samples were obtained. The precursor polymer for D-$(CH)_x$ was prepared as described by Edwards and Feast [6]. *trans*-D-$(CH)_x$ was obtained by heating powder of this precursor at 100 °C for 24 h under dynamic high vacuum ($<10^{-4}$ Torr).

ESR absorption lines are the picture of the spin Hamiltonian. However, in most of the π-organic materials, only one line is observed. This line can be the envelope of an unresolved hyperfine structure and is called 'inhomogeneous' (Fig. 1). In the case of diffusive spins in pure and undoped systems, local fields are averaged by the exchange interactions

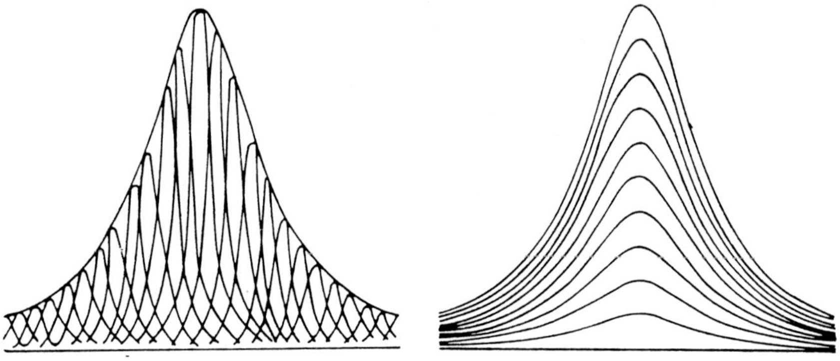

(a) (b)

Fig. 1. (a) Inhomogeneous and (b) homogeneous broadened ESR lines.

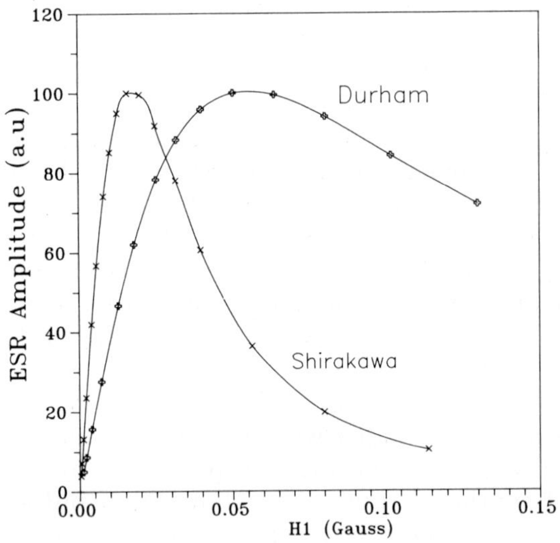

Fig. 2. Saturation curves of D-$(CH)_x$ and S-$(CH)_x$.

and all the spins have the same Larmor frequency. This distinction between homogeneous and inhomogeneous lines can be made by saturation experiments in CW ESR.

These experiments consist in irradiating the sample with an increasing power of the hyperfrequency field leading to an equalization of the Zeeman populations at high power for a homogeneous line, i.e., a zero net magnetization. In the case of an inhomogeneous line, this situation cannot be attained because of the widths of the energy levels (Fig. 2). Another way to show the difference between the two types of lines is to conduct an ESE experiment (Fig. 3). This experiment consists in applying a short intense hyperfrequency $\pi/2$ pulse to the sample. When the pulse is

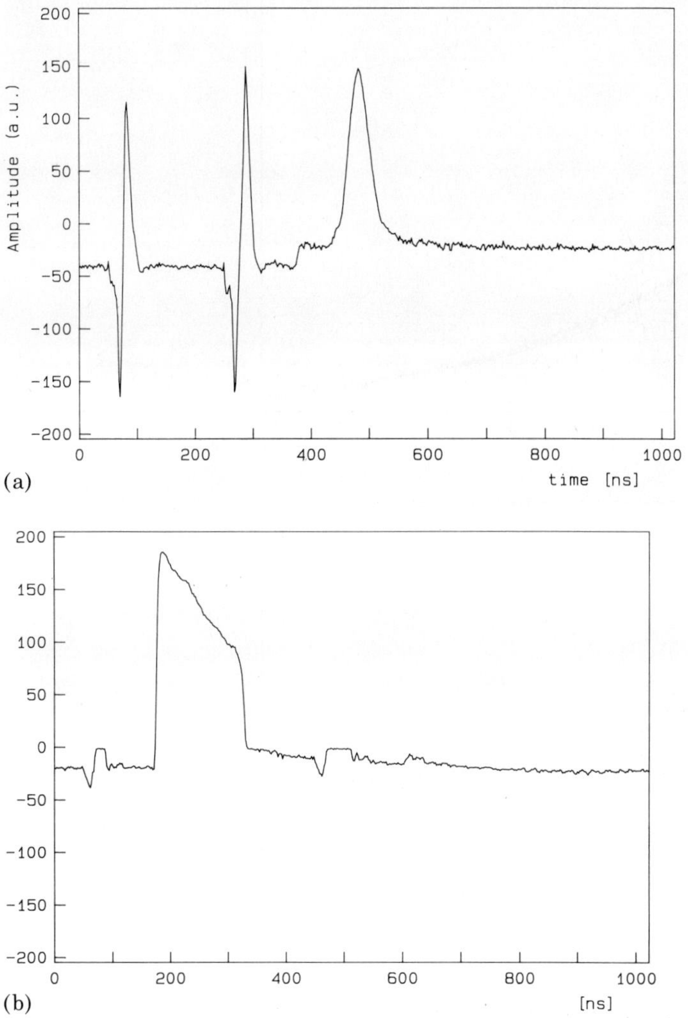

Fig. 3. ESE experiments on (a) D-$(CH)_x$ and (b) S-$(CH)_x$.

cut off, all the spin packets will precess in the rotating frame around the Zeeman field (z direction) at their own frequency, determined by the local fields tested by these spin packets. A π pulse follows the $\pi/2$ one after a time τ leading to the refocusing of the spin packets at time 2τ in the xy plane. This observation of an ESE cannot be made for homogeneous lines because no dephasing occurs between spins after the first $\pi/2$ pulse. Thus, an inhomogeneous line will lead to a spin echo, a homogeneous line will not. Figure 3 shows ESE experiments on D-$(CH)_x$ and S-$(CH)_x$. For the D-$(CH)_x$ sample, an intense echo is observed whereas no ESE is detectable at RT for the S-$(CH)_x$ sample. On D-$(CH)_x$, no free induction decay (FID)

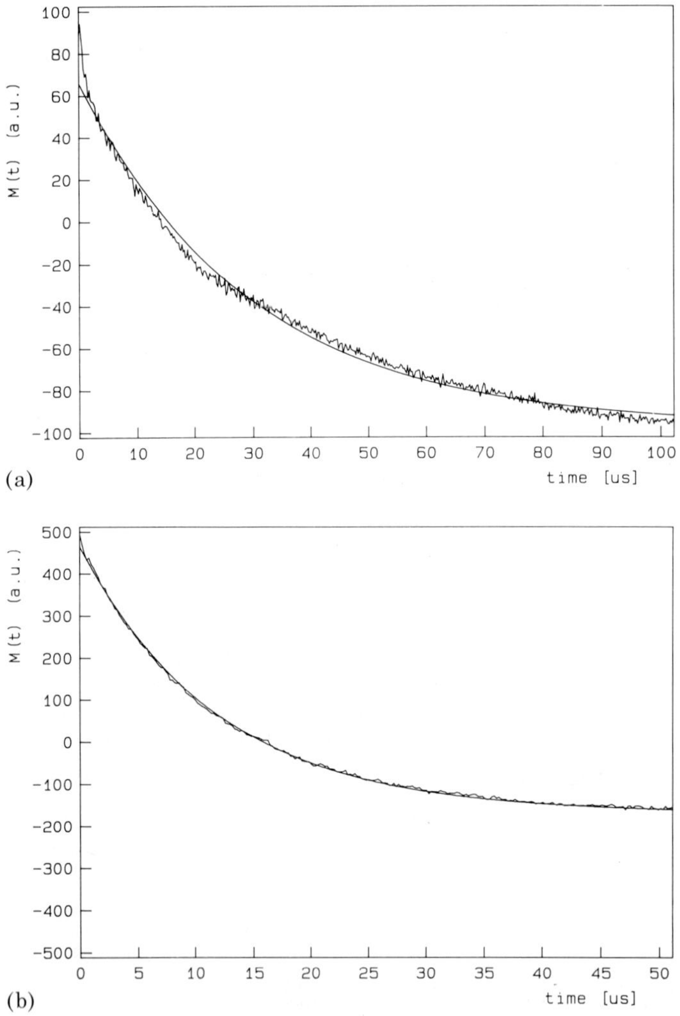

(a) (b)

Fig. 4. Decay of the magnetization of (a) D-$(CH)_x$ and (b) S-$(CH)_x$. Solid lines are monoexponential fits.

is observable indicating that if a homogeneous line was present, its relaxation time T_2 will be $T_2 \ll T_d$, where $T_d = 100$ ns is the dead time of the spectrometer, i.e., a CW ESR linewidth much larger than 0.7 G. In the case of S-$(CH)_x$, however, an intense FID of $T_2 = 180$ ns is seen. These first ESE experiments are confirmed by CW ESR saturation experiments shown in Fig. 2 where the decay of the magnetization is observed only for S-$(CH)_x$.

The drastic difference between both samples is explained by the fact that in D-$(CH)_x$ many chemical defects, such as chain bends, twists, crosslinks and terminations, resulting from the synthetic route, are present [8]. At the ESR timescale, solitons are trapped by these types of chemical defects of high energy so that the solitons test fluctuations of local magnetic fields. In the case of S-$(CH)_x$, no observed ESE at RT indicates:

(i) The inhomogeneity of local fields has no influence on the electronic spins. This is the situation for highly diffusive spins in pure materials. In this case local fields are averaged, leading to a unique effective magnetic environment.

(ii) There are very few chemical defects, mainly chain ends and residual *cis* units, in this material and thus the amount of localized spins is negligible as compared to diffusive ones.

Other experimental evidence of the discrepancy between these two materials lies in the relaxation of their magnetization characterized by a spin lattice relaxation time T_1. Whereas the decay of the magnetization is monoexponential in S-$(CH)_x$ at RT, it is multi-exponential in D-$(CH)_x$ (Fig. 4). This multiple exponential decay is due to the energy distribution of the chemical defects which leads to a distribution in relaxation times.

Conclusions

We have shown by ESE experiments that solitons in *trans*-$(CH)_x$ are very sensitive to the purity of the system. In the case of S-$(CH)_x$, conjugation length is high and very few chemical defects are present, leading to highly mobile spins, as has already been extensively demonstrated. In the case of D-$(CH)_x$ it is clear, from the distribution of the spin lattice relaxation time, that ESE is due to the chemical defects and not to the protons as has been proposed [4]; it is reasonable to assume that all the protons of the chain are equivalent in energy. These chemical defects reduce the length of the conjugated sequences so that the solitons are trapped in this system.

Acknowledgements

The authors are grateful to Professor G. Wegner and Dr R. Weizenhöfer from the Max-Planck-Institut für Polymerforschung, Mainz, Germany, for providing the Durham precursor polymer.

References

1 J. A. Pople and S. H. Walmsley, *Mol. Phys.*, *5* (1962) 15.
2 W. P. Su, J. R. Schrieffer and A. J. Heeger, *Phys. Rev. Lett.*, *42* (1979) 1698; W. P. Su, J. R. Schrieffer and A. J. Heeger, *Phys. Rev. B*, *22* (1980) 2099.
3 M. Nechtschein, F. Devreux, R. L. Greene, T. C. Clarke and G. B. Street, *Phys. Rev. Lett.*, *44* (1980) 356; K. Holczer, J.-P. Boucher, F. Devreux and M. Nechtschein, *Phys. Rev. B*, *23* (1981) 1051; M. Nechtschein, F. Devreux, F. Genoud, M. Guglielmi and K. Holczer, *Phys. Rev. B*, *27* (1983) 61.
4 H. Thomann and H. Jin, in C. P. Keijers, E. J. Reijerse and J. Schmidt (eds.), *Pulsed EPR: A New Field of Applications*, North-Holland, Amsterdam, 1989.
5 J. Ito, H. Shirakawa and S. Ikeda, *J. Polym. Sci., Polym. Chem. Ed.*, *12* (1974) 11.
6 J. H. Edwards and W. J. Feast, *Polymer*, *21* (1980) 595.
7 B. François, M. Bernard and J. J. André, *J. Chem. Phys.*, *75* (1981) 4142.
8 D. C. Bott, C. S. Brown, C. K. Chai, N. S. Walker, W. J. Feast, P. J. S. Foot, P. D. Calvert, N. C. Billingham and R. H. Friend, *Synth. Met.*, *14* (1986) 245.

3a. Photoexciations and Fast Phenomena

Ultrafast relaxation in conjugated polymers

Takayoshi Kobayashi
Department of Physics, Faculty of Science, University of Tokyo, Hongo 7-3-1, Bunkyo-ku, Tokyo 113 (Japan)

Abstract

Femtosecond pump–probe spectroscopy has been studied for polydiacetylenes (PDA-3BCMU (poly[4,6-decadiyne-1,10-diol-bis(n-butoxycarbonylmethylurethane)] (blue phase)) and PDA-4BCMU (poly[5,7-dodecadiyne-1,12-diol-bis(n-butoxycarbonylmethylurethane)] (red and blue phases)), and PDA-DFMP (poly[1,4-bis(2.5-bistrifluoromethylphenyl)-1.3-butadiyne]) and poly-thiophenes (P3MT (poly(3-methylthiophene)) and P3DT (poly(3-dodecylthiophene))) using a CPM (colliding-pulse mode-locked) laser with pulse width of 100 fs. The decay kinetics of the induced absorption intensity due to the self-trapped (ST) excitons in nonfluorescent polymers (PDA-3BCMU and PDA-4BCMU in the blue phase) is approximately represented by a single-exponential function after long-life components due to triplet excitons, polarons and/or bipolarons are subtracted. The exponential formation and decay time constants of ST excitons in nonfluorescent PDAs range between 70 and 150 fs and 1 and 3 ps, respectively, at any temperature between 10 and 290 K. Both the formation and decay times are only weakly temperature dependent. The decay kinetics of the ST exciton in fluorescent polymers (PDA-4BCMU in the red phase, P3MT and P3DT) substantially deviates from the single-exponential function. An adiabatic potential curve model is proposed to explain this dynamic behavior of ST excitons. The ST excition formation corresponds to the spontaneous geometrical relaxation of the free exciton in the one-dimensional system. The decay channel of the ST exciton is the tunneling process from the potential curve of the ST exciton to that of the ground state.

1. Introduction

Nonlinear optical materials are under extensive study for future practical applications in nonlinear optical devices, such as optical switches and memories because of their large nonlinearity, ultrafast response and large varieties of main chains and side groups [1–13]. Conjugated polymers are supposed to be a group of promising materials for application [14].

In order to realize nonlinear optical devices the mechanism of the ultrafast optical nonlinearity must be clarified. For example, the factors determining the size of the third-order susceptibility and relaxation time must be clarified to obtain a guiding principle of materials search and preparation.

In the present study the ultrafast optical responses in various polymers were investigated by femtosecond absorption and picosecond luminescence spectroscopies. The third-order nonlinear optical susceptibi-

0379-6779/93/$6.00

lities of various processes were also determined from the experimental results. The mechanism of the ultrafast relaxation in conjugated polymers is discussed in terms of self-trapped excitons (STEs).

2. Experimental

Several polydiacetylenes (PDAs) and polythiophenes (PTs) were studied. Cast films of blue-phase poly[4,6-decadiyne-1,10-diol-bis(n-butoxycarbonylmethylurethane)] (PDA-3BCMU) and red-phase poly[5,7-dodecadiyne-1,12-diol-bis(n-butoxycarbonylmethylurethane)] (PDA-4BCMU) were prepared by spin coating. Oriented films of PDA-4BCMU were prepared by the rubbing method similar to those used in liquid crystal studies. Evaporated thin films of PDA-3BCMU on a KCl single crystal surface (100) were also used. The sample films are composed of many single crystalline regions aligned either in the [110] or [1$\bar{1}$0] direction and amorphous domains. Electrochemically prepared films of poly(3-methylthiophene) (P3MT), poly(3-dodecylthiophene) (P3DT) were also used as samples.

A CPM laser with a four-stage dye amplifier pumped by a 10 Hz Q-switched Nd:YAG laser was used for femtosecond spectroscopy. The width, energy and central wavelength (photon energy) of the amplified femtosecond pulse are typically 80–100 fs, 0.2–0.1 mJ and 630–628 nm (1.97 eV), respectively.

3. Experimental results of femtosecond spectroscopy

The experimental results of the femtosecond time-resolved spectroscopy of several polydiacetylenes (PDAs) and polythiophenes (PTs) are described below.

3.1. Blue-phase PDA-3BCMU (cast film)

Figure 1 shows the femtosecond photoinduced absorption spectra in a cast film of PDA-3BCMU at 10 K [1]. The photon energy of the fundamental femtosecond pulse (1.97 eV) is very close to the absorption peak of the ^1B_u exciton (1.92 eV at 10 K and about 1.95 eV at 290 K) in blue-phase PDA-3BCMU. An oscillatory structure around 1.97 eV observed at negative delay times is due to the perturbed free induction decay observed also for several other materials [15–17]. The sharp bleaching peak at 1.97 eV is due to hole burning sharpened by the effect owing to coherent pump polarization coupling.

Two small, but highly reproducible, downward peaks at 1.79 and 1.71 eV observed at zero delay time are due to the probe amplification owing to the stimulated Raman gain [1, 2] corresponding to the Stokes Raman shifts of 1450 cm^{-1} (C=C stretching) and 2100 cm^{-1} (C≡C stretch-

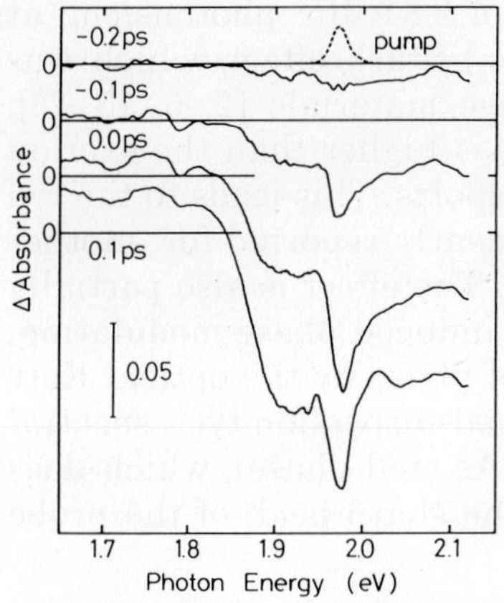

Fig. 1. Transient photoinduced absorption spectra of PDA-3BCMU cast film in the blue phase at 10 K and spectrum of the pump pulse (dotted curve). The downward peak at 1.97 eV between −100 and 100 fs is due to the hole burning and coherent interaction in the four-wave mixing process. The downward peaks around 1.71 and 1.80 eV are due to Raman gain with 2090 and 1450 cm^{-1} shift, respectively, from a pump laser photon energy at 1.97 eV. The excitation photon density is 9.5×10^{14} photons/cm^2.

ing), respectively [1–5]. The vibrational modes are coupled strongly with the exciton transition. They were also found in the inverse Raman of PDA-TS (poly[2,4-hexadiyne-1,6-diol-bis(p-toluene sulfonate)]) [5, 6].

The asymmetric bleaching near 1.97 eV in the transient spectrum of PDA-3BCMU cast film in the blue phase at 0.0 ps (shown in Fig. 2) is

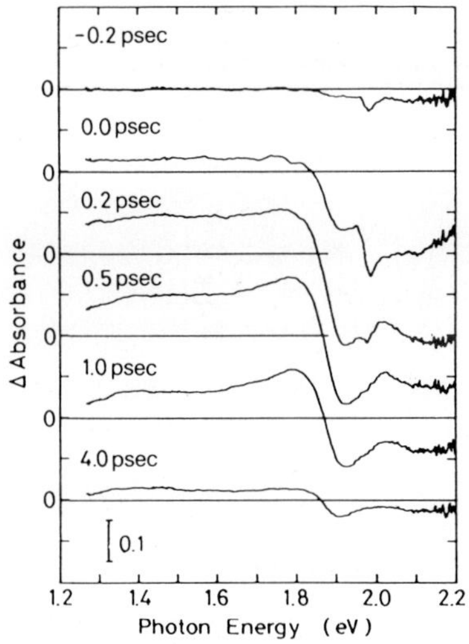

Fig. 2. Transient photoinduced absorption spectra of PDA-3BCMU cast film in the blue phase induced by a 1.97 eV pump pulse at 10 K. The excitation photon density is 3.8×10^{15} photons/cm^2.

induced by a high excitation photon density of 3.8×10^{15} photons/cm^2 at 10 K. This is partially due to the optical (a.c.) Stark effect, which was recently discussed for several condensed-phase materials [2, 6, 15–23]. The pump photon energy in the present study is higher than the exciton transition energy in contrast to the previous reports. This leads to the red shift [2] as theoretically expected and was recently reported for another system of a semiconductor by Hulin *et al.* [24]. The effect is also partially due to the shift of the probe spectrum by the induced phase modulation, which is caused by the ultrafast modulation of phase by the optical Kerr effect [25]. This can be verified by the observed dispersion-type spectral change similar to the a.c. Stark effect for PDAs (red phase), which does not have an exciton peak near 2 eV, due to the sharp peak of the probe spectrum [4].

3.2. Blue-phase PDA-4BCMU (oriented film)

The stationary and transient photoinduced absorption spectra of the blue-phase oriented PDA-4BCMU film at the delay times when both pump and probe pulses overlap in time in the sample are shown in Fig. 3. The polarizations of the pump and probe pulses are parallel to the polymer main chain. The bleaching observed at −0.2 and −0.1 ps at the pump photon energy at 290 K becomes broader with time from −0.1 to 0.2 ps and the bleaching peak shifts to 1.96 eV. At 10 K the bleaching peak at 1.97 eV due to coherent coupling of the pump polarization appears when the pump and probe pulses overlap and another bleaching peak appears at

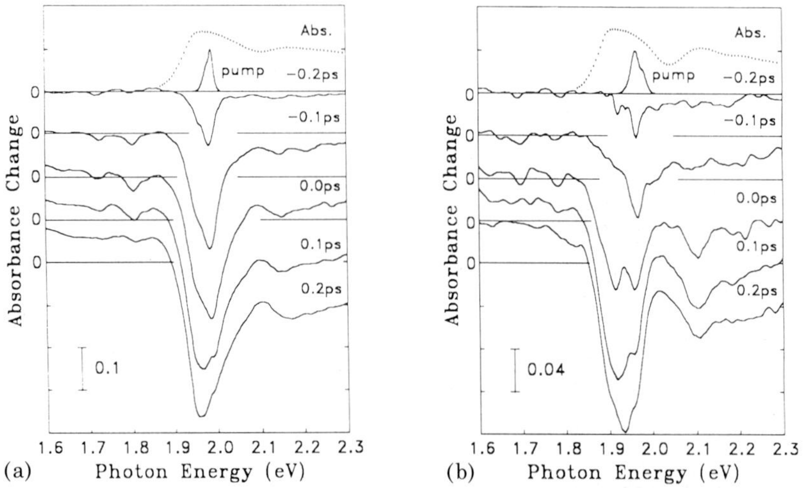

Fig. 3. Transient absorption spectra of a blue-phse oriented PDA-4BCMU film (a) at 290 K with the excitation energy of 1.98 eV and (b) at 10 K with that of 1.97 eV. The pump and probe pulses are polarized parallel to the polymer chain. The absorption (dotted curves) and pump spectra are shown together. The excitation photon density is 2.0×10^{15} photons/cm^2 in both cases.

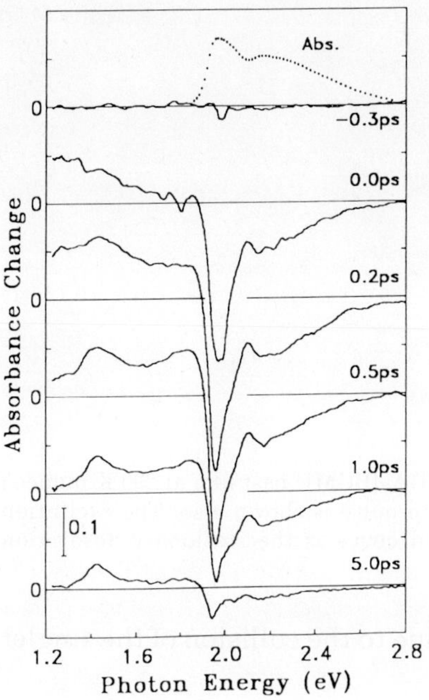

Fig. 4. Transient absorption spectra of the blue-phase oriented PDA-4BCMU film at 290 K induced by the 1.98 eV pump pulse. The excitation photon density is 2.0×10^{15} photons/cm^2. The dotted curve is the stationary absorption spectrum of the sample film.

1.91 eV due to the saturation of the excitonic absorption. The absorbance changes below the pump photon energy have two minima at 1.80 and 1.72 eV at 290 K, and 1.79 and 1.71 eV at 10 K. Figure 4 shows the transient absorption spectra of the oriented PDA-4BCMU film in the blue phase at 290 K up to 5 ps in the spectral range of 1.2–2.8 eV. At 0.0 ps bleaching above 1.9 eV and absorption below 1.7 eV are observed. The photoinduced absorption at 0.0 ps has a peak at the photon energy lower than 1.4 eV. Then the absorption spectrum shifts to higher energy with time from 0.0 to 0.5 ps. The absorption at 0.5 ps has a peak at 1.8 eV.

The absorption at 1.4 eV, which has a much longer life than 100 ps, is assigned to triplet excitons [27, 28]. The pump intensity dependence of the singlet excitons shows a linear response [1, 2, 28], while the long-lived component increases superlinearly with pump intensity. The pump intensity dependence of the triplet excitons was studied in another polydiacetylene, poly[1-(3-methylaminophenyl)-4-(3,5-bis(trifluoromethyl)-phenyl)-1,3-butadiyne] (PDA-MADF) in detail [28]. The result shows that the population of the triplet excitons is proportional to the square of the singlet exciton population even at higher-pump intensities, where the singlet excitons exhibit saturation. The formation mechanism of the higher excited singlet excitons which are the precursors of the triplet excitons generated by the 1.97 eV pump pulse was concluded to be not due

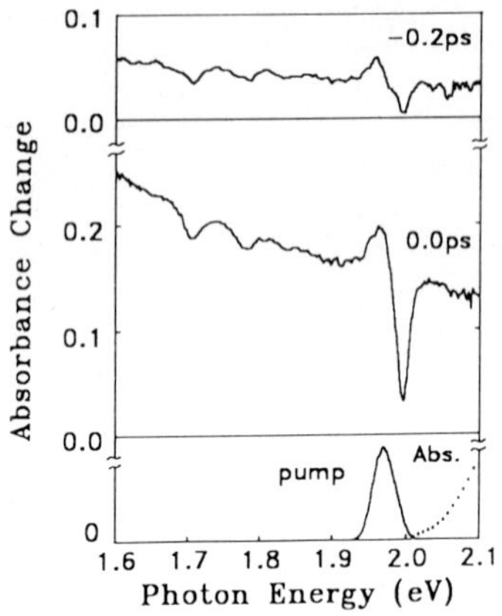

Fig. 5. Transient absorption spectra of a red-phase PDA-4BCMU cast film at 290 K induced by the 1.97 eV pump pulse. The spectrum of the pump pulse is shown also. The excitation photon density is 4.0×10^{16} photons/cm^2. The dotted curve is the stationary absorption spectrum.

to the two-step two-photon absorption but due to the collision of the singlet excitons [28].

3.3. Red-phase PDA-4BCMU (cast film)

Figure 5 shows the transient absorption spectra of the red-phase PDA-4BCMU cast film at 290 K excited by the 1.97 eV pump pulse. The observed absorbance change is positive in the whole photon energy region in the Figure. Since the absorption edge of the cast film is 2.01 eV, the 1.97 eV pump pulse cannot directly create the exciton. The pump intensity dependence of the absorbance change is represented by I^a with $a = 1.9 \pm 0.1$. The observed transient absorption is induced by the two-photon absorption of the 1.97 eV pump pulse. The absorbance change at -0.2 ps around 1.97 eV in the cast film has an antisymmetric oscillatory structure. The observed antisymmetric structure agrees with the differential spectra of the probe's white continuum pulse generated from the 1.97 eV pulse by the cross-phase modulation. The amplitude of the oscillation is proportional to pump intensity. The structure around 1.97 eV can be explained in terms of induced-frequency shift of the probe light [24, 29]. The induced phase modulation (IPM) can be verified by changing the spectra of the probe pulse. The change in the probe pulse spectrum due to the induced frequency shift was clearly found in GaAs microcrystallites [24].

The absorbance change spectrum at 0.0 ps near 1.97 eV is asymmetric. This is due to both the IPM and the pump polarization coupling. Since the refractive index change causing IPM is due to the formation of the free exciton and self-trapped exciton, the effect is even larger at a 0.0 ps delay time than -0.2 ps. If the refractive index change is the virtual effect as in the a.c. Stark effect, IPM disappears at 0.0 ps and the absorbance change reverses its sign at later delay times.

TABLE 1

Time constants τ_{ij} in the relaxation processes of excitons in polymers (i and j correspond to the states in Figs. 6, 8 and 9)

Polymer[a] (phase and form)	$\tau_{2\rightarrow 3}$ (fs) (290 K)	$\tau_{2\rightarrow G}$ (ps) (10 K)	$\tau_{2\rightarrow G}$ (ps) (290 K)	$\tau_{3,4\rightarrow G}$ (ps) (10 K)	$\tau_{3,4\rightarrow G}$ (ps) (290 K)	Emission property[b]
PDA-3BCMU (b, oriented film[c])	150 ± 40			1.9 ± 0.2	1.6 ± 0.1	n
PDA-3BCMU (b, cast film)	150 ± 50	<1.0	<1.0	2.0 ± 0.2	1.5 ± 0.2	n
PDA-DFMP (b, single crystal)	100 ± 30				1.6 ± 0.1	n
PDA-4BCMU (r, oriented film)	140 ± 40	<1.0	<1.0	2.1 ± 0.1	1.6 ± 0.1	n
PDA-4BCMU (r, oriented film)	120 ± 60					f
PDA-4BCMU (b, cast)	<200	0.9 ± 0.1	0.7 ± 0.1	5.6 ± 0.1	4.4 ± 0.5	f
P3MT (electrochem. prep.[d])	70 ± 50	0.6 ± 0.1		>5		f
P3DT (electrochem. prep.[d])	100 ± 50	0.5 ± 0.2	1.0 ± 0.2	>5	4.7 ± 1.2	f

[a]b is blue phase; r is red phase.

[b]Polymers with higher and lower fluorescence quantum efficiency than about 10^{-5} are indicated by f (fluorescent) and n (nonfluorescent), respectively.

[c]The sample was prepared by the polymerization of evaporated monomer film of PDA-3BCMU on a KCl crystal. The sample was composed of small crystals and amorphous regions.

[d]The samples were prepared by the method of electrochemical reaction.

The time constants of relaxations of STEs in the red-phase PDA-4BCMU are listed in Table 1.

3.4. P3DT (cast film)

The absorbance change in poly(3-dodecylthiophene) has similar properties to red-phase PDA-4BCMU. Bleaching due to absorption saturation and the broad absorption below 2 eV were observed. The decay kinetics of the absorbance change can be empirically fitted to the biexponential function and the time constants are obtained at 0.3 ± 0.1 and 4.7 ± 1.6 ps at 2.2 eV.

4. Discussion

4.1. Mechanism of exciton self-trapping in the one-dimensional system

Since there is no barrier in the potential curve between free and self-trapped (ST) excitons in the one-dimensional system as shown in Fig. 6, the formation of the ST excitons is expected to take place within a period of the coupled phonon cycle $T_M = 2\pi/\omega_M$. The experimental results of the absorption spectra, the Raman gain [1, 2], the resonance Raman scattering and the phonon-mediated optical nonlinearity [5, 6] show that the singlet excitons in polydiacetylenes are coupled with the C=C and C≡C stretching modes. The coupled stretching mode frequencies are about 1500 cm^{-1} (C=C stretching) and 2100 cm^{-1} (C≡C stretching) and correspond to oscillation periods of 20 and 15 fs, respectively. Both are much shorter than the experimentally observed appearance times of ST excitons, which are 150 fs in PDA-3BCMU, 100 fs in PDA-4BCMU, 70 fs in P3MT and 100 fs in P3DT.

In the following, 'self-trapping' is used in two ways. One is the change in the potential curve of excitons which have the same energy as the free

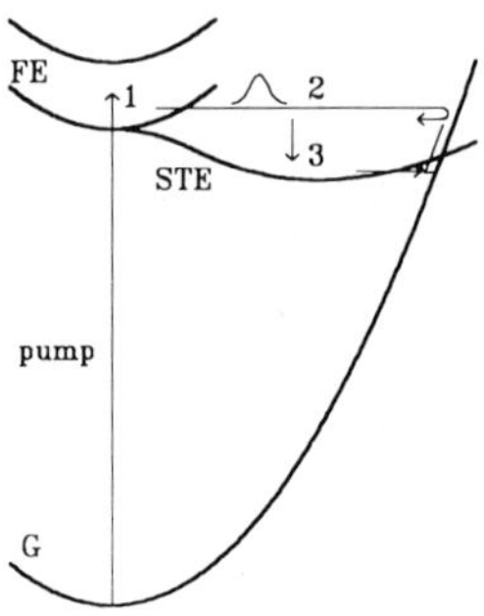

Fig. 6. Potential curves of the ground state (G), free exciton (FE) and self-trapped exciton (STE) of the nonfluorescent polydiacetylenes. There is no barrier between the free-exciton (FE) and self-trapped-exciton potential curves. Numbers 1, 2 and 3 indicate FE, nonthermal STE and quasi-thermal STE, respectively. A part of this Figure is enlarged and shown in Fig. 8.

excitons. The self-trapping time estimated above from the oscillation period is used in this meaning. The other is the quasi-thermalization associated with the intramolecular vibrational relaxation. After this process the exciton energy is lower than that of the free exciton and the population of the ST excitons is distributed near but not at the bottom of the 'ST' exciton (STE) potential in Fig. 6.

According to Jortner [30], the rate of the self-trapping process of the second meaning in rare gas solids such as Ne and Ar at low temperatures is given by $k_{ST} = \omega_M \exp(-\omega_M/\omega_D)$, ω_M and ω_D being the frequencies of the most strongly coupled mode and the Debye frequency, respectively. Using the values of ω_D in the literature and ω_M given above, the calculated rate k_{ST} is several orders of magnitude smaller than the observed rate in the polymers [1]. This is because of the simple phonon structure in the model being only suited to the rare gas solids. Much longer self-trapping time can be explained by the geometrical reorganization (relaxation) involving the lower frequency modes than by the oscillation period of the strongly coupled modes. The lower frequency modes may include those associated with the structural change in the bulky side chains, which are lacking in the rare gas solids.

There is a difference in the formation of ST excitons, namely 'self-trapping' time of the second meaning, between PDAs and PTs. This formation time corresponds to $\tau_{2\to3}$ representing the time constant of the process $2\to3$ in Fig. 6. The formation times are 150 fs in PDA-3BCMU and 140 fs in PDA-4BCMU in the blue phase, and 70 fs in P3MT and 100 fs in P3DT. This difference can be explained by the difference in the chemical structures of these two groups of polymers. PDA-3BCMU and PDA-4BCMU have side groups with hydrogen bonds connecting neighboring side chains, while P3MT and P3DT have rigid structures of thiophene rings conjugated with each other. The van der Waals interaction between the neighboring side chains of PTs results in the helical noncoplanar structure due to the steric hindrance among the side groups [31, 32]. Van der Waals interaction between the side chains and that between main chain and side chain is stronger in P3DT than in P3MT because of the longer side group in the former. This explains the longer formation time in the former. The self-trapping (quasi-thermalization of the intramolecular vibrational energy) time constant of excitons in P3MT is 70 fs [3], which is different from 150 fs in PDA-3BCMU by only a factor of two even though the differences in the structure and size of the side chains are very large. This may be due to the following three possible mechanisms. Mechanism (1): Finite time is needed for the self-trapped exciton to slide down along the potential curve because the potential curve of the self-trapped exciton deviates from the free exciton potential bottom, which is flat with a null derivative [33, 34]. Mechanism (2): According to the theory of Sumi *et al.* concerning the self-trapping process in reduced dimensionalities [34], the spontaneous geometrical relaxation, which takes place from the infinitely extending free exciton to the self-trapped exciton, follows the reaction pathway with a very small slope along the reaction coordinate. Mechanism (3): Because of the interchain interaction, the conjugated polymers may not be an ideal one-dimensional system. Then there may be a low barrier between two minima in the potential curves of the ground state and the ST exciton state.

4.2. Decay mechanism of self-trapped excitons in conjugated polymers

In this section the following properties are addressed concerning the excited state dynamics in conjugated polymers, especially polydiacetylenes:

(1) Many conjugated polymers are nonfluorescent or only weakly fluorescent. In particular, the quantum efficiency of the nonfluorescent blue-phase polydiacetylenes is estimated to be lower than 10^{-5} and that of the fluorescent red-phase polydiacetylenes is only of the order of 10^{-4}.

(2) There are common features in the transient absorption spectra of both PDAs (blue-phase PDA-3BCMU and PDA-4BCMU and red-phase PDA-4BCMU) and PTs either in random films or in oriented films. One feature is very broad absorption spectra in the 1.35–1.9 eV region observed just at excitation (0 ps). The broad width indicates the transition to a continuum state, i.e., the free exciton or ST exciton to the conduction band [1, 2].

(3) Triplet excitons cannot be formed directly from singlet excitons. They are formed only at a high density excitation of exciton transition or by creation of electrons and holes by the interband transition.

(4) The time constants of the relaxation processes of excitons and photoluminescence properties of several PDAs and PTs studied by our group are summarized in Table 1.

Figure 7 shows the time dependence of the absorbance change due to the self-trapped excitons in several PDAs and PTs, after the component with very long lifetimes is subtracted. As can clearly be seen in the Figure, the decays of nonfluorescent PDA-3BCMU, PDA-4BCMU and PDA-DFMP all in the blue phase are approximately given by an exponential function, except the early stage within 500 fs. The other fluorescent polymers have nonexponential decay from a very early delay time until 30 ps. The initial decay times, defined by the slope of the semilogarithmic plot of the absorbance change against delay time just after excitation, are determined as 900 ± 100 fs, 600 ± 100 fs and 500 ± 200 fs in PDA-4BCMU in the red phase, P3MT and P3DT, respectively, at 10 K. The decay becomes slower and slower at longer delay times because of the less efficient tunneling due to the thicker barrier width that the population of excitons must experience. The final time constants determined from the slope of the semilogarithmic plot of the absorbance change after the long-life component ($\gg 10$ ps) is subtracted are also listed. The slope was determined in the time range of 2–10 ps.

The lifetime of the self-trapped ^1B_u excitons in PDA-3BCMU is much shorter than the radiative life (a few ns) estimated from the oscillator

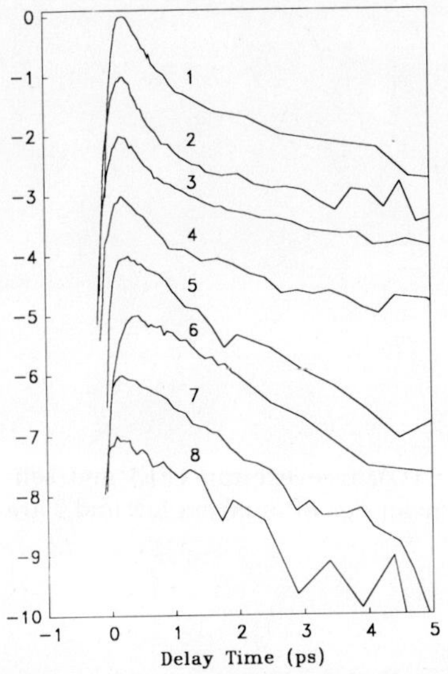

Fig. 7. Time dependence of the absorbance change due to the self-trapped excitons in several PDAs and PTs after the very long-lived component is subtracted: (1), P3MT (observed at 10 K at the probe photon energy of 1.63 eV); (2), P3DT (10 K, 2.34 eV); (3), PDA-4BCMU (red-phase cast film, 10 K, 1.77 eV); (4), PDA-4BCMU (red-phase oriented film, 10 K, 1.77 eV); (5), PDA-4BCMU (blue-phase oriented film, 290 K, 1.77 eV); (6), PDA-3BCMU (blue-phase cast film, 290 K, 1.77 eV); (7), PDA-3BCMU (oriented single crystals with smaller amorphous domains prepared by evaporation on a KCl single crystal surface (100), 290 K, 1.77 eV); and (8), PDA-DFMP (single crystals, 297 K, 1.98 eV). The excitation pulse width and energy are 100 fs and 1.97 eV, respectively. The samples of PDA-4BCMU in the red phase are excited by the two-photon absorption process.

strength of the transition. This means predominant radiationless relaxation in the exciton decay process. Excitons in many semiconductors and organic molecular crystals have lifetimes between several hundred picoseconds and several hundred nanoseconds at low temperatures. More than one order of magnitude differences in the lifetime of these excitons are usually found between low temperatures and room temperature. The activation potential barrier is very often invoked to explain the temperature dependence in most excitons in semiconductors, organic molecular crystals and also in ionic crystals.

Photoexcitation relaxation in polymers can be described using PDAs and PTs as examples [1–3, 35]. The differences in the relaxation kinetics of excitons in the nonfluorescent PDA-3BCMU and PDA-4BCMU in the blue phase and fluorescent PDA-4BCMU in the red phase and P3MT and P3MT described in Section 3.1 are discussed. The discussion is based on Toyozawa's theory [36, 37]. Figures 8 and 9 show the adiabatic potential curves of the ground state (G) and self-trapped exciton (STE), and free exciton (FE) band of the nonfluorescent (Fig. 8) and fluorescent (Fig. 9) polymers, respectively. The curvature radius of the potential of STE is

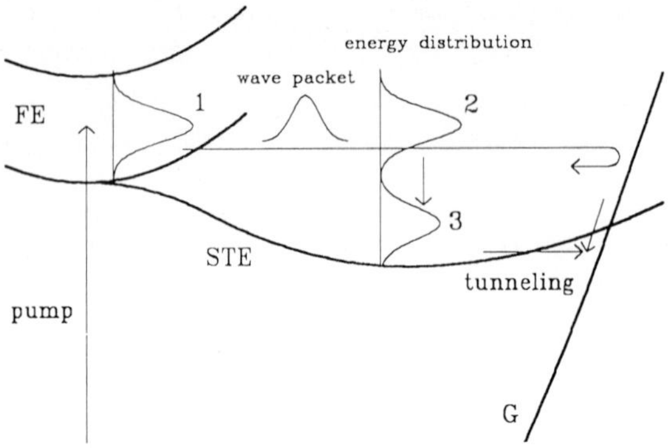

Fig. 8. Adiabatic potential curves of the ground state (G), free exciton (FE) and self-trapped exciton (STE) of the nonfluorescent PDAs. The meanings of numbers 1, 2 and 3 are the same as in Fig. 6.

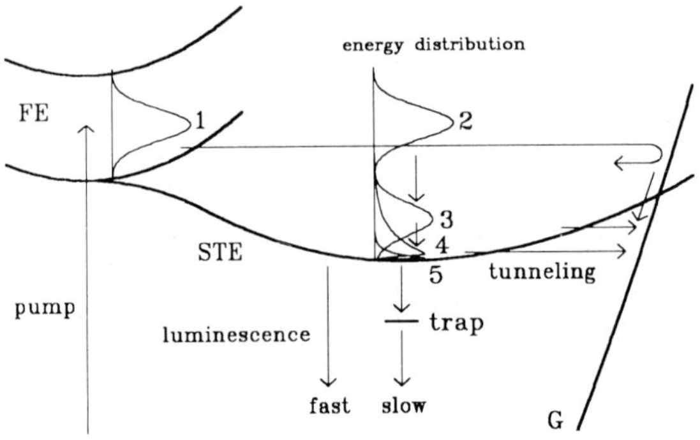

Fig. 9. Potential curves of the ground state (G), free exciton (FE) band and self-trapped exciton (STE) of the fluorescent polymers. Numbers 1, 2 and 3 are the same as in Fig. 6. Number 4 means STE after thermal equilibration among intramolecular vibrational modes to a local temperature, which is higher than the bulk experimental temperature of the samples. Number 5 indicates the STE after the local temperature is lowered to the bulk experimental temperature by the thermal diffusion process.

larger than those of G and FE states in order to take into account the discussion in the reaction coordinate by Sumi *et al.* [34] and of the small Stokes shift observed for the fluorescent polymers. The electronic structures of the polymer systems are too complicated to be represented by the one-dimensional potential curve. The small Stokes shift may also be explained by the ground state potential having two minima with a small barrier in-between corresponding to the acetylenic and butatrienic configurations or to the aromatic and quinoid configurations in the cases of PDAs and PTs, respectively.

In Figs. 8 and 9, the self-trapping process from FE (indicated by 1 in the Figures) to 'hot' STE (indicated by 2 in the Figures) is the process $1 \rightarrow 2$, followed by the intramolecular vibrational relaxation. The STEs after the vibrational process still remain in the unthermalized states, called 'quasi-thermal' STE in the present paper (indicated by 3 in Figs. 8 and 9). The quasi-thermal STEs (3) are coupled with intramolecular vibrations of low frequencies and then thermalize to states in which temperature of the intramolecular vibration can be defined. The thermalization process ($3 \rightarrow 4$) is observed as the spectral change at delay times from 0.5 to 5.0 ps. The time constant of the thermalization can be determined by the time dependence of absorbance change ratio among several different wavelengths. The obtained time constant of the thermalization process is about 1 ps in both red-phase PDAs and P3DT. However, the temperature of the STEs defined from equilibrated vibrational modes after this 1 ps thermalization process may be still higher than the bulk temperature of the polymer sample. Therefore, the temperature-decreasing process ($4 \rightarrow 5$) with a time constant in the nanosecond–microsecond range is expected to take place via thermal diffusion after the observed thermalization process. However, the spectral change due to the cooling process of the STEs could not be clearly detected in this study because of limited signal-to-noise ratio of the spectral data.

The change of the decay rate observed in the red-phase PDA-4BCMU and P3DT can be explained by the competition between the thermalization and tunneling processes. The STEs relax to the ground state mainly by tunneling through the barrier between the STEs and ground-state potentials. The initial fast decay is the relaxation from the nonthermal and quasi-thermal STEs to the ground state (indicated by $2 \rightarrow G$ and $3 \rightarrow G$, respectively, in Fig. 9) with the time constant of about 1 ps or slightly less. After the quasi-thermalization and subsequent thermalization, the STEs come down close to the bottom of the potential and the loss rate by tunneling becomes slower because of the increase in the barrier height and thickness. The decay time constant of about 5 ps is due to the tunneling from the thermal STE (indicated by 4 in Fig. 9). The thermalization process of the STE is estimated as about 1 ps from the spectral change. The wavelength dependence of the decay curve can be explained by the thermalization process. The absorbance change at 2.00 eV in the red-phase PDAs is mainly due to the quasi-thermal STEs and the thermal STEs have an absorption peak at 1.6 eV. Therefore, the decay curve at 2.00 eV can be due to a single exponential function with a time constant of about 1 ps. The time constant of the absorbance change below 1.8 eV is longer than that of the bleaching, because the quasi-thermal STEs thermalize with a time constant of 1 ps and the absorption peak due to the thermal STEs appears at 1.6 eV. The observed data for P3MT and P3DT can also be explained in the same way.

The decay of the quasi-thermalized (namely after the relaxation of intramolecular vibration with large frequency) STE in blue-phase PDAs is faster than in red-phase PDAs, because the crossing point is lower than the point in red-phase PDAs. The major part of the STEs relax to the ground state before thermalization. The decay rate from the quasi-thermal STEs to the ground state is not much slower than the decay rate from the nonthermal STEs. Therefore, the decay curves in the blue-phase PDAs can be fitted to single-exponential functions. The wavelength dependence of the time constant is due to the mixed signals of the STE states. The bleaching signal exhibits the nonthermal and quasi-thermal (i.e. after the relaxation of intramolecular vibrational relaxation with large frequency) STEs, while the absorption around 1.8 eV is mainly due to the quasi-thermal STEs. Therefore, the observed decay time constant of the bleaching is shorter.

The difference between fluorescent and nonfluorescent polymers is the time constant of the tunneling from the unthermalized and thermalized STEs to the ground state. The time constants in the blue-phase PDAs are between 1 and 3 ps and they are shorter than that in the red-phase PDAs. Therefore, the population of the remaining STE at the bottom of the STE potential curves in the blue-phase PDA is smaller and the luminescence could not be detected. The fast component of the luminescence in the red-phase PDAs is considered to be due to the free excitons and/or STEs and the slow component is due to the traps in the polymer chains. The detailed study using higly oriented samples is also consistent with the above model [38].

The proposed relaxation model predicts that the decay rate of the STE in conjugated polymers depends on the exciton energy. Conjugated polymers with large exciton energy are considered to have excitons with long lifetime and to be more fluorescent. For example poly(*p*-phenylene-vinylene) (PPV), which has an absorption peak at 2.5 eV, has strong luminescence with a quantum yield of several per cent [39]. Recently, a Langmuir–Blodgett film of PDA-(12,8), which has an exciton peak at 1.88 eV, has been studied and the decay time constant of the STE is estimated as 1.3 ± 0.1 ps at 290 K [40]. It is shorter than that in other blue-phase PDAs which have an exciton peak at 1.97 eV. These results are consistent with the above-proposed model. However, PDA under hydrostatic pressure has different decay kintics. When the pressure increases, the absorption edge of red-phase PDA-4BCMU shifts to lower energy but the decay kinetics becomes slower [41]. This pressure effect is probably due to a subsequent three-dimensional distortion of the polymer chain. The decay kinetics under the pressure may be different from that in the one-dimensional system.

Thus, low fluorescence quantum efficiency of the one-dimensional polymers is systematically explained by the proposed model shown by Figs. 8 and 9. While in fluorescent PDA-4BCMU in the red phase and P3MT and P3DT, the crossing points are higher than in PDA-3BCMU and PDA-4BCMU in the blue phase because of the higher energy of free and ST excitons. In the red-phase PDA-4BCMU at larger fraction (probably 10 to several tens of a per cent) of free excitons is relaxed to ST excitons with a 3 ps lifetime at 10 K. This is represented in Figs. 8 and 9 as fast

TABLE 2

Time constants of the relaxation processes of excitons at 290 K

Relaxation process[a]	PDA-4BCMU (blue) (oriented film)	PDA-4BCMU (red) (oriented film)	P3DT
$1 \rightarrow 2$ (self-trapping)[b]	(10–20 fs)[f]	(10–20 fs)[f]	(10–20 fs)
$2 \rightarrow 3$ (intramolecular vibrational relaxation)	140 ± 40 fs	120 ± 60 fs	100 ± 50 fs
$3 \rightarrow 4$ (intramolecular vibrational relaxation)	[c]	1.1 ± 0.1 ps	1.0 ± 0.2 ps
$2 \rightarrow G$ (tunneling from nonthermal STE)	<1.0 ps	0.7 ± 0.1 ps[d]	0.3 ± 0.1 ps[d]
$3 \rightarrow G$ (tunneling from quasi-thermal STE)	1.6 ± 0.1 ps		
$4 \rightarrow G$ (tunneling from thermal STE)	[e]	4.4 ± 0.5 ps	4.7 ± 1.2 ps

[a]1–4 and G are the states shown in Figs. 8 and 9.
[b]The time constants of the self-trapping processes could not be time-resolved in this study.
[c]The intramolecular vibrational relaxation process could not be detected in the blue-phase PDA.
[d]The tunneling processes from the nonthermal STE and from the quasi-thermal STE could not be separated.
[e]The tunneling from thermal STE in the blue-phase PDA-4BCMU could not be observed because the STEs disappear before thermalization.
[f]The value corresponds to the oscillation period of the C=C and C≡C stretching modes.

tunneling through and passing over the crossing point of the potential curve between the ST exciton and the ground state. The ratio of the contribution of the tunneling and activation processes must be determined for each polymer by detailed experiments of the temperature dependence of the primary decay kinetics. The time constants of the various relaxation processes in three polymers are listed in Table 2.

5. Conclusions

The mechanism of the low fluorescence quantum yield and the lack of triplet exciton formation from the singlet exciton have been clarified. This is because of the very short lifetime of the singlet exciton in the conjugated polymers. The short lifetime is caused by the electron–phonon coupling in the one-dimensional system, which is explained as follows. Since the conjugated polymers are one-dimensional, the spontaneous geometrical relaxation takes place without activation barrier in the potential curves between the minima of the free exciton and the self-trapped exciton. Consequently, adiabatic potentials of the self-trapped exciton and the ground state cross at a point lower than ordinary three-dimensional systems.

The present paper has already been published partially in refs. 1–3. The detailed study of the oriented films of PDA-3BCMU and PDA-4BCMU in the blue phase, the difference in decay kinetics between the amorphous cast film and the evaporated film on KCl of PDA-3BCMU and the triplet exciton formation in PDA-MADF are described in refs. 4, 42 and 28, respectively.

Acknowledgements

The author would like to thank Dr M. Yoshizawa for his contribution to the study and also Drs K. Ichimura, S. Halle and U. Stamm and Messrs M. Taiji and A. Yasuda for their collaboration in the femtosecond time-resolved experiment. The author is also indebted to Professor K. Yoshino for providing polythiophene samples and to Messrs T. Uemiya and Y. Hattori for providing PDA-3BCMU/KCl samples. The author thanks Professors Y. Toyozawa, E. Hanamura and H. Sumi and Dr S. Abe for their valuable discussions.

References

1 T. Kobayashi, M. Yoshizawa, U. Stamm, M. Taiji and M. Hasegawa, *J. Opt. Soc. Am., Part B, 7* (1990) 1558.
2 M. Yoshizawa, M. Taiji and T. Kobayashi, *IEEE J. Quantum Electron., QE-25* (1989) 2532.
3 U. Stamm, M. Taiji, M. Yoshizawa, T. Kobayashi and K. Yoshino, *Mol. Cryst. Liq. Cryst. 182A* (1990) 147.
4 M. Yoshizawa, A. Yasuda and T. Kobayashi, *Appl. Phys. B, 53* (1991) 296.
5 B. I. Greene, J. F. Mueller, J. Orenstein, D. H. Rapkine, S. Schmitt-Rink and M. Thakur, *Phys. Rev. Lett., 61* (1988) 325.
6 G. J. Blanchard, J. P. Heritage, A. C. von Lehmen, M. K. Kelly, G. L. Baker and S. Etemad, *Phys. Rev. Lett., 63* (1989) 887.
7 G. M. Carter, *J. Opt. Soc. Am., Part B, 4* (1987) 1018.
8 L. Y. Yang, R. Dornsville, Q. Z. Wang, W. K. Zou, P. P. Ho, N. L. Yang, R. R. Alfano, R. Zamboni, R. Danieli, G. Ruani and C. Taliani, *J. Opt. Soc. Am., Part B, 6* (1989) 753.
9 J. M. Nunzi and D. Frec, *J. Appl. Phys., 62* (1987) 2198.
10 F. Kajzar, L. Rothberg, S. Etemad, P. A. Chollet, D. Grec, A. Boudet and T. Jedju, *Opt. Commun., 66* (1988) 55.
11 W. M. Dennis, W. Blau and D. J. Bradley, *Appl. Phys. Lett., 47* (1985) 200.
12 G. I. Stegeman, R. Zanoni, E. M. Wright, N. Finlayson, K. Rochford, J. Ehrilich and C. T. Seaton in D. Bloor and R. A. Hann (eds.), *Organic Molecules for Nonlinear Optics*, Royal Society of Chemistry, London, 1989, pp. 369–381.
13 S. D. Halle, M. Yoshizawa, H. Matsuda, S. Okada, H. Nakanishi and T. Kobayashi, *Synth. Met., 49–50* (1992) 429.
14 See, e.g., T. Kobayashi (ed.), *Nonlinear Optics of Organics and Semiconductors*, Springer, Berlin, 1989.
15 M. Joffre, H. Hulin, A. Migus, A. Antonetti, C. Benoit à la Guillaume, N. Peyghambarian, M. Lindberg and S. W. Koch, *Opt. Lett., 13* (1988) 276.
16 J. P. Sokoloff, M. Joffre, B. Fluegel, D. Hulin, M. Lindberg, S. W. Koch, A. Migus, A. Antonetti and N. Peyghambarian, *Phys. Rev. B, 38* (1988) 7615.

17 C. H. Brito Cruz, J. P. Gordon, P. C. Becker, R. L. Fork and C. V. Shank, *IEEE J. Quantum Electron.*, *QE-24* (1988) 261.
18 M. Lindberg and S. W. Koch, *J. Opt. Soc. Am.*, *Part B*, *5* (1988) 139.
19 F. Charra and J. M. Nunzi, in J. Messier, F. Kajzar, P. Prasad and D. Ulrich (eds.), *Nonlinear Optical Effects in Organic Polymers*, Kluwer, Dordrecht, 1989, pp. 369–374.
20 A. N. Bonch-Bruevich and V. A. Khodovoi, *Usp. Fiz. Nauk*, *93* (1967) 71 (*Sov. Phys. Usp.*, *10* (1968) 637).
21 D. Fröhlich, A. Nothem and K. Reimann, *Phys. Rev. Lett.*, *55* (1985) 1335.
22 A. Mysyrowicz, D. Hulin, A. Antonetti, A. Migus, W. R. Masselink and H. M. Morkoc, *Phys. Rev. Lett.*, *56* (1986) 2748.
23 G. J. Blanchard and J. P. Heritage, *J. Chem. Phys.*, *93* (1990) 4377.
24 H. Hulin, A. Migus, M. Joffre and A. Antonetti, *Proc. Nonlinear Optics '90, Hawaii, USA, 1990*, p. 160.
25 E. Tokunaga, A. Terasaki and T. Kobayashi, *Opt. Lett.*, *17* (1992) 1131.
26 J. Orenstein, S. Etemad and G. L. Baker, *J. Phys. C: Solid State Phys.*, *17* (1984) L297.
27 L. Robins, J. Orenstein and R. Superfine, *Phys. Rev. Lett.*, *56* (1986) 1850.
28 K. Ichimura, M. Yoshizawa, H. Matsuda, S. Okada, M. Osugi, S. Hourai, H. Nakanishi and T. Kobayashi, *J. Chem. Phys.*, to be published.
29 P. L. Baldeck, R. R. Alfano and G. P. Agrawal, *Appl. Phys. Lett.*, *52* (1988) 1939.
30 J. Jortner, in E. I. Rashba and M. D. Sturge (eds.), *Excitons (Selected Chapters)*, North-Holland, Amsterdam, 1987, pp. 273–332.
31 C. X. Cui and M. Kertesz, *Phys. Rev. B*, *40* (1989) 9661.
32 G. Gustafson, O. Inganäs, H. Österholm and J. Laakso, to be published.
33 K. Nasu, *J. Lumin.*, *38* (1987) 90.
34 H. Sumi, M. Georgier and A. Sumi, *Rev. Solid State Sci.*, *4* (1990) 209.
35 T. Kobayashi, in J. Zyss (ed.), *Molecular Optoelectronics: Materials, Physics, and Devices*, Academic Press, Orlando, FL, 1992.
36 Y. Toyazawa, *J. Phys. Soc. Jpn*, *58* (1989) 2626.
37 Y. Toyozawa, *Tech. Rep. ISSP*, Institute for Solid State Physics, University of Tokyo, 1963.
38 M. Yoshizawa, A. Yasuda and T. Kobayashi, *Appl. Phys. B*, *53* (1991) 296.
39 R. H. Friend, D. D. C. Bradley and P. D. Townsend, *J. Phys. D: Appl. Phys.*, *20* (1987) 1367.
40 M. Yoshizawa, K. Nishiyama, M. Fujihira and T. Kobayashi, *Chem. Phys. Lett.*, to be published.
41 B. C. Hess, G. S. Kanned, Z. V. Vardeny and G. L. Baker, *Phys. Rev. Lett.*, *66* (1991) 2364.
42 M. Yoshizawa, T. Uemiya, Y. Hattori and T. Kobayashi, *Phys. Rev. B*, in press.

Synthetic Metals, 54 (1993) 93–98

Picosecond transient resonant Raman scattering in partially isomerized polyacetylene films

G. Lanzani*
Dipartimento di Fisica 'A. VOLTA' Universitá di Pavia, Pavia (Italy)

L. X. Zheng**, R. E. Benner and Z. V. Vardeny
Department of Physics and Electrical Engineering, University of Utah, Salt Lake City, UT 84112 (USA)

Abstract

Photoexcitation dynamics in the picosecond time domain in partially isomerized $(CH)_x$ films has been measured at room temperature, using the technique of transient photoinduced resonant Raman scattering (TPRRS). We show that TPRRS gives considerably more information than transient photoinduced absorption spectroscopy. Following the distinctively different decay kinetics of *cis* and *trans* resonantly enhanced vibrations, we have measured the time evolution of photoexcitations in *trans* and *cis* segments, separately. We observed a migration of excitons from *cis* to *trans* segments, with a time constant between 110 and 215 ps depending on the *cis*/*trans* ratio in the films.

Introduction

Polyacetylene ($(CH)_x$) is the simplest of conducting polymers and for this reason it has been extensively studied [1]. Two isomers of this polymer, *cis* and *trans*, coexist in a partially isomerized sample before complete isomerization to the more stable *trans* isomer which is achieved by temperature annealing (~180 °C). In *trans*-$(CH)_x$ the lowest lying electronic excitations are solitons, topological defects with unusual quantum numbers [1] which can be photogenerated with the above-gap photon absorption [2]. For this reason many recent experimental studies of $(CH)_x$ have focused on photospectroscopy [3]. In particular, picosecond transient photoinduced absorption (PA) measurements have played a central role in the study of the formation and dynamics of solitons [4, 5]. Unlike the *trans* isomer, the *cis* isomer does not support soliton excitations because of a lack of degeneracy in its backbone structure. It has been assumed, therefore, that photoexcitations in *cis* segments are self-trapped

*Present address: Istituto di Spettroscopia Molecolare, CNR, BO, Italy.
**Present address: Rochester, NY, USA.

0379-6779/93/$6.00

singlet excitons which quickly recombine, accompanied by a moderately strong photoluminescence at 1.96 eV [6].

In this paper we introduce a new transient pump and probe technique [7], namely picosecond transient resonant Raman scattering (TPRRS). One- and two-color excitations have been used to characterize separately the time evolution of photoexcitation density in both isomers. We demonstrate the existence of a fast migration of excitons from *cis* segments to the more energetically favorable *trans* segments.

Experimental and results

In TPRRS experiments a pump pulse (of about 3 ps) excites the sample, while a delayed pulse is used to measure the changes in the Stokes RRS cross section [8]. For one-color (degenerate) experiments the two laser pulses are split off the same beam, whereas two-color excitation is carried out by simultaneously pumping two different dye lasers.

In Fig. 1(a) we show the RRS spectrum of a *cis*-rich $(CH)_x$ film (50% *cis*) excited at 580 nm ($\hbar\omega = 2.14$ eV) in the 850–1300 wavenumber range. It contains two resonantly enhanced *cis* phonon lines (labeled C) at 900 and 1250 cm^{-1}, respectively, and a partially resonant *trans* line (denoted T) at 1070 cm^{-1}.

Figure 1(b) is the TPRRS spectrum at pump–probe delay time $\tau = 0$, as obtained by a suitable subtraction between pump-on and pump-off Raman scattering spectra [8]. $\Delta I(\tau)$, the change in the RRS intensity, is negative and its maximum value for $\tau = 0$ is about 2.5% for the *cis* phonon

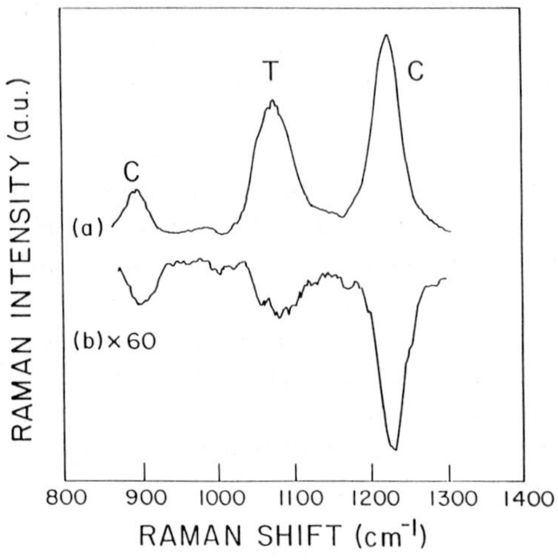

Fig. 1. (a) The cw and (b) normalized transient photoinduced RRS spectra of *cis*-rich $(CH)_x$ at room temperature, excited and probed at 580 nm. The pump–probe delay is $\tau = 0$ ps; *cis* (C) and *trans* (T) phonon lines are denoted.

lines, whereas ΔI is about 1.2% for the *trans* line. The TPRRS sign is accounted for by a reduction in the RRS cross-section rate due to the (incoherent) excited state population induced by the pump photons and it is a consequence of the Pauli exclusion principle: states that are already occupied are no longer accessible for optical transition [9]. The observed difference in the 'bleaching' magnitude between *cis* and *trans* lines can be accounted for by the Raman excitation profiles of the two isomers, which peak at different wavelengths. The photoinduced change in the RRS cross section is about one order of magnitude larger than the PA signal ($\Delta T/T$ where T is the transmission and ΔT are the changes in T) as measured in similar experimental conditions [7, 8]. This can be explained using the phase space filling model (PSF) to describe the $\chi^{(3)}$ response in resonance [9]. Within this model we have shown [8] that $\Delta I/I = (2/OD)\Delta T/T$, where $OD$is the optical density; since for our thin films $OD \ll 1$ at 750 nm, this explains the larger $\Delta I/I$ signal for TPRRS.

For $\tau > 0$ $\Delta I(\tau)$ decays. In Fig. 2 we report a comparison between a PA and TPRRS decay measured on the same sample (a full-*trans* $(CH)_x$ film) and in the same experimental configuration: pump at 580 nm and probe 750 nm. In PA (left panel) $\Delta T > 0$ and it shows a fast decay up to 10 ps followed by an oscillating structure which can be observed for hundreds of picoseconds (see inset). These oscillations can be accounted for by strain waves which bounce back and forth between the film surfaces [10]; they are not correlated with the photoexcitation dynamics. The PTRRS signal, which refers to the *trans* phonon band at 1100 cm^{-1}, is shown in the right panel of Fig. 2. The decay curve fits a power law $t^{-\alpha}\alpha = 0.17$, consistent with previous PA measurements [11] and does not contain any strain wave effect. This comparison shows that PA is not related to a single process and its interpretation is not straightforward, whereas TPRRS is a good probe of photoexcitation recombination kinetics, since the transient response of individual resonantly enhanced phonons in the system can be separately monitored free of artifact.

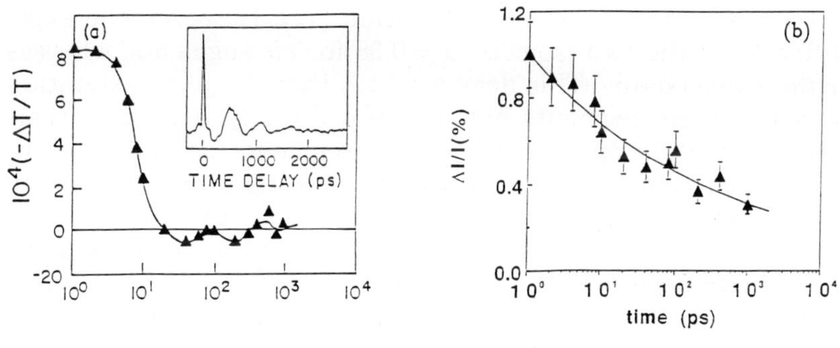

Fig. 2. (a) Photoinduced absorption decay of fully isomerized *trans*-$(CH)_x$ at room temperature, excited at 580 nm and probed at 750 nm. The inset shows the same data on a linear scale. (b) TPRRS decay at the same pump and probe conditions and for the same sample as (a).

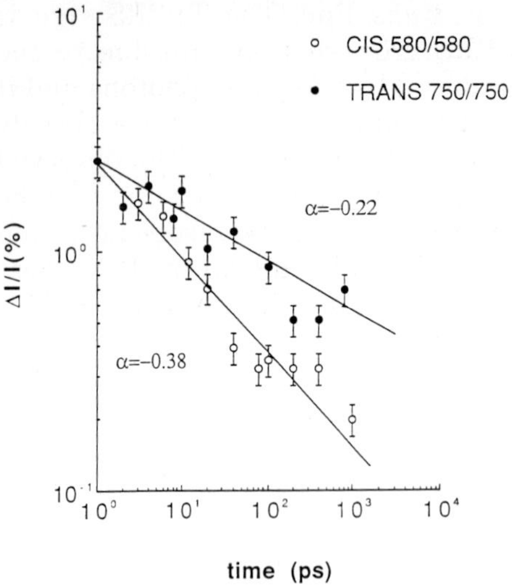

Fig. 3. TPRRS decays of the *trans* phonon line at 1070 cm^{-1} (solid circles) pumped and probed at 750 nm, and the *cis* phonon line at 1250 cm^{-1} (open circles) pumped and probed at 580 nm. The solid lines are the power law ($t^{-\alpha}$) fits with a $\alpha = 0.22$ for *trans* and $\alpha = 0.38$ for *cis*.

Since the strongly coupled resonant vibrations in *cis*- and *trans*-$(CH)_x$ have different frequencies and since their intensities in RRS can be influenced separately by photoexcitations in *cis* and *trans* segments, respectively, we have been able to monitor separately the photoexcitation dynamics in *cis* and *trans* segments by one-color TPRRS experiments.

We have measured TPRRS of the C–C stretching mode at 1100 cm^{-1} in *trans*-$(CH)_x$ by a one-color experiment at 750 nm (Fig. 3, solid circles), whereas the TPRRS decay for the *cis* phonon line at 1250 cm^{-1} was measured by pump and probe beams at 580 nm (Fig. 3, open circles). In both cases we obtained a power law ($t^{-\alpha}$) decay, consistent with one-dimensional random walk models for exciton recombination [12]. However, the exponent α is different in the two isomers: $\alpha = 0.38$ for *cis* segments, whereas $\alpha = 0.22$ for the *trans* chains. This demonstrates that the photoexcitation dynamic is faster in *cis* segments but it is not directly related to their lifetime since the excitons migrate to the *trans* segments, as shown below.

Two-color experiments have been used to investigate *cis-trans* exciton migration by pumping at 580 nm, mainly into the *cis* LUMO state, and probing in resonance with *trans* segments at 750 nm. In Fig. 4 we show the cw RRS and TPRRS decays of the C–C line in two partially isomerized $(CH)_x$ films. From cw RRS spectra we have obtained the *cis*/*trans* content ratios. It is 50% *cis* in film no. 1, whereas film no. 2 shows 37% *cis* with an uncertainty of about 7%. The TPRRS decays (Figs. 4(b) and 4(c), respectively) substantially deviate from the power law decay observed with one-color experiments. ΔI stops decaying after about 30 ps for both films

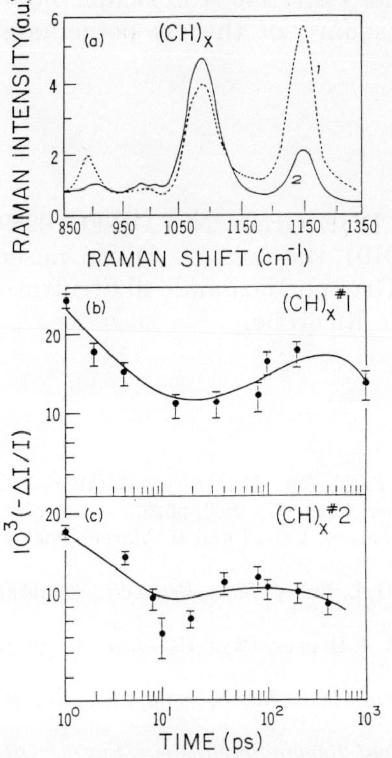

Fig. 4. (a) The cw RRS spectra of two partially isomerized $(CH)_x$ films (sample nos. 1 and 2) excited at 580 nm. TPRRS decays of $(CH)_x$ for (b) sample no. 1 and (c) sample no. 2 excited at 580 nm and probed at 750 nm.

and, contrary to expectation, begins to increase forming an apparent peak at $\tau_1 = 210$ ps in sample no. 1 and at $\tau_2 = 110$ ps in sample no. 2. We note that the TPRRS decay remains in the form of $t^{-\alpha}$ for both films when excited and probed at 750 nm. Moreover in Fig. 2 (lower panel) it is shown that the decay of $\Delta I(\tau)$ in full *trans*-$(CH)_x$ is a smooth power law even with two-color excitation, in agreement with the lack of *cis* segments in this film. This rules out exciton diffusion from short to long *trans* segments [7].

Based on the TPRRS decays of Fig. 4 we believe that when excitons are generated in both isomers a fast exciton migration occurs from *cis* to *trans* segments [13]. This happens since *trans* chains have a lower energy gap ($\simeq$1.8 eV) than *cis* chains ($\simeq$2.1 eV) and therefore are more energetically favorable. The time dependence of $\Delta I/I$ has been fitted by an hopping model [13] in which excitons undergo a one-dimensional random walk and eventually hop to neighboring chains with lower energy. The process is consistent with a diffusion of photoexcitations in *cis* segments such that: $\langle L\rangle^2 = D\tau_i$ where $\langle L\rangle$ is the average length of the *cis* segments and D is the exciton diffusion constant, and $\langle L\rangle$ is proportional to the *cis*/*trans* ratio (%*cis*). Using the proposed value for D[1] the mean free path of excitons is about 200 Å in sample no. 1 and 140 Å in sample no. 2 while the hopping times scale with the square of the *cis* percentage content.

Acknowledgements

This work was supported in part by DOE Grant No. DE-FG 0289 ER45409, ONR Contract No. N00014-91-C-0104, the John A. Dixon Laser Institute at the University of Utah and by Gruppo Nazionale di Struttura della Materia del Consiglio Nazionale delle Ricerche.

References

1 A. J. Heeger, S. Kivelson, J. R. Schrieffer and W. P. Su, *Rev. Mod. Phys.*, *60* (1988) 781.
2 W. P. Su and J. R. Schrieffer, *Proc. Natl. Acad. Sci. USA*, *77* (1980) 5626.
3 T. A. Skotheim (ed.), *Handbook of Conducting Polymers*, Vols. I and II, Marcel Dekker New York, 1986.
4 C. V. Shank, R. Yen, R. L. Fork, J. Orenstein and G. L. Baker, *Phys. Rev. Lett.*, *49* (1982) 1660.
5 Z. Vardeny, J. Strait, D. Moses, T. C. Chung and A. J. Heeger, *Phys. Rev. Lett.*, *49* (1982) 1657.
6 L. Lauchlan, S. Etemad, T. C. Chung, A. J. Heeger and A. G. MacDiarmid, *Phys. Rev. B*, *24* (1981) 3701.
7 D. L. Weidman and D. B. Fitchen, *Proc. 10th Int. Conf. Raman Spectroscopy, Eugene, OR, USA, 1986*, p. 12; D. L. Weidman, *Ph.D. Thesis*, Cornell University, 1987.
8 G. Lanzani, L. X. Zheng, R. E. Benner and Z. V. Vardeny, *Synth. Met.*, *49* (1992) 321.
9 B. I. Greene, J. Orenstein and S. Schmitt-Rink, *Science*, *247* (1990) 679.
10 G. S. Kanner, Z. Vardeny and B. C. Hess, *Phys. Rev. B*, *42* (1990) 5403.
11 Z. Vardeny, *Physica B*, *127* (1984) 338.
12 G. S. Kanner, *Ph.D. Thesis*, University of Utah (SLC), 1990.
13 G. Lanzani, L. X. Zheng, G. Figari, R. E. Benner and Z. V. Vardeny, *Phys. Rev. Lett.*, *68* (1992) 3104.

Studies of photoexcitations in conducting polymers by the absorption-detected magnetic resonance technique

Z. V. Vardeny and X. Wei
Department of Physics, University of Utah, Salt Lake City, UT 84112 (USA)

Abstract

The nature of photoexcitations in conducting polymers has been elucidated by the absorption-detected magnetic resonance (ADMR) technique. In this technique we measure changes in the photoinduced absorption (PA) spectrum due to changes in spin-dependent recombination rates induced by resonant absorption of μ-waves in magnetic fields. We have used the ADMR technique at 3 GHz to study spin states in *trans*-$(CH)_x$, oriented poly(*para*-phenylene vinylene) (PPV), polythiophene (PT) and polydiethynylsilane (PDES). We have verified that the charged excitations are solitons ($S^{\mp}$) in *trans*-$(CH)_x$ and PDES, and polarons ($P^{\mp}$) and bipolarons ($BP^{2\mp}$) in other nondegenerate ground state polymers. The long-lived neutral excitations, on the other hand, are self-trapped triplet excitons in all polymers, except for *trans*-$(CH)_x$ and PDES in which the triplets dissociate into neutral soliton (S^0) pairs due to their backbone degenerate ground state structure.

Introduction

Various powerful techniques have been used to investigate the photoexcitation properties in conducting polymers [1]. The most useful techniques are [1–8]: photoluminescence (PL), which studies radiative recombination of photoexcitations; photoconductivity (PC), which is sensitive to photogenerated mobile charge carriers; photomodulation (PM), which is capable of measuring photoexcitation energy levels in the gap and their recombination dynamics; and light-induced ESR (LESR), which has the potential to measure the spin state of the photoexcitations. Among these techniques LESR has been the most controversial [3, 9]. First, the sensitivity of ESR machines is usually too low for LESR: the overall minimum number of spins that can be measured is 10^{11}. A back of the envelope calculation using realistic dimensions of the type of μ-wave cavities used in LESR and the high absorbance of conducting polymer films (light penetration depth of ~ 1000 Å) limits the minimum steady state density of spin-carrying photoexcitations that can be detected in LESR to 10^{17} cm^{-3}. This relatively high photoexcitation density can be photogenerated if their lifetime is longer than about 10^{-2} s and the generation quantum efficiency is higher than about 10%. These conditions are not easy to satisfy for most photoexcitations in conducting polymers and, if attainable, it further requires high laser excitation intensity and this

0379-6779/93/$6.00

leads to sample heating, which is the second problem in LESR [3, 9]. Conducting polymers usually contain spin $\frac{1}{2}$ defects (dangling bonds, ends of chains, etc.) of substantial density, of order $10^{17}\,cm^{-3}$ or larger. The modulation of the laser excitation intensity in LESR measurements causes temperature modulation, which in turn modulates the Curie susceptibility leading to temperature artifacts in the photoinduced magnetic resonance spectra. Double modulation, where the μ-wave intensity and the laser excitation intensity are simultaneously modulated, does not help to reduce the heating problem since the LESR and the thermal lifetimes are similar [3]. A sophisticated scheme was tried in *trans*-$(CH)_x$ [9] to avoid laser heating effects in LESR, but the results were not conclusive due to lack of spectral information associated with the LESR signal.

In this paper we discuss the novel technique of photoabsorption-detected magnetic resonance (ADMR) [10], in which spin states and their correlated spectra in PM can be simultaneously measured. The main idea in ADMR is the detection of changes in steady state photoexcitation density induced by μ-wave absorption in resonance with Zeeman split sublevel electronic states. Since the changes are induced in the density of photoexcitations rather than in μ-wave absorption, the effect can be measured with detectors in the visible to near IR spectral ranges, with improved sensitivity of up to five orders of magnitude over that of μ-wave detectors [11]. Moreover, spectral information can be also gained in ADMR measurements, since the optical probe can be easily tuned.

Experimental

The ADMR experimental setup is shown in Fig. 1 [10]: A cw pump beam (from an Ar^+ laser) and probe beam (from a tungsten lamp) con-

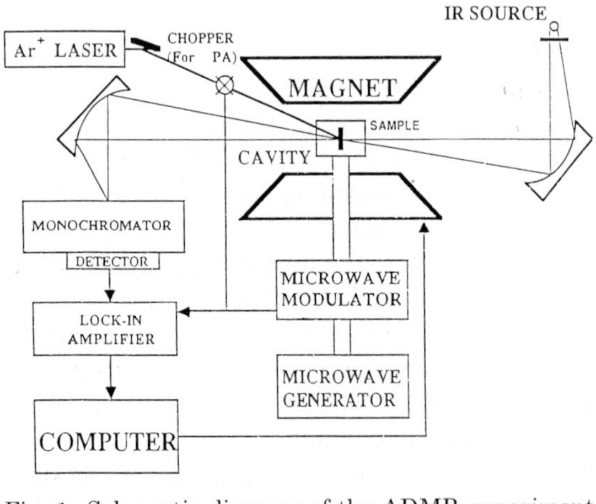

Fig. 1. Schematic diagram of the ADMR experimental setup.

stantly illuminate the sample, which is mounted in a high Q microwave cavity (at 3 GHz) equipped with optical windows and a superconducting magnet producing a field $\boldsymbol{H}$. Microwave (μ-wave) resonant absorption, modulated at 500 Hz, leads to small changes δT in the probe transmission T proportional to δn, the change in the photoexcitation density n produced by the pump; δn is induced by transitions in the μ-wave range that change spin-dependent recombination rates. Two types of ADMR spectra are usually obtained: the H-ADMR spectrum in which δT is measured at a fixed probe wavelength λ while sweeping $\boldsymbol{H}$, and the P-ADMR spectrum, in which δT is measured at a constant field, in resonance, while λ(probe) is changed. With suitable signal averaging, the system $\delta T/T$ sensitivity was 3×10^{-8} in the visible to near IR range and 3×10^{-7} in the mid-IR spectral range.

The polymer samples used in our ADMR studies were in the form of thin films ($d \simeq 1000$ Å) polymerized (or spin cast) on sapphire substrates, which do not reduce the Q value of the μ-wave cavity. Several conducting polymers have been measured: *trans*-$(CH)_x$, polydiethynylsilane (PDES), oriented and unoriented poly(phenylene vinylene) (PPV) and polythiophene (PT) and its derivative (P(3O)T).

Spin polarization

The ADMR is an excited state dynamical technique, which is not based on photoexcitations in thermal equilibrium. In this respect it is not similar to ESR, of which signal follows the Boltzmann statistics. In ADMR we must actively provide spin polarization or different population of spin sublevels for the photoexcitations. This can be achieved either during the generation process, or the recombination, or both. We will discuss spin $\frac{1}{2}$ and spin 1 cases seperately, under the assumption that the spin–lattice relaxation time is much longer than the photoexcitation lifetime.

Spin $\frac{1}{2}$ (doublets)

Two spin $\frac{1}{2}$ particles produce pairs with spins either parallel (P) or antiparallel (AP) to each other. In the case of a 'geminate' pair, the photoexcited electron–hole pair is correlated after photon absorption, hence the spins are in an AP configuration. The reason is that the ground state (GS) is spin singlet and the photon absorption process conserves spin. Spin polarization is achieved, therefore, by generation [11]: μ-wave absorption flips the spin of one carrier *reducing* the recombination rate of the pair to the GS. Consequently $\delta n > 0$ and the H-ADMR may contain two δn bands, if the respective g values for the electron and hole are sufficiently different from each other. In the case of uncorrelated carriers [11], usually with bimolecular recombination kinetics, there are pairs with P or AP spin configuration, initially with equal probability. With time, the pairs with P configuration prevail, since their recombination rate to the GS is smaller. Spin polarization is therefore achieved by recombination and any μ-wave-induced spin-flip increases recombination and consequently $\delta n < 0$. Again two δn bands can be detected in resonance at magnetic fields $\boldsymbol{H}$ close to $H_0 \simeq 1070$ G, which is the resonant field related to species with $g = 2$ and spin $\frac{1}{2}$, using μ-waves at ~ 3 GHz.

Spin 1 (triplets)

The spin Hamiltonian for spin 1 (triplet) states is more complex than that of doublets due to spin dipole interaction [12]:

$$\mathscr{H} = \mathscr{H}_0 + g\beta \boldsymbol{H} \cdot \boldsymbol{S} + D(S_z^2 - \tfrac{1}{3}S^2) + E(S_x^2 - S_y^2) \tag{1}$$

where $\mathscr{H}_0$ denotes the atomic Hamiltonian, $g\beta \boldsymbol{H} \cdot \boldsymbol{S}$ is the magnetic Zeeman splitting term, and D and E are the zero-field splitting (ZFS) parameters. The spin states are hence split into three sublevels (denoted x, y and z) even in the absence of $\boldsymbol{H}$.

Spin-triplet excitons cannot be generated directly from the singlet GS by photon absorption. They are produced, therefore, by intersystem crossing from the singlet manifold due to the spin–orbit (LS) coupling. LS coupling is usually different for each of the triplet sublevel states, x, y and z, causing spin polarization by generation. Then, μ-wave absorption among the triplet sublevels causes ADMR signals with either $\delta n > 0$ or $\delta n < 0$ (or $\delta n = 0$) depending on whether the recombination rate R decreases or increases (or remains constant), respectively. Even if the generation of the triplet sublevels is homogeneous, spin polarization can still be achieved by recombination [12], since the sublevel recombination rates to the GS are different ($R_x \neq R_y \neq R_z$). In this case, only ADMR signal with $\delta n < 0$ is observed, similar to the case discussed above for homogeneous generation of spin $\frac{1}{2}$ particles.

Applications of ADMR spectroscopy to conducting polymers

A considerable number of experimental studies, mainly using the PM, PL and LESR techniques, have been completed to elucidate the properties of photoexcitations in conducting polymers [1]. In general the predicted photoexcitation properties depend on whether the polymer has a degenerate or nondegenerate ground state backbone structure. Charged solitons ($S^{\mp}$) and neutral solitons (S^0) with a single absorption band in the gap are predicted for a degenerate ground state (DGS) polymer such as *trans*-$(CH)_x$ and PDES [13], whereas for the class of nondegenerate (NDGS) polymers, such as PT and PPV, the predicted charged excitations are polarons ($P^{\mp}$) with three optical transitions and bipolarons ($BP^{2\mp}$) with two optical transitions, respectively [14], below the gap. Triplet excitons were also predicted to exist in NDGS conducting polymers [15, 16], but the interest in them has never reached that given to the charged photoexcitations.

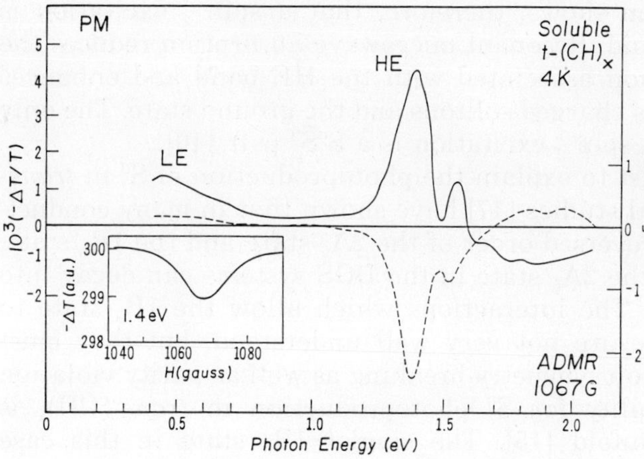

Fig. 2. PM spectrum (solid line) and P-ADMR spectrum (broken line) at 1067 G of soluble *trans*-$(CH)_x$ at 4 K. The inset shows the H-ADMR spectrum at 1.35 eV (HE peak). LE and HE PA bands are assigned. The $\delta T/T$ S/N ratio is 50 for $\hbar\omega > 1.1$ eV and 5 for $\hbar\omega < 0.7$ eV.

trans-Polyacetylene (trans-$(CH)_x$)

The PM spectrum of *trans*-$(CH)_x$ consists of two photoinduced absorption (PA) bands below the gap [2]: a LE band at 0.45 eV and a HE band at 1.35 eV. The LE band was identified as due to spinless charged solitons S^+ and S^- [4]; correlated with it is a prominent oscillation above 1.45 eV, identified [3] as due to electroabsorption (EA) caused by the photogenerated $S^{\pm}$. The HE band, which is associated with neutral photoexcitations, has remained a mystery for over a decade. Recent careful LESR studies of *trans*-$(CH)_x$ have conclusively shown [9] that spin $\frac{1}{2}$ excitations are also photogenerated. It was speculated [15, 16] that they are neutral solitons S^0 associated with the HE band.

The PM and P-ADMR spectra of soluble *trans*-$(CH)_x$ are shown in Fig. 2. The inset shows the H-ADMR at 1.4 eV (at the peak of the HE band in soluble *trans*-$(CH)_x$). A single symmetrical band, with a reduction of PA ($\delta n < 0$) is observed at 1067 G ($s = \frac{1}{2}$, $g \approx 2.003$). The P-ADMR spectrum at a fixed field of 1067 G consists of the following features [10] (Fig. 2):
(i) a high energy band, with a relatively large δn reduction: $\delta n/n \approx -5.7 \times 10^{-3}$, with the same spectral shape as the HE band in the PM spectrum;
(ii) a smaller increase in the $S^{\pm}$ population at the LE band: $\delta n/n \approx 10^{-3}$ and a corresponding small EA oscillation above 1.4 eV;
(iii) a small reduction of the PB band, suggesting enhanced recombination to the GS.

δn(HE) is correlated with δn(LE) as verified [10] by their identical dependences on the pump excitation intensity I_L, the microwave modulation frequency f and sample temperature θ. We note that in the PM spectrum, the HE and the LE PA bands do not decrease with θ in the same way [3]; their δn signals, however, decrease together with θ [10].

The ADMR spectrum shows, therefore, that a spin $\frac{1}{2}$ excitation is involved with the HE band. Resonant microwave absorption reduces the photoexcitation population associated with the HE band and enhances the population of spinless charged solitons and the ground state. The only viable candidate for this spin $\frac{1}{2}$ excitation is a $S^0\bar{S}^0$ pair [10].

Two possibilities exist to explain the photoproduction of S^0 in *trans*-$(CH)_x$: several theoretical studies [17] have shown that in many conducting polymers there is a reversed order of the $2A_g$ state and the $1B_u$ state. It has been shown that the $2A_g$ state in the DGS systems can decay into neutral solitons [18, 19]. The interactions which allow the $1B_u$ state to decay into the $2A_g$ state are not very well understood, but they must involve charge conjugation symmetry breaking as well as parity violation [16]. The second possibility for S^0 photoproduction in *trans*-$(CH)_x$ is through the triplet manifold [15]. The singlet $1B_u$ state in this case undergoes an intersystem crossing into the triplet $1B_u$ state, which consequently decays into $S^0\bar{S}^0$ pairs in a DGS polymer [15]. The first mechanism of S^0 photoproduction would lead to $\delta n > 0$, since the A_g state decomposes into a $S^0\bar{S}^0$ pair with AP spins (as in the geminate recombination case discussed above), whereas the second mechanism leads to $S^0\bar{S}^0$ pairs with P spin configuration and consequently gives $\delta n < 0$ upon μ-wave absorption. Our results favor, therefore, S^0 photogeneration via the triplet manifold. The reason why the intersystem crossing in *trans*-$(CH)_x$ is so fast (< 100 fs [20]) is not understood at this time.

Polydiethynylsilane (PDES)

The PDES polymer was thought to be the second candidate of the class of DGS polymers [21]. However, careful ^{13}C NMR and RRS measurements have shown that the real material is composed of many short chains; in such a case the GS degeneracy may be lifted [22]. Our PM and ADMR results conclusively show the existence of two kinds of long-lived photoexcitations in PDES. We detected $BP^{2\pm}$ in short chains whereas $S^{\mp}$ and S^0 are generated in long chains.

The PM spectrum of PDES at 4 K is shown in Fig. 3. It contains three prominent PA bands: LE_1 at 0.4 eV, LE_2 at 0.9 eV and an HE band at 1.45 eV; a small shoulder at 0.7 eV is also observed at the low-energy side of LE_2. By measuring the photoinduced IR active vibrations (IRAV) [4, 23], we identified LE_1 and LE_2 as due to charged excitations, whereas HE is due to neutral excitations. We also know that the shoulder at 0.7 eV and LE_1, LE_2 do not share a common origin due to different f, θ and I_L dependences [23].

The P-ADMR at 1067 G and the H-ADMR at 1.45 eV are shown in Fig. 3. We measured a single H-ADMR band at 1067 G ($g \simeq 2$, $s = \frac{1}{2}$) with $\delta n < 0$ and FWHM of 15 G, showing spin $\frac{1}{2}$ excitations. The P-ADMR spectrum, however, is not merely an image reflection of the PM spectrum, indicating that different photoexcitations have been separated. LE_2 in P-ADMR is much narrower and relatively weaker than its counterpart in the PM

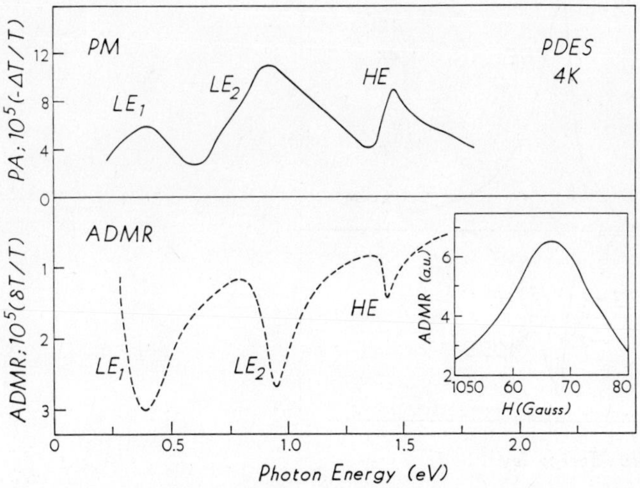

Fig. 3. PM spectrum (solid line) and P-ADMR spectrum at 1067 G of PDES at 4 K. The inset shows the H-ADMR spectrum at 1.45 eV (HE band). LE$_1$, LE$_2$ and HE PA bands are assigned. The S/N ratio in $\delta T/T$ is 50 in the whole spectral range.

spectrum (Fig. 3). Also the shoulder at 0.7 eV has completely disappeared from the P-ADMR spectrum. Moreover, the HE band in the P-ADMR spectrum is much weaker than in the PM spectrum. Based on these results we tentatively attribute the LE$_1$ and LE$_2$ PA bands to $BP^{2\mp}$ transitions in short PDES chains (also see below for $P^{\mp}$ and $BP^{2\mp}$ in PPV), whereas the HE and the shoulder at 0.7 eV are due to S^0 and $S^{\mp}$, respectively, in long PDES chains. This conclusion is in agreement with the PA band excitation spectra [23]. We found that for $\hbar\omega < 2$ eV, only the HE and the shoulder at 0.7 eV are photogenerated; the LE$_1$ and LE$_2$ bands can be photogenerated only at $\hbar\omega \geqslant 2.3$ eV, presumably due to shorter chains with higher E_g.

Poly(para-phenylene vinylene) (PPV)

The PM spectrum of PPV, an important member of the class of NDGS polymers, is shown in Fig. 4. It contains two prominent PA bands LE$_1$ at 0.3 eV and HE at 1.35 eV; a shoulder LE$_2$ can be also observed at around 1.2 eV. From studies of photoinduced IRAV it has been concluded [4, 24] that the LE$_1$ and LE$_2$ bands are due to charged photoexcitations, most likely $BP^{2\mp}$, whereas the HE is due to neutral photoexcitations, speculated to be due to triplet excitons from LESR measurements on a short oligomer of PPV [25].

We have measured ADMR spectra of oriented PPV (Figs. 4 and 5). Figure 5 shows the H-ADMR at 1.36 eV (peak of HE band in PM) and the H-PDMR of the strong PL band observed at 2 eV. In ADMR we identified two types of excitations, both with $\delta n < 0$: the line D at 1067 G is due to spin $\frac{1}{2}$ carriers, whereas the three lines denoted T are resonances associated

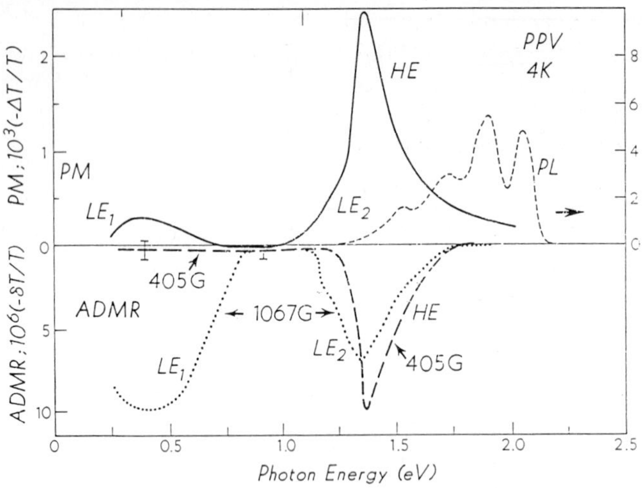

Fig. 4. PM spectrum (solid line) and P-ADMR spectra at 427 G (broken line) and at 1067 G (dotted line), of oriented PPV sample with the chain orientation parallel to the magnetic field. The S/N ratio in $\delta T/T$ is 50 for $\hbar\omega > 1.1$ eV and 10 for $\hbar\omega < 1.1$ eV.

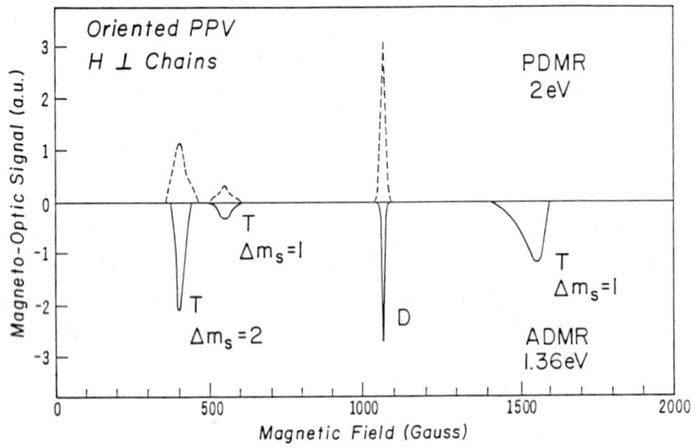

Fig. 5. H-ADMR spectrum at 1.36 eV (solid line) and H-PDMR spectrum at 2 eV (broken line) of oriented PPV perpendicular to $\boldsymbol{H}$ measured at 4 K. D is associated with spin $\frac{1}{2}$ excitations, whereas T are transitions associated with $s = 1$ excitations.

with μ-wave transitions among triplet sublevels. From their resonance value for chains parallel and perpendicular to $\boldsymbol{H}$ we calculated [26] the triplet ZFS parameters D and E and its angle β with respect to the chain direction: $D = 630$ G (0.059 cm^{-1}), $E = 92$ G (0.008 cm^{-1}) and $\beta = 15°$. Except for a weak line at high field, the H-PDMR shows a similar spectrum of a doublet and a triplet (Fig. 5), and we conclude that the PL at 2 eV is correlated with both singlet and triplet excitons, respectively.

The P-ADMR spectra of oriented PPV taken at 1067 G ($s = \frac{1}{2}$) and at 427 G ($s = 1$) are shown in Fig. 4. The triplet P-ADMR spectrum contains a single band peaked at 1.36 eV with the same spectral shape as the HE band in the PM spectrum. The spin $\frac{1}{2}$ ADMR spectrum, however, contains two bands peaked at 0.3 and 1.3 eV which are equivalent to the LE_1 band and LE_2 shoulder, respectively, in the PM spectrum. We therefore identify the LE_1 and LE_2 PA bands as due to optical transitions associated with photogenerated bipolarons ($BP^{2\pm}$), at somewhat lower energies compared to the doping induced BP [24]. The HE band, on the other hand, is due to optical transitions in the triplet manifold, associated with photogenerated triplet excitons in PPV.

The photoexcitation dynamics in PPV, similar to that of *trans*-$(CH)_x$, also shows two competing processes associated with the distinct B_u and $2A_g$ states (or singlet and triplet manifolds). An excited B_u state can decay into spin $\frac{1}{2}$ polarons ($P^{\mp}$), which are not topological defects and are therefore relatively mobile in the sample. Although we cannot discern a PA band in the PM spectrum associated with $P^{\mp}$, their $BP^{2\mp}$ by-products are observable in the PM spectrum. Since $\delta n < 0$, we identify the spin polarization needed in ADMR as due to the recombination process. Polaron recombination is a spin-dependent process with faster recombination for polaron pairs with AP spins [8]. This leads to excess $P^{\mp}$ pair population with P spins, which is partially transformed by μ-wave resonant absorption into $P^{\mp}$ pairs with AP spins with enhanced recombination. The enhanced $P^{\mp}$ recombination explains the positive PDMR signal (more PL) and the negative ADMR signal for $BP^{2\mp}$ (LE_1 and LE_2 in Figs. 4 and 5). This originates from the smaller $P^{\mp}$ steady state density under the μ-wave resonance conditions. On the other hand, if an intersystem crossing from B_u to $2A_g$ (or from the singlet to the triplet manifold) occurs [15, 16], triplet excitons are formed with a strong triplet–triplet transition at 1.36 eV. These excitons become spin polarized due to different recombination dynamics of each of the triplet sublevels [12]. Under μ-wave absorption, transitions among the triplet sublevels enhance recombination. This results in PDMR enhancement and ADMR reduction due to a decrease in the triplet steady state population; both changes are in agreement with the measurements presented in Figs. 4 and 5.

Polythiophene (PT)

PT is the first material among the class of NDGS polymers of which photoexcitations have been thoroughly investigated [7]. $P^{\mp}$ and $BP^{2\mp}$ excitations were tentatively identified by the PM and LESR techniques [27, 28], but a clear distinction between them is not available. Moreover, triplet excitons were identified to contribute in the PL process through PDMR [8], but the correlated optical transition in the triplet manifold has not yet been observed in PM spectra.

The PM spectrum of electrochemically polymerized PT film at 4 K is shown in Fig. 6 [28]. It clearly shows three PA bands: ω_1 at 0.45 eV, ω_2 at 1.25 eV and HE at 1.8 eV. All of these bands are due to charged excitations as verified by IRAV studies [7]. Although the PA bands have the

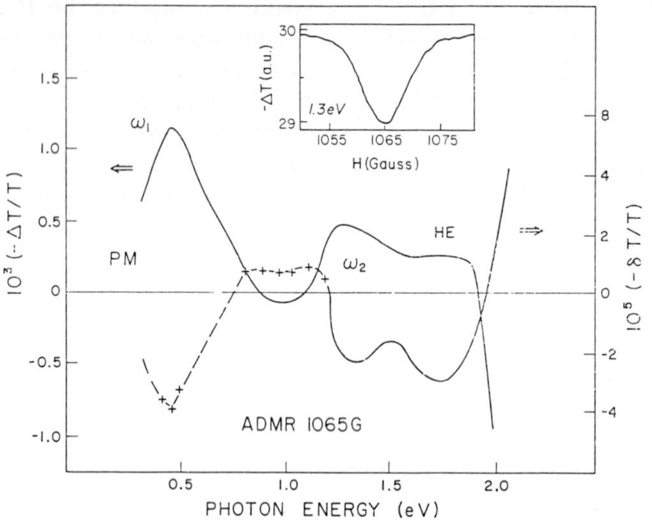

Fig. 6. PM (full line) and P-ADMR (full and dashed line) spectra at 1065 G of PT at 4 K. The dashed line part of the ADMR spectrum is a guide to the eye, whereas the full line part was continuously measured. The inset shows the H-ADMR spectrum at 1.3 eV (ω_2 PA band). The bands ω_1, ω_2 and HE are assigned.

same laser excitation spectra and I_L dependences [27], they do not share a common origin due to the distinctly different dependences on the modulation frequency f and temperature θ [28]. The HE band does not depend on f up to 100 KHz, whereas ω_1 and ω_2 decay with f starting from 200 Hz [28]. This shows that the lifetime of the photoexcitations associated with the HE band is much shorter than that of ω_1 and ω_2 which, however, share a common origin. Based on these experimental facts we have tentatively identified [28] the HE band due to $P^{\mp}$, whereas ω_1 and ω_2 are due to $BP^{2\mp}$.

The identification of the HE PA band with polarons is verified by ADMR, as shown in Fig. 6 by the P-ADMR spectrum at 1065 G and the H-ADMR spectrum at ω_2 (1.3 eV), shown in the inset. We note that the ADMR signal is negative, showing a decrease in photoexcitation density induced by μ-wave absorption. We also measured the PDMR spectrum of the same PT films under identical experimental conditions. The PDMR signal is positive (more PL) showing enhanced μ-wave-induced recombination [7]. Both spin-dependent spectra show an asymmetric band at 1065 G ($g \simeq 2.003$). Since we were unable to find in ADMR another resonance at $g \simeq 4$ due to triplets [7, 8], we conclude that the ADMR signal is due to spin $\frac{1}{2}$ excitations.

The P-ADMR spectrum at 1065 G (Fig. 6) contains the same three bands as in the PM spectrum, but with different intensities. The relative μ-wave-induced change in the photoexcitation density $\delta n/n$ is -3×10^{-2} for both BP bands ω_1 and ω_2, whereas $\delta n/n$ is -7×10^{-2} for the polaron band at HE. Recent careful light-induced ESR (LESR) studies of PT showed conclusively [27] that spin $\frac{1}{2}$ charged carriers are photogenerated in the polymer chains. It was then speculated that the LESR signal is due to photogenerated polarons [27, 28]; our ADMR results provide the experimental proof and verify that the HE band is due to charged polarons. In fact the ADMR shows that this band is due to spin $\frac{1}{2}$ photoexcitations that are correlated with BP, since the HE and the BP bands have the same H-ADMR spectrum, the same dependences on the μ-wave modulation frequency and on the laser excitation intensity. We identify, therefore, the HE band with the optical transition between the localized polaron levels in the gap [14].

The polaron-enhanced recombination in resonance can be explained by distant pair recombination [11], which leads to spin polarization. As discussed before for PPV, for times shorter than the spin–lattice relaxation, there are more polaron pairs with P spins than those with AP spins, due to the faster recombination rate of the latter. The μ-wave absorption with $\Delta m_s = \pm 1$ leads to a randomization of the spins, thereby increasing the density of pairs with antiparallel spins, which consequently enhances recombination. This, in turn, leads to a reduction in the polaron density ($\Delta n < 0$), causing a decrease in BP density and an increase in PL ($\delta L > 0$). That $\delta n/n(\mathrm{P}) \simeq 2\delta n/n(\mathrm{BP})$ shows that BP production ($\mathrm{P}^{\mp} + \mathrm{P}^{\mp} \rightarrow \mathrm{BP}^{2\mp}$) is only one of two channels of recombination of polaron pairs with AP spins, the other channel being direct recombination to the GS for polaron pairs with opposite charges ($\mathrm{P}^{\mp} + \mathrm{P}^{\mp} \rightarrow \mathrm{GS}$).

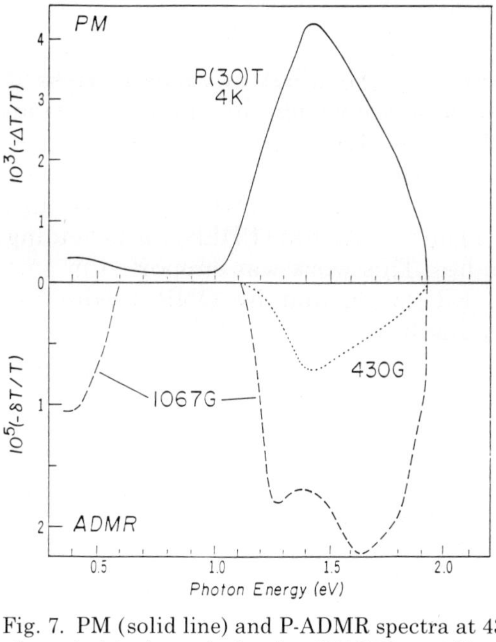

Fig. 7. PM (solid line) and P-ADMR spectra at 430 G (dotted line) and 1067 G (broken line) of a polythiophene derivative, P(3O)T, at 4 K.

The nature of photoexcitations in PT is completed by ADMR spectroscopy in a PT derivative (P(3O)T), as shown in Fig. 7. The PM spectrum shows an intense band at 1.4 eV, whereas the H-ADMR spectrum shows both spin $\frac{1}{2}$ (at 1067 G) and spin 1 (at 430 G). Similar to Fig. 6, we identify in Fig. 7, the $P^{\mp}$ and $BP^{2\mp}$ transitions at 1.7, 0.4 and 1.25 eV, respectively, using the P-ADMR spectrum at 1067 G ($s = \frac{1}{2}$). The P-ADMR spectrum at 430 G ($s = 1$), however, contains only a single PA band at 1.4 eV, which we therefore identify as due to triplets, similar to PPV (Fig. 4). From our results we conclude that the triplet state in PT is not different from that in PPV; it has a similar ZFS value and the intense optical transition in the triplet manifold occurs at about the same energy.

Conclusions

In this paper we have discussed the principles of ADMR spectroscopy and given several examples of studies completed in conducting polymers. In the class of DGS polymers such as *trans*-$(CH)_x$ and PDES, we identified $S^{\mp}$ and S^0 as the main long-lived charged and neutral photoexcitations, respectively. The spectroscopy of NDGS polymers, however, is richer. The charged excitations are $P^{\mp}$ and $BP^{2\mp}$, whereas the neutral photoexcitations are singlet and triplet excitons, some of which radiatively recombine giving rise to a relatively intense PL band.

Acknowledgements

The ADMR technique was successfully tried first in PT by B. Hess at the University of Utah (1989). We also acknowledge his major contribution to the ADMR measurements of *trans*-$(CH)_x$. We thank G. Kanner, P. C. Taylor, K. S. Wong and J. Shinar for useful discussions, and W. Ohlsen and J. M. Leng for technical assistance with the ADMR setup. We also thank F. Wudl (PPV), C. Taliani (PT) and T. Barton (PDES) for providing us with the samples used in our studies. This work was supported in part by the DOE, Grant DE-FG-02-89, ER 45409, and by ONR Grant No. N00014-91-C-0104 at the Utah Laser Institute.

References

1 A. J. Heeger, S. Kivelson, J. R. Schrieffer and W. P. Su, *Rev. Mod. Phys.*, *60* (1988) 781.
2 J. Orenstein and G. L. Baker, *Phys. Rev. Lett.*, *49* (1982) 1043.
3 J. Orenstein, in T. A. Skotheim (ed.), *Handbook of Conducting Polymers*, Marcel Dekker, New York, 1986.
4 Z. Vardeny, J. Orenstein and G. L. Baker, *Phys. Rev. Lett.*, *50* (1983) 2032.
5 L. Lauchlan, S. Etemad, T.-C. Chung, A. J. Heeger and A. G. MacDiarmid, *Phy. Rev. B*, *24* (1981) 3701.

6 Z. Vardeny, E. Ehrenfreund, J. Shinar and F. Wudl, *Phys. Rev. B*, *35* (1987) 2498.
7 Z. Vardeny, E. Ehrenfreund, O. Brafman, M. Nowak, H. Schaffer, A. J. Heeger and F. Wudl, *Phys. Rev. Lett.*, *56* (1986) 671.
8 L. S. Swanson, J. Shinar and K. Yoshino, *Phys. Rev. Lett.*, *65* (1990) 1140.
9 C. G. Levey, D. V. Lang, S. Etemad, G. L. Baker and J. Orenstein, *Synth. Met.*, *17* (1987) 569.
10 X. Wei, B. C. Hess, Z. V. Vardeny and F. Wudl, *Phys. Rev. Lett.*, *68* (1992) 666.
11 B. C. Cavenett. *Adv. Phys.*, *30* (1981) 475.
12 L. Robins, J. Orenstein and R. Superfine, *Phys. Rev. Lett.*, *56* (1986) 1850.
13 W. P. Su and J. R. Schrieffer, *Proc. Natl. Acad. Sci.*, *77* (1980) 5626.
14 K. Fesser, A. R. Bishop and D. K. Campbell, *Phys. Rev. B*, *27* (1983) 4804.
15 W. P. Su, *Phys. Rev. B*, *34* (1986) 1988; *36* (1987) 6040.
16 S. Kivelson and W.-K. Wu, *Phys. Rev. B*, *34* (1986) 5423.
17 B. S. Hudson, B. E. Kohler and K. Schulten, *Excited States*, *6* (1982).
18 P. Tavan and K. Shulten, *Phys. Rev. B*, *36* (1986) 4337.
19 J. Takimoto and M. Sasai, *Phys. Rev. B*, *39* (1989) 8511.
20 C. V. Shank, R. L. Fork, B. Greene, J. Orenstein and G. L. Baker, *Phys. Rev. Lett.*, *49* (1982) 1953.
21 X. Wei, L. X. Zheng, B. C. Hess, K. S. Wong, Z. V. Vardeny, J. Shinar, T. J. Barton and S. Grigoras, *Synth. Met.*, *42* (1991) 1583.
22 Z. Vardeny, E. Ehrenfreund and O. Brafman, *Synth Met.*, *17* (1987) 349.
23 Z. V. Vardeny, X. Wei, S. G. Han, K. S. Wong, L. X. Zheng, G. S. Kanner, J. Shinar, Y. Pang, S. Ijadi-Maghsoodi, T. J. Barton, S. Grigoras and B. Parbhoo, *Synth. Met.*, *50* (1992) 453.
24 K. F. Voss, C. M. Foster, L. Smilowitz, D. Mikailovic, S. Askari, G. Srdanov, Z. Ni, S. Shi, A. J. Heeger and F. Wudl, *Phys. Rev. B*, *43* (1991) 5109.
25 N. F. Colaneri, D. D. C. Bradley, R. H. Friend, P. L. Burn, A. B. Holmes and C. W. Spangler, *Phys. Rev. B*, *42* (1990) 11 671.
26 X. Wei, *Ph.D. Thesis*, University of Utah, 1992.
27 J. Poplawski, E. Ehrenfreund, K. Cromack, A. J. Epstein and A. J. Franck, *Synth. Met.*, *43* (1991) 1225.
28 G. S. Kanner, X. Wei, B. C. Hess, L. R. Chen and Z. V. Vardeny, *Phys. Rev. Lett.*, *69* (1992) 538.

Intrachain photogeneration of polarons in poly(3-methylthiophene)

J. Poplawski[a,b], E. Ehrenfreund[a], Z. V. Vardeny[c] and A. J. Epstein[d]

[a]*Physics Department and Solid State Institute, Technion-Israel Institute of Technology, Haifa 32000 (Israel)*

[b]*Max-Planck-Institut für Festkorperförschung, 1 Heisenbergstrasse, D-7000 Stuttgart 80 (Germany)*

[c]*Department of Physics, University of Utah, Salt Lake City, UT 84112 (USA)*

[d]*Department of Physics, Ohio State University, Columbus, OH 43210 (USA)*

Abstract

We report detailed measurements of the excitation profile of the absorption spectrum due to photoexcited polarons and bipolarons in the non-degenerate ground state conjugated polymer poly(3-methylthiophene). Two new sharp peaks at 2.10 and 2.25 eV are observed. The separation of the two peaks, which corresponds to an intrachain vibration, is taken as evidence for a channel of photogeneration of polarons via an intrachain process in poly(3-methylthiophene).

Introduction

The formation of self-localized polaron-like photoexcitation is a unique feature of conjugated polymers [1]. The nature of these photogenerated defects in polythiophene and its alkyl-substituted derivatives has been intensively studied. Steady state photomodulation spectroscopy studies of polythiophene [2] have revealed absorption features characteristic of bipolarons; namely, two intragap electronic bands and a set of infrared-active vibrational (IRAV) lines. Moreover, steady state light-induced electron-spin resonance (LESR) measurements done on the same samples indicated that the above photoinduced absorption (PA) features are due to spinless defects. These long-lived photoinduced defects were thus identified as bipolarons with charge $\pm 2e$ and no spin. Subsequent photoinduced absorption studies [3, 4] of poly(3-hexylthiophene) have also concluded that bipolarons are the predominant long-lived photogenerated defects.

Bipolarons are believed not to be directly produced by the photoexcited electron–hole pair, but rather are formed by two like-charge polarons residing on the same chain [2, 4, 5]. The initial electron–hole pairs rapidly decay into pairs of negative and positive polarons, most of which recombine, but due to the interchain interaction, some of these polarons transfer to different chains forming eventually bipolarons. Therefore, the efficiency of the bipolaron production could be sample dependent, as the

0379-6779/93/$6.00

importance of the interchain interaction may vary in different samples. This was demonstrated, for instance, in the photoinduced absorption study of poly(3-hexylthiophene) in solution [5]. Consequently, the rate at which polarons are converted into bipolarons becomes slower as the interchain interaction weakens and, thus, polarons may be observed in nearly steady state photoinduced absorption measurements. Indeed, by measuring both the PA and LESR absorptions in poly(3-methylthiophene) (P3MT) it has been demonstrated that both polarons and bipolarons, which have a common origin, coexist on a timescale of the order of 1 ms [6]. Recently, using time-resolved photoinduced absorption and photoinduced absorption-detected magnetic resonance (ADMR) techniques, Kanner *et al.* [7] showed that bipolarons are the secondary product of the primary photogenerated polarons in polythiophene. In this study, the typical nearly steady state (characteristic time longer than $\simeq 1$ ms) bipolaron absorption was shown to evolve dynamically from polarons with shorter lifetime. The steady state ADMR measurements provided the spin assignments of the photoinduced defects.

The quantum efficiency for the primary photogenerated carriers can be studied by measuring the excitation profile of the photoinduced absorption bands. We emphasize here that regardless of the long-time nature of the photogenerated defects (either polarons or bipolarons), the excitation profile is a measure of the initial primary process.

In addition to the observed photoinduced absorption bands, due to polarons and bipolarons, conjugated polymers from the polythiophene family are characterized by relatively strong photoluminescence (PL). It has been suggested that the photoluminescence is mostly due to exciton recombination. The excitons in these conjugated polymers are accompanied by a weak polaron-like distortion in the backbone dimerization pattern (polaron–exciton). The observed PL features a well-resolved vibronic structure, with vibronic sub-band separation corresponding to the strong C–C backbone vibration [8, 9]. This is, therefore, an indication that the excitons involved in the photoluminescence are intrachain.

In this work we report our measurements of the excitation profiles of both the photoinduced absorption and photoluminescence. The two profiles are strikingly dissimilar making it possible to separate the mechanisms by which polarons and excitons are photogenerated.

Experimental details and results

The samples of P3MT used for this study were described previously [6]. The PA experiments were done using the pump and probe technique. Its excitation profile (PAE) was measured utilizing Ar^+ laser lines followed by various dyes to cover the range 1.8 to 2.7 eV for the pump beam. The photoluminescence excitation (PLE) profile was measured by pumping with a tungsten lamp dispersed by a monochromator covering the same excitation range.

Figure 1 shows the photoinduced absorption spectrum of P3MT. It consists of two broad asymmetric electronic bands peaked at 0.39 and 1.28 eV followed by a shoulder around 1.6–1.8 eV, and a set of photoinduced IRAV lines below 0.2 eV. The intensity of the IRAV band centered at 1280 cm^{-1} versus the photon energy of the pump beam is shown in Fig. 2. This excitation profile is an extension of a previously reported measurement [6]. The IRAV intensity rises quickly as the pump energy increases above $\simeq$1.8 eV, features two sharp peaks at 2.10 and 2.25 eV, followed by a nearly energy-independent region above 2.4 eV.

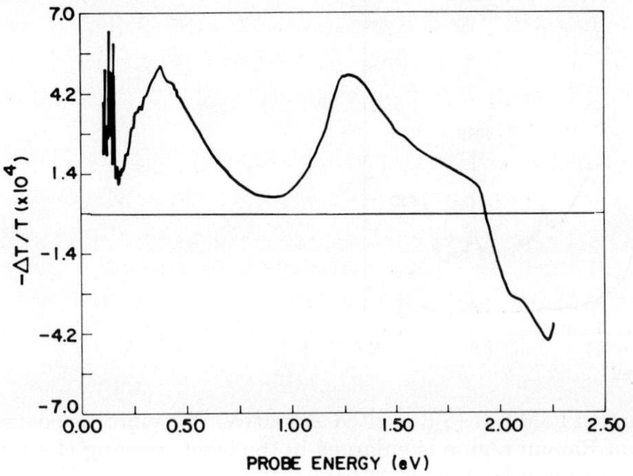

Fig. 1. Photoinduced absorption spectrum of P3MT, excited at 2.6 eV.

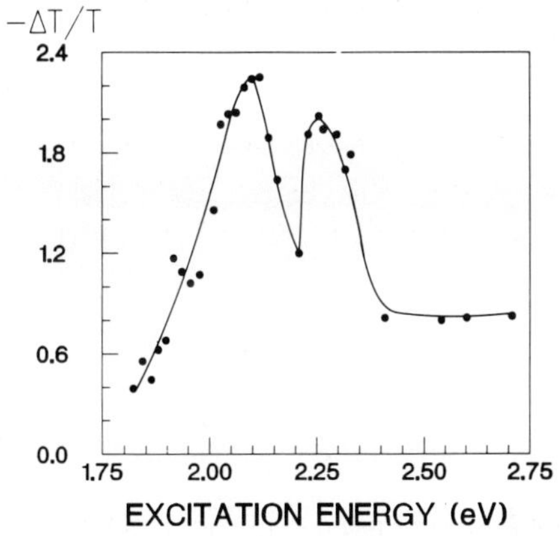

Fig. 2. Photoinduced absorption excitation (PAE) profile of the 1280 cm^{-1} IRAV band of Fig. 1. The PAE is normalized to a fixed number of absorbed photons: (●), data points; (——), a guide to the eye.

Figure 3 shows the PL spectrum of P3MT. The spectrum is composed of several phonon side bands superimposed on a much broader PL band. The energy separation of the phonon sidebands is $\simeq 0.17$ eV, which is nearly equal to the strongest resonantly scattered Raman line shown in the inset (0.18 eV, assigned to the C–C stretch). The zero-order PL emission at 15 K peaks at 1.90 eV. The PLE profile was measured at various probe energies in the PL band. A typical spectrum is shown in Fig. 4 for

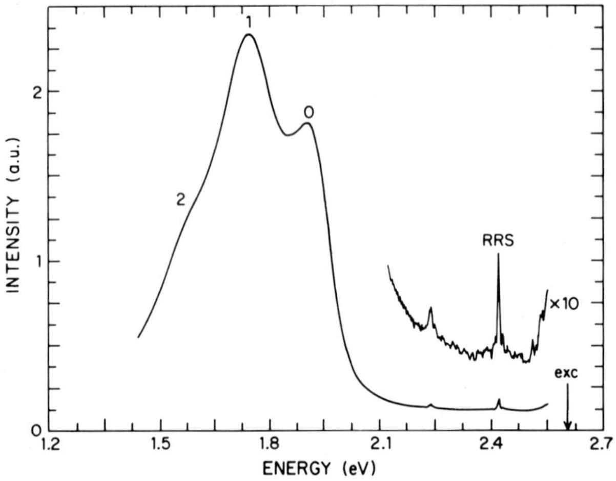

Fig. 3. Photoluminescence spectrum of P3MT at 15 K excited at 2.60 eV. The vibronic peaks are marked 0, 1 and 2. The resonant Raman region is enlarged in the inset showing clearly the first-order line at 2.42 eV and second-order line at 2.24 eV.

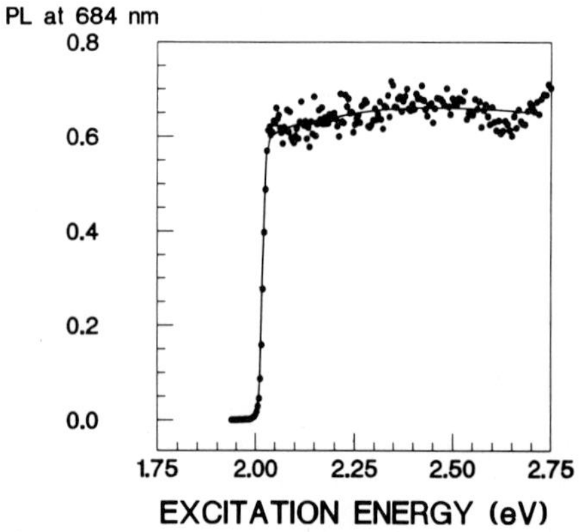

Fig. 4. Photoluminescence excitation (PLE) profile probed at 684 nm, taken at 15 K. Similar profiles were observed for various probe energies in the range 652–760 nm (1.63–1.90 eV). The PLE is normalized to a fixed number of absorbed photons: (●), data points; (——), a guide to the eye.

probe energy of 1.81 eV. Note the sharp onset at 2.00 eV and nearly flat response above 2.05 eV, in contrast to the double-peak structure observed for the PAE profile shown in Fig. 2.

Discussion

The excitation profiles presented in Figs. 2 and 4 measure the quantum efficiency for the primary photogenerated species responsible for the PA and PL, respectively. As mentioned above, although the spectrum shown in Fig. 1 is typical of bipolarons, the initial photogenerated carriers are polarons and therefore the PAE profile (Fig. 2) yields the action spectrum for the polarons. This profile shows that the spectrum of the quantum efficiency by which the absorbed photons create polarons has two peaks at 2.10 and 2.25 eV. The energy separation of these peaks ($\simeq$0.15 eV) is $\simeq$10–15% smaller than the 0.18 eV ground state A_g vibration observed by Raman scattering [9] or the vibronic separation of $\simeq$0.17 eV observed in the PL spectrum (Fig. 3). We tentatively assign, thus, the PAE spectral shape to a vibronic structure and the energy difference of $\simeq$0.15 eV to the intrachain A_g vibration in the excited state, which is typically somewhat lower than the ground state vibration. The PAE profile provides, therefore, an indication for the intrachain mechanism for the photoproduction of polarons. Previous photoconductivity studies of poly(3-hexylthiophene) also provided evidence for a similar intrachain mechanism [10]. By measuring the polarization dependence of the transient (subnanosecond) photoconductivity, these authors concluded that polarons formed via photoexcitation of electrons and holes on the same chain rather than on different chains are mostly responsible for the transient photoconductivity. More recently, following the evolution of photogenerated polarons and bipolarons on a time domain from 0.1 ps to 20 ms, Kanner *et al.* [7] have shown that in polythiophene, polaron–excitons are formed (with quantum efficiency close to 1) within 0.3 ps, whereas bipolarons are formed only after 3 to 250 ns. These findings led the authors to conclude that the fast polaron–exciton formation is due to intrachain photogeneration, which evolves in time to yield the bipolarons observed at much longer delay times. There are, thus, increasing experimental indications that in non-degenerate ground state conjugated polymers (at least from the polythiophene family) polaron formation is due mainly to an intrachain photogeneration process rather than an interchain process.

Theoretically, it was argued that the photoproduction of intrachain polaron pairs is forbidden because two close polarons have the same A_g symmetry as the ground state [11, 12]. However, recently, Gartstein [13] observed that even in ideal, infinitely long chains parity violation of the optical selection rules occurs when the electron–phonon interaction is treated non-classically. He then further argues that, in such a case, the probability of photogeneration of (free) polaron pairs is comparable to that of soliton–anti-soliton pairs (which is the allowed process in the classical lattice). In real systems, parity violation may perhaps occur due to the presence of defects on the chain. The longer the chain the larger is the probability of having a defect on it and, in turn, larger quantum efficiency for photogeneration of charges. In such a case, intrachain polaron pair formation occurs preferentially in longer chains whereas exciton–polarons are more evenly photoproduced in long and short chains. The photoselection of the longer chains (with defects) should result in a polaron quantum efficiency spectrum which is peaked at the B_u level of the longer chains, as seen in Fig. 2. The double-peak structure observed in Fig. 2 is interpreted as the long chain B_u absorption at $\simeq$2.10 eV and its vibrational replica at $\simeq$2.25 eV, as mentioned above. Since there is no photoselection of chains in the neutral exciton–polaron formation, any similar structure in the quantum efficiency spectrum of the photoluminescence should be masked out due to the continuous chain length distribution. The sharp onset at $\simeq$2.00 eV and the nearly flat response above $\simeq$2.05 eV of the PL quantum efficiency (Fig. 4) are therefore not unexpected.

Conclusions

We have observed a double-peak structure in the PAE profile of P3MT and argued that it shows that the intrachain photogeneration of polarons is possible. This conclusion is corroborated by earlier photoconductivity and time-resolved photoinduced absorption studies, which point to the same conclusion, and by recent calculations.

Acknowledgements

The work at the Technion is supported by the US–Israel Binational Science Foundation (BSF 89-322), Jerusalem, Israel. The work at the University of Utah is supported in part by the DOE, Grant DE-FG-02-89, ER 45409.

References

1 A. J. Heeger, S. Kivelson, J. R. Schrieffer and W. P. Su, *Rev. Mod. Phys.*, *60* (1988) 781.
2 Z. Vardeny, E. Ehrenfreund, O. Brafman, M. Nowak, H. Schaffer, A. J. Heeger and F. Wudl, *Phys. Rev. Lett.*, *56* (1986) 671.
3 Y. H. Kim, D. Spiegel, S. Hotta and A. J. Heeger, *Phys. Rev. B*, *38* (1988) 5490.
4 J. Rühe, N. F. Colaneri, D. D. C. Bradley, R. H. Friend and G. Wegner, *J. Phys.: Condens. Matter*, *2* (1990) 5465.

5 D. Spiegel, P. Pincus and A. J. Heeger, *Synth. Met.*, *28* (1989) C385.
6 J. Poplawski, E. Ehrenfreund, K. Cromack, A. J. Epstein and A. J. Frank, *Synth. Met.*, *41* (1991) 1225.
7 G. S. Kanner, X. Wei, C. Hess, L. R. Chen and Z. V. Vardeny, *Phys. Rev. Lett.*, *69* (1992) 538.
8 B. C. Hess, J. Shinar, Z. Vardeny, E. Ehrenfreund and F. Wudl, *Mater. Res. Soc. Symp. Proc.*, *109* (1988) 363.
9 J. Poplawski, E. Ehrenfreund, S. Glenis and A. J. Frank, *Synth. Met.*, *28* (1989) C335.
10 G. Yu, S. D. Phillips, H. Tomozawa and A. J. Heeger, *Phys. Rev. B*, *42* (1990) 3004.
11 G. W. Hayden and E. J. Mele, *Phys. Rev. B*, *34* (1986) 5484.
12 J. Yu, H. Matsuoka and W. P. Su, *Phys. Rev. B*, *37* (1988) 10 367.
13 Yu. N. Gartstein, *Phys. Lett. A*, *144* (1990) 487.

3b. Nonlinear Optical Properties

Synthetic Metals, *54* (1993) 123–137

New advances in molecular engineering for quadratic nonlinear optics

Isabelle Ledoux
Centre National d'Etudes des Télécommunications, 196 Avenue Henri Ravera, 92220 Bagneux (France)

Abstract

Optimization of organic molecules for quadratic nonlinear optics is investigated in the perspective of nonlinearity/transparency trade-off, using intramolecular charge-transfer compounds with limited conjugation, such as oligomers of polysilane or polyphenyl. A large increase of the transparency in the blue and near UV spectral range is obtained for silane derivatives, but at the cost of a limitation of the nonlinear optical response at the molecular scale. On the contrary, polyphenyls appear to be 'optimized' materials, their nonlinear optical performances, both at the microscopic and macroscopic scale, being much better than those of paranitroanilines, with a significant improvement of the transparency. On the other hand, organometallic complexes containing ferrocene moieties conjugated to Ni, Pd or Pt metals through an organic polarizable unit display interesting nonlinear optical properties: the role of complexation appears to be quite important for ferrocenyl-containing ligands. A metal-to-metal intramolecular charge transfer is clearly evidenced. The role of the nature of the complexing metal and the influence of the nature of the ligand are also investigated. The moderate static hyperpolarizability values (of the order of 20×10^{-30} e.s.u.) may be partially accounted for by the defavourable projection factor between the ground state dipole moment and the vector part of the molecular hyperpolarizability.

1. Introduction

The huge development of organic materials for quadratic nonlinear optics has given rise, during the last decade, to a wide extension of the so-called 'molecular engineering' rules, previously limited to classical intramolecular charge-transfer (ICT) conjugated molecules, such as benzene or stilbene derivatives. An increasing demand for applications of these materials in the field of optoelectronic devices has oriented the research effort towards specific directions for which organic compounds could be of practical use for achieving optical functions, such as frequency conversion, electro-optic (EO) modulation, all-optical switching and so on [1–4].

More precisely, the recent investigations towards optimization of molecular materials in terms of nonlinear efficiency, transparency range, processability, etc., have been oriented in two complementary directions [5]. First, a quite attractive application of quadratic nonlinear materials is frequency-doubling of laser diodes in order to increase the capacity of

0379-6779/93/$6.00

optical disks, by reduction of the minimal size of the writing and reading beams emitted by laser diodes around 850 nm and focused onto the disk. In that case, the demand for highly efficient materials must be generally moderated by the limits of the transparency domain of the nonlinear compound, especially in the near UV and blue spectral range. Therefore, a delicate compromise must be found between linear (transparency) and nonlinear properties, opening the way to a special part of molecular engineering, specifically oriented towards a 'double' optimization of the spectral properties of the material. Secondly, the possible use of highly nonlinear and strongly polar organic molecules inside poled transparent polymers for elaboration of electro-optic devices has led to a large extension of the organic synthesis towards elaboration of strongly conjugated molecules, without any consideration of their absorption spectra in the visible range, the typical wavelengths for optical telecommunication systems being 1.3 or 1.5 μm. The additional limitations to the practical use of such 'active' molecules in EO components are more related to their thermal and (photo)chemical stability.

In the present work, these complementary orientations will be illustrated by various examples. In the search for an optimal nonlinearity/transparency trade-off, oligomers of polysilane and polyphenyl are investigated, on the basis of the limitation of the polarizability or the conjugation of the intermediate electrons separating the donor from the acceptor group, using sp^3 hybridization (for silanes) or steric constraints (for polyphenyls). In the field of molecular optimization for EO applications, where the constraints on the transparency are widely relaxed, we will focus on metallo-organic compounds and investigate the role of complexation and metal–metal intramolecular charge transfer on the possible increase of the dipole moment and of the first-order hyperpolarizability β of organic molecules.

Most nonlinear optical results reported here were obtained using the electric field-induced second-harmonic generation (EFISH) [6] experiment for β determination, the fundamental wavelength being generally 1.34 μm, and sometimes 1.9 or 1.06 μm. The β values deduced from these measurements are usually strongly affected by resonance effects. Therefore, it is more significant to compare the 'static' hyperpolarizabilities $\beta(0)$ of various molecules, deduced from β using a two-level model [7]:

$$\beta(0) = \beta\left[1 - \left(\frac{\lambda_0}{\lambda}\right)^2\right]\left[1 - \left(\frac{2\lambda_0}{\lambda}\right)^2\right]$$

2. Sp^3 hybridization: silane derivatives

A strong increase of the blue-UV transparency is obtained using sp^3 hybridized heteroatoms of the third row of the Periodic Table. The expected blue shift could be related to the electron-isolating properties of

σ bonds as compared to π ones, and to the relatively poor overlap of the external 3s or 3p orbitals of the heteroatoms with the 2s and 2p of the neighbouring carbons. However, the polarizability of the corresponding Si–Si σ bonds is significantly higher than that of a C–C one, owing to the increased electronegativity of the atoms with larger atomic numbers.

2.1. Si as part of the conjugation path

The role of the $Si(CH_3)_2$ unit was investigated by testing the linear and nonlinear optical properties of the series:

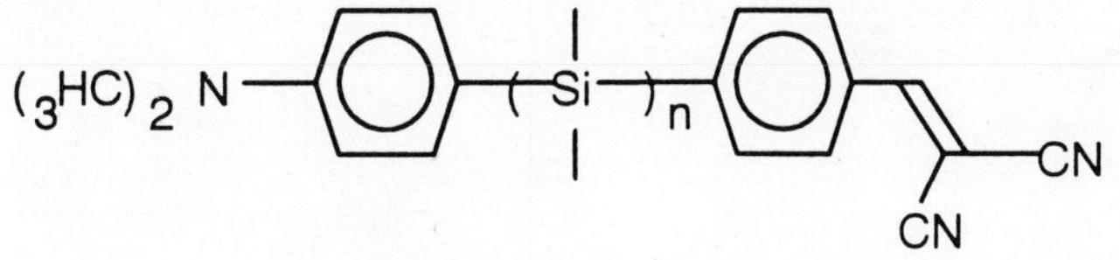

where $n = 1, 2, 6$ [8].

The dipole moment, maximum absorption wavelength and hyperpolarizability of these molecules are reported in Table 1. The ability of the dimethylsilane groups to 'transmit' the intramolecular charge transfer (ICT) from the donor to the acceptor part of the molecule is estimated comparing the measured $\beta(0)$ of the entire molecule (i.e. deduced from the experimental β using a two-level model) to the sum of the measured $\beta(0)$ values of the two separate parts of the molecule:

$\beta^D(0)$

and $\beta^A(0)$

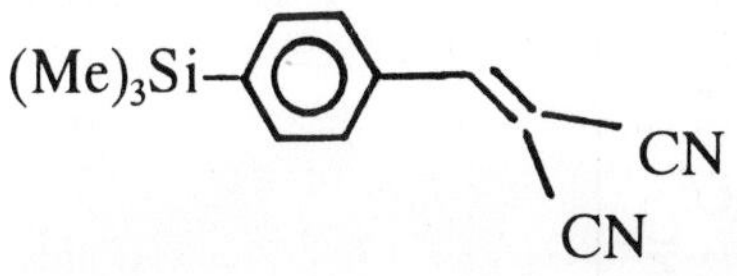

This sum corresponds to a description of the nonlinear response of the molecules in terms of an additive model, where the resulting hyperpolarizability is $\beta_{add}(0) = \beta^D(0) + \beta^A(0)$.

For $n = 1$, no ICT could be detected. This clearly shows that one Si atom is unable to act as a polarizable connection between the phenyl rings. The ICT processes involved here remain located within each subunit; in fact, the two corresponding individual ICT axes are not aligned, they make an angle of approximately 109°, owing to sp^3 hybridization of Si. Such a bent structure is not favourable to an ICT over the whole molecule, the two phenyl rings, owing to steric constraints, not being coplanar [9].

TABLE 1

Dipole moment μ, maximum absorption wavelength λ_{max}, static first-order hyperpolarizability deduced from experimental data ($\beta_{exp}(0)$), from an additive model ($\beta_{add}(0)$) and an intramolecular charge transfer contribution ($\beta_{CT}(0)$) to $\beta(0)$, for silane derivatives with various numbers of methylsilane units between the electron-donating and the electron-accepting parts of the molecule ($n = 6$ for the last molecule)

Compound	μ (D)	λ_{max} (nm)	$\beta_{exp}(0)$ (10^{-30} e.s.u.)	$\beta_{add}(0)$ (10^{-30} e.s.u.)	$\beta_{CT}(0)$ (10^{-30} e.s.u.)
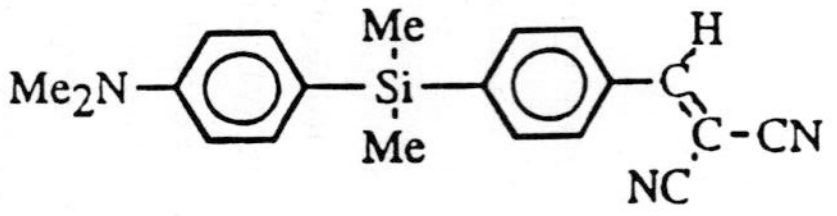	6	320	11	11	0
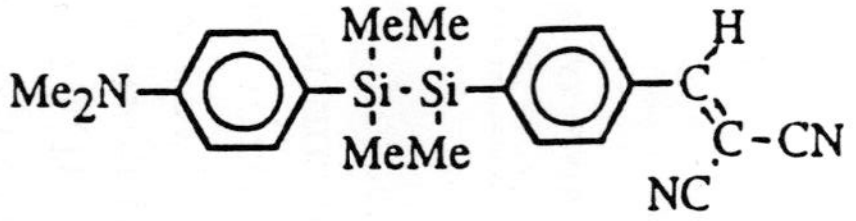	7	334	16	13	3 ± 2
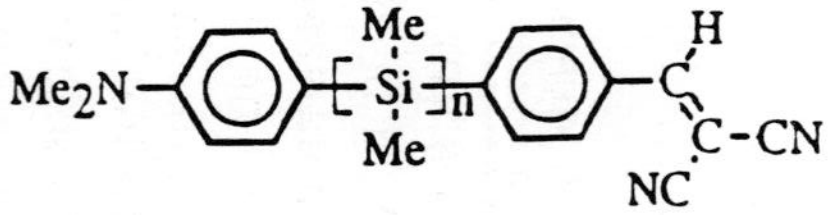	6.8	385	26	13	13

For $n = 2$, $\beta(0)$ is significantly higher, but the possible ICT contribution remains within the experimental error range. The increase of $\beta(0)$ as compared to $n = 1$ is mainly accounted for by the all-*trans* configuration of the molecule along the Si–Si bond. Crystallographic studies have shown a better coplanarity between the benzene rings, the corresponding dihedral angle being only 11°. In addition, the good parallelism between the two vector parts of the $\beta(0)$ values of each subunit maximizes the cosine projection factor and hence the $\beta(0)$ of the whole molecule [9].

For $n = 6$, the $\beta(0)$ value is largely improved with respect to smaller n values [8]. This may be accounted for by a significant delocalization of the σ electrons along the Si–Si chain. The ICT becomes large (50% of the total $\beta(0)$ vlaue). In addition, the compound exhibits an interesting nonlinearity/transparency trade-off: $\beta(0)$ is almost three times larger than that of paranitroanilines for a very similar transparency. However, the molecule is also approximately three times longer. The σ delocalization over the whole Si–Si backbone shows the ability of silicon-based molecules to increase the optimization of the nonlinearity/transparency compromise.

2.2. Role of substituents in ICT through disilane units

The scale of the electron-donating strength of various substituents: $H < F < OCH_3 < N(CH_3)_2$ was clearly established from nonlinear optical experiments or from various other physical or chemical properties when connected to standard π-electron systems, such as benzene or stilbene derivatives, and conjugated to acceptors such as CN. A quantitative estimate of these electron-donating properties may be given by Hammet's constants. Similarly, it has been well established that the vinyltricyano group exhibits a stronger accepting character than the vinyldicyano one. Therefore, a higher nonlinear optical response is expected for compounds having the strongest electron-donor or electron-acceptor substituents.

In Table 2 are given the experimental data for five disilane molecules, four of them bearing the same vinyldicyano acceptor group for different donors; a significant (and unexpected) decrease in $\beta(0)$ is observed when replacing the methoxy group by the dimethylamino one, although the latter is supposed to be a stronger donor. The same surprising phenomenon occurs when replacing the vinyldicyano group by the vinyltricyano one, while keeping the same methoxy donor. One should also note that the $\beta(0)$ values are well described in terms of an additive model for donor substituents, like H, F or dimethylamino, but that is not the case for the corresponding methoxy derivatives, where a significant ICT is evidenced [10, 11].

Figure 1 plots the $\beta(0)$ values of the four disilane derivatives bearing different donor substituents as a function of their corresponding Hammet constants. The behaviour is consistent with the model proposed by Marder *et al.* [12] and Beratan [13], where β does not increase indefinitely when increasing the relative strengths of the donor and acceptor substituents, but exhibits a maximal value. In fact, for an acentric molecule

TABLE 2

Dipole moment μ, maximum absorption wavelength λ_{max}, static first-order hyperpolarizability deduced from experimental data ($\beta_{exp}(0)$), from an additive model ($\beta_{add}(0)$) and an intramolecular charge transfer contribution ($\beta_{CT}(0)$) to $\beta(0)$, for disilane derivatives with various couples of donor–acceptor substituents

Compound	μ (D)	λ_{max} (nm)	$\beta_{exp}(0)$ (10^{-30} e.s.u.)	$\beta_{add}(0)$ (10^{-30} e.s.u.)	$\beta_{CT}(0)$ (10^{-30} e.s.u.)
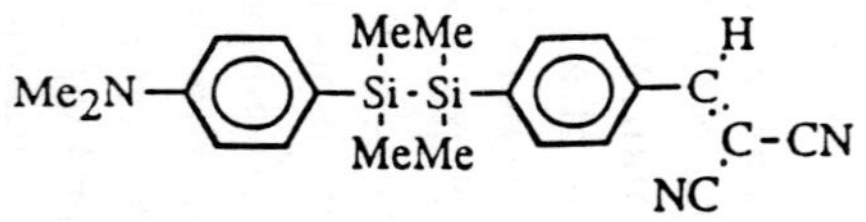	7	334	16	13	3 ± 2
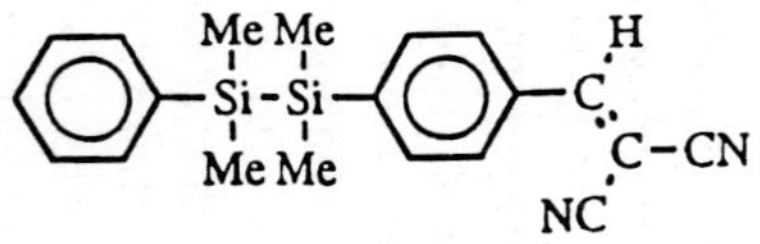	5.1	345	8.5	9	0
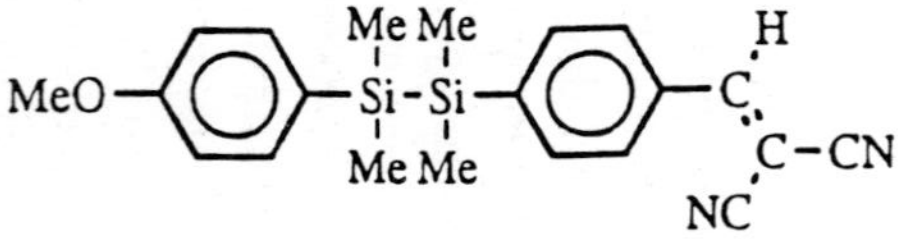	5	322	22.1	9.3	12.8
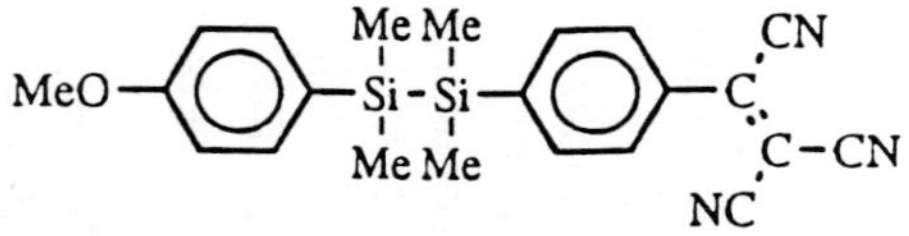	6.2	367	16.1	17.5	0
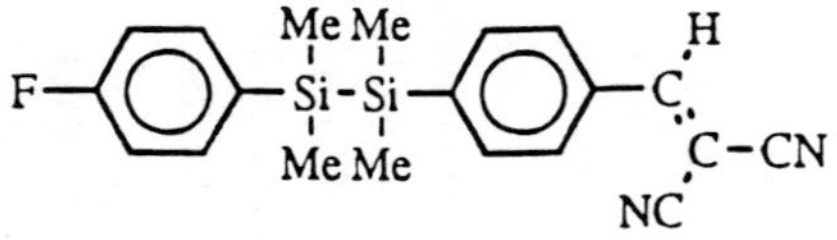	5.8	344	9.1		

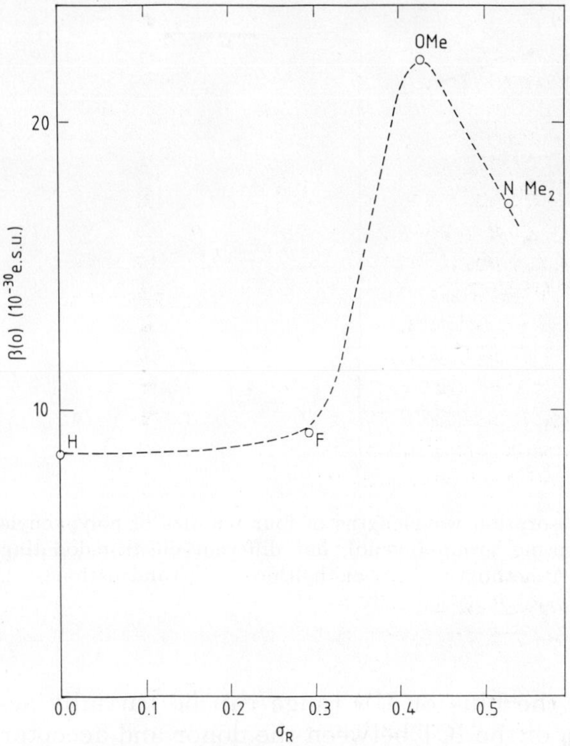

Fig. 1. Static first-order hyperpolarizabilities $\beta(0)$ for vinyldicyanodisilane derivatives substituted by various electron-donor groups, H, F, MeO and Me_2N, as a function of Hammet's constants of these substituents.

substituted with very strong donor and acceptor groups, the polarizable electrons are strongly linked to the electroactive substituents, so that ground and excited state dipole moments are equal. This should be particularly the case for weakly polarizable electrons as occurring in the Si–Si σ bonds. In that case, only the methoxy group exhibits the appropriate electronic properties (when associated to the vinyldicyano group) to induce a significant ICT. Weaker substituents, such as F or H, are not active enough to withdraw any substantial electronic charge towards the acceptor through the disilane chain. This experimental behaviour clearly evidences the possibility of a decrease of the quadratic nonlinear response of an electronic system substituted by quite strong donor and acceptor units, as already predicted for inorganic compounds.

3. Hybridized carbon atoms: polyphenyls

An efficient way to increase the transparency, in keeping with a reasonably large β value, has been reported for polyphenyls [14–17].

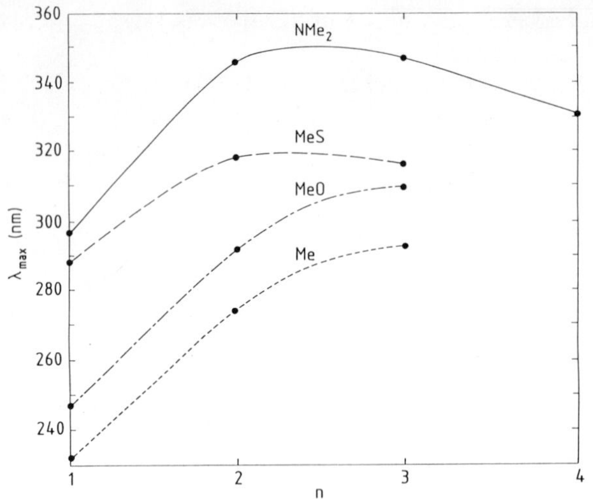

Fig. 2. Plot of the the maximal absorption wavelengths of four families of polyphenyls exhibiting the same electron-accepting group (cyano), but different electron-donating groups, such as dimethylamino (——), methoxy (–·–·–), methylthio (– – –) and methyl (- - -), vs. the number n of phenyl rings.

Their good transparency in the blue or UV range can be partially accounted for by the limitation of the ICT between the donor and acceptor groups, owing to the large torsion angle (40° for dimethylaminocyanobiphenyl) between the phenyl rings.

3.1. Linear optical properties

Figure 2 gives the maximum absorption wavelengths λ_{max} as a function of n in the series:

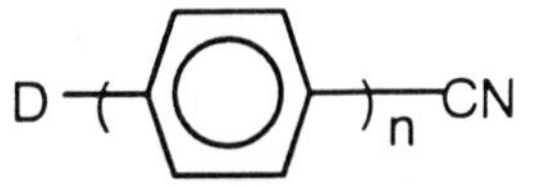

for four different electron-donating groups with increased electron-releasing character when going from the methyl to dimethylamino one. As expected, λ_{max} increases with polarizability of the donor group. The relative increase of λ_{max} is directly correlated to the relative electronegativity of the heteroatoms N, S, O and C, and thus to the corresponding polarizability of the nonbonding electrons. For a given donor substituent, the large increase of λ_{max}, when going from $n = 1$ to $n = 2$, results from the additional number of π electrons involved in the ICT process, which largely overcomes the decrease of the electronic polarizability due to the large twist angle between the benzene rings. For $n = 3$, the additional increase of the number of π electrons contributes less to the ICT, owing to the presence of two torsion angles between the phenyl rings; λ_{max} saturates or even decreases in some cases.

It must be pointed out that, for any molecule of this series, even for the strongest donor group and the largest π-electron system, λ_{max} is significantly smaller than for paranitroaniline derivatives. All compounds are white or very pale yellow in the solid state.

3.2. Nonlinear optical properties

The $\beta(0)$ values of the four molecular series are plotted in Fig. 3 as a function of the number of phenyl rings n. It is noticeable that the electron-donating character of S–Me is slightly weaker than that of O–Me, although the polarizability of the surrounding non-bonding electrons is higher in the case of the sulfur atom. This is consistent with dipole moment values, which display a similar relative behaviour. This could be accounted for by the larger atomic volume of sulfur, the orbital overlap integral between the neighbouring sp^2 hybridized atom and the sulfur one being smaller than with the oxygen–carbon pair. The ICT is then stronger for the methoxy substituent.

It should be pointed out that, in fact, the relevant parameter is the quadratic hyperpolarizability per volume (or length) unit; taking into account these volume effects, the actual efficiency of the polyphenyls displays a maximal value for $n = 3$ and then slightly decreases; this

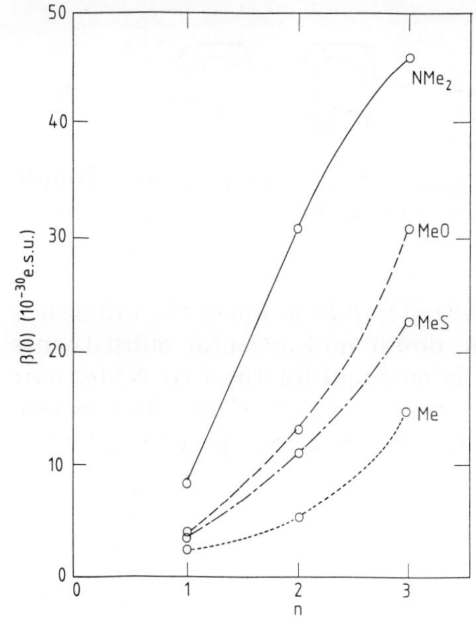

Fig. 3. Plot of $\beta(0)$ of four families of polyphenyls exhibiting the same electron-accepting group (cyano) but different donating groups.

observation is consistent with the theoretical predictions made on similar compounds by Morley *et al.* [18].

Saturation of $\beta(0)$ values is observed only for a terphenyl molecule containing the most efficient donor group. This phenomenon corresponds to a saturation of the transmitting capacity of the conjugated chain linking the donor and acceptor groups. The limitation of the ICT has a purely electronic origin, already predicted by theoretical studies for idealized fully planar systems and a steric origin, due to the torsion angles between the adjacent phenyl rings. The two effects are self-consistently related: torsion angles have a significant influence on the electronic properties of the whole π-electron system while, conversely, the torsion angles will depend on the strength of the donor–acceptor interaction. Such a 'saturation' of the $\beta(0)$ values is not observed for the weaker electron-donating groups. This behaviour is probably linked to the electronic properties of ICT molecules such as described in the model of Marder *et al.* [12] and Beratan [13]. The strong electron-donating character of dimethylamino induces a large ICT in both ground and first excited states, then limiting the relative increase of $\Delta\mu$ with n as compared to molecules bearing more moderate electron-donor substituents. It is even possible, for a given n value ($n = 2$), to decrease the $\beta(0)$ value when using very strong donor and acceptor substituents, such as O^- as a donor and a pyridinium as an acceptor. In this case $\beta(0)$ becomes even strongly negative and the molecule displays a huge red shift ($\lambda_{max} = 481$ nm) as compared to the previous ones. This is due to a large contribution of the quinonic form in the ground state, preferably to the aromatic one:

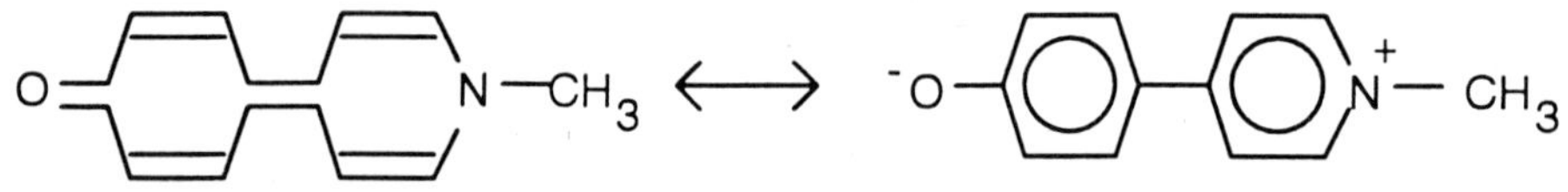

As predicted in ref. 12, this predominance of the quinonic form strongly decreases the energy gap between the ground and the first excited states, and simultaneously leads to negative $\Delta\mu$ values. Therefore, the sign of $\beta(0)$ is reversed as compared to that of classical substituents. Figure 4 exemplifies this phenomenon; $\beta(0)$ is plotted as a function of the difference between the Hammet constants of the donor and acceptor substituents $\sigma^D - \sigma^A$ for biphenyl derivatives. $\beta(0)$ is maximal for the CN–NMe_2 pair and then strongly decreases towards a large negative value. The experimental data seem to be consistent with the behaviour predicted by the model proposed in ref. 12.

4. Metallo-organic molecules

The search for molecules possessing large second-order polarizabilities has led to the investigation of organometallic coordination complexes

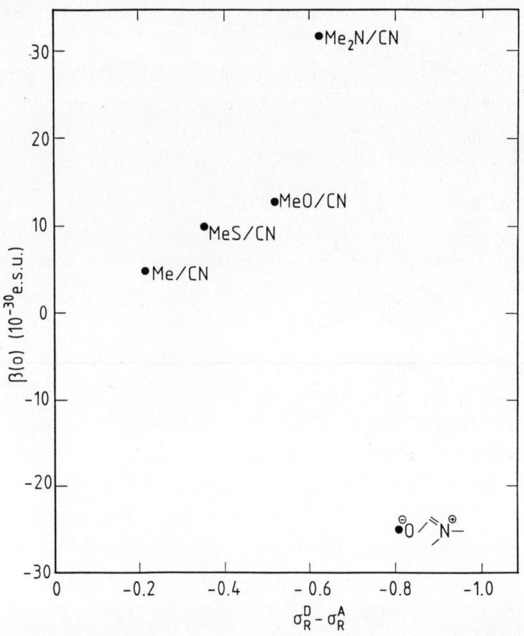

Fig. 4. $\beta(0)$ values of various biphenyl derivatives bearing donor and acceptor substituents with various strengths as a function of the difference of the corresponding Hammet constants.

[19, 20]. Strong hyperpolarizabilities β on ferrocenyl derivatives have been recently shown [21, 22]. Some novel ferrocene derivatives and their corresponding palladium, platinum or nickel complexes that exhibit promising nonlinear optical properties [23, 24] are described here. The compounds investigated in this study are shown in Fig. 5. The influence of the nature of the ligand, of the complexation and of the nature of the complexing metals is studied.

The absorption spectra of ferrocene derivatives usually exhibit a broad band with a maximum around 450 nm and a higher energy band around 330 nm. It may be considered, as in ref. 20, that the low energy transition is assigned to a metal-ligand charge-transfer (MLCT) band in the ferrocenyl group, the other transitions at higher energy being attributed to intraligand $\pi-\pi^*$ transition, with some metal character. The peak wavelengths of the MLCT band are listed in Table 3 for all compounds. This band exhibits a bathochromic shift upon complexation by a second metal. This demonstrates the involvement of transition metal orbitals in the MLCT transition. It should be noted that the most electropositive metal (Pt) leads to the larger shift. The substitution of S by SO or SO_2 slightly increases the shift. The addition of a double bond also leads to an expected red shift. On the contrary, an hypsochromic shift is observed with the complex **3** which has a triple bond. The bathochromic shift for the higher energy bands is quite small. All the compounds (except **1′Pt** and **2′Pd**) exhibit a slight positive solvatochromic shift of the MLCT band.

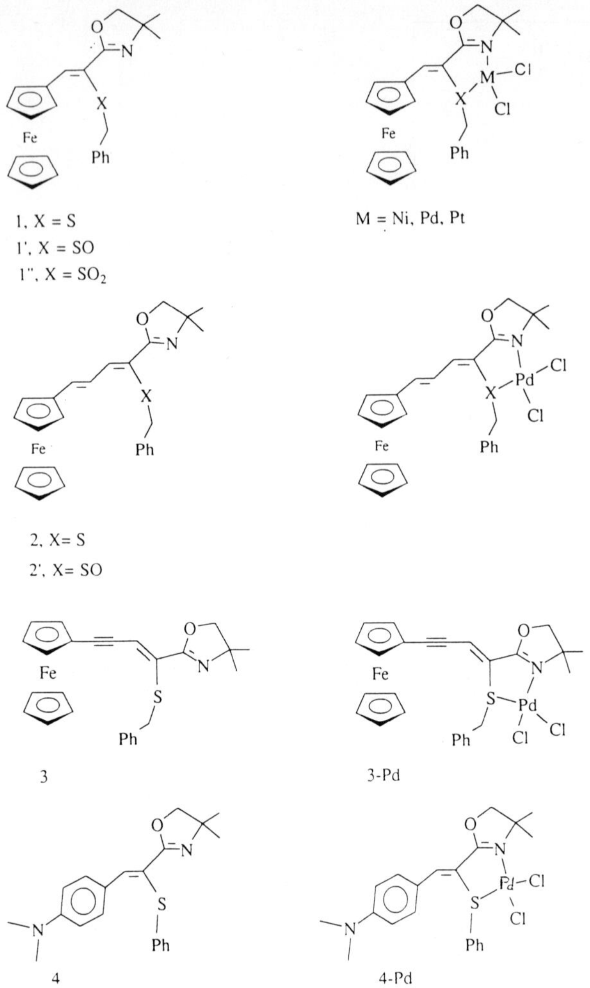

Fig. 5. Ferrocenyl derivatives investigated in the present work.

EFISH measurements were carried out at 1.34 μm for all compounds and at 1.9 μm for some of them. The level taken into account in the two-level model corresponds to the MLCT transition. As $\beta(0)$ values deduced from measurements at 1.34 and 1.9 μm are in agreement, it can be concluded that the contribution of the MLCT level to the hyperpolarizability is predominant. Of course, one cannot rule out a contribution due to the $\pi-\pi^*$ ICT band, but this contribution should be weak.

Measurements of third-harmonic generation at 1.9 μm allow the negligible contribution due to the third-order polarizability γ in the EFISH experiment to be ruled out.

The most important feature of these results is the important increase of $\beta(0)$ values upon coordination. The increase is particularly large for

TABLE 3

Peak positions of the MLCT band (λ_{MLCT}) in chloroform, ground state dipole moments (μ_g), difference $\Delta\mu$ between excited and ground state dipole moments and hyperpolarizability coefficients β

Compound	λ_{MLCT} (nm)[a]	μ_g (D)[b]	$\Delta\mu$ (D)	β (10^{-30} e.s.u.)[c]	β_0 (10^{-30} e.s.u.)[d]	β (10^{-30} e.s.u.)[e]	β_0 (10^{-30} e.s.u.)[f]
1	472	3.85	2	4.5	2.0	2.9	2.07
1′	470	5.7	2.1	8.4	3.7		
1″	477	6.45	2.5	6.3	1.6		
1-Ni	500 (511)[g]	14 (14)[g]	8.3	10.3 (7.14)[g]	3.92 (2.55)[g]		
1-Pd	530	13.2	3	12.2	4.13	9	5.8
1-Pt	523	11.6	3	16.0	5.2		
1′Ni	498 (494)[g]	14 (14)[g]	11.8	15.8 (5.85)[g]	6.0 (2.3)[g]		
1′-Pd	540	12.2	6	30.2	7.6		
1′-Pt	548	11.3	0.2	28.0	7.7		
2	482	3.22	21	13.3	5.6	8.8	6.14
2′	494	6.65	2.9	11.0	4.6		
2-Pd	567	13.7	0.3	58	13.5	18.2	10.8
2′-Pd	602	13.5	1.4	123.5	19		
3	460	5.55	7.7	2.6	1.2		
3-Pd	558	13.6	1.9	33.6	8.5		
4	371	3.5	17	23.0	15.0		
4-Pd	442	13.5	9	37.0	20.0		

[a]λ_{MLCT} was determined in chloroform solutions.
[b]The ground state dipole moment μ_g was determined in chloroform solutions.
[c]β was measured by the EFISHG technique (solvent: chloroform) at 1.34 μm.
[d]β_0 is the frequency-independent nonresonant first-order hyperpolarizability calculated on the basis of a two-level system with a transition energy given by the MLCT peak position in chloroform deduced from β measured at 1.34 μm.
[e]β was measured by the EFISHG technique (solvent: chloroform) at 1.9 μm.
[f]β_0 was deduced from β measured at 1.9 μm.
[g]The figures in brackets are for the complex formed with bromine ions (instead of chlorine ions).

compounds with two double bonds or one triple and one double bond in the conjugated linkage between the two complexes (compare compounds **2** and **2-Pd**, and compounds **3** and **3-Pd**). The enhancement factor can reach 7 for the **3** derivatives. On the contrary, compounds **4** and **4-Pd**, where the ferrocenyl group has been replaced by a *p*-*N*,*N*-dimethylanilino group, is less affected by metal complexation [24]. Reasonably large $\beta(0)$ values could be obtained (20×10^{-30} e.s.u. for compound **4-Pd**).

The nature of the second transition metal has a weak influence on β values. The slight increase of β observed in the series **1-Ni**, **1-Pd**, **1-Pt** and also **1′-Ni**, **1′-Pd** and **1′-Pt** correlates to the increase in the bathochromic shift of the MLCT band upon coordination to the second metal. This can be a consequence of the decrease of electronegativity from Ni to Pt [25].

In the two-level model, the charge-transfer part of $\beta(0)$ can be evaluated from the oscillator strength of the transition and the difference $\Delta\mu$ between ground and excited state dipole moments. As far as this two-level model is valid, it is possible to correlate $\Delta\mu$ and $\beta(0)$ values. It should be noted that this correlation holds only for ferrocenyl derivatives without a second metal. The presence of the second metal does not lead to a systematic increase of the measured $\Delta\mu$ in relationship with $\beta(0)$. This can be, at least partially, accounted for by the poor validity of the solvatochromic experiments on large molecules [26].

The substitution of the thio group by a sulfoxy group leads to an increase in both $\Delta\mu$ and $\beta(0)$, as a consequence of the increase of the electron-attracting effect; further replacement of the sulfoxy group by a sulfonyl one leads to a decrease of $\beta(0)$. This can be due either to an increase of the angle between the ground state dipole moment and the vector part of the hyperpolarizability, which approaches 90°, or to a kind of ‘saturation’ effect on β similar to that observed in disilanes or polyphenyls in the presence of strongly donating or accepting groups.

In summary, investigation of the nonlinear optical properties of a wide range of bimetallic complexes has made possible evidence of interesting properties in terms of the influence of the complexation on the molecular hyperpolarizability. The absolute $\beta(0)$ values are not yet very large; this could be due to a large projection factor between the dipole moment and the vector part of the hyperpolarizability, as observed on the sulfoxy derivative. Some progress should then be made towards a better optimization of this projection factor, together with investigation of novel ligands and complexing metals for β optimization.

5. Conclusions

Two complementary directions of molecular engineering towards optimization of quadratic nonlinear optical response of organic molecules have been investigated. A quite satisfactory nonlinearity/transparency trade-off was found for polyphenyls, especially for dimethylamino-cyanobiphenyl (DMACB) which exhibits a large nonlinear optical response (higher than that of 3-methyl-4-nitropyridine-1-oxide) in the solid state, with an increased transparency range [17]. On the other hand, bimetallic complexes open a new and interesting way towards optimization of organic molecules for incorporation of oriented electro-optics polymers, but some improvements have still to be found in terms of optimization of the ICT and of the projection factor between the dipole moment and the vector part of the quadratic hyperpolarizability.

References

1 D. S. Chemla and J. Zyss (eds.), *Nonlinear Optical Properties of Molecules and Crystals*, Academic Press, Orlando, FL, 1987.
2 S. R. Marder, J. E. Sohn and G. D. Stucky, (eds.), *Materials for Nonlinear Optics—Chemical Perspectives, ACS Symp. Ser., 455* (1991).
3 K. Singer (ed.), *Proc. SPIE, 1560* (1991).
4 J. Messier, F. Kajzar and P. Prasad, (eds.), *Organic Molecules for Nonlinear Optics and Photonics, NATO ASI Ser., 194* (1991).
5 J. Zyss, in J. L. Brédas and R. R. Chance (eds.), *Conjugated Materials: Opportunities in Electronics, Optoelectronics and Microelectronics*, Kluwer, Dordrecht, 1990.
6 I. Ledoux and J. Zyss, *Chem. Phys. 73* (1982) 203.
7 J.-L. Oudar, *J. Chem. Phys., 67* (1977) 446.
8 G. Mignani, A. Krämer, G. Puccetti, I. Ledoux, G. Soula, J. Zyss and R. Meyrueix, *Organometallics, 9* (1990) 2640.
9 G. Mignani, M. Barzoukas, J. Zyss, G. Soula, F. Balegroune, D. Granjean and D. Josse, *Organometallics, 10* (1991) 3660.
10 G. Mignani, A. Krämer, G. Puccetti, I. Ledoux, G. Soula and J. Zyss, *Mol. Eng., 1* (1991) 11.
11 G. Mignani, A. Krämer, G. Puccetti, I. Ledoux, J. Zyss and G. Soula, *Organometallics, 10* (1991) 3656.
12 S. R. Marder, D. N. Beratan and L. T. Cheng, *Science, 252* (1991) 103.
13 D. N. Beratan, in S. R. Marder *et al.* (eds.), *Materials for Nonlinear Optics—Chemical Perspectives, ACS Symp. Ser., 455* (1991) 89.
14 G. Berkovic, Y. R. Shen and M. Shadt, *Mol. Cryst. Liq. Cryst., 150B* (1987) 607.
15 C. Combellas, H. Gauthier, J. Simon, A. Thiébault, F. Tournilhac, M. Barzoukas, D. Josse, I. Ledoux, C. Amatore and J.-N. Verpeaux, *J. Chem. Soc., Chem. Commun*, (1988) 203.
16. I. Ledoux, J. Zyss, A. Jutand and C. Amatore, *Chem. Phys., 150* (1991) 117.
17 J. Zyss, I. Ledoux, M. Bertault and E. Toupet, *Chem. Phys., 150* (1991) 125.
18 J. O. Morley, V. J. Dogherty and D. Pugh, *J. Chem. Soc., Perkin Trans.* 2, (1987) 359.
19 L. H. M. Green, S. R. Marder, M. E. Thompson, J. A. Bandy, D. Bloor, P. V. Kolinsky and R. J. Jones, *Nature (London), 330* (1987) 360.
20 J. C. Calabrese, L. T. Cheng, J. C. Green, S. R. Marder and W. Tam, *J. Am. Chem. Soc., 113* (1991) 7227.
21 D. R. Kanis, M. A. Ratner, T. J. Marks and M. C. Zerner, *Chem. Mater., 19* (1991) 3.
22 D. R. Kanis, M. A. Ratner and T. J. Marks, *J. Am. Chem. Soc., 112* (1990) 8203.
23 G. Doisneau, G. Balavoine, T. Fillebeen-Khan, J.-C. Clinet, R. Loucif, J. Delaire, I. Ledoux and G. Puccetti, *J. Organomet. Chem., 421* (1991) 299.
24 R. Loucif-Saïbi, J. Delaire, L. Bonazzola, G. Doisneau, G. Balavoine, T. Fillebeen-Khan, I. Ledoux and G. Puccetti, *Chem. Phys.* submitted for publication.
25 K. F. Purcell and J. C. Kotz, *An Introduction to Inorganic Chemistry*, Holt-Sanders International Editions, Tokyo, 1982.
26 C. Bosshard, G. Knöpfle, P. Prêtre and P. Günter, *J. Appl. Phys., 71* (1992) 1594.

Quadratic nonlinear optical properties of thermally crosslinkable polymers

S. Muller, E. Chastaing, P. Le Barny and P. Robin
Groupe de Chimie pour l'Électronique, Laboratoire Central de Recherche, THOMSON-CSF, Orsay (France)

F. Pellé
CNRS, Meudon (France)

Abstract

Amorphous polymeric films including active NLO moieties frozen in an acentric alignment have proved to be the most promising candidates for electro-optic application. However, these systems generally suffer from a low stability of their NLO properties. Until now, the best results in terms of stability and efficiency have been obtained by forming epoxy–amine networks. Following this concept, two new types of epoxy-ended NLO compounds, among which are two different NLO units, have been synthesized at THOMSON. Corresponding poled and crosslinked films have been elaborated and their EO properties characterized. The realization of good quality films possessing a thickness ranging from 1 to 1.5 μm required a formulation optimization work. The curing process was characterized by T_g measurements and IR spectroscopy, therefore allowing for each system the determination of the best curing/poling procedure. The r_{33} coefficients were determined for each system before and after a thermal aging treatment. A long-term thermal stability of the EO properties at elevated temperatures (ranging from 80 to 130 °C) was demonstrated for these systems. A high T_g associated with the fixation of the chromophores by both ends has proved to be a major factor of stability.

Introduction

The search for new and efficient materials in which to carry out second-order nonlinear optical (NLO) processes, such as electro-optic (EO) modulation or second-harmonic generation (SHG), has become increasingly important in recent years. Polymers, which combine easy processability into guided wave structure and flexibility of organic chemistry, seem to be the most promising NLO materials. However, the necessary acentric alignment of the chromophores provided by the application of a d.c. electric field is a thermodynamically unstable state. This results in a decay of the NLO response with time due to a relaxation of the induced alignment.

Recent successful efforts to achieve thermal stability in EO polymeric materials include the elaboration of a network by epoxy–amine crosslink-

0379-6779/93/$6.00

ing during the poling [1, 2]. EO thermal stability at 80 °C is reported. Nevertheless, two comments can be made on that work:
— Aromatic amines are used as curing agent. Due to their lower reactivity, a very long curing time is required to obtain the fully cured network.
— Active moieties are covalently attached at the network junction points, but only by their donor group, hence allowing possible rotations.

In this communication, we report on the optimization procedure and EO properties of two new types of epoxy–amine polymeric networks. The first type is based on the use of an electroactive copolymer, whereas the second one includes a low molecular weight NLO molecules as the starting material. Two slightly different chromophores, **A** and **B**, have been investigated. These compounds have been specially designed to obtain a very efficient locking of the orientational ordering after poling as well as to exhibit very good NLO properties. In addition, to decrease the curing time, we have used a commercially available aliphatic amine, triethylenetetramine (TETA), instead of aromatic ones as the curing agent.

Experimental

Synthesis of the NLO crosslinkable compounds, the molecular structure of which is depicted in Fig. 1, has been reported elsewhere [3, 4].

The first step of the elaboration procedure was to optimize the formulation of the solutions made from the three different crosslinkable systems (Copo **I**/TETA, **A**/TETA and **B**/TETA) to provide good optical quality films with thicknesses ranging from 1 to 1.5 μm. For all systems, a

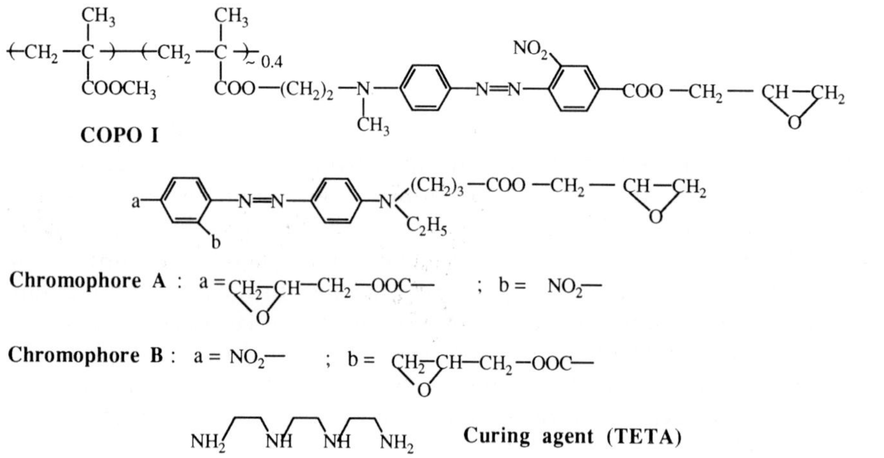

Fig. 1. Molecular structure of the main components of three different crosslinkable NLO materials.

stoichiometric amount of NH groups and epoxy groups was introduced in the mixtures. Chloroform ($CHCl_3$) was chosen as the solvent because of its solvent properties towards the different components, its well-adapted boiling point and its good wettability properties on glass + ITO substrate. In order to improve the film quality, namely, to avoid cracks in the case of the Copo **I**/TETA system, or to impart sufficient mechanical strength to the films as well as to increase their thickness in the case of films based on low molecular weight molecules, additives such as a flexibilizing agent (PMMA LG156) or a high molecular weight poly(methyl methacrylate) (PMMA 950 000) were introduced in the mixtures. $CHCl_3$ and TETA were both distilled just before the preparation of solutions. After filtration of the solutions on 0.2 μm Millipore filters, films were deposited by the conventional spin-coating technique on a glass + ITO substrate. Their thicknesses were measured by profilometry and their optical qualities were appreciated by a simple visual examination.

The thermal treatment applied to the samples to crosslink was chosen such that molecular movements would be possible, while avoiding too much fluidification of the material that would result in breakdowns during the poling. This means that the applied temperature must be at every moment within or just above the glass transition range of the material. A thermal treatment of 1 h at 50 °C + 1 h at 100 °C + several hours at 140 °C was thus chosen. Then, in order to determine as precisely as possible the most accurate moment for the poling field application during the curing process, as well as the total poling/curing procedure time, we followed the curing process in two different ways. The first one consisted in following the T_g evolution of the systems by DSC analysis (Perkin-Elmer DSC 4). The second one consisted in following the structural evolution of the system and, more particularly, the disappearance of epoxy or amino groups by infrared spectroscopy (Nicolet FT-IR).

The poling was performed using the conventional Corona poling technique. We then evaluated the EO thermal stability of each poled and crosslinked system by following the evolution of the electro-optic coefficient r_{33} (measured by the reflection method described by Teng and Man [5]) while submitting the material to a thermal aging treatment.

Results and discussion

Formulation optimization

Contrary to other systems, films made from the **B**/TETA system exhibited a very bad surface aspect, due to a problem of segregation between the curing agent and epoxy molecules. In order to avoid this problem, a precuring was done by submitting the solution to a thermal treatment of 40 min at 50 °C + 20 min at 100 °C before spin-coating. This treatment initiates the condensation processes between the amino and epoxy groups. This prevents, at the same time, segregation between the

TABLE 1

Formulation and film optimization procedure for the three systems

System formulation	Precuring of the solution	Spinning speed (thickness)	Optical quality
COPO **I**: 80 mg TETA[a]: 3.43 mg PMMA LG156: 5.32 mg $CHCl_3$[a]: 1 ml	no	1000 rpm (1 μm)	good
A: 100 mg TETA[a]: 9.5 mg PMMA LG156: 7 mg $CHCl_3$[a]: 1 ml	no	2000 rpm (0.2–0.4 μm)	excellent
B: 100 mg TETA[a]: 9.5 mg PMMA LG156: 14 mg $CHCl_3$[a]: 1.2 ml	40 min at 50 °C +20 min at 100 °C	1000 rpm (1.5 μm)	excellent

[a]Freshly distilled.

two types of molecules and the natural tendency of the chromophores to form associations.

The formulation and the procedure we adopted for the three different systems to obtain a good film are given in Table 1.

Curing process characterization

T_g measurements

Evolution of T_g as a function of curing time is given for the three systems in Fig. 2. At the beginning of the curing, a more or less rapid increase in T_g was observed depending on the system. Then, after a certain curing time τ, a plateau was reached for Copo **I**/TETA and **B**/TETA systems, whereas a very weak increase was still observed for the **A**/TETA system. After a curing time of about 20 h, we assumed that the maximum T_g value (T_g(max)) had been reached. T_g(max) and τ values are summarized for each system in Fig. 2.

For the Copo **I**/TETA system, the T_g value we observed at the very beginning of the curing process was much lower than that of Copo **I** (117 °C) due to the presence of the curing agent which was plasticizing the copolymer. For the **A**/TETA system, a very slow T_g increase was observed in the two first hours of treatment, followed by an acceleration as soon as the temperature reached 140 °C. On the contrary to the previous system, probably due to the precuring performed in the solution, the **B**/TETA system exhibited a rapid increase of T_g in the first three to four hours leading to a value of about 120 °C for T_g. After this time, as the sample was submitted to 140 °C, only a slight evolution was noticed for T_g.

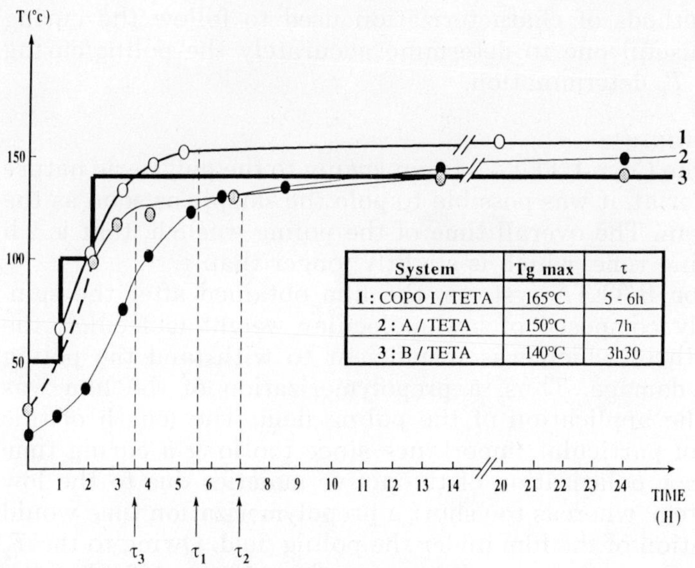

Fig. 2. Plots of T_g vs. curing time for Copo **I**/TETA, **A**/TETA and **B**/TETA systems.

IR characterization

IR spectra were recorded at different times of the curing process. In all cases only weak modifications of the whole spectra were observed. The most noticeable changes were due to the progressive disappearance of the epoxy vibration bands between 1000 and 700 cm^{-1}. In the case of the Copo **I**/TETA system, a rapid disappearance of the epoxy groups was observed (at 750 cm^{-1}) in the first two hours with an acceleration of the phenomenon as the temperature reached 100 °C. After 2.5 h, no more evolution was observed on the spectra, whereas the system exhibited a T_g value of about 110 °C, this is far below T_g(max) (see Fig. 2). This suggests that only a slight structural evolution occurred after 2.5 h of the curing process, leading, however, to an important increase in T_g. This T_g evolution can be first attributed to a limited number of reacting sites which further the polycondensation, hence reducing the number of plasticizing chain-ends. They can also be assigned to a 'thermal aging' effect [6] leading to a redistribution of free volume inside the material.

Concerning **A**/TETA and **B**/TETA systems, the epoxy band evolution appeared in good agreement with the evolution of T_g with the curing process. However, in the case of the **B**/TETA system, a typical band of primary amide appeared between 1660 and 1670 cm^{-1} after a very long curing time (28 h). This band could be the consequence of the reaction of the residual amino groups with the ester groups present in the system at the end of the crosslinking process. This reaction contributes to the liberation of low molecular weight compounds which plasticize the network, hence resulting in the **B**/TETA system having a lower T_g(max) value than the **A**/TETA system.

Of the two methods of characterization used to follow the curing process, the only useful one to determine accurately the poling/curing procedure was the T_g determination.

Poling/curing procedure

In the case of the Copo **I**/TETA system, owing to the polymeric nature of the starting material, it was possible to pole the sample as soon as the curing process began. The overall time of the poling was 8 h, that is 2 h more than the curing time, which is slightly longer than τ.

For **A**/TETA or **B**/TETA systems, the film obtained after the spin-coating was mainly composed of low molecular weight molecules, the mechanical strength of which was insufficient to withstand the poling field without any damage. Thus, a prepolymerization of the film was necessary before the application of the poling field. The length of this precuring time is of particular importance since too long a curing time would lead to a poor polarization of the active moieties due to the low mobility of the system, whereas too short a prepolymerization time would result in a degradation of the film under the poling field. Owing to the T_g measurements, we assumed that a T_g ranging from 100 to 120 °C of the films should allow molecular movements without risking the degradation of the film under Corona poling. Therefore, we fixed the time of the precuring treatment to 4.5 h for the **A**/TETA system and to 2.5 h for the **B**/TETA system. The additional curing time was 10 h for both systems, whereas the total time for the poling field application was 12 h.

EO thermal stability

The evolution of r_{33} coefficients with thermal aging treatment performed on the systems is given in Fig. 3. Table 2 summarizes the perfor-

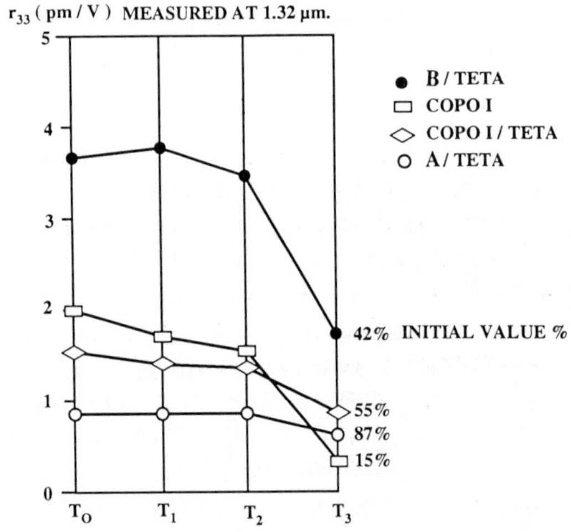

Fig. 3. Evolution of the electro-optic coefficients vs. thermal aging treatment performed on different NLO materials: T_0, without any thermal treatment; $T_1 = T_0 + 1$ h at 80 °C; $T_2 = T_1 + 2$ h at 80 °C; $T_3 = T_2 + 0.5$ h at 130 °C.

TABLE 2

Comparison of the EO properties exhibited by the different studied systems with the poling/curing procedure

Studied system (chromophore density (10^{20} molecules/cm^3))	Initial r_{33}[a] value of the poled and crosslinked system (pm/V)	Starting time of the poling field application	T_g after the poling/curing process (°C)	Residual r_{33}[a] after 3 h at 80 °C + 30 min at 130 °C (%)
COPO **I**/TETA (7.71)	1.55	as soon as the curing begins	150	55
COPO **I** alone (7.74)	2		117	15
A/TETA (8.07)	0.9	4.5 h after the beginning of the curing process ($T_g \approx 120$ °C)	145	85
B/TETA (7.61)	3.5	2.5 h after the beginning of the curing process ($T_g \approx 100$ °C)	135	42

[a]Measured at 1.32 μm.

mances (initial r_{33} value and relative stability) as a function of the poling/curing procedure adopted for each system studied. As a reference, a film made from the copolymer COPO **I** alone has been poled and its stability checked by the same method.

For the thermal stability of the different systems, the results can be well explained if we consider the two following parameters: the fixation of the chromophores by both their ends and the T_g of the cured system. In the case of Copo **I** alone, the combination of a relatively low T_g and the fixation of the chromophores by only one end results in a rapid decrease of the NLO properties. In the Copo **I**/TETA system, a 45% decrease of the r_{33} initial value is observed after the thermal treatment and this despite a T_g value of the system higher than 130 °C. This can be explained by the presence of nonreacted epoxy-ended chromophores inside 'cages' of the network that can easily lose their orientation. In comparison with the Copo **I**/TETA system, a better stability is observed in the **B**/TETA system even though both systems exhibit a T_g of the same order of magnitude. This can be assigned to a higher number of chromophores covalently linked by both ends to the network in the latter case. The lower stability of the **B**/TETA system in comparison with the **A**/TETA one can be simply attributed to their difference in T_g.

The initial r_{33} values are strongly dependent, for a similar content of active moieties, on the $\beta\mu$ value of the chromophores as well as on the orientational ordering imparted to the chromophores by the poling field.

EFISH measurements using a Nd–Yag laser source (1.06 μm) performed on the two different chromophores displayed $\beta\mu$ values of 4000 × 10^{-30} cm^5 D/e.s.u. for **B** compared with only 3200 × 10^{-30} cm^5 D/e.s.u. for **A**. This helps to explain the higher initial electro-optic coefficient value exhibited by the **B**/TETA system in comparison with the other systems. The difference of magnitude of the initial r_{33} value for the **A**/TETA and **B**/TETA systems is also due to the difference in the precuring time. The longer the precuring time, the lower is the $(r_{33})_0$. A long precuring time results in the crosslinking of a certain number of chromophores in a random orientation, hence reducing the final EO coefficient.

Among the materials including the **A**-type chromophoric unit (Copo **I**, Copo **I**/TETA and **A**/TETA), a higher electro-optic coefficient is obtained for the non-crosslinked system (Copo **I** alone). This result suggests a possible competition between the poling and the crosslinking process.

Conclusions

The optimization process and the study of EO properties of new epoxy–amine networks have led to the following conclusions:

An important increase in stability has been demonstrated for these systems in comparison with side-chain NLO polymers. Despite the use of an aliphatic amine instead of an aromatic one as the curing agent, long times are still required to obtain fully cured materials. The ease of fabrication is greatly improved when using a crosslinkable copolymer as the starting material since the general procedure is simplified and the curing time is decreased. However, systems based on low molecular weight molecules exhibit better optical quality films as well as improved thermal stability.

Even though they can be increased by using chromophores possessing a higher $\beta\mu$ value, r_{33} coefficients remain quite low. A better control of the curing process combined with a better knowledge of the Corona poling mechanisms should improve these results.

References

1 M. Eich, B. Reck, D. Y. Yoon, C. G. Wilson and G. C. Bjorklund, *J. Appl. Phys.*, *66* (1989) 3241.
2 D. Jungbauer, B. Reck, R. Twieg, D. Y. Yoon, C. G. Wilson and J. D. Swalen, *Appl. Phys. Lett.*, *56* (1990) 2610.
3 P. Le Barny, E. Chastaing, J. C. Dubois, S. Muller and F. Soyer, *Brevet No. 9 108 233.*
4 S. Muller, P. Le Barny and D. Broussoux, *Brevet No. 9 013 041.*
5 C. C. Teng and H. T. Man, *Appl. Phys. Lett.*, *56* (1990) 1734.
6 H. L. Hampsch, J. Yang, G. K. Wong and J. M. Torkelson, *Polym. Commun.*, *30* (1989) 40.

Synthetic Metals, *54* (1993) 147–153

Electro-optic modulator and second-harmonic generator with nonlinear polymers

V. Lemoine, M. Papuchon, J. P. Pocholle, P. Le Barny and P. Robin
Laboratoire Central de Recherche, THOMSON-CSF, Orsay (France)

Abstract

With a new waveguiding structure we have realized monomode waveguides with different nonlinear polymers. Electro-optic modulators were fabricated with a $V\pi$ value of 20 V cm. Frequency bandwidths from d.c. up to 200 MHz were achieved with classical lumped electrodes. Using a microstrip line as travelling-wave electrode, preliminary results show a flat frequency response up to 500 MHz. Second-harmonic generation from a 1.064 μm YAG laser beam was also demonstrated. The structure was phase-matched through modal dispersion. A 2×10^{-3} %/(W cm^2) second-harmonic generation efficiency was measured.

Introduction

Nonlinear optical polymers have been the focus of intensive fundamental and applied studies. We have at our disposal materials presenting very promising properties in the field of integrated optics: high nonlinear coefficients, possibility to spin-coat the polymer on nearly any kind of substrate in good optical quality thin films, possibility to modify the optical properties because of molecular engineering versatility, etc.

Among these materials we have chosen two different polymers. The first one is well adapted to the realization of an electro-optic modulator working at 1.32 μm; it is a copolymer of MMA and 4-(4′-nitro)-diethylaminoazobenzene (DR1) [1], with $r_{33} = 20$ pm/V at 1.32 μm (for a 100 V/μm poling field). The other one is used for the second-harmonic generation of a 1.064 μm YAG laser beam; it is a copolymer of MMA and 4-*N*-*N*-dialkylamino-4′-cyanobiphenyl [1] (referred to as PLBP80) with $d_{33} = 12$ pm/V at 1.064 μm (achieved by Corona poling) and $\lambda_{max} = 350$ nm.

In this contribution, after a short description of the bidimensional confinement method used to fabricate monomode waveguides, we will present the realization of a travelling-wave electro-optic modulator and some results concerning the possibility of achieving phase-matched second-harmonic generation in integrated optics.

0379-6779/93/$6.00

Bidimensional confinement method

The efficiency of nonlinear optic interactions depends on the optical power density in the material; consequently the possibility of easily fabricating low-loss bidimensional waveguides in an active polymer layer was the first major achievement.

We report a high-index buried-rib structure (shown in Fig. 1), very versatile and simple to fabricate with classical semiconductor technological means and presenting low intrinsic losses. The rib has to be very thin in order to confine most of the optical power in the active layer; it has been made with a high refractive-index photoresist material (either a polysilane or a classical photoresist). This structure was realized on a silicon substrate, the buffer layer being a PMMA-based crosslinkable polymer. The input and output sides of the waveguides were easily prepared by cleaving the substrate.

Monomode waveguides at 1.32 μm were fabricated with the copolymer of DR1 and MMA as active material; the mode size for a 3 μm rib width was comparable to those of classical optical fibres. The losses of this structure were estimated to be 7 dB/cm with a pessimistic Fabry–Perot technique [2] (the cleaved sides are not perfect mirrors).

The main advantage of this bidimensional confinement structure is the possibility of controlling very well the geometric definition of the rib (thickness and width) and, thus, the optical properties of the waveguide (effective indices).

Electro-optic modulator

Using the Pockels effect demonstration of many devices is possible with nonlinear polymers, from phase modulators to sophisticated directional couplers or to Mach Zehnder interferometers.

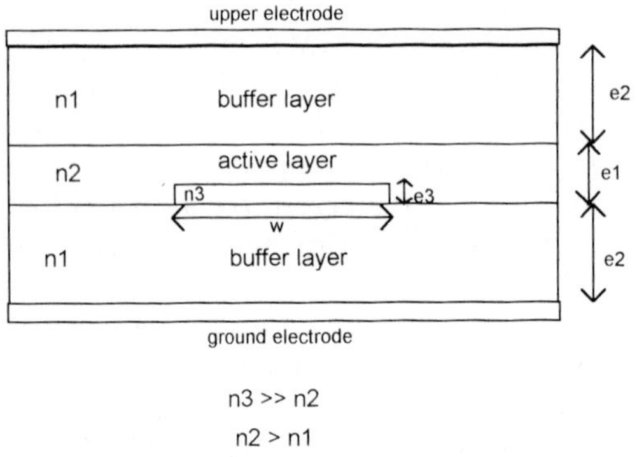

Fig. 1. Schematic view of a high-index buried-rib structure.

We chose to investigate the properties of a fast electro-optic modulator based on a phase modulator mounted between crossed polarizers the axis of which forms an angle of 45° with the structure axis. The phase modulation induced by the electric field $E = V/d$ depends on the polarization of the electromagnetic wave involved (d is the distance between the two electrodes):

$$\Delta\Phi_{\mathrm{TE}} = \frac{2\pi}{\lambda} L \frac{n^4 r_{13} V}{2\beta_{\mathrm{TE}} d} \theta$$

and

$$\Delta\Phi_{\mathrm{TM}} = \frac{2\pi}{\lambda} L \frac{n^4 r_{33} V}{2\beta_{\mathrm{TM}} d} \theta$$

with

$$\theta = \frac{\iint_{\mathrm{AL}} \varepsilon\varepsilon^* \, dx \, dy}{\int_{-\infty}^{+\infty}\int \varepsilon\varepsilon^* \, dx \, dy}$$

where L is the interaction length, β_{TE} and β_{TM} are the effective indices, n is the refractive index and ε the electromagnetic field; in the integral AL stands for the active layer. Between crossed polarizers the output intensity can be written as

$$I(V) \cong I_0 \left[1 - \cos\left(\frac{2\pi}{\lambda} L \frac{n^4 r_{\mathrm{eff}} V}{2\beta d} \theta \right) \right]$$

with $r_{\mathrm{eff}} = r_{33} - r_{13} = 2/3\, r_{33}$.

The first modulators were fabricated with classical lumped electrodes. The optical waveguides were sandwiched between two planar electrodes. The samples were 1.2 cm long, the upper electrode was 250 μm wide and the distance between the two electrodes was 5 μm; θ was calculated to be 60% for the realized monomode high-index buried-rib structure. The modulation depth was 80% as shown in Fig. 2. We measured a $V\pi$ of 20 V cm between d.c. and a few kHz. This $V\pi$ corresponds to an electro-optic coefficient $r_{\mathrm{eff}} = 13$ pm/V ($r_{33} = 20$ pm/V). We observed that the Kerr effect was negligible with this polymer. For this type of electro-optic modulator the electric frequency cut-off is fixed by the capacitance of the electrode system ($F_c = 1/2\,\pi\, RC$) and measured as 10 pF for our samples, implying (for $R = 50\ \Omega$) $F_c = 300$ MHz. Experimentally we observed a flat modulation efficiency from d.c. up to 200 MHz, probably limited by the contact circuit.

In order to increase the frequency bandwidth we designed the electrodes as for an electric transmission line (travelling-wave interactions).

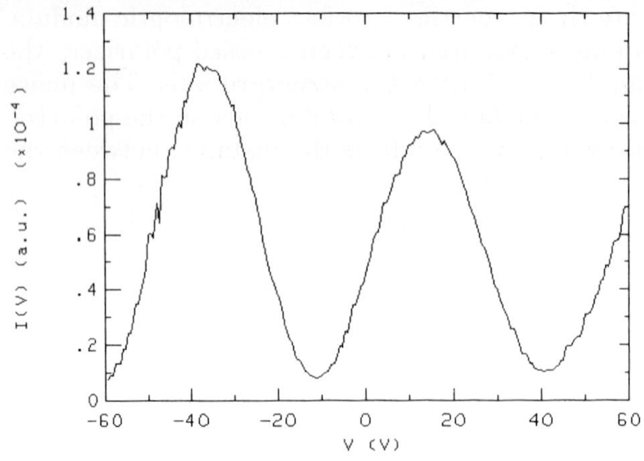

Fig. 2. Amplitude modulation.

The best line adapted to this type of structure is the microstrip line. A schematic diagram of the device is shown in Fig. 3. If the distance between the ground and signal electrodes is 5 μm, the characteristic impedance of the line is 50 Ω for a signal electrode 12 μm wide and 4 μm thick (the electrode should be this thick to limit ohmic losses) [3]. These parameters are compatible with classical technological means. The thick electrode was fabricated using an electroplating technique.

The characteristic impedance of the line at 1.5 GHz was measured as $Z_{char} = 40 + 12j\ \Omega$ and the voltage standing wave ratio (VSWR) is lower than 1.5 from d.c. up to 2 GHz (SMA connectors were used). Preliminary experimental results show a flat electro-optical response up to 500 MHz.

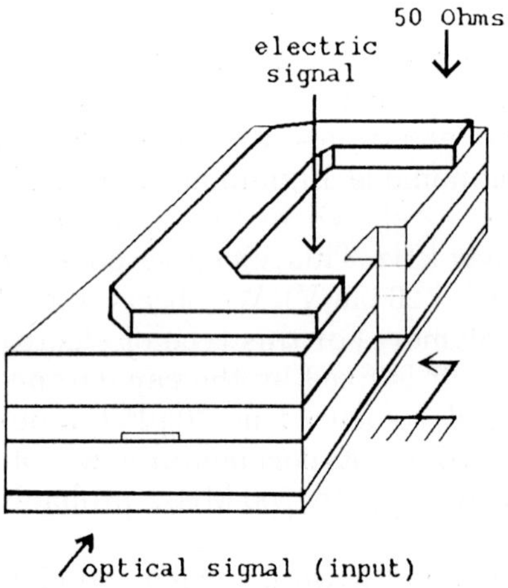

Fig. 3. Schematic view of the travelling-wave modulator.

Second-harmonic generator

The basic problem of efficient second-harmonic generation is related to the phase-matching of the nonlinear polarization wave (effective index is β^{ω}) and the second-harmonic wave (effective index is $\beta^{2\omega}$). In general, these two effective indices are different, partly because of the refractive index dispersion imposing $n^{2\omega}$ always greater than n^{ω}. The second-harmonic power $P^{2\omega}$ generated after a propagation length L in a nonlinear waveguide with a pump power P^{ω} is given by

$$P^{2\omega} \propto \left[P^{\omega} d_{\text{eff}} \sin c\left(\frac{\Delta\beta L}{2}\right)\right]^2 A^2$$

where A^2 is the overlap integral between the fundamental and harmonic fields on the nonlinear layer(s) and $\Delta\beta = 2\beta^{\omega} - \beta^{2\omega}$. This phase mismatch ($\Delta\beta \neq 0$) results in a constructive interaction length limited to the coherence length:

$$L_c = \frac{\pi}{\Delta\beta}$$

All the studies on second-harmonic generation look for an increase of this coherence length L_c either by a quasi-phase-matching technique such as periodic modulation of the nonlinear coefficient [4], or by real phase-matching (for example the refractive index dispersion can be compensated by modal dispersion).

The refractive indices at both fundamental (1.06 μm) and harmonic (0.53 μm) wavelengths of a poled PLBP80 film have been measured precisely for a beam polarization along the poling axis (TM mode corresponding to the highest nonlinear coefficient: $d_{\text{eff}} = d_{33}$) using the m-line method on a highly multimode waveguide. The first device tested was simply a planar waveguide consisting of the active layer deposited on a well-known buffer layer. From the refractive index values a chart giving the effective indices of the planar waveguide at both wavelengths as a function of PLBP80 layer thickness was computed. The crossing points in Fig. 4 show that the phase-matching condition by modal dispersion is achieved for two thicknesses of the PLBP80 layer: for $E_{\text{al}} = 0.6$ μm the TM_0 mode at ω and the TM_1 mode at 2ω are phase-matched, for $E_{\text{al}} = 1.37$ μm TM_0 at ω and TM_2 at 2ω are the phase-matched modes. In both cases the overlap integral A^2 is an hundred times weaker than it could be between two fundamental modes (the interactive modes have not the same order). A PLBP80 layer nearly 1.37 μm thick was spin-coated on a 2.5 μm thick buffer layer. The second-harmonic power generated for a 5 mm long sample was measured to be 30 nW for a 80 mW pump power in the waveguide, corresponding to a second-harmonic generation efficiency of

$$\frac{P^{2\omega}}{(P^{\omega}L)^2} = 2 \times 10^{-3} \qquad (\text{in } \%/(\text{W cm}^2))$$

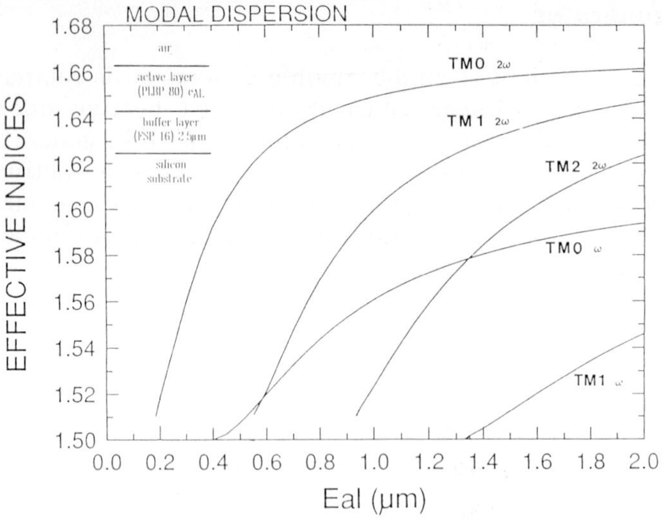

Fig. 4. TM mode dispersion curves of the fundamental (1.06 μm) and harmonic waves (530 nm).

This value is far from the theoretical efficiency of a phase-matched structure over the whole sample length (it is nearly a thousand times lower) but the output power was considerably larger than the one which could be obtained for a non-phase-matched structure presenting a 5 μm coherence length. This efficiency could be obtained with a structure having a 300 μm coherence length. This could be due either to a deviation of the layer thicknesses from those required for phase-matching, to fluctuations in the opto-geometrical parameters of the waveguide or more probably to both. Further investigations to achieve less critical phase-matching conditions should lead to higher conversion efficiency.

Conclusions

We have achieved the fabrication of an amplitude modulator at 1.32 μm putting a channel waveguide between crossed polarizers. The $V\pi$ of the structure was measured to be 20 V cm. To achieve a large frequency bandwidth ($>$1 GHz) we have fabricated travelling-wave electrodes based on a microstrip line design. Preliminary results show a flat modulation efficiency from d.c. up to 500 MHz. The fabrication of a Mach Zehnder interferometer should be the next step towards the realization of an efficient electro-optic modulator.

Second-harmonic generation is a very interesting field of application for optically nonlinear polymers. They offer various solutions to the phase-matching problem. We have studied the possibility of achieving the phase-matching conditions by compensating the refractive index dispersion through modal dispersion. The structure we used was a simple slab waveguide; through modal dispersion a strong increase of the second-harmonic generated power was observed when the thickness of the active layer was close to the theoretical phase-matching value. We did not succeed in adjusting exactly this theoretical value because the precision needed for a 1 cm efficient interaction length is unrealistic for this structure. More sophisticated structures fabricated on the basis of less critical phase-matching conditions will be the next step of this study and should lead to high conversion efficiency.

Acknowledgements

A part of this work was supported by ECC under ESPRIT 2284. The authors are grateful so S. Muller and E. Dupont for the fruitful discussions and to J. Lehoux, J. P. Truffer, R. Luseaux and D. Macreaux for the technological advise.

References

1 V. Lemoine, J. P. Pocholle, P. Le Barny and P. Robin, in J. Zyss (ed.), *Integrated Optics with Non Linear Polymers*, Academic Press, Orlando, FL, submitted for publication.
2 R. G. Walker, *Electron. Lett.*, *21* (1985) 581.
3 M. V. Schneider, Microstrip lines for micro-wave integrated circuits, *Bell. Syst. Tech. J.*, *48* (1969) 1421–1444.
4 N. Bloembergen and A. J. Sievers, *Appl. Phys. Lett.*, *17* (1970) 483.

Recent advances in the synthesis and characterization of nonlinear optical materials: second-order materials

L. R. Dalton*, L. P. Yu, M. Chen, L. S. Sapochak and C. Xu
Department of Chemistry, University of Southern California, Los Angeles, CA 90089-1062 (USA)

Abstract

A major problem in utilizing second-order organic materials for device applications is the requirement for the assembly of these chromophores into a stable, noncentrosymmetric lattice. For certain applications, such as frequency doubling, a periodic noncentrosymmetric lattice may be desirable. In this presentation, we discuss the synthesis of such organic polymeric lattices and evaluation of modulators and doublers fabricated from resulting materials. The general approach is to develop soluble, processible precursor polymers which can be transformed into stable noncentrosymmetric lattices by electric field poling and subsequent photoinduced or thermally induced solid state crosslinking reactions. Photo-processing is utilized to generate periodic noncentrosymmetric lattices for quasi-phase-matched second-harmonic generation.

Introduction

Second- and third-order organic nonlinear optical materials are at quite different stages of development. With third-order materials, 'nonresonant' nonlinear susceptibilities are not of sufficient magnitude at this time to consider device development; thus, research focuses upon the identification of materials and mechanisms with improved optical nonlinearity [1–6]. Our research has focused upon simultaneous measurement of the real and imaginary components of optical nonlinearity over significant spectral bandwidths and upon defining the temporal dependence of contributions to optical nonlinearity from various mechanisms [1–6]. We have recently investigated a number of systems characterized by delocalized electron excitations including phenylpolyenes, anthracenylpolyenes, thienylpolyenes, ferrocenylpolyenes, ladder polymers, metallated macrocycles, mixed valence materials and polymers incorporating rare earth metals [1–6]. Many currently available organic second-order materials, on the other hand, exhibit adequate 'nonresonant' optical nonlinearity to permit focus upon development of prototype devices including phase modulators and frequency doublers [1]. A major problem with such materials is that overall noncentrosymmetric symmetry is required for activity.

*Author to whom correspondence should be addressed.

0379-6779/93/$6.00

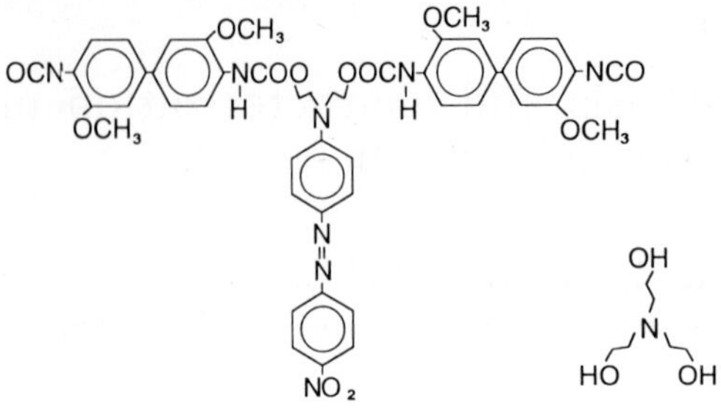

(a) (b)

Fig. 1. Monomer (**a**) used to prepare precursor polymer by isocyanate condensation; crosslinking reagent (**b**).

Indeed, much of the research activity involving organic second-order chromophores has focused upon exploiting various methods for molecular self-assembly into noncentrosymmetric lattices [1]. In addition to the well-known techniques of crystal growth and sequential synthesis by Langmuir–Blodgett methods, sequential syntheses involving covalent [1, 7] and ionic [1, 8] interactions have been demonstrated. Another approach to the desired macroscopic symmetry has involved incorporation of polar molecules as guests into host materials with appropriate cavities; again both organic [1, 9] and inorganic hosts [10] have been considered. However, one of the most popular routes to ordered lattices has been electric field poling followed by treatment to stabilize poling-induced order. It was quickly realized that chromophores incorporated into polymers to form composite materials rapidly lost poling-induced order [1]. Even chromophores covalently attached to a polymer backbone at a single point of attachment have been found to relax too rapidly for device application [1]. To improve the stability of poling-induced order, a number of investigators have pursued the development of bifunctional chromophores [1, 11, 12]. We have had considerable success in stabilizing poling-induced order employing the heavily crosslinked polymer based upon the monomer **a** and crosslinking reagent **b** shown in Fig. 1. Our approach is to prepare a precursor polymer based upon the self-condensation of monomer **a** which is processed into an appropriate polymer film, then heated above the glass transition temperature and poled. The poling-induced noncentrosymmetric symmetry is stabilized by thermally induced crosslinking involving **a** and **b**. In this manner we have produced materials which exhibit greater than 95% retention of second-order activity for periods exceeding 5000 h. The following second-harmonic generation and electro-optic coefficients have been realized (when the disperse red 19, DR-19, chromophore is employed): $d_{33} = 150–200$ pm/V; $r_{33} = 40–60$ pm/V; and $r_{13} = 13–20$ pm/V.

In this communication, we provide further evidence of the utility of this approach to the stabilization of poling-induced order and of photo-processing to produce periodic lattices appropriate for 'quasi-phase-matched' second-harmonic generation [13].

Results and discussion

Second-harmonic generation employing quasi-phase-matching

Several routes are available for achieving phase-matching in waveguides. These include (i) phase-matching by periodic spatial modulation of either the refractive index or the nonlinear optical susceptibility [14], (ii) phase-matching by anomalous dispersion [15], (iii) phase-matching by conversion of a guided fundamental mode into a second-harmonic radiation mode in the form of Cerenkov radiation [16], and (iv) phase-matching by conversion of a guided fundamental mode into a guided second-harmonic mode using modal waveguide dispersion and birefringence [17].

A disadvantage of the last technique is that the thickness has to be perfectly adjusted for phase-matching and the overlap integral is difficult to optimize. One effective route around this limitation is to use the quasi-phase-matching scheme of Bloembergen and co-workers [13] which involves creating a structure where the index of refraction or $\chi^{(2)}$ is modulated (with a periodicity of the phase mismatch) along the direction of propagation. If we define the periodicity for the variation of d_{33} or n as Λ, then we can write

$$\Delta\boldsymbol{k} = 4\pi[N(2\omega)_{\text{eff}} - N(\omega)_{\text{eff}}]/\lambda = 2\pi/\Lambda = \boldsymbol{k}(2\omega) - 2\boldsymbol{k}(\omega)$$

where N is the mode index in the waveguide and $\boldsymbol{k}(\omega)$ is the wavevector associated with the guided wave at the fundamental frequency. The second-harmonic intensity produced in a slab waveguide is related to the Fourier transform of the second-order susceptibility, $\chi^{(2\omega)}$, along the waveguide length x, specifically:

$$P(2\omega) = \{128\pi^5[P(\omega)]^2/WN(2\omega)_{\text{eff}}N^2(\omega)_{\text{eff}}\lambda^2 c\}(F_{112})^2 \int |\chi^{(2)}(x)\exp(\mathrm{i}\Delta\boldsymbol{k}x)\,\mathrm{d}x|^2$$

where W is the spatial extent of the fundamental in the y direction, L is the sample length and F_{112} is the normalized overlap integral. Obviously, if a periodic modulation of $\chi^{(2)}$ exists along x such that

$$\chi^{(2)}(x) = \chi^{(2)}(a) + \chi^{(2)}(b)\cos(2\pi x/\Lambda)$$

then the integral will have the nonvanishing value of $\chi^{(2)}(b)/2$.

The phase-matching length, L_{pm}, can be calculated as follows:

$$[P(2\omega)/P_0(2\omega)] = \sin c^2(\Delta\boldsymbol{k}L_{\text{pm}}/2) = 1/2$$

$$L_{\text{pm}} = 1.4/\Delta\boldsymbol{k} = (1.4\Lambda^2)/(2\pi\Delta\Lambda) = 2.3\ \text{mm}$$

If a polymer system such as shown in Fig. 2 is exposed to light which has passed through a photomask (see Fig. 3) then an appropriate grating can be written in the material. By rotating the grating pattern with respect to

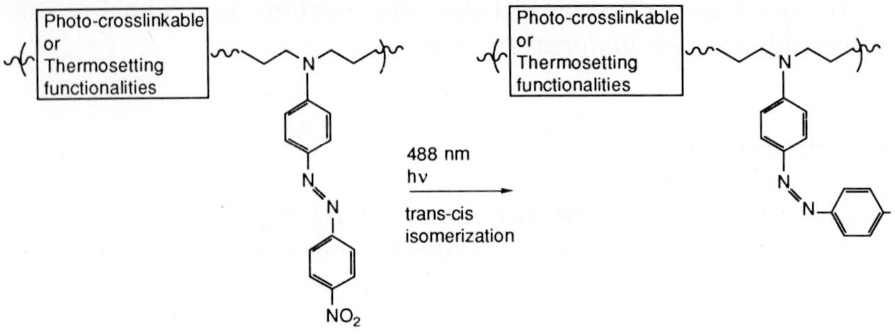

Fig. 2. Illustration of a photoinduced reaction which can be used to alter refractive index and second-harmonic generation coefficient, d_{33}, of a polymer.

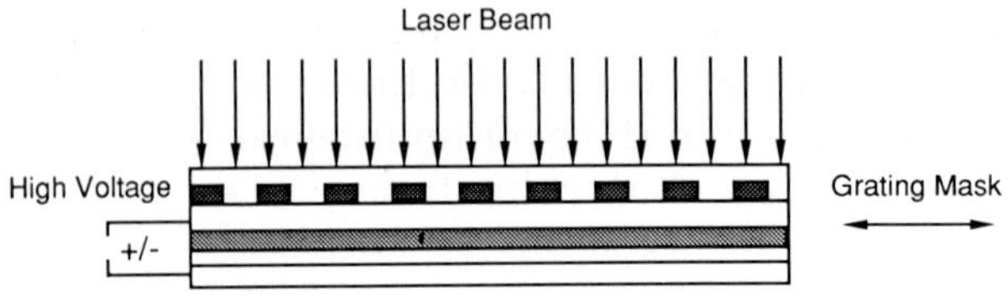

Fig. 3. Schematic diagram of apparatus used for electric field poling and photochemical generation of a periodic second-order lattice.

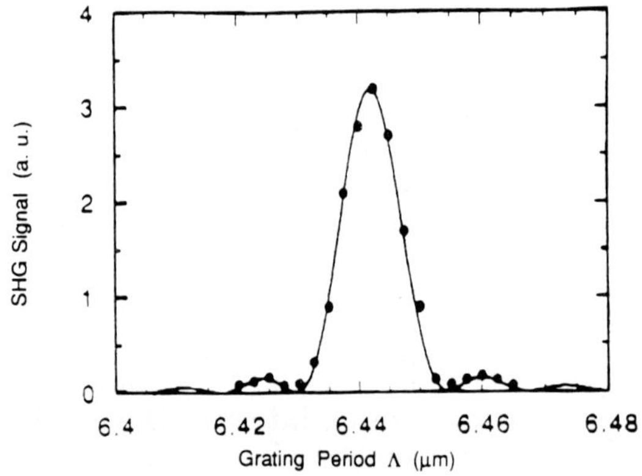

Fig. 4. Second-harmonic signal as a function of grating period.

the laser input beam, one can easily measure second-harmonic signal as a function of grating period (see Fig. 4). We have also observed a linear relationship between input (fundamental) and output (second-harmonic) power as required when effective phase-matching has been achieved.

Photo-processing such as we have described here can also be employed in the development of integrated optical circuits as discussed elsewhere [18, 19].

Electro-optic modulation including development of an ultrahigh frequency (100 GHz) infrared modulator

Highly crosslinked polyurethane polymers (see Fig. 1) have been used for phase modulation of infrared (633 nm to 1.8 μm) light at frequencies ranging from a few kilohertz (using a Mach-Zehnder interferometer [20]) to 95 GHz using the ultrahigh frequency phase modulator shown in Fig. 5. The performance of the various devices over the frequency range 1 kHz to 100 GHz is consistent with theory and the magnitude of electro-optic coefficients cited above. In Table 1, we compare the performance of the DR-19/polyurethane crosslinked polymer system with representative inorganic materials.

A key requirement of the polymeric material used in realizing the hybrid device structure shown in Fig. 5 is stability at elevated temperatures. No weight loss in the themogravimetric analysis of DR-19/polyurethane polymer is observed up to 310 °C. This is compatible with the temperature required for fabricating the hybrid structure. With ladder-type chromophores [1] even greater temperature stability can be realized.

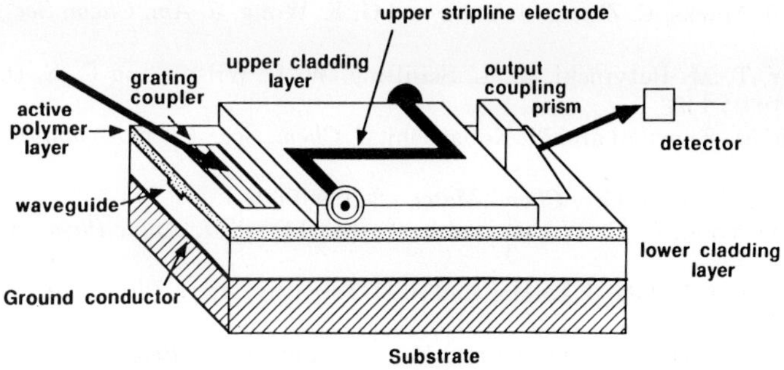

Fig. 5. Schematic diagram of 100 GHz infrared phase modulator.

TABLE 1

Comparison of electro-optic materials for millimeter wave phase modulation

	Polymer	GaAs	$LiNbO_3$
Velocity mismatch $L_{max}f_m = c/2(n - \varepsilon_{eff}^{1/2})$ (in GHz cm)	130	40	6
Modulation efficiency $(\Phi/E_m)(\lambda_0/\lambda_m) = n^2r/(n - \varepsilon_{eff}^{1/2})$	12×10^{-4}	1.7×10^{-4}	1.3×10^{-4}

Acknowledgements

This research was supported by the Air Force Office of Scientific Research under Contracts F49620-91-0270 and F49620-90-0210 and by National Science Foundation Grant DMR-9107806.

References

1 L. R. Dalton, L. S. Sapochak and L. P. Yu, in K. Sienicki (ed.), *Molecular Electronics and Molecular Electronic Devices*, Vol. 2, CRC Press, Boca Raton, FL, 1992.
2 L. R. Dalton, L. S. Sapochak, M. R. McLean, L. P. Yu, M. Chen and C. W. Spangler, in P. N. Prasad and J. K. Nigam (eds.), *Frontiers of Polymer Research*, Plenum, New York, 1991, p. 115.
3 A. N. Bain, L. P. Yu, M. Chen, L. S. Sapochak and L. R. Dalton, *Spectrosc. Int. J.*, *8* (1990) 73.
4 L. P. Yu, D. W. Polis, M. R. McLean and L. R. Dalton, in T. A. Skotheim (ed.), *Electroresponsive Molecular and Polymeric Systems*, Marcel Dekker, New York, 1991, p. 113.
5 L. P. Yu, D. W. Polis, L. S. Sapochak and L. R. Dalton, in J. Messier, F. Kajzar and P. Prasad (eds.), *Organic Molecules for Nonlinear Optics and Photonics*, Kluwer, Dordrecht, 1991, p. 273.
6 L. S. Sapochak, M. R. McLean, M. Chen, L. R. Dalton and L. P. Yu, *Proc. SPIE*, *1626* (1992) 431.
7 D. Li, M. A. Ratner, T. J. Marks, C. Zhang, J. Yang and G. K. Wong, *J. Am. Chem. Soc.*, *112* (1990) 7389.
8 H. E. Katz, G. Scheller, T. M. Putvinski, M. L. Schilling, W. L. Wilson and C. E. D. Chidsey, *Science*, *254* (1991) 1486.
9 S. Tomaru, S. Zembutsu, M. Kawachi and M. Kobayashi, *J. Chem. Soc., Chem. Commun.*, (1984) 1207.
10 S. D. Cox, T. E. Gier and G. D. Stucky, *Chem. Mater.*, *2* (1990) 609.
11 M. Eich, B. Reck, D. Y. Yoon, C. G. Wilson and G. C. Bjorklund, *J. Appl. Phys.*, *66* (1989).
12 B. K. Mandal, Y. M. Chen, J. Y. Lee, J. Kumar and S. K. Tripathy, *Appl. Phys. Lett.*, *58* (1991) 2459.
13 J. A. Armstrong, N. Bloembergen, N. Ducuing and P. S. Pershan, *Phys. Rev.*, *127* (1962) 1918.
14 G. Khanarian, R. A. Norwood, D. Haas, B. Feuer and D. Karim, *Appl. Phys. Lett.*, *57* (1990) 977.
15 P. A. Cahill, D. R. Tallant, T. C. Kowalcyzk and K. D. Singer, *Proc. SPIE*, *1560* (1991) 130.
16 P. K. Tien, R. Ulrich and R. J. Martin, *Appl. Phys. Lett.*, *17* (1970) 447.
17 G. I. Stegeman, C. T. Seaton, J. Ariyasu, R. F. Wallis and A. A. Maradudin, *J. Appl. Phys.*, *58* (1985) 2453.
18 Y. Q. Shi, W. H. Steier, L. P. Yu, M. Chen and L. R. Dalton, *Appl. Phys. Lett.*, *58* (1991) 1131.
19 Y. Q. Shi, W. H. Steier, L. P. Yu, M. Chen and L. R. Dalton, *Appl. Phys. Lett.*, *59* (1991) 2935.
20 Y. Shi, W. H. Steier, M. Chen, L. P. Yu and L. R. Dalton, *Appl. Phys. Lett.*, *60* (1992) 2577.

4. Poster Session: Synthetic Materials for Molecular Electronics

Transport properties of Q1D metals (electron–electron scattering)

Eugene N. Dolgov and Dmitrii V. Dmitriev
Institute of Chemical Physics, Russian Academy of Sciences, Kosygin Street 4, 117977 Moscow V-334 (Russian Federation)

Abstract

In the framework of the Q1D model with a soft phonon mode it is shown that the transport properties may be determined by all phonons of the Brillouin Zone (BZ) (not only by $2p_F$ phonons). The transport problem is a problem of interaction of the subsystem 'electrons and $2p_F$ phonons' and other phonons. In the absence of the soft phonon mode this subsystem is mainly independent, but in the presence of a soft mode this interaction becomes essential. This so-called 'phonon drag' is determined by electron–electron interaction, phonon–phonon interaction etc. Therefore, the interpretation of the transport phenomena in terms of a simple τ-approximation for electrons and phonons is an oversimplification.

In this paper we propose a new approach to the transport problem of Q1D metals and discuss our preliminary qualitative results. A full investigation in the framework of these models needs tedious computer calculations.

For the sake of simplicity we shall describe only a high-temperature situation in the framework of the standard kinetic equation. This means that the Fermi surfaces (FS) are of a 3D feature, i.e. the transversal overlap integral $t_\perp$ is not too small ($t_\perp$ may be considered as a parameter of the 'corrugation' of the almost plane FS).

It was shown in refs. 1 and 2 that if $k_\parallel = 2p_F$ ($k_\parallel$ is the distance between two almost planar FS) is incommensurate with $\boldsymbol{K}$ (vector of reciprocal lattice), the electrons and $2p_F$ phonons form a subsystem, which is weakly coupled with the lattice ('phonon drag'). A simple estimation gives the T^2 or T^3 law for resistivity depending on 3D effects (Fig. 1) [1]. The experimentally observed T^2 law for resistivity may also be obtained from the electron–libron scattering mechanism [2], but this mechanism implies strong model assumptions concerning some phonon modes.

Undoubtedly, both mechanisms may play a crucial role in the transport properties of Q1D compounds, but their manifestation in, for example, conductivity of real materials is probably ambiguous because of the many experimental and theoretical difficulties.

The presence of a soft phonon mode, which is a precursor of the structural phase transition or Peierls transition, may change transport

0379-6779/93/$6.00

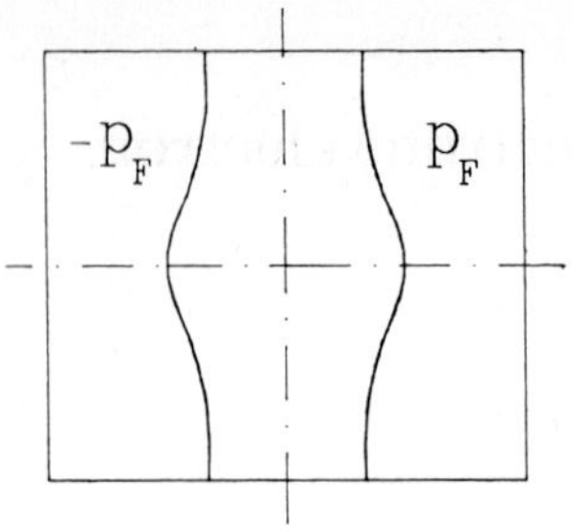

Fig. 1. Two parts of the almost planar FS in the BZ.

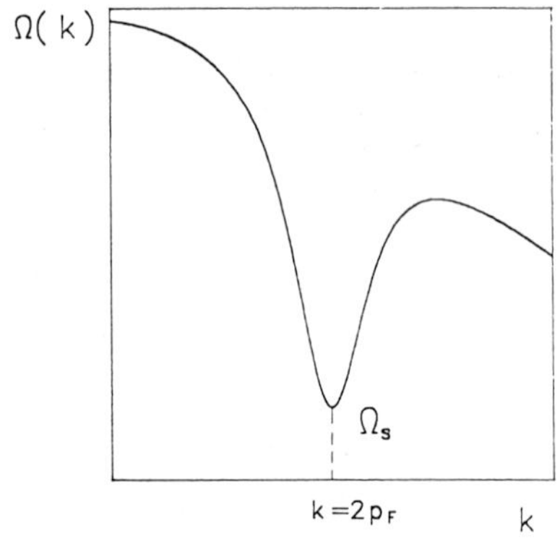

Fig. 2. A Kohn anomaly appears in the phonon spectrum.

phenomena in a crucial way. The simplest example of this manifestation was given in ref. 1. The conductivity anomalies near T may be explained by changing the phonons occupation numbers because of phonon softening. A conductivity maximum near T may be interpreted in terms of the soft mode.

The next important step was made in ref. 3. It was recognized that a soft phonon mode (a Kohn anomaly) caused a strong dependence of the carrier scattering amplitude on momentum transfer (this dependence may be exponential because of the exponential dependence of the phonon occupation number, see Fig. 2.). Such a strong dependence may change the kinetic coefficients of Q1D systems drastically. Quantitative results were obtained from the standard Ziman variational principle [4]. Transport properties were interpreted in terms of different groups of charge carriers characterized by different drift velocities in the electric field.

Unfortunately, the concept of a soft phonon mode is not generally accepted. A direct method of optical observation of this mode is proposed in the following paper [5]. It is obvious that soft mode manifestations in transport phenomena are very complex and ambiguous. In addition, their parameters depend strongly on all possible interactions in the system.

The Ziman variational principle is a version of the principle of a minimum of entropy production in a weakly non-equilibrium state. Since an external electric field can be considered as a source of momentum, this state should be characterized by a process of momentum transfer from electrons to lattice. In an incommensurate Q1D situation the relaxation process consists of the processes of momentum transfer between electron groups, from electrons to phonons, from phonons to lattice etc. The relaxation times for these processes differ significantly, and a soft phonon mode is an additional important source of this difference.

Thus, a minimum of entropy production should correspond to a very complicated current state. It is obvious that the region of k-space near the points of maximal softening corresponds to a minimal entropy production at a given momentum transfer. On the other hand, an effective width of $2p_F$ phonons should be wider to reduce the phonon–phonon contribution to the entropy production. So, this effective width is determined by 'competition' between these tendencies.

The variational principle may be derived from the usual kinetics equations for electrons and phonons (for details see refs. 1 and 6).

$$-ev_F\boldsymbol{E}\left(\frac{\delta n_0}{\delta \varepsilon}\right) = I_{e,\,ph}(\varphi, \chi) + I_{e,\,e}(\varphi)$$

$$0 = I_{ph,\,e}(\varphi, \chi) + I_{ph,\,ph}(\chi) \tag{1}$$

$$\delta n(\boldsymbol{p}) = -\left(\frac{\delta n_0}{\delta \varepsilon}\right)\varphi(\boldsymbol{p}), \quad \delta N(\boldsymbol{k}) = -\left(\frac{\delta N_0}{\delta \omega}\right)\chi(\boldsymbol{k})$$

$\delta n(\boldsymbol{p})$ and $\delta N(\boldsymbol{k})$ are linear corrections to electron and phonon occupation numbers, respectively. $I_{e,\,ph}$ and $I_{e,\,e}$ are linearized collision integrals for electron–phonon and electron–electron scattering. $I_{ph,\,e}$ corresponds to phonon–electron scattering processes and $I_{ph,\,ph}$ describes a phonon–phonon relaxation due to anharmonicity.

This phonon–phonon relaxation may be the only mechanism for momentum transfer in the Q1D situation. $I_{ph,\,ph}$ is the sum of Normal processes and Umklapp processes:

$$I_{ph,\,ph} = I^{N}_{ph,\,ph} + I^{U}_{ph,\,ph} \tag{2}$$

Note, that collision integrals depend on the parameters of a soft mode. The balance of a momentum is expressed by eqn. (3).

$$-e\boldsymbol{E}n_e = \sum_{g} \int \frac{d^3\boldsymbol{k}}{(2\pi)^3}\, \boldsymbol{k} I^{U}_{ph,\,ph}(\chi) \tag{3}$$

In the normal 3D situation one should integrate over all phonons of the BZ. After simple estimations one gets (in the 3D situation!): $\chi \ll \varphi$ (see ref. 6, §80). But a main problem in the Q1D situation is to estimate the region of integration in eqn. (3). The simplest approach [1] gives a too high conductivity magnitude.

The expression for entropy production for a weakly non-equilibrium state may be obtained from the kinetics equations (1). (This procedure for a phonon gas is described in ref. 6, §72.) After straightforward but tedious calculations one gets

$$\dot{S} = \dot{S}_{\mathrm{e,\,ph}} + \dot{S}_{\mathrm{ph,\,ph}} + \dot{S}_{\mathrm{e,\,e}} \tag{4}$$

Every contribution in eqn. (2) is an integral in phase space of the positively determined quadric form.

So, expression (4) is equivalent to system (1). But expression (4) may be minimized by an appropriate set of probe functions. The standard Ziman principle should be modified to describe systems with phonon drag. One should minimize expression (4) with additional relations, which describe a momentum transfer from the electrons and phonons and between different groups of electrons. A simple physical way is to use a probe function, corresponding to a definite group of electrons and phonons and to investigate the momentum transfer from the electric field to the electrons and phonons. Although electron–electron scattering does not introduce new Umklapp processes, it may transfer momentum between electron groups.

Thus, if we take a set of probe functions in the form [4, 6]

$$\varphi(\boldsymbol{p}) = \sum_{l} n_i \varphi_i(\boldsymbol{p}), \quad \chi(\boldsymbol{k}) = \sum_{j} \theta_j \chi_j(\boldsymbol{k}) \tag{5}$$

we obtain a quadric form and linear additional conditions. (For example, eqn. (3) may be considered as an additional condition. This condition is of the obvious form: $-eEn_{\mathrm{e}} = \Sigma_i A_i \theta_i$.) This minimum may be found in principle by well-known methods.

Unfortunately, this program cannot be realized by simple methods. A main difficulty is to write down the equation of the balance of momentum. The electron–electron scattering is of great importance, since it may determine the effective width of $2p_{\mathrm{F}}$ phonons in contrast to the electron–phonon scattering.

Simple estimations confirm results [1] if the soft mode is absent. It is also possible to obtain a wider effective width of $2p_{\mathrm{F}}$ phonons and reduce conductivity.

But in the presence of a soft mode the physical situation becomes more complex. Phonon parameters become changed in a characteristic scale $(T_{\mathrm{p}}/v_{\mathrm{F}}) \ll p_{\mathrm{F}}$ in the phase space. This means that one should consider the probe functions in eqn. (5) in relation to such a behavior. Therefore, the subsystem 'electrons and $2p_{\mathrm{F}}$ phonons' would be closely related with other phonons in the BZ. In other words, BZ is divided into many different groups of phonons and electrons with different drift velocities.

Of course, this problem may be solved only by computer methods. It is obvious that such a model would reveal unusual galvanomagnetic and frequency characteristics.

Acknowledgement

This work was financially supported in part by ELORMA.

References

1 L. P. Gor'kov, E. N. Dolgov and A. G. Lebed, *Zh. Eksp. Teor. Fiz.*, *55* (1982) 365.
2 M. Weger and H. Gutfreund, *Comments Solid State Phys.*, *8* (1978) 135.
3 E. N. Dolgov and E. S. Nikomarov, *Fiz. Tverd. Tela*, *32* (1990) 133.
4 J. M. Ziman, *Electrons and Phonons*, Clarendon, Oxford, 1960.
5 E. N. Dolgov and E. Ya. Sherman, *Synth. Met.*, *54* (1993) 169.
6 E. M. Lifshitz and L. P. Pitaevskii, *Physical Kinetics*, Nauka, Moscow, 1979.

Synthetic Metals, 54 (1993) 169–172

Two-phonon Raman scattering in Q1D conducting materials with soft phonon mode

Eugene N. Dolgov and Eugene Ya. Sherman

Institute of Chemical Physics, Russian Academy of Sciences, Kosygin Street 4, 117977 Moscow V-334 (Russian Federation)

Abstract

A new method for optical observation of a soft phonon mode based on the analysis of two-phonon Raman spectra is proposed. It is also shown that in a Q1D system resonant phenomena due to a topology of two almost planar Fermi surfaces can be observed. These phenomena occur in a wide range of temperature and frequency transfer.

Although properties of Q1D conductors have been intensively investigated, many problems still remain unresolved. For example, there is no general approach to the universal phenomena of Q1D physics—Peierls transition or structural transition.

In this paper we propose a new method for optical observation of a soft phonon mode based on the analysis of two-phonon Raman spectra. The term 'soft phonon mode' denotes a deep and abrupt minimum of the phonon dispersion curve $\tilde{\Omega}(\boldsymbol{k})$ (see Fig. 2 in ref. 1). This singularity is in fact a Kohn anomaly in a system with approximately planar Fermi surfaces (FS). A transition temperature T_{p} corresponds to $\tilde{\Omega}(\boldsymbol{k}_0)=0$, where $\boldsymbol{k}_0$ is a point of surfaces $k_{\parallel}=\pm 2p_{\mathrm{F}}$. In other words, the soft mode is a precursor of a structural transition.

There is strong experimental evidence in favor of the fact that a strong softening takes place in a small enough region of the Brillouin Zone (BZ) near the point $\boldsymbol{k}_0$, which does not correspond to the center of the BZ. The last fact excludes the possibility of observing such a soft mode in the one-phonon Raman spectra. But in the two-phonon scattering channel there are processes corresponding to emitting or absorption of two soft phonons with opposite momenta: $\boldsymbol{k}_1+\boldsymbol{k}_2=0$ (Fig. 1). At large enough softening the contribution of soft phonons near the singularity into the spectral density of Raman scattering $\rho(\Omega)$ may stay out of the wide two-phonon band, corresponding to the appropriate phonon mode ($2\tilde{\Omega}_{\mathrm{min}}<\Omega<2\tilde{\Omega}_{\mathrm{max}}$). This fact enables one to observe such a contribution. Two possible processes are represented in Fig. 1. A photon excites an electron from the valence band to the conductivity band, then the electron emits two phonons and photon with changed frequency.

0379-6779/93/$6.00

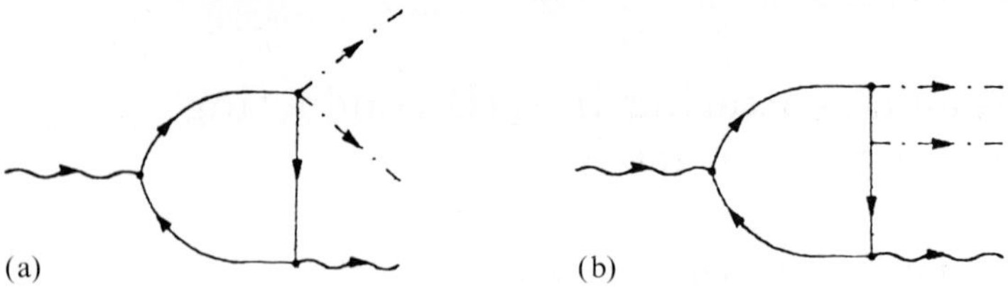

Fig. 1. A diagram representation of possible two-phonon processes. Wavy lines correspond to photons, solid lines to electrons, dotted lines to phonons. (a) Process of anharmonic electron–phonon interaction. (b) The second order process by electron–phonon interaction. Contributions of these diagrams are comparable by order of magnitude. Near the resonant frequency diagram (b) is more important.

The real shape of the Raman line is determined by the form of the phonon spectrum near the point of mimimal frequency. The calculation of these parameters is a complex problem. We shall demonstrate calculations in the framework of a simple physical model [1, 2]; we shall determine the spectral density by the Fermi Golden Rule at finite temperature [3]

$$\rho(\Omega) = \int \mathrm{d}^3\boldsymbol{k}_1 \mathrm{d}^3\boldsymbol{k}_2 |F(\boldsymbol{k}_1, \boldsymbol{k}_2)|^2 \delta(\Omega - \tilde{\Omega}(\boldsymbol{k}_1) - \tilde{\Omega}(\boldsymbol{k}_2))\delta(\boldsymbol{k}_1 + \boldsymbol{k}_2) \times (1+n_1)(1+n_2) \tag{1}$$

where $F(\boldsymbol{k}_1, \boldsymbol{k}_2)$ = effective matrix element, $\boldsymbol{k}_{1,2}$ = the phonons momenta, n_1, n_2 = their occupation numbers. Moreover, we shall neglect the damping of soft phonons, and adopt for a soft mode the following expression

$$\tilde{\Omega}(\delta, \boldsymbol{k}_\perp) = g_{\mathrm{ph}}\Omega_0 t^{1/2} + \Omega_0 \alpha\left(\frac{v_{\mathrm{F}}\delta}{T_{\mathrm{p}}}\right)^2 + \beta_1\Omega_0(k_{1\perp}a)^2 + \beta_2\Omega_0(k_{2\perp}a)^2 \tag{2}$$

$g_{\mathrm{ph}}^2 \ll 1$ is a dimensionless constant of an electron–phonon interaction, Ω_0 = Debye phonon frequency, $t = (T - T_{\mathrm{p}})/T_{\mathrm{p}}$, $\delta = k_\parallel - 2p_{\mathrm{F}}$, $\Omega_{\mathrm{S}} = g_{\mathrm{ph}}\Omega_0 t^{1/2}$, v_{F} = Fermi velocity, α, β = phonon spectrum parameters (~ 1), $\boldsymbol{k}_\perp$, a = phonon momentum component and lattice constant perpendicular to the c axis, $T \gg \Omega_{\mathrm{S}}$.

After a simple calculation one gets for the spectral density:

$$\rho(\Omega, T) = A\frac{1}{(\alpha\beta_1\beta_2)^{1/2}}\frac{(\Omega - 2\Omega_{\mathrm{S}})^{1/2}}{\Omega^4}\left(\frac{T_{\mathrm{p}}}{E_{\mathrm{F}}}\right)T^2 \tag{3}$$

where E_{F} = Fermi energy, A is a constant depending on the photon frequency and electronic properties of the crystal.

Thus, one may expect the following experimental situation. At some temperatures $T > T_{\mathrm{p}}$ it becomes possible to separate a small narrow peak from the wide two-phonon band (Fig. 2). Then this peak moves to zero frequencies. In the vicinity of the phase transition its intensity may

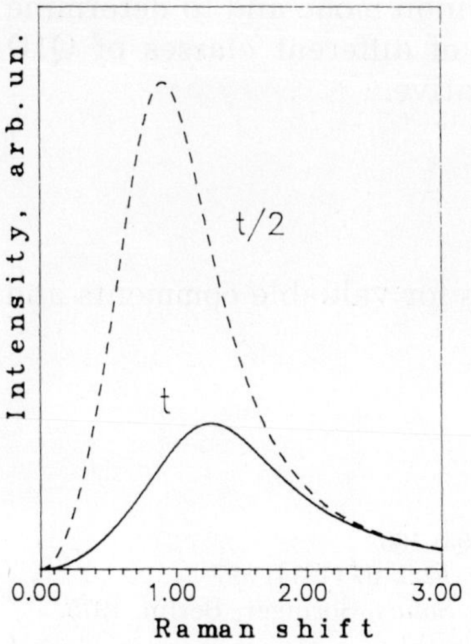

Fig. 2. A schematic representation of the spectral density of a soft phonon mode at two different temperatures.

diverge as $t^{-7/4}$, and its maximum position is proportional to $t^{1/2}$. Such a high critical index is due to large occupation numbers of the soft phonons in this two-phonon scattering process. A rapid increase of the intensity would be restricted by phonon damping. However, undoubtedly we may predict that in a wide enough temperature region the peak intensity would increase rapidly.

Note, that this effect in a metal is determined only by a soft phonon mode. However, in Q1D systems specific resonance phenomena may occur, which are closely related to the topology of almost planar FS. It is known that a resonant effect in metals or semiconductors may be observed if the frequency of the light is equal to the energy gap between a conductivity band and a valence band in some point of the BZ [3]. In a Q1D system this frequency may be resonant for the whole plane in the BZ near the almost plane isoenergy surface. Moreover, for the process of Fig. 1(b) three Green functions may be large for $2p_F$ phonons, i.e. this process is anomaly resonant. It is obvious that the frequency of a $2p_F$ phonon may be of a normal ($\sim \Omega_0$) magnitude, i.e. the resonance may be observed without phonon softening. Then the resonance frequencies are in the interval $(\Omega_{min}(2p_F), \Omega_{max}(2p_F))$, where $\Omega_{min}(2p_F)$ and $\Omega_{max}(2p_F)$ are minimal and maximal phonon energies in the plane $k_\parallel = 2p_F$, respectively. This resonance is difficult to distinguish from that in a 3D situation.

A phonon softening may spread anomaly in a resonance frequency interval. Although this question needs a more accurate analysis, one can expect some resonance peaks in the interval (Ω_S, Ω_{max}). This picture would depend on light frequency, temperature etc. So, it may be very informative.

Thus, one may hope to observe a soft phonon mode and to determine its parameters. A comparable investigation of different classes of Q1D compounds would be very useful and informative.

Acknowledgements

We wish to thank Professor V. E. Zubkus for valuable comments and ELORMA for financial support.

References

1 E. N. Dolgov and D. V. Dmitriev, *Synth. Met.*, *54* (1993) 163.
2 L. P. Gor'kov and I. E. Dzyloshinskii, *Zh. Eksp. Teor. Fiz.*, *67* (1974) 397.
3 M. Cardona, in M. Cardona (ed.), *Light Scattering in Solids*, Springer, Berlin, 1975.

Theoretical investigation on the effects of methyl substitution on the electronic properties of polysilane

R. Crespo*, M. C. Piqueras, E. Ortí and F. Tomás
Departamento Química Física, Universitat de València, Dr. Moliner 50, E-46100 Burjassot, Valencia (Spain)

Abstract

We present a valence effective Hamiltonian (VEH) investigation of the electronic structure of polysilane and its methyl derivatives. The main features observed in the band structure calculated for polysilane are preserved upon substitution with methyl groups. This substitution leaves almost unaffected the ionization potential but produces an important reduction of the band gap and a corresponding increase of the electron affinity as the degree of methylation increases. VEH results are in excellent agreement with photoemission and UV-absorption data.

Introduction

Polysilane polymers consist of a saturated backbone formed by concatenated silicon atoms with two organic sidegroups attached to each silicon atom [1]. Polysilanes are different from all other high polymers in that they exhibit σ-electron delocalization. This phenomenon leads to special physical properties such as strong electronic absorption, conductivity, photoconductivity, photosensitivity, etc., which determine many of the technological applications of polysilanes. The linear optical properties of polysilanes include intense absorption bands in the near UV and good transparency in the visible and near-IR. Experimental work has demonstrated that the wavelength of maximum absorption increases with increasing length of the silicon backbone as well as with increasing size of the substituents [2]. Furthermore, polysilanes with long alkyl substituents show dramatic changes in the absorption spectra as a function of temperature [3].

In this paper, we seek to investigate how the introduction of methyl groups attached to the silicon backbone affects the electronic structure and optical properties of polysilane. For this purpose, we focus on polysilane and its increasingly methylated derivatives, poly(methylsilylene-*co*-silylene), polymethylsilane and polydimethylsilane. These polymers have one, two and four methyl groups per translational unit cell, respectively. We have used the valence effective Hamiltonian (VEH) technique [4] to

*Author to whom correspondence should be addressed.

0379-6779/93/$6.00

calculate the one-dimensional electronic band structure of polysilane chains in the all-*trans* conformation. The VEH approach has been successfully applied to silicon containing polymers and has been shown to provide accurate estimates of essential electronic properties [5, 6]. It is completely non-empirical and has been parametrized to yield one-electron energies of *ab initio* double-zeta quality. We have used *ab initio* SCF-LCAO-MO methods to obtain reliable geometric structures for the unit cell of polysilane and its methylated derivatives. All the geometry optimizations have been performed using the MONSTERGAUSS-90 system of programs [7].

Results and discussion

Figure 1 shows the bond lengths and bond angles used to build up the unit cells employed in the electronic band structure calculations. The geometric parameters presented in Fig. 1 have been obtained by optimizing the geometries of the related molecules, trisilane, 1,2,3-trimethyltrisilane and 1,1′,2,2′,3,3′-hexamethyltrisilane at the *ab initio* level. The reason for the choice of trisilane compounds is that geometric optimizations previously reported for n-oligosilanes have shown that the bond lengths and bond angles obtained for trisilane are almost equal to those obtained for longer n-oligosilanes [8, 9].

All the geometry optimizations have been carried out using the 3-21G* split-valence basis set which has a complete set of six supplementary d Gaussians on silicon atoms. The 3-21G* basis set has been shown to

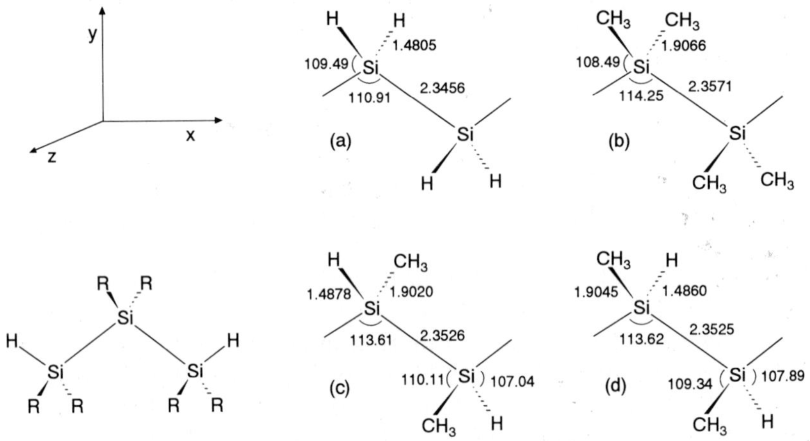

Fig. 1. Geometric parameters used to build up the unit cells of polysilane (a), polydimethylsilane (b), syndiotactic polymethylsilane (c) and isotactic polymethylsilane (d). Distances are in Å and angles in degrees. Carbon–hydrogen bond lengths have values between 1.087–1.088 Å and are not indicated.

yield calculated geometries in excellent agreement with experimental structures for molecules containing second-row elements [10].

The geometries of trisilane and 1,1′,2,2′,3,3′-hexamethyltrisilane have been fully optimized subject to a C_{2v} symmetry constraint and maintaining the three RSiR planes parallel to each other (see Fig. 1). These assumptions are taken in order to reproduce the environment of SiR_2 units in the polymer. For 1,2,3-trimethyltrisilane we have optimized the up–down–up and up–up–up isomers using the same constraint for the RSiR planes. These isomers only differ in the position of the methyl groups relative to the silicon backbone plane and reproduce adequately the environment of syndiotactic polymethylsilane and isotactic polymethylsilane, respectively. For each polymer, the parameters adopted to construct the unit cell correspond to those obtained for the central silicon atom of the corresponding molecule.

As can be seen from Fig. 1, the incorporation of methyl groups only produces a small increase in the Si–Si bond length passing from 2.3456 Å in polysilane to 2.3571 Å in polydimethylsilane. However the changes observed in the ∠SiSiSi bond angle are more accentuated (110.91–114.25°) because of the steric hindrance between methyl groups in 1 and 3 positions. The geometries obtained for the two isomers of 1,2,3-trimethyltrisilane are very similar.

All the VEH band structure calculations have been performed by assuming an all-*trans* conformation for the silicon backbone. The electronic band structures calculated for polysilane, syndiotactic polymethylsilane and polydimethylsilane have been obtained by taking into account the helical symmetry of these polymers (i.e. $-SiR_2-$) and are displayed in Fig. 2. The band structure of poly(methylsilylene-*co*-silylene) and isotactic polymethylsilane have been calculated using their respective translational unit cells constituted by two silicon atoms. That used for poly(methylsilylene-*co*-silylene) has been made up from the unit cells of polysilane and syndiotactic polymethylsilane. This way of building up the unit cell of the polymer has been adopted after stating that it yields reliable geometric parameters. The electronic bands depicted in Fig. 2 for polysilane and polydimethylsilane are classificed as σ- or π-like bands depending on whether one-electron wave functions are either symmetric or antisymmetric with respect to the reflection plane containing the silicon backbone. It should be emphasized that the band structure of isotactic polymethylsilane is very similar to that obtained for syndiotactic polymethylsilane when the translational cell is used for the latter. On the other hand, the diagram calculated for poly(methylsilylene-*co*-silylene) is a combination of both polysilane and polymethylsilane. These are the resons why these band structures are not displayed in Fig. 2.

The band structure calculated for all-*trans* polysilane is very similar to that previosuly reported using a standard geometry to build up the unit cell [5]. Going from the bottom to the top of the valence band structure, the lowest band is a σ band mainly built from 3s atomic orbitals of the

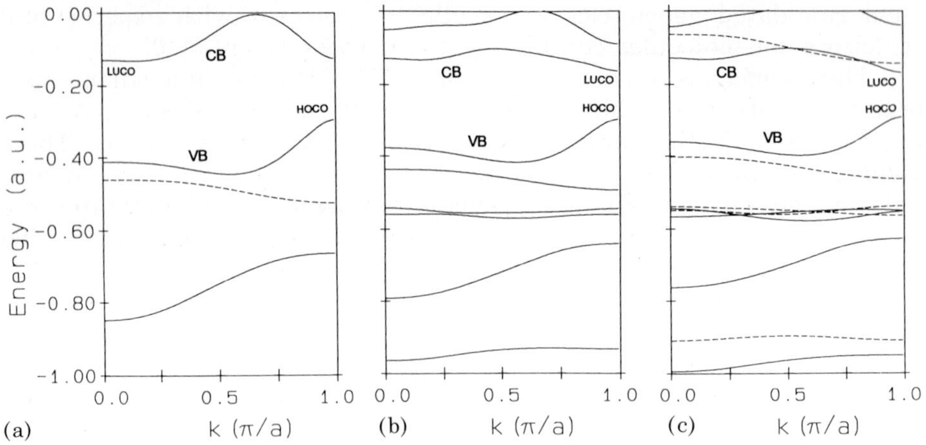

Fig. 2. VEH band structures calculated for polysilane (a), syndiotactic polymethylsilane (b) and polydimethylsilane (c). Energies are in atomic units. Unoccupied bands with energies larger than 0.0 a.u. are not shown. Solid lines indicate σ bands. Dashed lines indicate π bands. VB denotes valence band and CB denotes conduction band. HOCO and LUCO denote the highest occupied and lowest unoccupied crystal orbital, respectively.

silicon atoms but important contributions from hydrogen 1s orbitals and silicon $3p_y$ orbitals are also present (see Fig. 1 for axes assignment). The second occupied band is a π band resulting from the bonding interaction of the hydrogen 1s orbitals with the silicon $3p_z$ orbitals. The highest occupied or valence band is a σ band and is composed of hydrogen 1s and silicon $3p_y$ contributions at the center of the Brillouin zone ($k = 0$) but only silicon $3p_x$ contributions are present at the edges of the Brillouin zone ($k = \pm\pi/a$), where the HOCO (highest occupied crystal orbital) is located. The lowest unoccupied band or conduction band of polysilane is a σ band and results from silicon $3p_x$ atomic orbitals at $k = 0$, where the LUCO (lowest unoccupied crystal orbital) is located, while silicon 3s and $3p_y$ and hydrogen 1s contributions constitute this band at $k = \pm\pi/a$.

The band structure of syndiotactic polymethylsilane shows similar features to those observed for polysilane. The main difference is the appearance of C–H bands as a consequence of the attachment of methyl groups. The first of these bands is located at the bottom of the diagram and is mainly composed of carbon 2s and hydrogen 1s contributions, some contributions from silicon orbitals being also observed. The appearance of this band provokes a shift of the first band of polysilane to higher energies, which is more pronounced at the bottom of the band. The other two C–H bands are located around -15 eV. These bands are quasi degenerate and show very small bandwidths. The lower energy band is composed of hydrogen 1s and carbon $2p_y$ and $2p_z$ orbitals. The other band is built from 1s orbitals of the hydrogens lying out of the CSiH plane and carbon $2p_x$ orbitals. The highest two occupied bands of polysilane preserve their shape in syndiotactic polymethylsilane but are shifted to

TABLE 1

VEH values calculated for Koopmans' ionization potential (*IP*) and electron affinity (*EA*), direct band gap (E_g), bandwidth of the highest occupied or valence band (BW-VB) and bandwidth of the lowest unoccupied or conduction band (BW-CB). All the values are in eV

Polymer	*IP*[a]	*EA*[a]	E_g	BW-VB	BW-CB
Polysilane	8.01	3.58	4.50	4.10	3.45
Poly(methylsilylene-*co*-silylene)	8.04	3.89	4.15	3.54	1.13
Isotactic polymethylsilane	8.06	4.19	3.87	3.32	1.52
Syndiotactic polymethylsilane	8.06	4.46	3.60	3.30	1.69
Polydimethylsilane	7.92	4.58	3.34	2.93	1.79

[a]*IP* and *EA* values are those obtained for an isolated polymeric chain and no factor has been substracted to take into account the polarization energy of the lattice.

higher energies, in a similar way to that mentioned above for the first band of polysilane, due to the presence of C–H bands. It should be emphasized that the appearance of an unoccupied band below 0.0 eV implies a drastic reduction of the width of the conduction band. Now the LUCO corresponds to the crystal orbital located at the edges of the Brillouin zone ($k = \pm\pi/a$). The addition of a second methyl group to go to polydimethylsilane induces the same changes indicated for syndiotactic polymethylsilane but duplicates the number of C–H bands. Larger shifts to higher energies are now calculated for polysilane bands.

The VEH values obtained for Koopmans' ionization potentials and electron affinities, direct band gaps, and bandwidths of the valence and conduction bands of polysilane, poly(methylsilylene-*co*-silylene), polymethylsilane and polydimethylsilane are summarized in Table 1. As can be inferred from this Table, the ionization potential remains almost constant on going from polysilane to polydimethysilane. This is an expected result since the HOCO is mainly composed of silicon $3p_x$ orbitals and the attachment of methyl groups only introduces a weak contribution from the $2p_x$ orbitals of the carbon atoms and the 1s orbitals of the hydrogen atoms lying out of theCSiR planes. In order to correlate our theoretical results with experimental data, the VEH ionization potentials have to be corrected by a factor to account for the polarization energy of the solid. The correction factor of 2.3 eV applied in previous works for conjugated polymers is also used here [11]. The corrected value for the ionization potential of polydimethylsilane is therefore 5.62 eV. This value is in excellent agreement with the 5.9 ± 0.1 eV value reported by Seki *et al.* [12] from XPS and UPS data.

In contrast to the ionization potential, the value calculated for the electron affinity increases with the incorporation of methyl groups. Note that while in polysilane the LUCO results from the antibonding interaction of the $3p_x$ orbitals of silicon atoms, in the methylated polymers a bonding interaction between carbon 2s and silicon 3s orbitals takes place at the edges of the Brillouin zone leading to a stabilization of the crystal orbital at $k = \pm\pi/a$ that becomes the LUCO. The difference between the electron affinity calculated for isotactic and syndiotactic polymethylsilane arises from the fact that in the isotactic polymer there is a large contribution of silicon 3s orbitals which produces an increase in the antibonding interaction between silicon atoms along the chain. The high values obtained for the electron affinity (3.58–4.58 eV) suggest the possibility of obtaining electrical conductive n-doped polysilanes as we previously reported [5].

As can be seen from Table 1, the direct band gap calculated for polysilanes decreases in going from polysilane to polydimethylsilane (4.50–3.34 eV). The value calculated for polydimethylsilane (3.34 eV) is in very good agreement with the solid-state UV absorption maximum measured for the all-*trans* conformationally ordered phase of polydimethylsilane which appears at 342 nm (3.63 eV) [13].

In conclusion, the VEH electronic properties calculated for polysilanes are in excellent agreement with photoemission and UV absorption experimental data and support the reliability of the VEH method to deal with organopolysilanes.

Acknowledgements

We thank the Servei d'Informàtica de la Universitat de València for the use of their computing facilities. R.C. and M.C.P. are grateful to the Ministerio de Educación y Ciencia of Spain for a Doctoral Grant. This work has been supported by the Generalitat Valenciana and the DGYCIT Projects PS88-0112 and OP90-0042.

References

1 (a) R. D. Miller and J. Michl, *Chem. Rev.*, *89* (1989) 1359; (b) J. M. Zeigler and F. G. Fearon (eds.) *Silicon-Based Polymer Science. A Comprehensive Resource*, Advances in Chemistry Series 224, American Chemical Society, Washington, DC, 1990.
2 P. Trefonas III, R. West, R. D. Miller and D. Hofer, *J. Polym. Sci., Polym. Lett. Ed.*, *21* (1983) 823.
3 L. A. Harrad and J. M. Zeigler, *J. Polym. Sci., Polym. Lett. Ed.*, *23* (1985) 209; *Macromolecules*, *20* (1987) 601.
4 J. M. André, J. L. Brédas, J. Delhalle, D. J. Vanderveken, D. P. Vercauteren and J. G. Fripiat, in E. Clementi (ed.), *MOTECC-90 (Modern Techniques in Computational Chemistry)*, ESCOM, Leiden, 1990, Ch. 15, p. 745.
5 E. Ortí, R. Crespo and M. C. Piqueras, *Synth. Met.*, *42* (1991) 1575.
6 R. Crespo, M. C. Piqueras, E. Ortí and J. L. Brédas, *Synth. Met.*, *43* (1991) 3457.
7 M. R. Peterson and R. A. Poirier, *MONSTERGAUSS*, Dept. of Chemistry, University of Toronto, Ont., M5S 1A1, Canada.
8 J. T. Nelson and W. J. Pietro, *J. Phys. Chem.*, *92* (1988) 1365.
9 J. V. Ortiz and J. W. Mintmire, *J. Am. Chem. Soc.*, *110* (1988) 4522.

10 W. J. Pietro, M. M. Francl, W. J. Hehre, D. J. DeFrees, J. A. Pople and J. S. Binkley, *J. Am. Chem. Soc.*, *104* (1982) 5039.
11 E. Ortí, R. Crespo and M. C. Piqueras, *J. Mol. Struct.*, *254* (1992) 243.
12 K. Seki, T. Mori, H. Inokichi and K. Murano, *Bull. Chem. Soc. Jpn.*, *61* (1988) 351.
13 F. C. Schilling, F. A. Bovey, D. D. Davis, A. J. Lovinger and R. B. Macgregor, Jr., *Macromolecules*, *22* (1989) 4645.

Synthetic Metals, *54* (1993) 181–186

Electronic properties of poly(pyrrylene vinylenes): a theoretical approach

M. C. Piqueras*, R. Crespo, E. Ortí and F. Tomás

Departamento Química Física, Universitat de València, Dr. Moliner 50, *E-46100 Burjassot, Valencia (Spain)*

Abstract

We present a theoretical investigation of the electronic properties of poly(pyrrylene vinylene) and its methyl and methoxy derivatives by using the valence effective Hamiltonian (VEH) method. Theoretical calculations predict that the electronic properties of poly(pyrrylene vinylene) remain almost unaffected upon methyl substitution, while a noticeable lowering of $\sim$0.4 eV is obtained for the ionization potential and energy gap upon methoxy substitution. The VEH trends are in quantitative agreement with experimental data reported for poly(thienylene vinylene) derivatives.

Introduction

Among the great variety of conducting polymers nowadays known, only those which are able to combine good mechanical properties together with processability and environmental stability have chances to be used towards the production of electronic or optical devices. Poly(*p*-phenylene vinylenes) and poly(thienylene vinylenes) figure prominently in current research on conducting polymers [1] and are of interest since the successful synthesis of soluble precursor polymers for poly(*p*-phenylene vinylene) [2].

The vinylene polymers are synthesized via a high molecular weight, soluble precursor yielding high quality materials which can be easily formed into films and oriented to a high degree. The common backbone of these polymers is made up of a ring bridged by a vinylene linkage which not only reduces steric hindrance between backbone rings and groups attached to them but also has a beneficial effect on electronic properties as evidenced by experimental and theoretical data on both poly(*p*-phenylene vinylenes) [3, 4] and poly(thienylene vinylenes) [4, 5]. Furthermore, the electronic properties of poly(arylene vinylenes) can be adjusted by suitable selection of the arylene and substituent group, making them an excellent family of materials for experimental investigation [6].

*Author to whom correspondence should be addressed.

0379-6779/93/$6.00

Recently, a new conducting polymer, poly(bis-(2,5-*N*-methylpyrrylene) vinylene), with a conductivity intermediate between that of polypyrrole and that of poly(*N*-methylpyrrole) has been synthesized [7]. The properties displayed by this polymer have opened the way for the study of poly(pyrrylene vinylenes).

In this paper, we report a theoretical investigation on the electronic properties of poly(pyrrlene vinylene) (PPyV), poly(3-methylpyrrylene vinylene) (PMePyV) and poly(3-methoxypyrrylene vinylene) (PMxPyV). We investigate how the attachment of methyl and methoxy groups to the ring part of the polymer backbone affects both geometric and electronic structures. Band structure calculations have been performed using the valence effective Hamiltonian (VEH) technique [8]. The VEH method yields one-electron energies of *ab initio* double-zeta quality and has been demonstrated to provide accurate predictions of essential electronic properties in the context of conducting polymers [9]. We have performed geometry optimizations using *ab initio* SCF-LCAO-MO methods as implemented in MONSTERGAUSS-90 [10] to obtain reliable geometric structures for the monomers of PPyV, PMePyV and PMxPyV.

Results and discussion

Since no structural data exist for the polymers considered here, we have used *ab initio* methods to obtain the geometrical parameters of the unit cell of the polymer. The geometry of the closely related molecule *trans*-bis(2-pyrryl)ethylene and its 3,3′-dimethyl and 3,3′-dimethoxy derivatives has been optimized using the 3-21G basis set. The central part of these molecules has the same environment as the unit cell of the polymers, as shown in Fig. 1. The molecules were kept planar and subjected to a C_{2h} symmetry constraint. As can be seen from Fig. 1, the rings in the monomer unit are alternated in such a way that the nitrogen atoms in adjacent rings point to opposite directions in order to accomplish that all the vinyl groups are *trans* with respect to the double bonds on adjacent

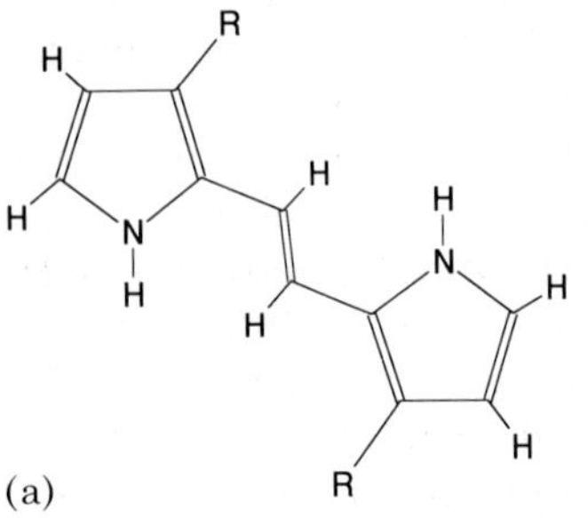
(a)

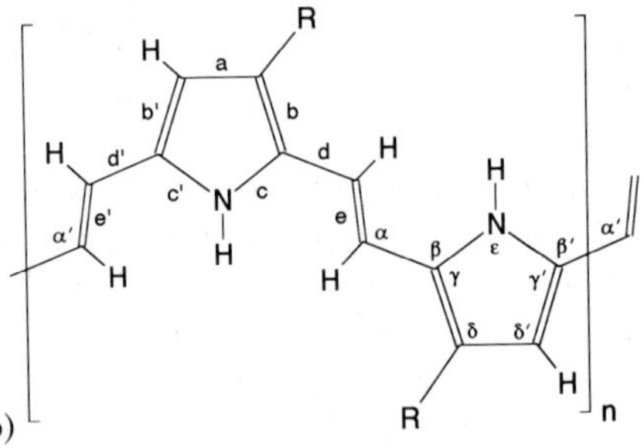
(b)

Fig. 1. Geometry of *trans*-bis(2-pyrryl)ethylenes (a) and poly(pyrrylene vinylenes) unit cell (b). See Table 1 for values of bond lengths and bond angles as optimized by *ab initio* calculations.

rings. This assumption has been taken on the basis of accurate *ab initio* calculations performed to investigate the conformational properties of 2-vinylpyrrole for which the *trans* rotamer is predicted as the most stable conformation [11]. Furthermore, the restriction to coplanarity is justified by the slight dependence of the electronic properties on the torsion angles at least for values smaller than 20 °. It is further justified by the fact that we are mainly interested in the relative changes that the electronic properties undergo upon substitution and not in their absolute values. In the case of the 3,3′-dimethoxy derivative, the methoxy groups are considered to be coplanar with the rings. This assumption is supported by X-ray data reported for 3,3′- and 4,4′-dimethoxy-2,2′-bithienyl molecules [12] and by accurate *ab initio* calculations performed by us on 3-methoxy-2-vinylpyrrole. In all cases, the methoxy groups are found to lie in the plane of the molecule.

Table 1 summarizes the main geometrical parameters employed to define the unit cells used as input for the VEH band structure calculations. The unit cells have been built up from the central part of the corresponding molecules in order to minimize end group effects and to reproduce more adequately the environment of the unit cell in the polymer. This approach yielded good ground state geometries for VEH band structure calculations in the past. As can be seen in Table 1, the incorporation of methyl groups introduces minor changes in the geometry of the polymer repeat unit. In contrast, the addition of methoxy groups lowers the single and double bond character of the bonds located in the interannular regions where substituents lie. As was expected, the 3,3′-dimethoxy-*trans*-bis(2-pyrryl)ethylene is slightly more delocalized than the parent *trans*-bis(2-pyrryl)ethylene.

TABLE 1

Geometries of polymer unit cell for poly(pyrrylene vinylene) (PPyV), poly(3-methylpyrrylene vinylene) (PMePyV) and poly(3-methoxypyrrylene vinylene) (PMxPyV) calculated at the *ab initio* double-zeta level. Bond lengths are in angstroms. Bond angles are in degrees. See Fig. 1 for the labeling of bond lengths and bond angles

Bond length	R group			Bond angle	R group		
	–H	$-CH_3$	$-OCH_3$		–H	$-CH_3$	$-OCH_3$
a	1.4035	1.4117	1.3989	α	126.52	126.59	126.49
b	1.3687	1.3721	1.3672	α'	126.52	126.52	126.52
b′	1.3687	1.3687	1.3687	β	124.91	124.50	126.54
c	1.3773	1.3798	1.3819	β'	124.91	124.91	124.91
c′	1.3773	1.3773	1.3773	γ	107.13	107.46	106.56
d	1.4501	1.4491	1.4426	γ'	107.13	107.13	107.13
d′	1.4501	1.4501	1.4501	δ	108.05	107.43	108.68
e	1.3278	1.3293	1.3284	δ'	108.05	108.35	107.87
e′	1.3278	1.3278	1.3278	ε	109.65	109.64	109.76

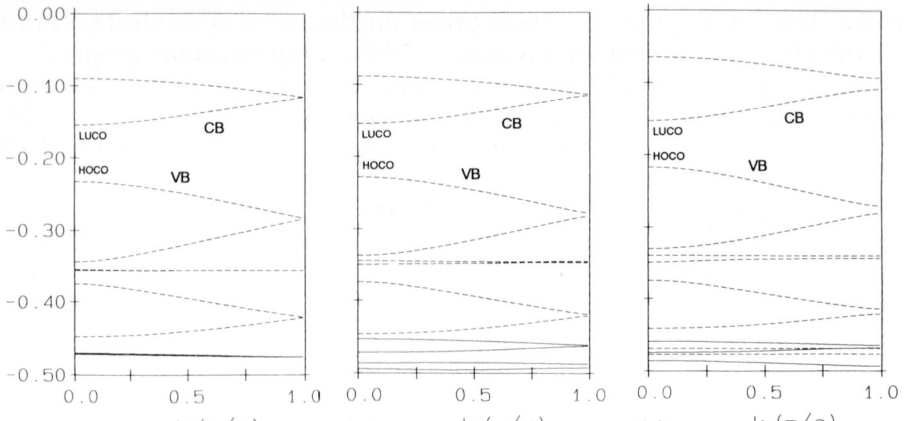

(a) (b) (c)

Fig. 2. VEH band structures calculated for poly(pyrrylene vinylene) (a), poly(3-methylpyrrylene vinylene) (b), and poly(3-methoxypyrrylene vinylene) (c). Only bands lying between −0.50 and 0.00 atomic units are displayed. Solid lines indicate σ bands. Dashed lines indicate π bands. HOCO and LUCO denote the highest occupied and lowest unoccupied crystal orbital, respectively. VB denotes valence band and CB denotes conduction band.

The VEH electronic band structures calculated for PPyV, PMePyV and PMxPyV are displayed in Fig. 2. The valence band and the conduction band of PPyV, PMePyV and PMxPyV are, as expected, π bands. The HOCO (highest occupied crystal orbital), which is located at the top of the valence band, corresponds to a wave function delocalized over the carbon backbone with bonding interactions between carbons forming double bonds and antibonding interactions between carbons forming single bonds, and with essentially no contribution from the nitrogen atoms. Methyl carbon atoms have no contribution to the HOCO in PMePyV but methoxy oxygen atoms have an antibonding interaction with the carbon to which they are attached to in PMxPyV. Compared to the HOCO, the LUCO (lowest unoccupied crystal orbital), which corresponds to the orbital lying on the bottom of the conduction band, shows opposite interactions between carbon atoms along the conjugated chain and important antibonding contributions from the nitrogen atoms. It should be emphasized that no contribution from methyl or methoxy groups is detected in PMePyV and PMxPyV, respectively.

The VEH values calculated for the ionization potentials, electron affinities, band gaps and bandwidths of PPyV, PMePyV and PMxPyV are collected in Table 2. Ionization potentials (*IP*)and electron affininities (*EA*) are obtained using Koopmans' theorem. We should point out that the calculated values are corrected downwards by 2.3 eV to account for the solid-state polarization. The band gaps are calculated as the energy difference between the top of the valence band and the bottom of the conduction band ($E_g = IP - EA$) and correspond to direct band gaps at $k = 0$.

TABLE 2

VEH values calculated for the ionization potentials (*IP*), electron affinities (*EA*), band gaps (E_g), bandwidths of the highest occupied or valence band (BW-VB) and bandwidths of the lowest unoccupied or conduction band (BW-CB) of poly(pyrrylene vinylene) (PPyV), poly(3-methylpyrrylene vinylene) (PMePyV) and poly(3-methoxypyrrylene vinylene) (PMxPyV). All values are in eV

Polymer	*IP*[a]	*EA*[a]	E_g	BW-VB	BW-CB
PPyV	4.01	1.90	2.11	3.05	1.75
PMePyV	3.90	1.89	2.00	1.35	1.06
PMxPyV	3.60	1.88	1.72	1.45	1.22

[a] *IP* and *EA* values are those obtained by substracting a constant factor of 2.3 eV to take into account the polarization energy of the lattice.

As can be seen from Table 2, almost identical values are predicted for the electronic properties of PPyV and PMePyV. Methyl substitution only produces a small lowering of 0.11 eV in the ionization potential. This result is in quantitative agreement with the experimental decrease obtained from electrochemical potential spectroscopy data when passing from poly(thienylene vinylene) (PTV) (4.76 eV) to poly(3-methylthienylene vinylene) (PMeTV) (4.64 eV) [4, 5]. As mentioned above, the methoxy group contributes more strongly to the HOCO and produces a larger shift (0.41 eV) towards lower binding energies in the ionization potential. This shift is in very good accord with the lowering of the ionization potential observed for PMxTV with respect to PTV (0.4 eV) [4, 5]. Substitution with both methyl or methoxy groups induces no changes in the values calculated for the electron affinity (0.01–0.02 eV). This fact is totally understandable since the LUCO is not affected by the introduction of methyl or methoxy groups. As a consequence, the band gap of PMePyV (2.00 eV) remains almost unaffected compared with that obtained for PPyV (2.11 eV) and in fact the variation is the same as that calculated for the *IP*. The same feature is observed when passing from PPyV to PMxPyV and a lowering of 0.39 eV is calculated for the value of the band gap. The VEH predictions concerning the effect of methyl and methoxy groups on the band gap are in perfect accord with the experimental data reported for poly(3-methylthienylene) PMeT and PMxTV. For methyl substitution a small lowering of the optical band gap of the order of 0.0–0.1 eV is found on going from PT to PMeT [5]. In contrast, an important reduction of the optical band (about 0.4 eV) is measured when passing from PTV to PMxTV [4, 5].

It is important to note that the bandwidths calculated for PMePyV and PMxPyV are approximately half of that obtained for PPyV because of the non-degeneracy of the bands at the edges of the Brillouin zone in PMePyV and PMxPyV. This non-degeneracy is due to the lower symmetry of PMePyV and PMxPyV.

In summary, we have studied the effects of methyl and methoxy substitution on the electronic structure of poly(pyrrylene vinylene) at the VEH level. We find that methyl substitution induces a small lowering (0.11 eV) of the *IP* and the E_g. Substitution with a methoxy group has a more accentuated effect on both the *IP* and the E_g and reduces these values by about 0.4 eV. The calculated trends are in excellent quantitative agreement with available experimental data and accord with those previously reported for poly(thienylene vinylenes) and poly(furylene vinylenes) [13]. These results seem to suggest that through careful design of polymer structure (a proper choice of substituents and backbone structure) one can adjust the electronic properties of these materials over a wide range.

Acknowledgements

We thank the Servei d'Informàtica de la Universitat de València for the use of their computing facilities. M.C.P. and R.C. are grateful to the Ministerio de Educación y Ciencia of Spain for a Doctoral Grant. This work has been supported by the Generalitat Valenciana and the DGYCIT Projects PS88-0112 and OP90-0042.

References

1 *Proc. ICSM '90 (Int. Conf. Science and Technology of Synthetic Metals)*; *Synth. Met.*, *41* (1991), *42* (1991), *43* (1991).
2 F. E. Karasz, J. D. Capistran, D. R. Gagnon and R. W. Lenz, *Mol. Cryst. Liq. Cryst. 118* (1985) 327; I. Murase, T. Ohnishi, T. Noguchi, M. Hirooka and S. Murakami, *Mol. Cryst. Liq. Cryst.*, *118* (1985) 333.
3 G. Kossmehl, *Ber. Bunsenges. Phys. Chem.*, *83* (1979) 417, and refs. therein.
4 H. Eckhardt, L. W. Shacklette, K. Y. Jen and R. L. Elsenbaumer, *J. Chem. Phys. 91* (1989) 1303.
5 K. Y. Jen, R. Jow, L. W. Shacklette, M. Maxfield, H. Eckhardt and R. L. Elsenbaumer, *Mol. Cryst. Liq. Cryst.*, *160* (1988) 69.
6 A. J. Brassett, N. F. Colaneri, D. D. C. Bradley, R. A. Lawrence, R. H. Friend, H. Murata, S. Tokito, T. Tsutsui and S. Saito, *Phys. Rev. B*, *41* (1990) 10 586 and refs. therein.
7 M. A. Takassi, W. Chen, R. E. Niziurski, M. P. Cava and R. M. Metzger, *Synth. Met.*, *36* (1990) 55.
8 J. M. André, J. L. Brédas, J. Delhalle, D. J. Vanderveken, D. P. Vercauteren and J. G. Fripiat, in E. Clementi (ed.), *MOTECC-90 (Modern Techniques in Computational Chemistry)*, ESCOM, Leiden, 1990, Ch. 15, p. 745.
9 J. L. Brédas, in T. A. Skotheim (ed.), *Handbook of Conducting Polymers*, Marcel Dekker, New York, 1986, Ch. 25, p. 859.
10 M. R. Peterson and R. A. Poirier, *MONSTERGAUSS*, Dept. of Chemistry, University of Toronto, Ont., M5S 1A1, Canada.
11 E. Ortí, M. C. Piqueras, R. Crespo and F. Tomás, *NATO ASI Series*, *Ser. B*, *248* (1991) 447.
12 E. F. Paulus, R. Dammel, G. Kämpf and P. Negener, *Acta Crystallogr.*, *Sect. B*, *44* (1988) 509.
13 E. Ortí, R. Crespo and M. C. Piqueras, *J. Mol. Struct.*, *254* (1992) 243.

Synthetic Metals, 54 (1993) 187–194

New conductive polymeric systems

J. Kubát[a], R. Kuželʼ[*,b], I. Křivka[b], P. Bengtsson[a], J. Prokeš[b] and O. Stefan[b]

[a]*Department of Polymeric Materials, Chalmers University of Technology, S-412 96 Gothenburg (Sweden)*

[b]*Faculty of Mathematics and Physics, Charles University, Ke Karlovu 5, 121 16 Prague 2 (Czech Republic)*

Abstract

Some new polymeric composites exemplified by polyethylene or polystyrene filled with precious metal particles, especially with spherical palladium, silver and gold powders, of various diameters are discussed. In addition to bulk samples, thick films are prepared and studied. The determined critical parameters describing resistivity–content dependence are strongly influenced both by the polymer and by the shape and conductive particle dimensions. The percolation threshold varies from about 10 to 31 vol.%. All the composites exhibit a large positive temperature coefficient of resistance in a wide temperature range. Dielectric measurements show also a strong temperature dependence of capacitance up to the melting point. The results are compared with those obtained on polymers filled with carbon black of various types.

1. Introduction

Various kinds of polymeric composites have been studied by many authors for their wide possibilities and varieties of applications. Until now the composites have mostly been polymers filled with carbon black of various types. We have studied new types of polymeric composites containing precious metal powders or oxides.

Bulk samples or thick films can be prepared by various methods. Their resistivities can cover a wide range of values ranging from about about 0.01 to 10^{12} Ω cm or more. The composites were studied by measuring d.c. and a.c. conductivities and the capacitances of capacitors at various temperatures and frequencies. The results are compared to the dependences achieved with polymeric systems containing various kinds of carbon black.

2. Sample preparation

2.1. Materials

Four polymer types were used for the composite preparation: low density polyethylene (LDPE), linear low density polyethylene (LLDPE), polystyrene (PS) and chloroprene rubber (CR).

*Author to whom correspondence should be addressed.

0379-6779/93/$6.00

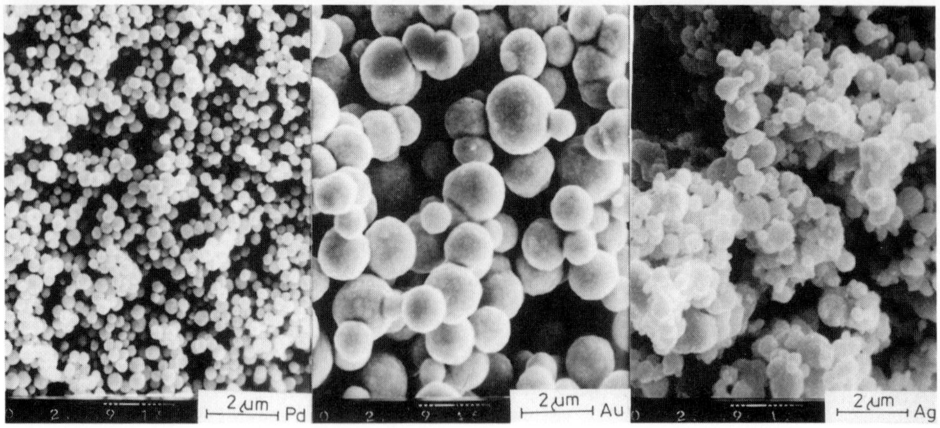

Fig. 1. SEM photomicrographs of Pd, Au and Ag powders.

The following conductive ingredients were selected for the filling of polymers: ruthenium dioxide powder (type Ru-109) characterized by small, perfect single crystals 52 nm long and 30 nm wide; palladium powder consisting of uniform spheres of about 0.30 μm diameter (type Pd-213); silver and gold powders containing spherical particles of diameters ranging from about 0.2 to 2.5 μm (type Ag-023 and Au-317) (see Fig. 1) produced by Shoei Chemical Inc. and carbon black (Vulcan P and Flammruss 101).

2.2. Composite preparation

Small sized samples were prepared in the case of precious metals. The diameter of the pellets was about 12 mm and their thickness 1–2 mm.

A mixture of a given polymer with conductive particles was first mixed and homogenized at elevated temperatures from about 130 to 160 °C depending on the polymer. Then the compounds were compressed at elevated temperatures under a pressure of about 100 MPa and cooled slowly.

PS films of 40 μm thickness were prepared by a screen-printing technique applying the composition PS plus a conductive powder dissolved in toluene on alumina substrates at room temperature. The alumina substrates were provided with silver terminations. The electrodes of bulk samples were prepared by evaporating gold in vacuum or by painting the planar surfaces with a thin layer of a colloidal silver dispersion.

3. Experimental and results

3.1. Electrical resistivity — blending curves

One of the primary characteristics of conductive composites is the variation of resistivity with the content of the conductive ingredient, the

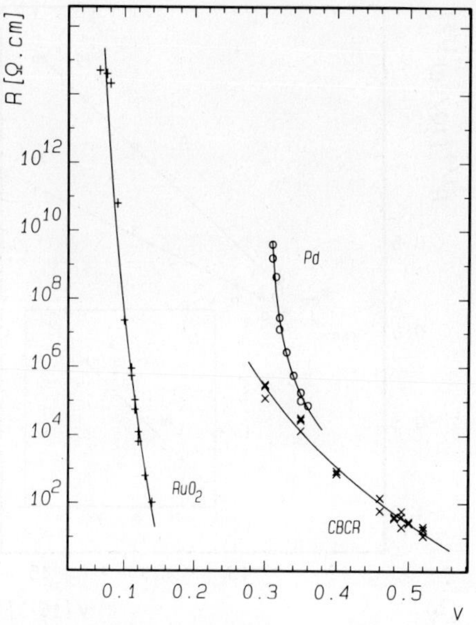

Fig. 2. Dependence of the resistivity R on the volume fraction of the conductive component for polystyrene filled with RuO_2 or Pd and for carbon black-filled chloroprene rubber (CBCR).

so-called blending curves. They can be described by means of percolation theory using the equation:

$$R = R_o(v - v_c)^s \tag{1}$$

where v is the volume fraction of conductive phase, v_c is the critical volume fraction (the threshold), s is the critical exponent and R_o is the material parameter.

For the PS system, based on RuO_2 or Pd, the data are shown in Fig. 2(for comparison, data of the carbon black-filled chloroprene rubber are plotted). The percolation threshold determination is difficult in this case and it is better to use the equation of straight line:

$$R^{1/s} = R_o^{1/s}(v - v_c) \tag{2}$$

with two fitted parameters v_c and R_o for different fixed values of the exponent s. The solid lines in Fig. 3(a) represent blending curves obtained from the least-squares fit for the measured dependences. These dependences for other systems (PS–Ag, PS–Au and PS–Pd composites) in the forms of thick films are shown in Fig. 3(b). The fits of the straight lines can be considered very good and the linear correlation coefficient is very close to unity. The unknown parameters, v_c and s, are given in Table 1.

The value of critical volume predicted by percolation theory ($v_c = 0.16$) approximately applies for two systems: Au–PS and Au–PS composites. The experimental values of v_c of the other composites differentiate slightly

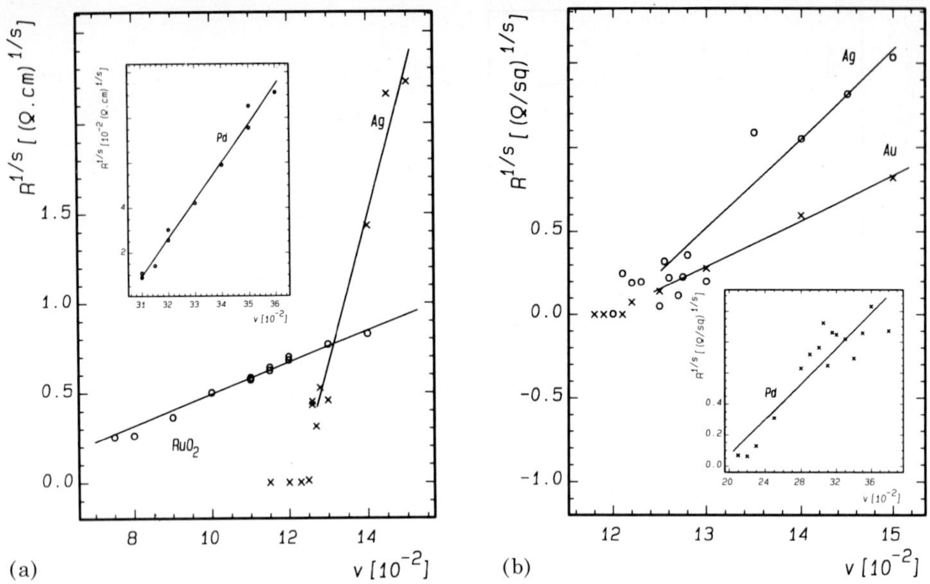

Fig. 3. Dependence of $R^{1/s}$ on the volume fraction for polystyrene filled with RuO_2, Pd, Au and Ag for (a) bulk samples and (b) thick films.

TABLE 1

Critical parameters in eqn. (1) for polymers filled with precious metal powders, RuO_2 and carbon black

Conductive component	Polymer/bulk (B) or thick film (TF)	Critical volume fraction, v_c	Critical exponent, s
RuO_2	PS/B	0.044 ± 0.005	-24.5
Pd	PS/B	0.31 ± 0.01	-4.7
Au	PS/B	0.11 ± 0.01	
Ag	PS/B	0.12 ± 0.01	-3.7
Pd	PS/TF	0.19 ± 0.03	-2.9
Au	PS/TF	0.12 ± 0.01	-1.8
Ag	PS/TF	0.12 ± 0.01	-1.3
Ag	LDPE/B	0.10 ± 0.01	
Ag	LLDPE/B	0.10 ± 0.01	
CB (Flammruss)	CR/B	0.10 ± 0.01	-13.2
CB (Vulcan P)	LLDPE/B	0.24 ± 0.05	-5.0
CB (Flammruss 101)	LLDPE/B	0.24 ± 0.02	-1.8

in comparison to the theoretical ones. The lower values of v_c were explained by McLachlan [1]. The explanation can be found in the inhomogeneous distribution of the conductive powder particles in the composites and in segregating structures creating long conductive chains, especially on the low conductive powder content side of the percolation threshold.

The elongated form of the particles (in case of RuO_2) can also influence the resistivity of the composites.

On the other hand, the critical exponent s is mostly higher than the theory predicts (-1.9) and varies widely from about -1.3 to -20 or more. The theoretical value corresponds only with composites containing gold powder. The reason for this behaviour is not clear and it will not be easy to explain. We do not know of any author who has explained the high absolute values of the exponent s, as for example in the case of CB–CR composites with $s = -13.2$ or RuO_2–PS systems with $s = -24.5$. Regarding other composites, the value of s is around -5.0. This means that the critical exponent is influenced both by the polymer type and the conductive ingredient. A higher absolute value of $|s| = 2.4$ was explained by Miyazima *et al.* [2]. The critical volume fraction of the thick films containing Ag or Au powder is close to the v_c value of bulk samples. However, the critical volume of the Pd–PS system is lower (0.19) than that for bulk samples (0.305). The reason may be in the different technology of the thick films prepared without a pressure step. This method enables the preparation of resistors with sheet resistance ranging from about 0.1 to 10^{13} $\Omega/\square$.

3.2. Temperature dependence of d.c. resistivity

A further important characteristic of conductive composites is the variation of their resistivity with temperature. The results shown in Fig. 4 relate to the temperature dependence of the d.c. resistivity measured for the PS composite containing RuO_2 from about 20 K up to room temperature. The composite is interesting for two reasons:

(1) The temperature coefficient of resistance is positive and reaches the high value of 8.6×10^{-3} /K at room temperature.

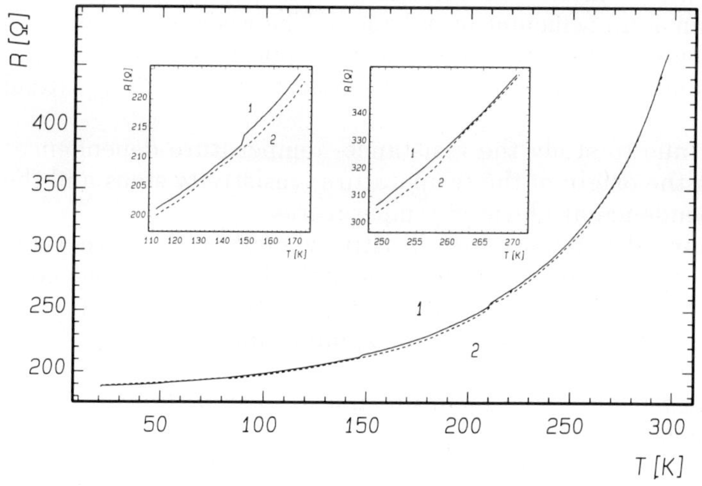

Fig. 4. Resistance vs. temperature for polystyrene filled with 14 vol.% of RuO_2 (low temperatures).

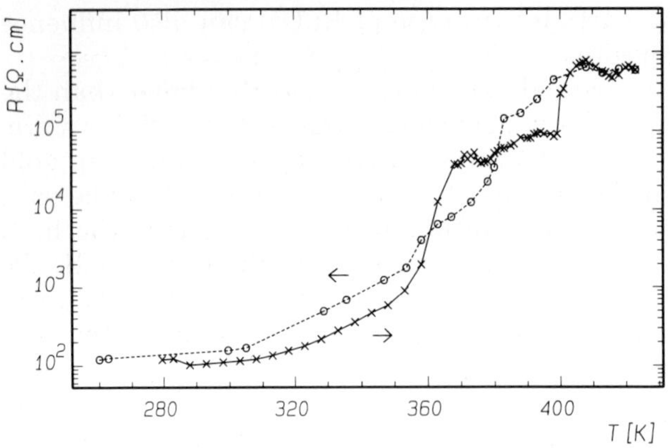

Fig. 5. Resistivity vs. temperature for polystyrene filled with 14 vol.% of RuO_2 (elevated temperatures).

(2) Two steps in resistivity dependence on temperature are observed when cooling: one is between 212 and 213 K and the other is between 147 and 148 K. The reason for these steps in not clear and is the subject of further investigation. When heated, there is only one step at a higher temperature of about 258 K. These dependences could be reproduced.

A strong variation of resistivity with temperature is observed especially above room temperature (see Fig. 5). The dependence exhibits three strong maxima at 373, 393 and 408 K. The highest resistivity is indicated close to and above the melting point. Different dependences of resistance versus temperature are obtained upon cooling. However, the resistivity value at room temperature increases by about 25% only. The positive temperature coefficient of resistance can be explained by the large thermal expansion coefficient of polymers. The resistance increases very rapidly especially when the temperature approaches the melting point and the distances between individual particles of the filling extend substantially.

We shall continue to study the resistance–temperature dependences in order to explain the origin of the temperature–resistivity steps and the nature of this dependence at elevated temperatures.

The temperature dependences of resistivity measured above room temperature show the same characteristics for all the measured composites. They differ only in some details and their structure. Our interest is now centred on three details and their explanation.

3.3. Temperature dependence of capacitance and a.c. conductivity

The dielectric behaviour of capacitors filled with polymeric composite at varying temperatures, as shown in Fig. 6, is similar to that observed for the temperature–resistivity dependence, but with reverse behaviour. The capacitance decreases with increasing temperature. The

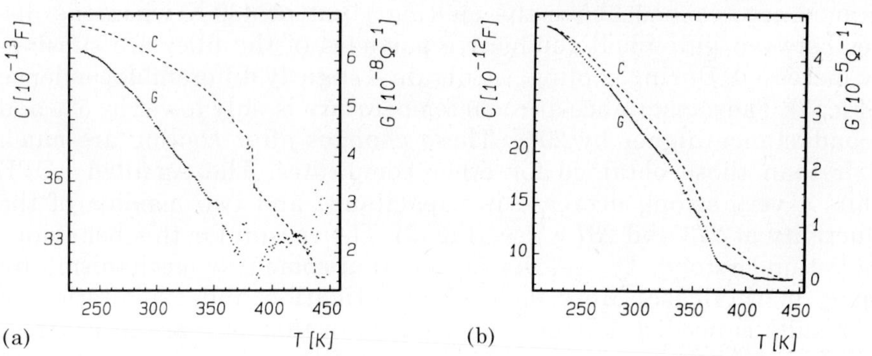

Fig. 6. Capacitance (C) and conductance (G) vs. temperature for polystyrene filled with (a) 25 vol.% and (b) 34 vol.% of Pd powder for 100 kHz.

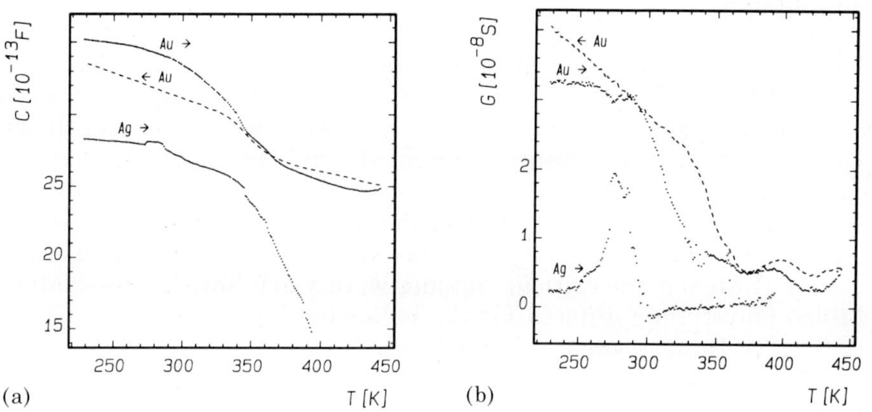

Fig. 7. (a) Capacitance and (b) conductance vs. temperature for polystyrene filled with 11.8 vol.% of Au powder and for LDPE filled with 9.5 vol.% of Ag powder for 100 kHz.

minimum is reached close to the melting point. The explanation is the same as in the case of resistivity. As the temperature is increased the distances between individual particles of the filling increased and the capacitance of the individual microcapacitors decreased, causing a decrease in the resulting total capacitance of the composite.

There are some composites, like the Pd–PS systems, below the threshold that exhibit a sudden two-step decrease in capacitance and corresponding sharp conductivity maxima at 380 and 440 K, respectively (see Fig. 6(a)). This behaviour is probably caused by material changes in the system. It is not observed for the Pd–PS systems above the threshold (high filling) (see Fig. 6(b)).

Figure 7 shows two examples of temperature dependence of capacitance (a) and conductivity (b) for polystyrene filled with Au powder and for LDPE filled with Ag particles for 100 kHz. The course of the curves is typical. The lowest capacitance of the Au–PS system and the highest resistance are reached above the melting point at 430 K when the distances between individual conductive particles of the filler are substantially increased. During cooling, we obtain a slightly different dependence (see Fig. 7). The capacitance at room temperature is only lower by 5% and the conductance higher by 20%. These changes after cooling are much smaller than those obtained for other composites. The Ag-filled LDPE exhibits a very strong decrease in capacitance and two maxima of the conductivity at 273 and 287 K (see Fig. 7). The reason for this behaviour is not yet understood. The results suggest that some new mechanisms are involved in this case. More detailed investigation and exploitation of these results is needed. Further experiments in this area are in progress and will be published soon.

4. Conclusions

New types of composites in the form of bulk samples and screen-printed thick films were developed. Polystyrene or polyethylene were used as matrices, and gold, silver and palladium powder were used as fillings.

The evaluation of resistance–content dependences showed that the critical parameters of blending curves were almost the same both for the bulk samples and for the thick films in the case of gold and silver fillers. The critical exponent and critical volumes were very close to the theoretical values. However, the critical volumes were much lower. These values of Pd-filled polystyrene differed for the bulk samples and the thick films. The reason for these results is the different particle sizes of the conductive powder. The results were compared with the corresponding data on polymeric systems containing various kinds of carbon black.

The positive temperature dependence of resistivity over a wide temperature range was explained as due to the large thermal expansion coefficient of polymers. The various steps observed in resistance and capacitance in respective temperature dependences are the subject of further investigation.

The experimental and theoretical results make it possible to suggest several applications of the developed polymeric systems, especially thick films in some fields of electronics.

References

1 D. S. McLachlan, *J. Appl. Phys.*, *70* (1991) 3681.
2 S. Miyazima, K. Maruyama and K. Okumura, *J. Phys. Soc. Jpn*, *60* (1991) 2805.

Electrochemical behaviour and structural characterization of highly crystalline poly(*p*-phenylene) films obtained by oxidation of benzene in a concentrated sulfuric acid emulsion

M. D. Levi*, E. Yu. Pisarevskaya, E. B. Molodkina and A. I. Danilov
Frumkin Institute of Electrochemistry of the Russian Academy of Sciences, Leninsky Prospect 31, Moscow 117071 (Russian Federation)

Abstract

Poly(*p*-phenylene) (PPP) films deposited from benzene in concentrated H_2SO_4 emulsion reveal a highly crystalline structure under TEM, SEM and XDA in contrast to nearly completely amorphous films deposited from $C_6H_5NO_2$ medium. Possible effects of solvent and electrolyte specificity on the morphology and mechanical properties of PPP films are discussed in view of these results.

Introduction

Among different conducting polymers it is poly(*p*-phenylene) (PPP) that attracts much attention due to both the very convenient procedure of grafting the films onto an electrode surface during anodic oxidation of benzene [1–6] and its rather high chemical stability, in particular, with respect to oxygen of the air. Similar to polyacetylene and polythiophene, PPP can be doped by anions (p-doping) and cations (n-doping) [7] thus yielding a rechargeable battery with both PPP cathode and anode. In addition, the polymer is extremely suitable as a model of p-conjugated polymers.

Solvent nature as well as electrolyte solution composition markedly influence the morphology, conductivity and electrochemical behaviour of resulting films. Owing to the relatively high reactivity of intermediate benzene radical-cations the optimum medium was shown to consist of a low donor number solvent, such as CH_2Cl_2 or $C_6H_5NO_2$, and slightly nucleophilic salts, typically with tetraalkylammonium cations, in the presence of some Broensted or Lewis acids [8]. The latter catalysts form a complex with the benzene molecules, facilitating the oxidation due to some decrease in conjugation energy of the aromatic ring [8].

Liquid SO_2 at -75 °C with different acid additions [4, 9] and concentrated H_2SO_4 in the presence of $AlCl_3$ [5] were shown to be extremely

*Present address: C.E.N.G., DRFMC/SESAM/EM, 85X 38041 Grenoble Cédex, France.

0379-6779/93/$6.00

suitable for benzene oxidation resulting in homogeneous PPP coatings. One of the characteristic features of the electrochemically prepared PPP films is their poor crystallinity as compared to those obtained chemically, the latter being increased up to 60% by annealing at rather high temperatures [10].

It is the purpose of this publication to demonstrate that regular well-faceted crystals as well as homogeneous PPP film consisting of very small crystals can immediately be produced on Pt, Au, Ni, Ti etc. electrodes by electrochemical oxidation of benzene in concentrated H_2SO_4 without adding Lewis catalysts. Preliminary results on this new method of obtaining PPP have recently been published [11]. Here we would like to give more information on the conditions of the electrochemical procedure leading to highly crystalline films and their structural and electrochemical characterization.

Experimental

The deposition of PPP films was carried out in either one- or two-compartment cells containing benzene in H_2SO_4 emulsion (volume ratio 1:25) at 8 °C with continuous mixing, where Au, Pt, Ni, Ti and glassy carbon were used as working electrodes (visible surface area was typically 0.13 cm^2). A piece of graphite textile or platinum sheet was used as a counter electrode. All the potentials were measured relative to a mercury sulfate electrode in concentrated H_2SO_4. Other experimental details have already been reported [11, 12] or will be published in the future [13].

Results and discussion

Effect of electric regime of benzene oxidation on the electrochemical behaviour of the resulting PPP films

Three different regimes of electrosynthesis, namely, potentiostatic, rectangular pulse stepped and cyclic voltammetric, have been used for preparation of the films. Figure 1(a) shows that potentiostatic synthesis proceeds effectively at electrode potential value $E = 0.91$ V, whereas application of higher potential used, $E = 0.95$ V, gives rise to a rather rapid decrease in current due to concurrent degradation of the grafted film. The lowest potential used, $E = 0.85$ V, as can be easily seen from Fig. 1(a), is insufficiently large to ensure the formation of the film at a reasonably high rate.

The rectangular pulse stepped method was shown to be as effective as the potentiostatic one (see Fig. 1(b)). The increase in current values with the number of pulses corresponds well with the shape of the chronoamperometric curve during potentiostatic synthesis at the same time interval. Cyclic linear potential sweeping between −0.2 and 0.91 V results in PPP

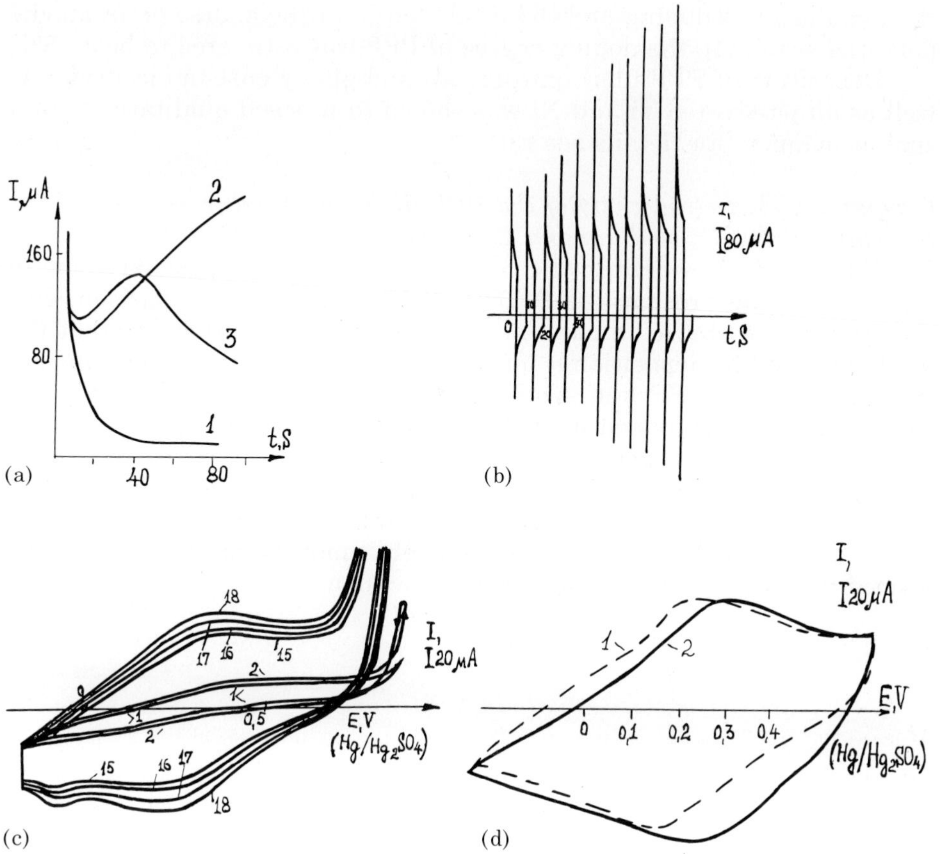

Fig. 1. Chronoamperometric curves in the course of potentiostatic (a) and pulse stepped (b) electrolysis of benzene in concentrated H_2SO_4 at different potentials (potentiostatic regime) E (V): 1, 0.85; 2, 0.91: 3, 0.95. Pulse duration and amplitude are indicated in Fig. 1(b). (c) Cyclic voltammograms for oxidation of benzene in concentrated H_2SO_4 at potential scan rate $v = 50$ mV/s. (d) Cyclic voltammograms of PPP doping in concentrated H_2SO_4. Films were obtained in potentiostatic (1) and pulse stepped regime (2).

films (Fig. 1(c)) with nearly the same voltammetric response as in the case of the potentiostatic (or pulse stepped) regime (Fig. 1(d)). We therefore came to the conclusion that the redox-activity of the formed PPP films depends primarily on the appropriate choice of the upper anodic potential rather than on the electric mode of the electrosynthesis. Therefore, the deposition mechanism of PPP is practically the same (for rather thin films) irrespective of the potential changing mode, as is clearly seen from the typical nucleation loops in the early cyclic voltammetric curves (Fig. 1(c)) which correspond well to the rising position of the appropriate chronoamperometric response (Fig. 1(a)).

Based on the relation between the equilibrium charge extracted from the film during dedoping and the total charge in the course of an anodic potential scan [14], the doping degree of PPP was estimated to be *c.* 18%.

Deposition of PPP films onto Pt, Au and glassy carbon electrodes as well as on passivated Ti and Ni was shown to proceed qualitatively in a similar manner (for details see ref. 13).

Comparison of electroactivity of the PPP films deposited from concentrated H_2SO_4 medium and $C_6H_5NO_2$

Figure 2(a) shows the cyclic voltammograms for benzene oxidation in $C_6H_5NO_2$ in the presence of $BF_3O(C_2H_5)_2$ (10 vol.%). Comparison with the appropriate curves in concentrated H_2SO_4 (Fig. 1(c)) shows that the mechanism of film deposition is qualitatively very similar. The voltammetric response of a PPP film synthesized in $C_6H_5NO_2$ reveals exactly the same features when doped in H_2SO_4 as another film obtained in concentrated sulfuric acid (compare Figs. 1(d) and 2(b)).

Surprisingly, drastic variations in the structure of the resulting films have been found, namely, the highly crystalline one when deposited in H_2SO_4 as opposed to the almost entirely amorphous structure when formed in $C_6H_5NO_2$ or another organic medium.

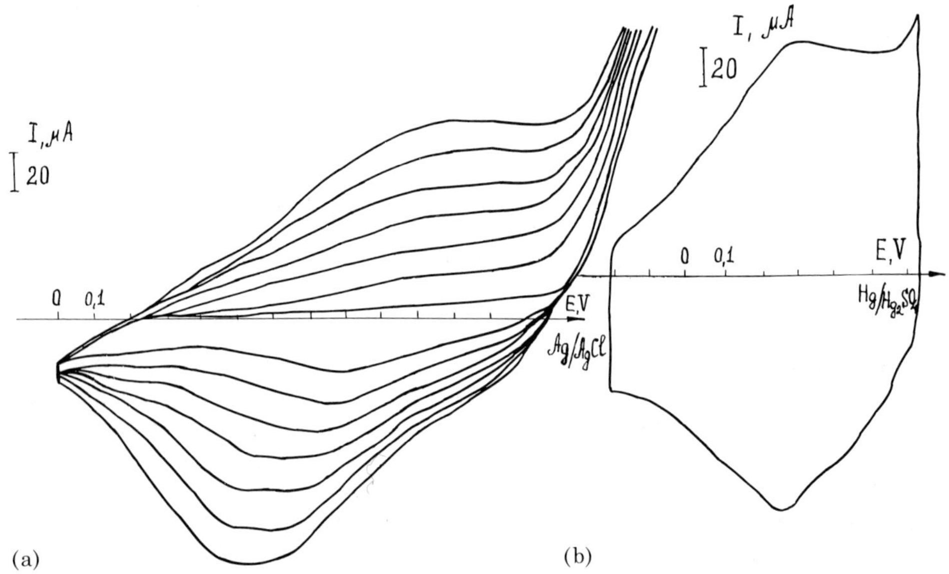

(a) (b)

Fig. 2. Cyclic voltammograms for oxidation of benzene in $C_6H_5NO_2$ in the presence of $BF_3O(C_2H_5)_2$ (10 vol.%) (a) and corresponding doping–dedoping curves of the resulting PPP films when immersed in concentrated H_2SO_4 (b).

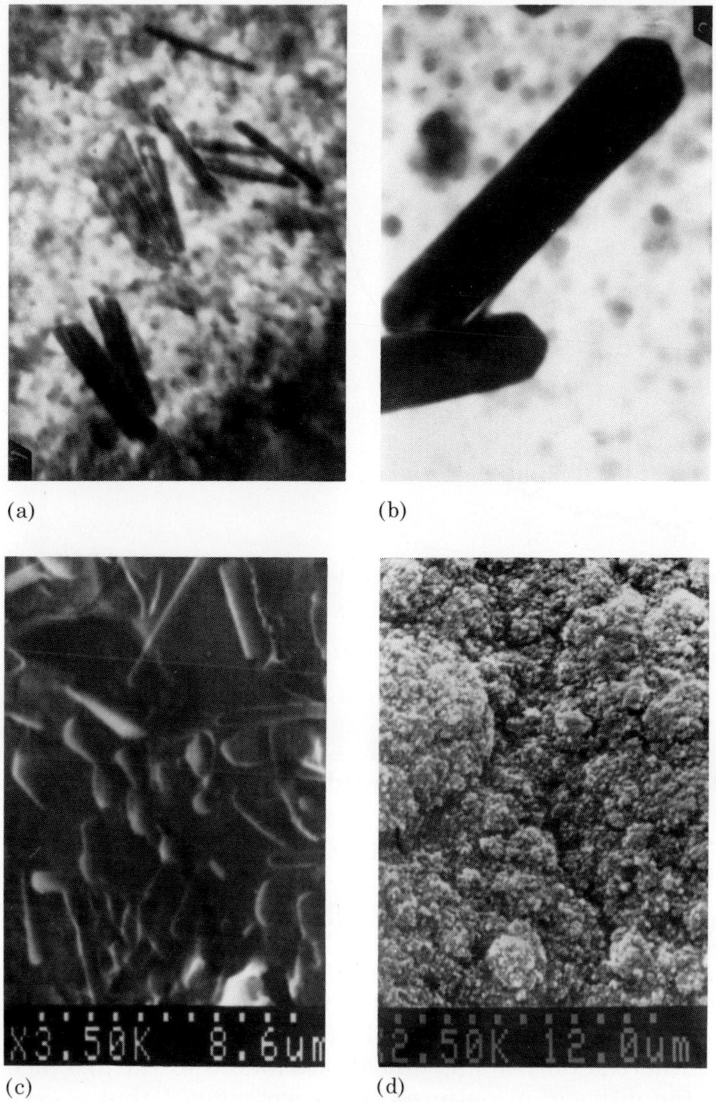

Fig. 3. Crystals of the PPP film under TEM (80 kV, (a) ×15 000, (b) ×31 250). (c) Crystalline 'needles' on the inner side of the PPP film under SEM. (d) Globe-shaped structure of the outside of the PPP film under SEM. ((a) and (b) reduced 83% in reproduction and (c) and (d) 55%.)

Structural characterization

Figure 3(a) shows a large number of regular well-faceted crystals of different sizes and thicknesses (under TEM) located on the inner side of the homogeneous PPP film (synthesis in H_2SO_4). We observed typical electron diffraction patterns from these crystals [11]. The location of the crystals was confirmed using SEM. Figure 3(b) shows particles of an oblong shape on the inner side of the film corresponding to the crystalline

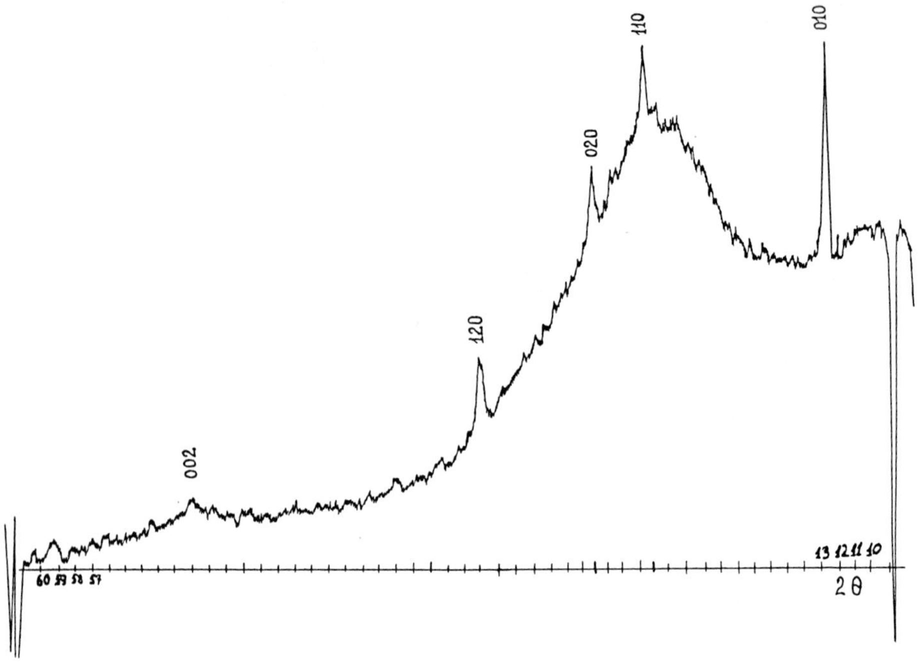

Fig. 4. X-ray diffraction patterns of the PPP films obtained in concentrated H_2SO_4 (Co Kα). Reflecting faces are indicated near the peaks.

'needles' in Fig. 3(a). The outer (directed towards the solution) side of the surface reveals a typical globular structure (Fig. 3(c)).

The highly crystalline structure of the powdered film was confirmed by X-ray diffraction analysis, XDA, see Fig. 4. We have indexed the reflections using a monoclinic unit cell with parameters $a = 0.787$, $b = 0.517$, $c = 0.432$ nm and $\beta = 105.7°$.

We observed also a marked dependence of the number of peaks in the X-ray diffraction patterns on the intensity of mixing of the benzene/sulfuric acid emulsion. At an increased rate of mixing, crystal sizes were shown to decrease whereas only a single peak attributed to the 120 face was prominent on the diffraction patterns. Thus we suggest that a very fine emulsion gives rise to a textured PPP film with the *c* axis directed along the electrode surface.

PPP films obtained from H_2SO_4 were rather brittle. These mechanical properties also seem to be in reasonable agreement with our recent preliminary data on their porous structure [15] investigated by the standard porosity method which was based on a comparison of the structure of the standard porous samples and that under study (on capillary equilibrium conditions).

PPP films synthesized in aprotic solvents revealed an amorphous structure under XDA and none of the separate crystals can be seen under TEM and SEM.

To account for such a different behaviour we have to take into consideration at least two points. First, it is well known that the crystal structure of the conducting polymer film could be a matter of the nature of the doping ion. There was no indication of a crystalline structure of PPP obtained in CH_2Cl_2 in the presence of oleum [6] though such a result can in principle be a consequence of the excess of Bu_4NBF_4 background electrolyte. We believe the other point which could be of crucial importance, is the solvent specificity. The formation of microcrystals at the onset of electrochemical synthesis could be favoured by the very low solubility of benzene as well as oligophenylenes in concentrated H_2SO_4 as opposed to that in aprotic solvents. It would be of interest to find a strong correlation between the latter quantity and the nature of the solvent during synthesis. Bearing in mind this hypothesis we are currently undertaking a new line of research.

Conclusions

It has been shown that PPP films deposited from benzene in concentrated H_2SO_4 emulsion reveal a highly crystalline structure under TEM, SEM and XDA. The resulting films are rather brittle and totally incompressible which can also be regarded as indirect evidence of their crystalline structure. In contrast, PPP films deposited from $C_6H_5NO_2$ and some other aprotic solvents are flexible as well as completely amorphous.

References

1 M. Satoh, M. Tabata, K. Kaneto and K. Yoshino, *J. Electroanal. Chem.*, *195* (1985) 203.
2 T. Onsawa and K. Yoshino, *Synth. Met.*, *17* (1987) 601.
3 S. Aeiyach, J. E. Dubois and P. C. Lacaze, *J. Chem. Soc., Chem. Commun.*, (1986) 1668.
4 S. Soubiran, S. Aeiyach, J. J. Aaron, M. Delamar and P. C. Lacaze, *J. Electroanal. Chem.*, *251* (1988) 89.
5 J. H. Ye, Y. Z. Chen and Z. W. Tian, *J. Electroanal. Chem.*, *229* (1987) 215.
6 L. M. Goldenberg, I. B. Nazarova, O. N. Efimov, R. N. Lyubovskaya, O. S. Roschupkina and A. N. Titkov, *Synth. Met.*, *44* (1991) 133.
7 R. Baunhman, L. Shacklette, N. Murthny, G. Miller and R. Elsenbaumer, *Mol. Cryst. Liq. Cryst.*, *118* (1985) 253.
8 K. Yamamoto, T. Asada, H. Niscede and E. Tsuchida, *Bull. Chem. Soc. Jpn.*, *61* (1988) 1731.
9 S. Aeiyach, S. Soubiran, P. C. Lacaze, J. Froyer and Y. Pelous, *Synth. Met.*, *32* (1989) 103.
10 P. Kovacic and M. B. Jones, *Chem. Rev.*, *87* (1987) 357.
11 M. D. Levi, E. Yu. Pisarevskaya, E. B. Molodkina and A. I. Danilov, *J. Chem. Soc., Chem. Commun.*, (1992) 149.
12 E. Yu. Pisarevskaya and M. D. Levi, *Sov. Electrochem.*, *27* (1991) 496.
13 M. D. Levi and E. Yu. Pisarevskaya, to be published.
14 E. M. Geniès and C. Tsintavits. *J. Electroanal. Chem.*, *200* (1986) 127.
15 Yu. M. Vol'fkovich, M. D. Levi, T. K. Zolotova and E. Yu. Pisarevskaya, *Polym. Commun.*, submitted for publication.

Thermal dedoping of poly(3-n-hexylthiophene) films

R. H. M. van de Leur*, B. de Ruiter and J. Breen

TNO Plastics and Rubber Research Institute, P.O. Box 6031, NL-2600 JA Delft (Netherlands)

Abstract

It is shown that the conductivity of undoped poly(3-n-hexylthiophene) is reversibly reduced a few orders of magnitude by heating beyond the glass–rubber transition (room temperature) in a nitrogen ambient. Impedance spectroscopy reveals the presence of localized and delocalized charge carriers and shows the influence of oxygen on the conductivity.

Introduction

The good processing properties of the conducting poly(alkylthiophene)s are promising with respect to possible applications. To allow widespread application the polymers should be stable when exposed to air. However, it has been shown that at least some of these polymers (methyl and butyl substituted) are sensitive to atmospheric oxygen or water vapour [1, 2]. The aim of the present work was to study the effect of oxygen on the dielectric properties of undoped poly(3-n-hexylthiophene) (PHT).

Experimental

PHT was synthesized via Grignard coupling reactions. 3-Hexylthiophene was prepared from 3-bromothiophene and the appropriate Grignard reagent in the presence of a nickel catalyst [3, 4]. 3-Hexylthiophene was converted into its 3-hexyl-2,5-diiodo analog by reaction with iodine in the presence of nitric acid [5, 6]. The products were purified by means of column chromatography (SiO_2/hexane). Polymerization was carried out with Mg in THF in the presence of a nickel catalyst [6]. THF was removed by evaporation under reduced pressure; the solid residue was vigorously shaken with an HCl/MeOH mixture and filtered off. The product was further purified by washing procedures, extensive extractions, and finally dried under vacuum at 40 °C. All synthesized materials were characterized by means of ^{1}H NMR spectroscopy.

*Author to whom correspondence should be addressed.

0379-6779/93/$6.00

The molecular weight (M_n) as obtained by GPC on polystyrene standards was 10×10^3. This value is probably not too different from the actual molecular weight [7]. The dispersion factor M_w/M_n of the polymer is 2.5.

For the impedance spectroscopy (IS) and the dynamic mechanical analysis (DMA) PHT was dissolved in chloroform (at room temperature) to obtain 10 (IS) and 40 (DMA) μm thick free-standing films via solution casting. Gold electrodes with an area of 1.1 cm^2 were vacuum deposited on both sides of the films for the IS at frequencies between 0.1 Hz and 10 kHz with an amplitude of the applied voltage equal to 0.1 or 1 V. The DMA was performed with a harmonic tensile strain with an amplitude of 0.03% superposed on a static tensile strain of 0.3%. The temperature was swept between -160 and $+160$ °C with a heating or cooling rate of 0.25 °C/min. The measurements were performed at intervals of 10 °C in complete darkness.

Glass–rubber transition

The DMA reveals a strong decrease of the elastic modulus at room temperature (see Fig. 1). The glass–rubber transition temperature T_g (peak in tan δ) is observed to vary between 10 °C at 0.1 Hz and 30 °C at 10 Hz. No difference was observed between the experiments during heating and subsequent cooling. There is an indication of a second (β) transition at approximately -90 °C which is in accordance with DMA on octyl- and dodecylthiophenes [8].

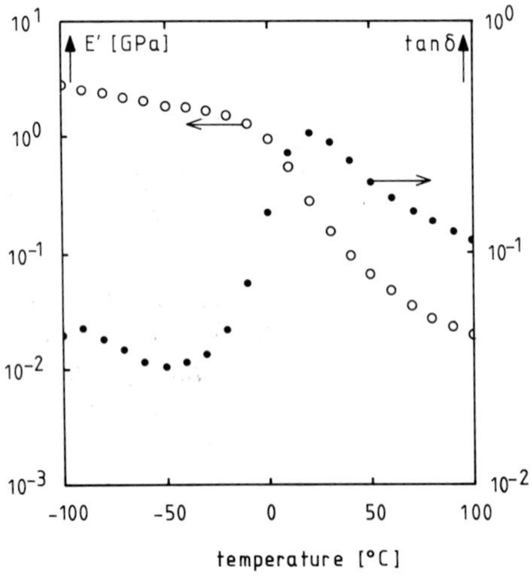

Fig. 1. The elastic modulus E' (○) and the loss tangent tan δ (●) as a function of temperature at a frequency of 1 Hz.

Thermal dedoping in a nitrogen ambient

The d.c. conductivity of PHT exhibits a local maximum on heating at approximately 30 °C if the film was exposed to normal environmental conditions or dry air for some minutes preceding the experiment (see Fig. 2). Up to T_g the conductivity is thermally activated with an activation energy $E_{act} = 0.29$ eV. With the heating rate chosen, the film seems to be completely dedoped at 120 °C. The conductivity of the dedoped film is thermally activated, with $E_{act} = 0.55$ eV, in the whole temperature range investigated. Provided that the film is kept in nitrogen, the conductivity does not change during subsequent cooling and heating runs.

Experiments in a nitrogen ambient at three different temperatures (see Fig. 3) indicate that below T_g the dedoping rate is at least a few orders of magnitude lower than at or above this transition temperature.

The doping (with oxygen) and dedoping processes are observed to be reversible, in accordance with experiments on poly(methylthiophene) [1]. In contrast to the dedoping, the doping by exposure to air is observed to be rather fast; it occurs within a few minutes.

The dedoping process at temperature above T_g might originate from a reduction of the conjugation length due to torsional defects in the polymeric backbone. This reduction possibly leads to a back transfer of electrons from the ionized oxygen. Due to the enhanced diffusion of the oxygen at these temperatures the atomic oxygen will leave the polymer rather fast.

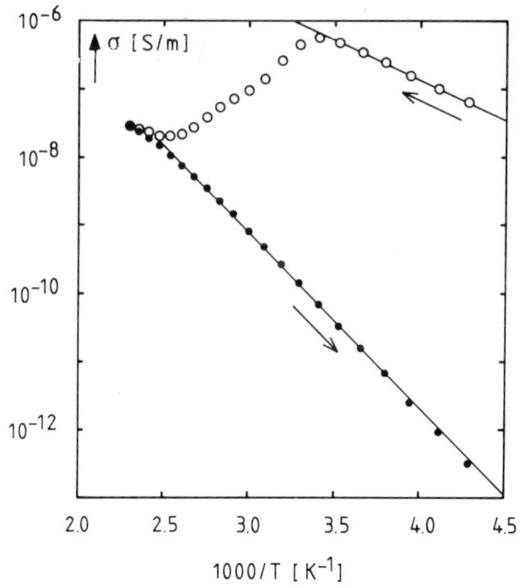

Fig. 2. The d.c. conductivity σ as a function of the reciprocal temperature during heating (○) and cooling (●) at a rate of 0.25 °C/min in a dry nitrogen ambient.

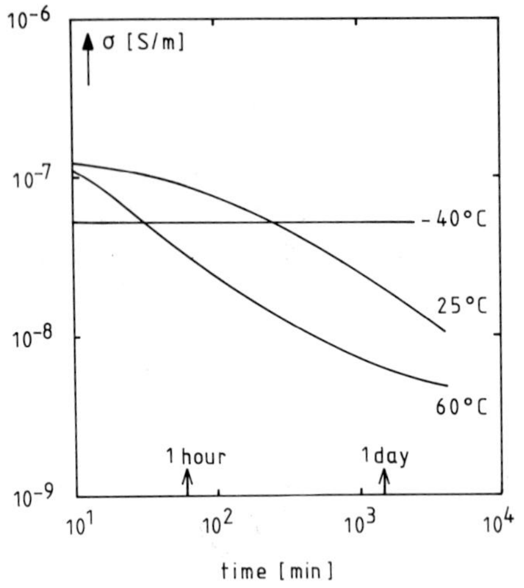

Fig. 3. The d.c. conductivity σ of PHT films at three different temperatures in a dry nitrogen ambient as a function of time. Prior to these experiments the samples were exposed to normal environmental conditions.

Dielectric response

The impedance of the PHT films discussed in this paper can be described by the equivalent circuit shown in Fig. 4. At high frequencies the impedance is largely determined by electronic polarization (represented by C_0) resulting in a dielectric constant $\varepsilon' = 2.3$. This value corresponds rather well to 2.23, calculated on the basis of the structural units of the polymer [9]. The higher polarization observed earlier in powder samples and films synthesized by electropolymerization [9, 10] can possibly be explained by the microstructure (rough surface) of these two types of PHT samples.

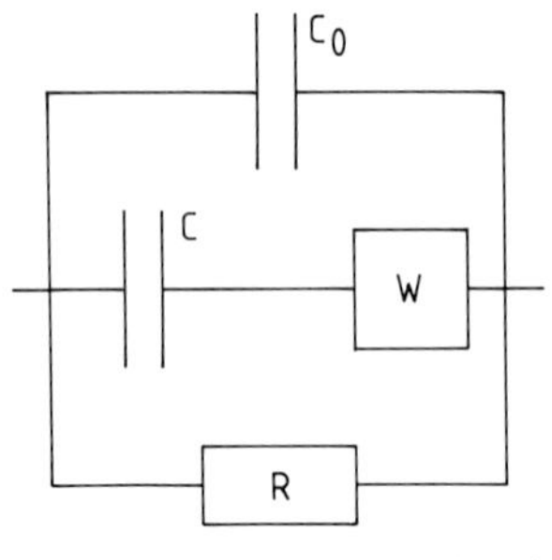

Fig. 4. An equivalent circuit for the dielectric response of the PHT samples. The capacitor C_0 represents electronic polarization, the CW combination a hopping process for localized charge carriers, and R the d.c. conduction of delocalized charge carriers.

TABLE 1

Thermally activated processes

	Hopping		D.c.	
	E_{act} (eV)	W_0 (μS/m)	E_{act} (eV)	$1/R_0$ (S/m)
Before dedoping	0.15	0.6	0.29	0.02
After dedoping	0.26	0.2	0.55	0.12

Two different thermally activated processes are observed in parallel with the electronic polarization, viz. the transport of localized and delocalized charge carriers. In the doped state the charge carriers are likely to be (bi)polarons but also ionized oxygen might contribute to the conduction [11]. In the equivalent circuit of Fig. 4 the localized charge transport is represented by the capacitor C and the Warburg element W in series. The admittance of a Warburg element is given by $Y_w = W(j\omega)^{0.5}$ with $j = \sqrt{(-1)}$. This element is often associated with a hopping (or diffusion) process. It should be noticed that the equivalent circuit of the capacitor C_0 parallel with a second capacitor and the Warburg element in series is a special case of the Cole–Cole relation which is often used to describe dipole orientation in polymers. In fact, the dielectric response of localized charge transport is identical to that of dipole orientation.

The activation energy for the d.c. conduction (delocalized charge transport) is observed to be twice as large as the one for the hopping process both before and after dedoping (see Table 1 and Fig. 5). The

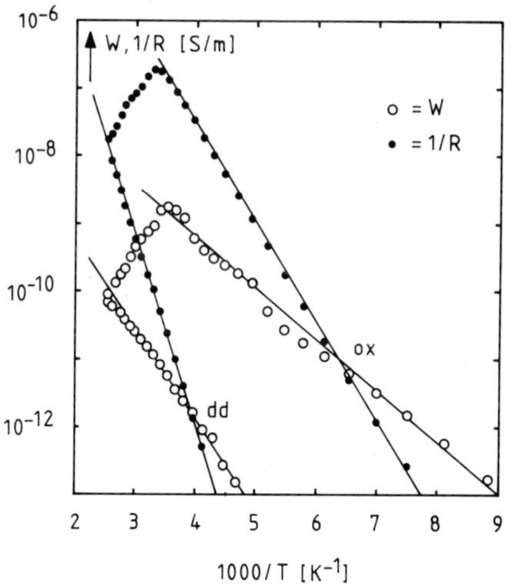

Fig. 5. The d.c. conductivity $1/R$ (●) and the admittance W (○) representing the hopping process before (ox) and after dedoping (dd).

admittance is given by $Y = Y_0 \exp(-E_{act}/kT)$ with $Y_0 = W_0$ and $Y_0 = 1/R_0$ for the hopping process and the d.c. conduction, respectively. Note that the admittance corresponding with the hopping process is six orders of magnitude lower than that of the d.c. conduction.

Conclusions

The experiments reported in this paper and those reported earlier on PHT obtained via oxidative coupling and electropolymerization [9] lead towards the following conclusions.

(i) The glass–rubber transition and the dedoping of oxygen in a nitrogen ambient occur at the same temperature.

(ii) The dielectric response of PHT can be described by a combination of electronic polarization and localized and delocalized charge transport.

(iii) The activation energy for the charge transport is increased by dedoping.

(iv) For the d.c. conduction the activation energy is twice as large as for the hopping conduction.

Acknowledgements

A. Buik, P. J. Droppert and M. Ringenaldus are gratefully acknowledged for performing the syntheses, the IS measurements and the DMA, respectively. This work is financially supported by the Dutch Ministry of Economic Affairs, Innovation Oriented Research Program on Polymer Composites and Special Polymers (IOP-PCBP project BP201).

References

1 D. M. Taylor, H. L. Gomes, A. E. Underhill, S. Edge and P. I. Clemenson, *J. Phys. D*, *24* (1991) 2032.
2 Y. Ohmori, H. Takahashi, K. Muro, M. Uchida, T. Kawai and K. Yoshino, *Jpn. J. Appl. Phys.*, *30* (1991) L1247.
3 K. Tamao, S. Kodama, I. Nakajima, M. Kumada, A. Minato and K. Suzuki, *Tetrahedron*, *22* (1982) 3347.
4 C. Van Pham, H. B. Mark, Jr. and H. Zimmer, *Synth. Commun.*, *16* (1986) 689.
5 J. M. Barker, P. R. Huddleston and M. L. Wood, *Synth. Commun.*, *5* (1975) 59.
6 J. Laakso, *Acta Polytech. Scand., Chem. Technol. Metall. Ser.*, *184* (1988) 1.
7 G. W. Heffner and D. S. Pearson, *Macromolecules*, *24* (1991) 6295.
8 S.-A. Chen and J.-M. Ni, *Makromol. Chem., Rapid Commun.*, *13* (1992) 31.
9 R. H. M. van de Leur and B. de Ruiter, *Synth. Met.*, *44* (1991) 327.
10 R. H. M. van de Leur, *Synth. Met.*, *51* (1992) 69.
11 N. T. Binh, M. Gailberger and H. Bässler, *Synth. Met.*, *47* (1992) 77.

Electron spectroscopic studies of models for PPP and PT on p- and n-doping

M. G. Ramsey*, D. Steinmüller and F. P. Netzer
Institut für Experimentalphysik, Karl-Franzens Universität, Graz, A-8010 (Austria)

Abstract

Condensed layers of sexiphenyl and bithiophene are investigated as models for polyparaphenylene (PPP) and polythiophene (PT). Bipolaron-like gap states are directly observed by ultraviolet photoemission spectroscopy (UPS) on the n-doping of sexiphenyl with Cs. Comparison with the results for Cs dosed biphenyl indicate that the dimer reflects the electronic modifications induced by doping. The *in situ* dosing of bithiophene with gaseous Br_2 is characterized by thermal desorption mass spectrometry (TDS), UPS and electron energy loss spectroscopy (EELS) and the formation of a Br–bithiophene complex is indicated. Comparison of the EELS spectra with optical absorption results for the p-doping of longer thiophene oligomers suggests that a bipolaron to polaron-like transformation can be achieved by thermal 'dedoping'.

Introduction

Polaron or bipolaron states arising from the geometric changes induced by the charge transfer upon doping in the non-degenerate ground state polymers are generally accepted. The experimental evidence for such states in the conducting polymers, has however been both indirect and not entirely unambiguous. This has led to studies of the oligomers as experimental models of the polymers which provide well defined systems that express the essential features of the electronic structure induced by doping. In earlier studies of biphenyl and bithiophene on 'n-doping' with Cs the gap states were directly observed by UPS and indirectly via EELS [1–3]. Here we report an extension of this work and present preliminary results for the n-doping of sexiphenyl and compare them to those of Cs dosed biphenyl and benzene to access how well the dimers model longer systems. In the case of thiophene systems the study of the oligomers has become intrinsically interesting due to their technologically exploitable characteristics [4]. We have extended therefore our n-doping experiments of bithiophene in this work to p-doping, and report here a combined UPS and EELS study of the valence band of bromine dosed bithiophene which compare favourably with studies of the p-doping of polythiophene and its longer oligomers.

---

*Author to whom correspondence should be addressed.

0379-6779/93/$6.00

Experimental

The experiments were performed in an ultrahigh vacuum (UHV base pressure $< 1 \times 10^{-10}$ mbar) VG ADES-400 angle resolving electron spectrometer with facilities for sample cooling and heating, UV photoemission and EELS with a resolution of 0.35 eV. A mass spectrometer mounted in the spectrometer chamber in line-of-sight of the sample allowed the performance of TDS following up to ten desorbing masses simultaneously. All thin films were applied *in situ* to a Ni single crystal substrate. Thin films of benzene, biphenyl and bithiophene were condensed from a pinhole effusion source on to the cooled (90 K) substrate. For sexiphenyl an evaporator within a liquid nitrogen cooled shroud was constructed so that it could be brought within close proximity to the surface or removed from the spectrometer chamber through a gate valve. This allowed sexiphenyl films to be grown, whilst maintaining ultrahigh vacuum conditions, with rates as low as a monolayer in 15 min. Cs was evaporated from an SAES getter source whilst Br_2 dosing from the gas phase was from the decomposition of a AgBr pellet in an electrochemical cell mounted within the chamber. All doses quoted here are given in units of monolayers (ML) estimated from TDS, UPS and work function measurements.

Results and discussion

In Fig. 1 a comparison is made between the highest lying valence band emission features of benzene, biphenyl and sexiphenyl for both the pristine films and the films after exposure to Cs. The spectra of the pure films have been aligned to the $1e_{1g}$ orbital of benzene and the non-bonding orbital emissions and display the breaking of the degeneracy and the evolution of the benzene derived $1e_{1g}$ orbitals in accordance with the results of Seki *et al.* [5]. On Cs exposure two new states appear in the gap for all three molecules. This is in agreement with the predictions of Brédas *et al.* [6] and is discussed elsewhere in detail for biphenyl [1]. The qualitative similarity between the features of biphenyl (and even benzene) and sexiphenyl reported here for the first time are noteworthy and demonstrate that even such short molecules as the dimers express already the essentials of the electronic modifications induced by doping. A more thorough discussion of this will be given elsewhere [7].

The p-doping of biphenyl was attempted with gaseous Br_2, however the molecules essentially ignored each other as evidenced in TDS, UPS and EELS and no indications for the formation of a charge transfer complex were found, in contrast to the results of bithiophene reported below. TDS spectra, which reveal the energetics of the layers, for four ML of bithiophene after exposure to four ML of Br_2 are displayed in Fig. 2 for the principal masses observed: bithiophene, Br_2 and Br. The desorption spectra of bithiophene and Br_2 are very similar to those of the individual

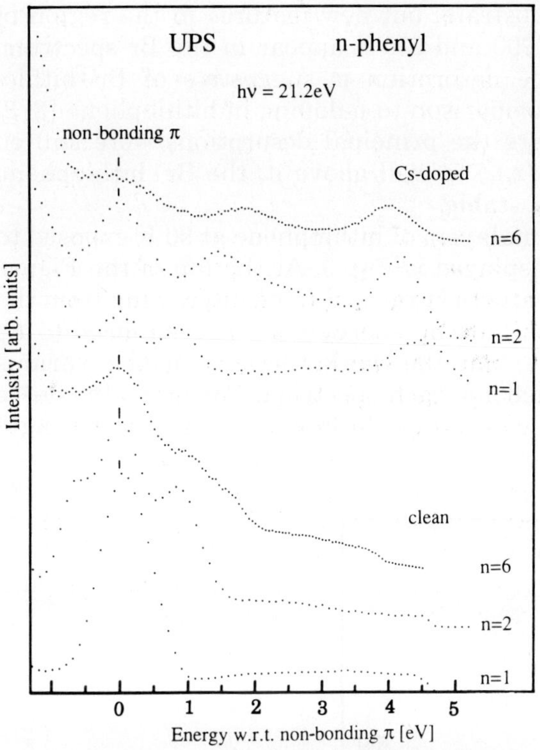

Fig. 1. UPS spectra of clean and Cs-doped *n*-phenyl for $n = 1, 2, 6$. The energy alignment is to the non-bonding π emission features.

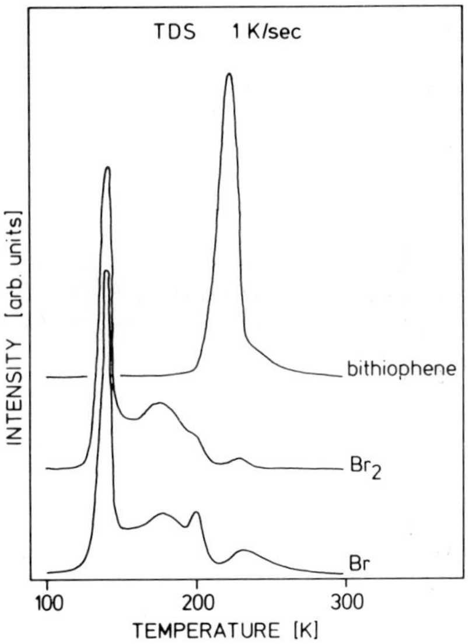

Fig. 2. TDS of bithiophene, Br_2 and Br for four monolayers of Br_2 condensed on four monolayers of bithiophene.

molecular films from the Ni substrate, but new features in the region of the bithiophene desorption at 200 and 230 K appear in the Br spectrum. This concerted Br–bithiophene desorption is suggestive of Br–bithiophene complex formation. In comparison to n-doping of bithiophene [3, 8] and biphenyl [2] with Cs, where the principal desorptions were shifted from below room temperature (r.t.) to well above it, the Br–bithiophene complex seems to be much less stable.

The UPS series for four monolayers of bithiophene at 90 K exposed to increasing amounts of Br_2 is displayed is Fig. 3. At the top of the Figure a heating series, for the temperatures suggested to be interesting from the TDS results, is also shown. The binding energy scale is referenced to E_F observed for the Ni substrate and the workfunctions of the various surfaces are indicated on the left for each spectrum. For up to Br_2 doses of 0.5 of an ML the valence band spectra of bithiophene undergoes a rigid

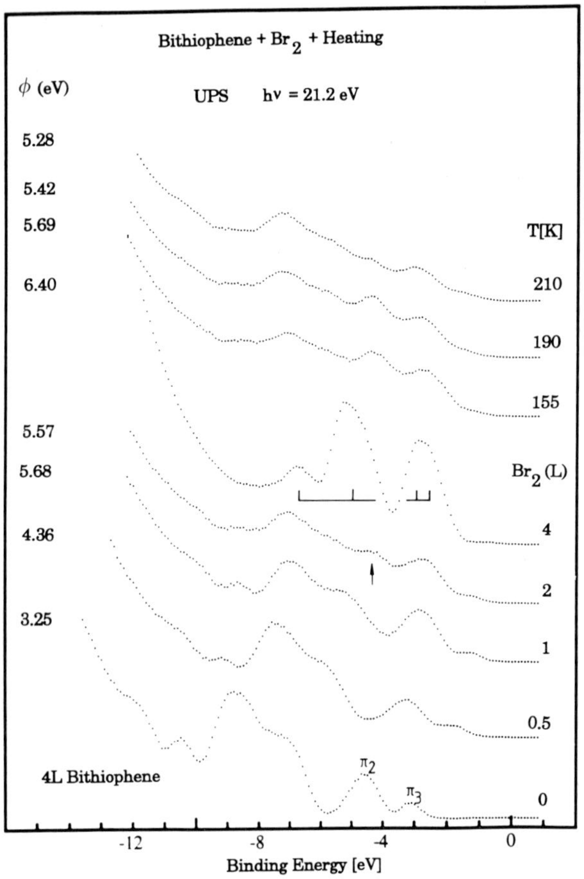

Fig. 3. Dosing and heating series of Br_2 dosed bithiophene. Br_2 exposure is quoted in monolayers (L). The workfunctions are indicated on the left, ϕ (eV).

shift to lower binding energy equivalent to the increase in the workfunction, with little change in the emission features. Between 0.5 and 1 ML Br_2 the spectra again shift to lower binding energy ($\Delta E = 0.5$ eV), however significantly less from the change in workfunction ($\Delta\phi = 1.3$ eV). Up to this point the only perceptible change in the emission features is a relative increase in band π_2 and a small relative decrease in the HOMO π_3 which is now observed as an indistinct step at 1.3 eV binding energy. Further dosing (2 ML) leads to the first noticeable bromine induced feature at 4.3 eV (arrow in Fig. 3). Subsequent exposures result in the removal of any recognisable bithiophene emissions and the three prominent peaks seen at 2.9, 5.2 and 6.8 eV can be attributed to condensed Br_2; the gas phase Br_2 emissions are indicated as ticks on the spectra (with their positions implying a 2.7 eV relaxation shift in condensed Br_2). Clearly, by four layers exposure the bithiophene has been buried under a condensed Br_2 layer. The bulk of the excess Br_2 is desorbed by warming to 155 K and the spectrum returns to a bithiophene-like spectrum with the additional 4.3 eV emission. On heating from 190 to 210 K through the Br desorption, the 4.3 eV emission all but disappears and there is a loss in intensity in the π_2 band at 2.6 eV relative to the prominent bithiophene emission at 7.4 eV. Careful inspection of the spectra also revealed a recovery in the π_3 HOMO emission at 1.5 eV.

Br–bithiophene complex formation therefore results in shifts to lower binding energy, a new 4.3 eV emission, some emission intensity in the π_2 band and perhaps a small decrease in intensity in the HOMO emission π_3. The energy shifts are similar to those reported for the p-doping of PT [9, 10] and, as pointed out by Lögdlund *et al.* [10] the relationship to $\Delta\phi$ changes can be attributed to shifts in E_F arising from the formation of empty (bi)polaron gap states. The result here that for high exposures the shifts are less than $\Delta\phi$ may be attributed to the lowering of the molecules screening ability (polarizability) on electron extraction: the equivalent but opposite effect has been observed for the n-doping of biphenyl and bithiophene with Cs [1, 3, 8].

The 4.3 eV peak is associated with dissociation of Br_2 and is assigned to an emission of Br^- in the complex: the emission from nickel bromide is observed in the region of the 4 to 6 eV binding energy. That it could be associated with Br_2 emission, not unreasonable when workfunction values are considered, is ruled out by the following EELS results where heating between 155 and 190 K removes all evidence of Br_2 from the excitation spectra.

The EELS spectral series taken concurrently with the UPS series is shown in Fig. 4. The primary energy of 20 eV is chosen to minimize substrate emission features and to provide a similar probing depth to the HeI UPS spectra. For pristine condensed bithiophene the EELS spectrum is similar in form to the optical absorption spectra of polythiohpene and its oligomers albeit with a higher energy $\pi \rightarrow \pi^*$ transition threshold: ≈ 3.5 eV compared to 3.12 and 2.97 eV for quarter- and sexithiophene,

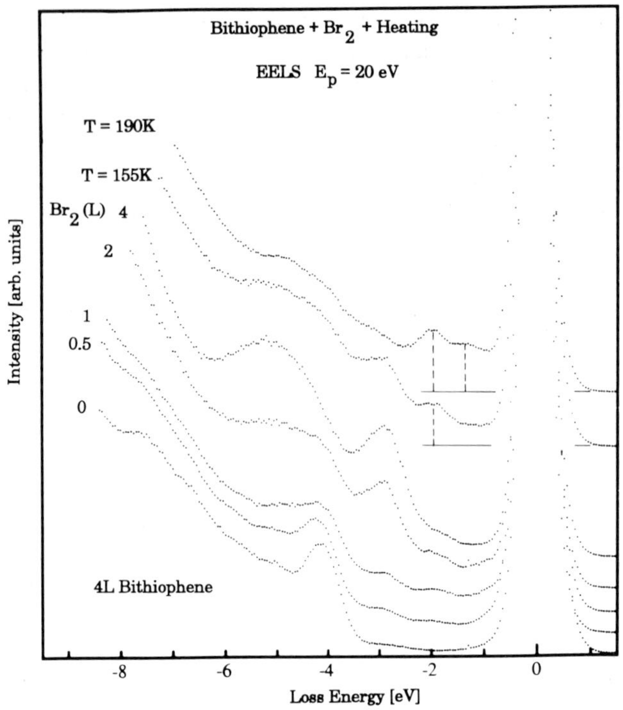

Fig. 4. EELS of equivalent surfaces to Fig. 3.

respectively [11, 12]. On increasing exposure to Br_2 the spectra in the region of the $\pi \rightarrow \pi^*$ transition gradually transforms to that of Br_2: pristine condensed Br_2 has its excitation threshold at ≈ 2.5 eV with prominent peaks at 2.9 and 5 eV. Apart from this gradual transformation a peak grows within the forbidden gap region at 2 eV loss energy; there is also increase in loss intensity at lower energy starting as a small shoulder on the elastic peak at ≈ 1 eV. Clearly bromine interaction induces states in the forbidden gap which can be accessed. The excitations to the gap states are poorly resolved in the dosing series, however on warming these features become more distinct. Heating past the Br_2 multilayer desorption temperature ($T = 155$ K) reduces the 2.8 eV Br_2 feature and leaves the 2 eV loss as the only prominent feature in the gap. Heating to 190 K almost entirely removes the loss features associated with Br_2 and bithiophene whilst increasing the 2 eV loss feature and creating a new feature at 1.2 eV. Further heating past the 200 K Br desorption (not shown) reduces the 2 eV loss peak whilst the low loss energy intensity around 1.2 eV remains. Both TDS and UPS imply that the heating can be considered as 'dedoping'. For high doping concentrations EELS shows one gap state excitation feature prominent at 2 eV whereas for lower concentrations two excitations at 1.9 to 1.2 eV exist. This result has its correspondence

TABLE 1

Gap state excitation energies for thiophene oligomers in eV

					Reference
$6T^{+}$	1.59	0.84	$6T^{2+}$	1.24	12
$5T^{+}$	1.7	0.98			12
$4T^{+}$	1.92	1.16			12
$4T^{+}$	1.85	1.1	$4T^{2+}$	1.7	11
$2T^{+}$	2	1.2	$2T^{2+}$	2	
$2T^{-}$	2.7	1.5	$2T^{2-}$	2.3	3

both in optical absorption results for the p-doping of PT and its longer oligomers and the EELS results for the n-doping of bithiophene. For sexithiophene Fichou *et al.* [12] observed at low $FeCl_3$ concentrations absorptions at 1.59 and 0.84 eV which for increasing concentrations were replaced by a prominent absorption at 1.24 eV. With additional ESR evidence this was assigned to transformation from polaron to bipolaron-like states. Waragai and Hotta [11] obtained qualitatively similar spectra for dimethyl quarterthiophene on $NOPF_6$ doping with low dosing features at 1.85 and 1.1 eV transforming to a prominent feature at 1.71 eV for higher doping levels. The bithiophene EELS results clearly fit a trend for oligomer length as expressed in Table 1. Also displayed in the Table are the results obtained for the gap state losses obtained in the Cs dosing of bithiophene which models the n-doping situation [3]. The n-doping is similar to the p-doping for increasing dopant concentration in that the excitation spectra transforms from two to one major feature. There are however significant energy differences between the (di)cation and (di)anion cases: clearly one cannot expect that excitations to the empty gap states created on p-doping to have the same energy as excitations from the filled states of n-doping. Furthermore, despite the fact that on n-doping the gap states are clearly observed directly by UPS the relationship between their excitation and ionization energies is not obvious, as a result of significant electron–hole interaction energies. Thus despite optical absorption or EELS results giving information regarding the gap states one cannot simply infer the position of the molecular orbitals within the gap as has often been the case in the literature in the past. In summary: p-doping of bithiophene was achieved with Br_2 from the gas phase with the formation of a Br–bithiophene complex. This was evidenced by the emptying of the HOMO observed in UPS and in EELS by excitations to the gap states formed. In the latter a transformation on dedoping by heating from bipolaron to polaron-like states was inferred. Similar p-doping of biphenyl could not be achieved presumably due to the inhibition of Br_2 dissociation.

Acknowledgements

This work has been supported by the Austrian Fonds zur Förderung der Wissenshaftlichen Forschung. G. Leising of the Technical University Graz is thanked for supplying the sexiphenyl.

References

1 M. G. Ramsey, D. Steinmüller and F. P. Netzer, *Phys. Rev. B*, *42* (1990) 5902.
2 F. P. Netzer, M. G. Ramsey and D. Steinmüller, *Synth. Met.*, 41–43 (1991) 1343.
3 M. G. Ramsey, D. Steinmüller and F. P. Netzer, *Phys. Rev. B*, in press.
4 G. Horowitz, D. Fichou, X. Peng and F. Garnier, *J. Appl. Phys.*, *67* (1990) 528.
5 K. Seki, U. Karlsson, R. Engelhardt, E. E. Koch and W. Schmidt, *Chem. Phys.*, *91* (1984) 459.
6 J. L. Brédas, B. Themans, J. G. Fripiat, J. M. Andre and R. R. Chance, *Phys. Rev. B*, *29* (1984) 4491.
7 M. G. Ramsey, D. Steinmüller, F. P. Netzer and D. Leising, to be published.
8 M. G. Ramsey, D. Steinmüller, F. P. Netzer and D. R. Lloyd, *J. Chem. Phys.*, *97* (1992) 4489.
9 Y. Jugnet, G. Tourillon and Tran Minh Duc, *Phys. Rev. Lett.*, *56* (1986) 1862.
10 M. Lögdlund, R. Lazzaroni, S. Stafsström, W. R. Salaneck and J.-L. Brédas, *Phys. Rev. Lett.*, *63* (1989) 1841.
11 K. Waragai and S. Hotta, *Synth. Met.*, *41–43* (1991) 519.
12 K. Fichou, B. Xu, G. Horowitz and F. Garnier, *Synth. Met.*, *41–43* (1991) 463.

Stability of thick polypyrrole films electrogenerated on stainless steel

T. F. Otero, E. Angulo, C. Santamaria and J. Rodriquez
U.P.V., Facultad de Química, Departamento de Ciencia y Tecnología de Polímeros, Laboratorio de Electroquímica, P.O. Box 1072, 20080 San Sebastián (Spain)

Abstract

Thick polypyrrole films (>10 μm) were electrogenerated on steel electrodes, and their stability was checked through the variation of electrical charge stored in the film under different anodic potential limits and in different solvents. 0.1 M $LiClO_4$ acetonitrile, water and propylene carbonate solutions were used. Consecutive potential waves (square and triangular) were employed in the cycling. Water and acetonitrile solutions give higher charge storage efficacies (25–30 A h kg^{-1}) than propylene carbonate solutions (0.7 A h kg^{-1}). The stability is lower when the anodic potential is higher: polypyrrole films lose half of their charge storage efficacy after 250 cycles between -500 and 1000 mV versus SCE, after 1100 cycles between -600 and 700 mV and after 1800 cycles between -600 and 400 mV.

Introduction

Electrogenerated conducting polymers such as polypyrrole and polythiophene have been successfully applied in electronic devices with a wide doping range [1]. The oxidation (doping) of the conducting polymers takes place by generation of positive charges along the polymeric chains. Those charges are compensated by negative ions coming from the solution.

The doping process, carried out easily with the aid of electrochemical techniques [2, 3], takes place in conjunction with a degradation reaction promoting irreversible variations of the polymer properties [4, 5] (conductivity, charge storage ability, electrochromism, . . .). Doping conditions must be optimized in order to diminish the degradation process and increase the stability of those materials.

The aim of this work is to study the electrochemical degradation process taking place when the polypyrrole is consecutively cycled between a cathodic and an anodic potential in different solvents.

Experimental

Polypyrrole films were electrogenerated and checked in a one compartment electrochemical cell with the aid of an M273 PAR potentiostat-galvanostat connected to a PC-AT IBM microcomputer through an

0379-6779/93/$6.00

interface permitting the registration of consecutive chronoamperograms or voltammograms. As counter electrodes, two platinum sheets 4 cm^2 each were used in the generation and controls. The reference electrode was a saturated calomel electrode (SCE) (Crison Instruments) and was introduced directly into the electrochemical cell.

Pyrrole, anhydrous lithium perchlorate and propylene carbonate were from Jansen. Acetonitrile was Lab Scan HPLC grade. All these products were used without previous purification. Ultrapure water was obtained in our laboratory with a Milli Q Reagent water system fed by a Milli RO4 water purification system. All the studies were made in an inert atmosphere by bubbling Ar continuously in the electrochemical cell.

Generation of films

The polypyrrole films checked in this work were electrogenerated on stainless steel electrodes (AISI 304) of 7 cm^2 surface area. The solution consisted of pyrrole 0.25 M and $LiClO_4$ 0.1 M in acetonitrile. Polymerizations were carried out by application of consecutive square waves of potential between −100 (3 s) and 800 (10 s) mV. This method was optimized to obtain good quality films on steel electrodes [6]: electrogeneration takes place by alternating between oxidation-polymerization and polymer oxidation, in the anodic part of the wave, and polymer reduction in the cathodic part of the wave. After polymerizations, the coated electrode was weighed in a microbalance (clean electrodes were weighed before polymerization).

Results and discussion

Polypyrrole films electrogenerated in acetonitrile solutions were checked in 0.1 M $LiClO_4$ solutions in different solvents (acetonitrile, water and propylene carbonate) by consecutive potential sweeps and potential steps. The general shape of the chronoamperograms of control can be observed in Fig. 1. In this Figure some of the chronoamperograms obtained from a polypyrrole film of 0.0131 g weight, 7 cm^2 area on a stainless steel electrode submitted to consecutive square waves of potential between −700 and 1100 mV versus SCE in a 0.1 M $LiClO_4$ acetonitrile solution are depicted.

In the anodic part of the wave, a 'nucleation' like process takes place [4, 7, 8], as can be deduced by the increase of the intensity in the first seconds. This process is related to the polymeric network opening as the oxidation occurs due to the entrance of counterions, the ClO_4^- anions, in the polymer [4, 9]. At the beginning of the oxidation, the polymeric network is closed and the process is very slow. As the anions enter into the film, the network is opened and the oxidation process is more and more easy, with the subsequent current increasing. In a few seconds, oxidation is starting to finish, and the current decays.

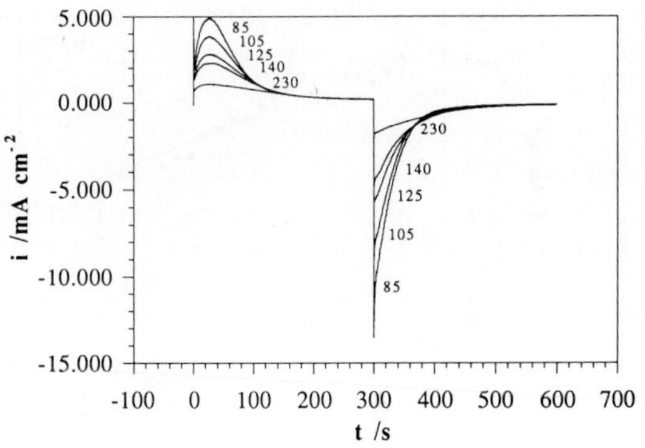

Fig.1 . Chronoamperograms obtained after cycling 85, 105, 125, 140 and 230 times a polypyrrole film 0.0133 g weight by consecutive square waves of potential between −700 (300 s) and 1100 (300 s) mV vs. SCE in a 0.1 M $LiClO_4$ acetonitrile solution.

As the number of cycles is increased the anodic and cathodic intensities and the charges involved in this oxidation/reduction process (obtained by integration of the chronoamperograms) become smaller. This loss in redox capacity with cycling is the variable studied in this work. During oxidation positive charges are stored along the polymeric molecules. Thus, the charge storage efficacy (stored charge per unit of polymer weight) was obtained for different electric potentials of polarization and solvents from the charge involved on the anodic part of the square wave and the weight of the polymer film measured after generation.

In Fig. 2 the evolution of this charge storage efficacy with the number of polarization cycles for experiments carried out in different media is shown. They correspond to consecutive potential square waves applied in acetonitrile (a), water (b) and propylene carbonate (c) solutions of 0.1 M $LiClO_4$, between −500 and 1000 mV versus SCE, with a period of 600 s. The shape of the three curves is very similar: the charge storage efficacy is smaller as the number of cycles increases. In the case of the propylene carbonate, the values of the charge are lower than in acetonitrile or water (0.7 versus 25 A h kg^{-1} for the first cycles). This fact can be due to a lower participation of the polymer in the oxidation/reduction process in propylene carbonate solution, under those conditions of polarization.

The stability of the polypyrrole films can be followed by these experiments through a parameter quantifying the decay in the charge storage efficacy, but taking into account that the values of these charges are different for each experiment. For example, a loss of 0.3 A h kg^{-1} in 200 cycles is an appreciable quantity in the case of propylene carbonate but it is negligible in acetonitrile or water. However, the number of cycles necessary to lose half of the initial charge storage efficacy ($Nc_{1/2}$) is a parameter taking this into account.

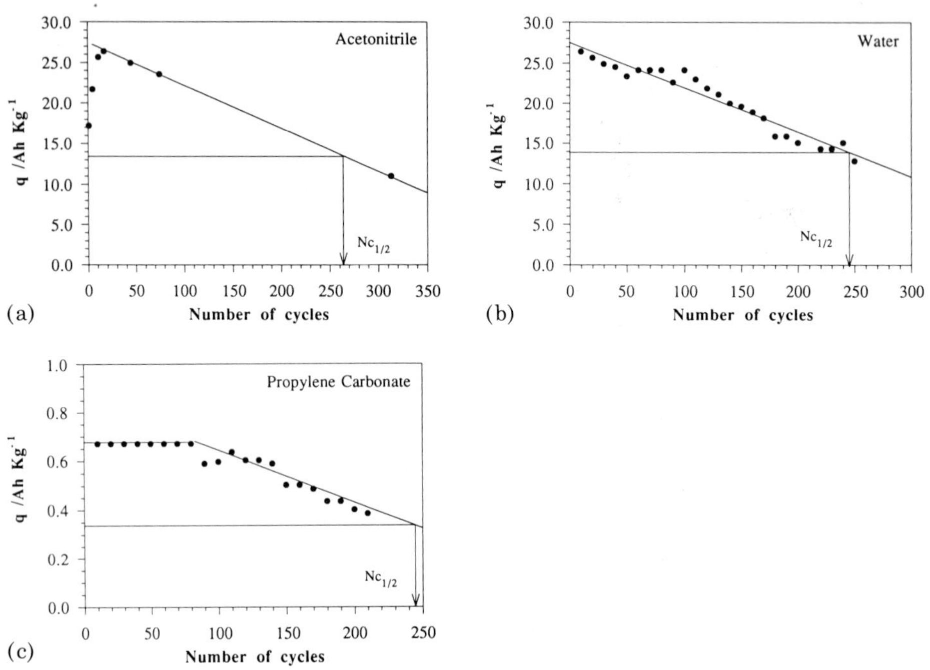

Fig. 2. Variation of the charge storage efficacy with the number of cycles for the application of a square wave of potential (600 s period) between −500 and 1000 mV vs. SCE using three polypyrrole films in 0.1 M $LiClO_4$ solutions: polypyrrole 0.0105 g weight in acetonitrile (a); polypyrrole 0.0184 g weight in water (b); and polypyrrole 0.0207 g weight in propylene carbonate (c). Charge storage efficacy values are obtained by dividing the anodic charge of each cycle by the weight of the polymer film.

In all the cases of Fig. 2 $Nc_{1/2}$ is around 250 cycles; the solvent seems to affect the value of the charge storage efficacy, but not the stability of the films. In these three solvents (acetonitrile, water and propylene carbonate) and in the studied potential range (−500 to 1000 mV versus SCE), 250 cycles (~42 h) are enough to lose half of the charge storage efficacy in polypyrrole films.

The loss of the redox activity of the polymer is related to the polarization at higher anodic potentials than the oxidation potential of the polypyrrole [4, 10]. In Fig. 3, another technique was employed: consecutive triangular waves of potential between −600 mV versus SCE and 700 (a) or 400 (b) mV at 10 mV s^{-1}. The solution consisted of 0.1 M $LiClO_4$ in acetonitrile. Under those conditions, the decrease of the charge storage efficacy with the number of cycles was lower than when square waves of potential were applied (Fig. 2). The reason can be related to the higher anodic potential used in square waves.

For consecutive potential sweeps at an anodic potential of 700 mV (Fig. 3(a)), $Nc_{1/2}$ is around 1100 cycles. When the anodic limit was 400 mV, $Nc_{1/2}$ is near 1800 cycles. In the last case, a stabilization at 1.0 A h kg^{-1}

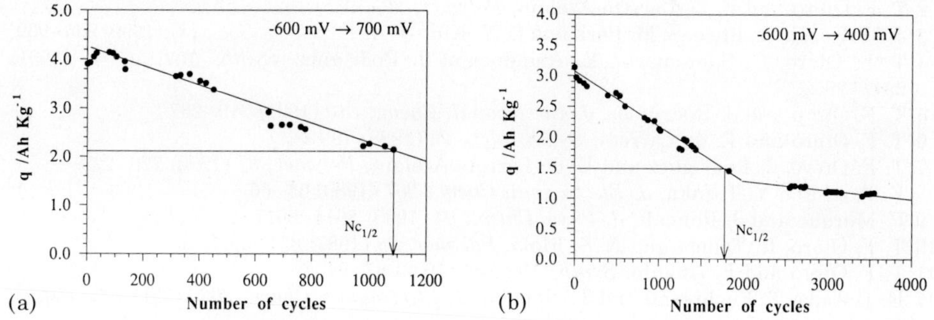

(a) (b)

Fig. 3. Variation of the charge storage efficacy with the number of cycles for the application of a triangular wave of potential (10 mV s^{-1}) in 0.1 M $LiClO_4$ acetonitrile solution, using two polypyrrole films of 0.012 g weight, between −600 mV vs. SCE and 700 mV (a); 400 mV (b). Charge storage efficacy values are obtained as above by dividing the anodic charge of each cycle by the weight of the polymer film.

above 2000 cycles is observed. The charge storage efficacy values in Fig. 3 are smaller than in Fig. 2(a) since the anodic potentials are also smaller.

All these results point to the presence of a polymeric degradation parallel to the oxidation/reduction process taking place in polypyrrole by potential cycling, being more important at higher anodic potentials. This means that the polypyrrole film oxidation occurs at the same time as a different anodic discharge [11, 12], promoting the 'degradation' of the polymer and the loss of its redox properties. This anodic process is not related to the solvent, because when the solvent varies, the degradation process does not change. Its relative influence varies with the anodic potential of polarization.

The loss in charge storage efficacy can arise from any alteration of the conjugated structure of the polymeric chains, making the polymer more and more non-oxidizable at each cycle. It can be due, as well, to a reticulation process promoting an irreversible closing of the polymer network, making the entry of the ion more difficult.

Acknowledgement

This work has been supported by the Commission of the European Communities (NAPOLEO program, BRITE-EURAM 0-148-C).

References

1 J. R. Ellis, in T. A. Skotheim (ed.), *Handbookf of Conducting Polymers*, Marcel Dekker, New York, 1986, pp. 489–505.

2 T. F. Otero and E. de Laretta-Azelain, *Polymer*, *29* (1988) 1522–1527.
3 J. M. Ko, H. W. Rhee, S. M. Park and C. Y. Kim, *J. Electrochem. Soc.*, *137* (1990) 905–909.
4 T. F. Otero, C. Santamaría, E. Angulo and J. Rodríguez, *Synth. Met.*, *41–43* (1991) 2947–2952.
5 T. F. Otero and J. Rodríguez, *J. Electroanal. Chem.*, *310* (1991) 219–237.
6 T. F. Otero and E. de Larreta, *Synth. Met.*, *26* (1988) 79–88.
7 T. F. Otero, J. Rodríguez and E. de Larreta-Azelain, *Polymer*, *31* (1990) 220–222.
8 K. Aoki and Y. Tezuka, *J. Electroanal. Chem.*, *267* (1989) 55–66.
9 P. Marque and J. Roncali, *J. Phys. Chem.*, *94* (1990) 8614–8617.
10 T. F. Otero, R. Tejada and A. S. Elola, *Polymer*, *28* (1987) 651–658.
11 T. F. Otero and E. Angulo, *Synth. Met.*, *51* (1992) 87–94.
12 H. Harada, T. Fuchigami and T. Nonaka, *J. Electroanal. Chem.*, *303* (1991) 139–150.

Synthetic Metals, *54* (1993) 223–228

Polymerization of pyrrole by laser induced photolysis of onium salts

W. Behnck and J. Bargon*
Institute of Physical Chemistry, University of Bonn, Bonn (Germany)

Abstract

Reactivity of photolyzed onium salts in thin layers of conventional polymer towards pyrrole polymerization was investigated. Introducing diphenyliodonium salts into matrices of common polymer without special functional groups, i.e. polystyrene, gives a transparent polymer composite sensitive to light in the near UV region. When irradiated with a KrF excimer laser or a high pressure mercury lamp in the presence of pyrrole vapor a black–brown coloured polymer results. The reactivity of polymerization depends largely on the anions (Cl, Br, I, BF_4, PF_6) of the diphenyliodonium salts used; the differences are outlined by spectroscopic methods. The resulting photoproduct is identified as a black polypyrrole with the appropriated anion.

Introduction

Two- and three-dimensional patterning of polymers plays an important role in modern electronics [1–3]. Among the large variety of polymeric materials oligomers and polymers of pyrrole and thiophene derivatives are of considerable interest because of their unusual electrical and non-linear optical properties [4–7].

It is known that these monomers can be polymerized in thin films either electrochemically or chemically by protonic or cationic support [8–11].

Earlier investigations of the acid catalyzed polymerization of pyrrole proved the formation of oligomeric or low molecular weight polymers [11–13]. We have already reported the 'oxopolymerization' of pyrrole by $FeCl_3$ [14, 15].

The aim of this paper is to show that pyrrole can be polymerized in polymer matrices of conventional, insulating polymers by UV irradiation. The basic idea is to generate polymerization-initiating species by laser induced photolysis of onium salts thereby polymerizing pyrrole in polymer matrices.

It is well known that onium salts can be used as photosensitive bivalent initiators both for free radical and cationic polymerization initia-

*Author to whom correspondence should be addressed.

0379-6779/93/$6.00

tion [16, 17]; typical examples are diarylhalonium (Ar_2Cl^+, Ar_2Br^+, Ar_2I^+), triarylsulfonium (Ar_3S^+) and triarylselenomium (Ar_3Se^+) salts.

On irradiation-induced decomposition (=photolysis) of diphenyliodonium salts cationic species and Brönsted acids are generated which can initiate polymerization of the monomer along a cationic pathway.

In this study it is shown that photolyzed diphenyliodonium salts can also start the polymerization of pyrrole in a polystyrene matrix. The effect can be applied in a variety of ways and it may be visualized by generating patterned black structures in a transparent polymer.

Experimental

Several diphenyliodonium salts were under investigation:

- $C_{12}H_{10}I^+Cl^-$, CARN [1483-72-3], purity $\gg$ 98%, m.p. 228–229 °C
- $C_{12}H_{10}I^+Br^-$, CARN [1483-73-4], purity 98+%, m.p. 240–247 °C
- $C_{12}H_{10}I^+PF_6^-$, CARN [58109-40-3], m.p. 138–139.5 °C
- $C_{12}H_{10}I^+BF_4^-$, CARN [313-39-3], m.p. 134–136 °C

Except for the last one, all salts were used without further purification. $C_{12}H_{10}I^+BF_4^-$ was synthesized by a metathesis as outlined in the literature [18].

The photolysis experiments were carried out with 5×10^{-3} molar solutions in tetrahydrofuran and acetonitrile. These solvents were of p.a. quality and were used without further purification.

Polymeric thin films of about 3 μm were prepared by spincoating 2 ml of a 10 wt.% solution of common polystyrene in tetrahydrofuran containing the desired onium salts as a 0.01 molar solution in the same solvent (THF).

Irradiation of the films was performed through a UV-transparent Suprasil-glass carrier from the back with either a KrF excimer laser (248 nm, typical intensity per pulse = 120 mJ/cm^2) or using a 1000 W Hg (Xe) lamp (typical efficiency = 10.8 mW/cm^2 at 252 nm). The patterning experiments were conducted in a specially designed irradiation apparatus filled with monomer vapor and the samples were irradiated through a UV-transparent window.

All measurements in the UV–Vis range were performed using a double-beam spectrophotometer (Shimadzu UV-1200 A) in 1 cm cells. The IR spectra were taken on a double-beam spectrophotometer (Philips/Pye Unicam SP1000) using the common sample techniques.

Results

The decomposition of the diphenyliodonium ions by irradiation at 254 nm is described in the literature as a combined electron-transfer and homolysis reaction [19]

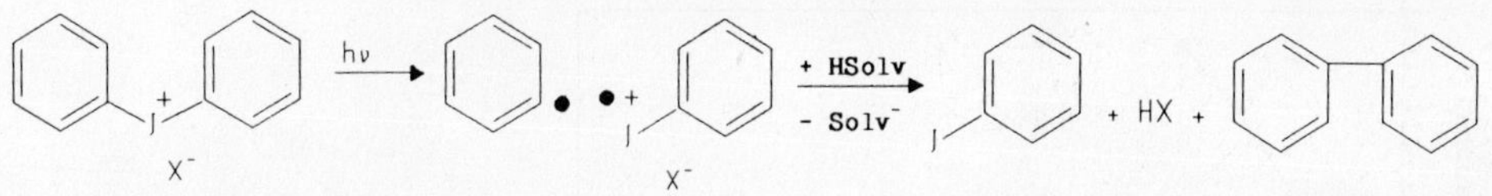

It is remarkable that all salts show a very low solubility in appropiate solvents for polymers and function even at concentrations of less than 5×10^{-3} mol/l as good polymerization initiators.

The photolysis of these salts in various solvents has been investigated. The following outlines the UV–Vis spectrum of the photolysis of diphenyliodonium chloride in tetrahydrofuran as a typical example. Similar spectra have been obtained by photolysis of diphenyliodonium bromide and diphenyliodonium hexafluorophosphate. Although the same cations are used in all cases there is no significant difference in location of the maximum of the absorption bands. However, there is a difference in the extent of absorption.

According to the expected decomposition scheme there should be a decay of the absorption band at 240 nm related to the undecomposed diphenyliodonium salt and an increase in the absorption bands at 295 and 365 nm due to photolysis products (Fig. 1).

The identification and characterization of the reaction products absorbing at 295 and 365 nm is still under investigation.

A comparison of the tested salts with different anions results in the statement that there is a significant difference in photolytic yield between

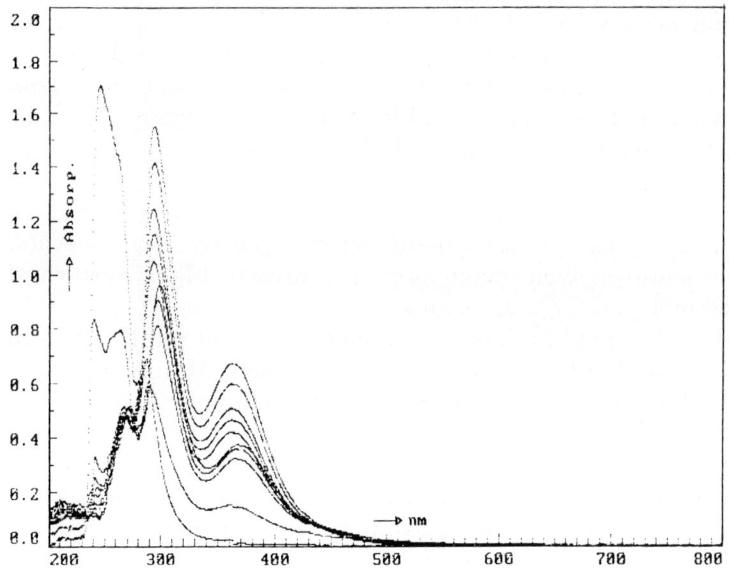

Fig. 1. Photolysis of diphenyliodonium chloride in tetrahydrofuran. UV–Vis spectra recorded after increasing time of irradiation.

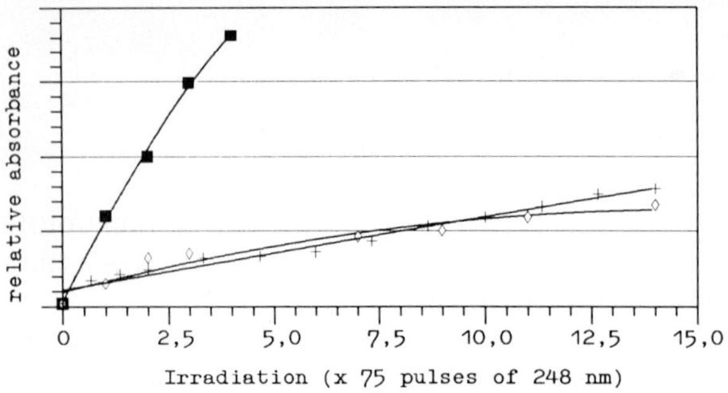

Fig. 2. Photolysis of the diphenyliodonium salts in THF as the result of KrF laser irradiation. Absorption at 365 nm of the unique photoproduct as a function of number of laser pulses: ■ diphenyliodonium hexafluorophosphate, ◇ diphenyliodonium chloride, + diphenyliodonium bromide.

onium salts with nucleophilic halogenides (Cl, Br, etc.) and non-nucleophilic complex anions (PF_6, BF_4, etc.).

Figure 2 shows a comparison of the onium salts. Although the absorption at 240 nm due to the undecomposed salts decreases for nucleophilic halogenides much faster with growing absorbed energy than for complex anions, the rate of formation of photoproduct is much faster in the case of complex anions.

In order to evaluate the feasibility of initiating the conversion of pyrrole into its electrically conducting polymer in the photoexposed regions, the following experiments were carried out.

Various onium salts were incorporated into a thin and transparent polystyrene matrix and irradiated in the presence of pyrrole vapor through a UV-transparent mask. Thereby these films become more and more black–brown coloured and insoluble in common organic liquids. This finding is common to all anions used. Again, the changes are more obvious and faster when complex anions like PF_6^- are used.

This effect can be used and visualized in the form of black patterns in otherwise transparent polymer films using laser exposure. This method yields a sharp and well-resolved image as a dark brown–black pattern of the mask down to micrometer dimensions.

In agreement with our photolysis experiments we find a difference in the resulting contrast depending on the anions used. Halogen anions produce a weak and less contrasted image relative to complex anions like hexafluorophosphate. These latter salts gave images of high contrast as brown to black patterns.

In order to identify the photoproducts we investigated the system pyrrole/diphenyliodonium hexafluorophosphate in more detail. When pyrrole and diphenyliodonium hexafluorophosphate are irradiated together, a black cloud appears which changes the colour of the solution into deep black within 5 min. An aqueous mixture of the resulting solu-

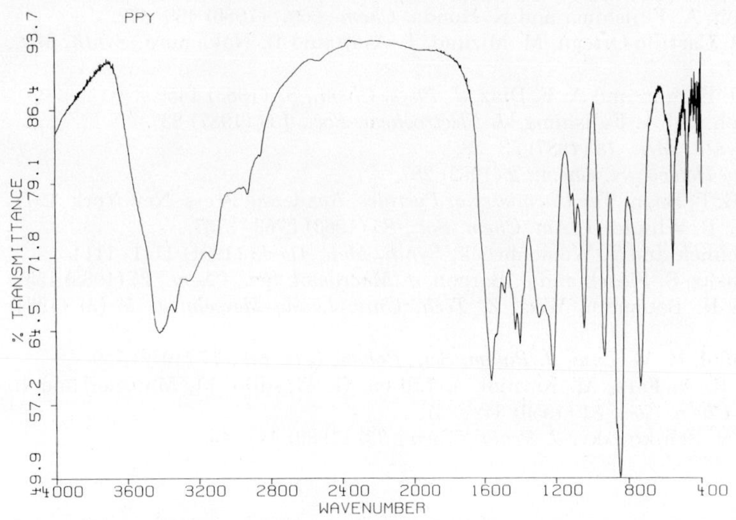

Fig. 3. FT-IR spectrum of the black precipitate as photoproduct.

tion is acidic (pH ~ 3); the methanolic solution shows in addition to the absorption at 250 nm a second weaker absorption at 550 nm. First TLC separations show more than four different products. Separation by HPLC yields the expected reaction byproducts: iodobenzene and biphenyl. Upon treatment with diisopropylether a black flocky powder precipitates; it could be isolated and identified as pyrrole black: it is unburnable, insoluble and, according to first chemical analysis, contains halogenide. The FT-IR spectrum (Fig. 3) is very similar to that of the pyrrole monomer. XRD and electron microscopical studies proved it to be an amorphous, partial conducting material. Further investigations are still in progress.

Acknowledgements

We thank Dr R. Baumann, Leipzig, for helpful discussions. This work has been supported by the Fonds der Chemischen Industrie.

References

1 A. Craievich, G. G. B. de Souza and V. Baranauskas (eds.), *Microlithography: High Integration in Microelectronics (Proc. First Workshop, Rio de Janeiro, Brazil, Aug. 28–Sept. 1, 1989)*, World Press, Singapore, 1990.
2 A. D. Ketley and F. Magnotta, *J. Radiat. Curing, 9* (1982) 11–18.
3 J. Bargon, *Methods and Materials in Microelectronic Technology*, Plenum, New York, 1984, pp. 181–241.
4 R. J. Waltman and J. Bargon, *Can. J. Chem., 64* (1986) 76–95.
5 J.-M. Kern and J.-P. Sauvage, *J. Chem. Soc., Chem Commun.*, (1989) 657–658.

6 M. Okano, K. Itoh, A. Fujishima and K. Honda, *Chem. Lett.*, (1986) 469–472.
7 M. Inoue, Ma. M. Castillo-Ortega, M. Mizuno, T. Asaji and D. Nakamura, *Synth. Met.*, *33* (1989) 355–364.
8 R. J. Waltman, J. Bargon and A. F. Diaz, *J. Phys. Chem.*, *87* (1983) 1459.
9 M. Okano, K. Itoh and A. Fujishima, *J. Electrochem. Soc.*, *134* (1987) 837.
10 M. Takakubo, *Synth. Met.*, *18* (1987) 53–58.
11 G. F. Smith, *Adv. Heterocycl. Chem.*, *2* (1963) 287.
12 R. A. Jones and G. P. Bean, *The Chemistry of Pyrroles*, Academic Press, New York, 1977.
13 Y. Chiang and E. B. Whiple, *J. Am. Chem. Soc.*, *85* (1963) 2763–2767.
14 J. Bargon, W. Behnck and T. Weidenbrück, *Synth. Met.*, *41–43* (1991) 1111–1114.
15 T. Ueno, H.-D. Arntz, S. Flesch and J. Bargon, *J. Macromol. Sci., Chem.*, *25* (1988) 1557.
16 H.-J. Timpe and H. Baumann, *Wiss. Z. Tech. Univ. Leuna-Merseburg*, *26* (3) (1984) 439–448.
17 J. V. Crivello and J. H. W. Lam, *J. Polym. Sci., Polym. Lett. Ed.*, *17* (1979) 759.
18 F. M. Beringer, R. A. Falk, M. Karniol, J. Lillien, G. Masullo, M. Mausner and E. Sommer, *J. Am. Chem. Soc.*, *81* (1959) 342–351.
19 H.-J. Timpe and V. Schikowsky, *J. Prakt. Chem.*, *331* (1989) 447–460.

Langmuir–Blodgett and vacuum sublimed films of terbium bisphthalocyanine

J. Souto* and J. A. de Saja
Física de la Materia Condensada, Universidad de Valladolid, Valladolid 47011 (Spain)

R. Aroca and M. L. Rodríguez
Department of Chemistry and Biochemistry, University of Windsor, Windsor, Ont., N9B 3P4 (Canada)

Abstract

Langmuir–Blodgett monolayers and evaporated thin solid films of terbium bisphthalocyanine ($TbPc_2$) have been prepared and characterized using Raman scattering, surface enhanced resonance Raman scattering, IR and UV–Vis absorption spectroscopy. The surface pressure/molecular area isotherm, measured for the floating Langmuir layer, was consistent with an edge-on molecular orientation of the macrocycle rings on the water subphase. Vibrational characterization and molecular organization in Langmuir–Blodgett layers and evaporated thin solid films are discussed.

Introduction

The unique electrochromic and semiconductive properties of lanthanide bisphthalocyanines ($LnPc_2$) have been the subject of a number of recent reports [1–3]. Terbium bisphthalocyanine is a member of this series of molecular semiconductors that has attracted much attention due to their potential applications. The starting material presented a good thermal stability allowing the preparation of evaporated films. However, several devices built with the Langmuir–Blodgett (LB) technique have demonstrated an enhanced performance when compared to those based on thin solid films obtained by high vacuum deposition [3, 4]. The formation of stable LB films of several lanthanide bisphthalocyanines has been recently reported [2, 5, 6]. The aim of the present work was to prepare Langmuir–Blodgett (LB) monolayers of $TbPc_2$, and to carry out the spectroscopic characterization of LB and evaporated films. The work will continue testing possible applications in the field of chemical sensing and electrochromic displays.

Experimental

Terbium bisphthalocyanine was synthesized by the method described by Kirin *et al.* [7]. The equipment used for thermal evaporation and LB deposition

*Author to whom correspondence should be addressed.

0379-6779/93/$6.00

has been described previously [6]. A 200 nm thin solid film of $TbPc_2$ was sublimed onto a KBr disk and Corning 7059 glass slides. LB monolayers of $TbPc_2$ were prepared at 10 °C and transferred to glass slides cleaned with chromosulfuric acid and treated with NaOH [8]. One LB monolayer was also transferred to glass slides coated with a gold island film. Monolayers were transferred by Z-type deposition at a constant speed of 3 mm/min, and at a surface pressure of 15 mN/m. Au island films were deposited by evaporation of 4 nm of gold onto glass slides kept at 200 °C. A Spectra Physics model 2020 Kr^+ ion laser was used to obtain Raman spectra. Raman shifts were measured with a Spex-1403 double spectrometer.

Results and discussion

Characterization of LB monolayers

The molecules were spread from a 5×10^{-5} M solution in toluene onto a triple distilled and deionized water subphase. The temperature of the subphase was a key factor to obtaining reproducible isotherms. Reproducible surface pressure/area isotherms were obtained at 10 °C. The isotherm is shown in Fig. 1. The hysteresis obtained on relaxation of the monolayer and a subsequent compression of the layer are also displayed. The relaxation and following compression of the monolayer revealed that some residual packing remained after the surface pressure went down to zero. This caused a slight delay in the rising of the curve when compressing the layer for a second time. The area per molecule obtained by linear extrapolation of the low compressibility region of the isotherm to 0 mN/m was 78 Å^2. Similar molecular areas on LB films have been obtained for $LuPc_2$, $YbPc_2$ and $PrPc_2$

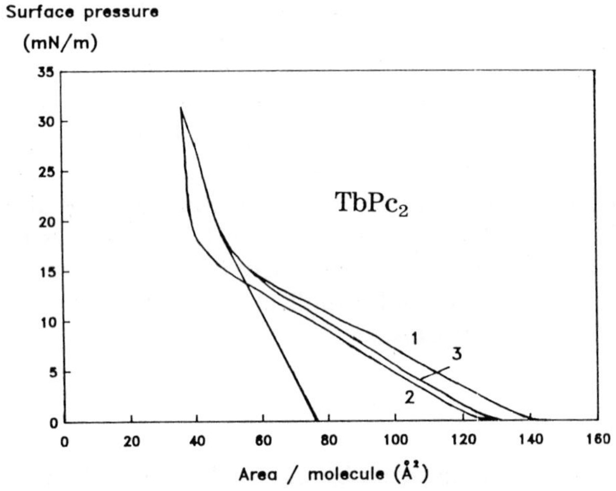

Fig. 1. Surface pressure vs. molecular area isotherm for terbium bisphthalocyanine (1), hysteresis curve on relaxation of the compressed layer (2) and subsequent compression of the floating layer (3).

[2, 5, 6]. The estimated values from space-filling models for edge-on and face-on configurations are about 65 and 170 Å^2, respectively. Therefore, an edge-on (tilted) configuration of the macrocycle in the floating monolayer with a tilt angle θ of about 5° would be compatible with the limiting area. LB monolayers were transferred to glass substrates with a near unity transfer ratio.

Electronic spectra

The electronic spectrum of $TbPc_2$ in solution is in correspondence with what has been reported for lanthanide bisphthalocyanines [9, 10]. It exhibits the characteristic Q-band with an absorption maximum at 668 nm and vibrational sidebands at 602 and 584 nm. The starting green material presented a weak band at 458 nm, which is a characteristic absorption of the Pc radical [11], and a Soret band with two maxima at 322 and 280 nm.

In the spectrum of two monolayers of $TbPc_2$ (one on each side of the substrate) the maximum of the Q-band is located at 658 nm, the weak band of the Pc radical has also shifted to 476 nm, and the Soret band is the strongest one. Figure 2 shows the spectra for LB monolayers, an LB film and for the evaporated film. Polarized light was used to probe organization in the monolayer. The absorbance at 658 nm is greater when the electric field vector is polarized parallel to the dipping direction than when it is perpendicular. The absorbance ratio (parallel divided by perpendicular) at this wavelength is almost 1.5. The latter result clearly indicated that the $TbPc_2$ molecules were not face-on (lying flat on the substrate), and that a preferential, though not perfect, ordering of the molecules with the planes of macrocycles parallel to the dipping direction was present [12]. The spectrum

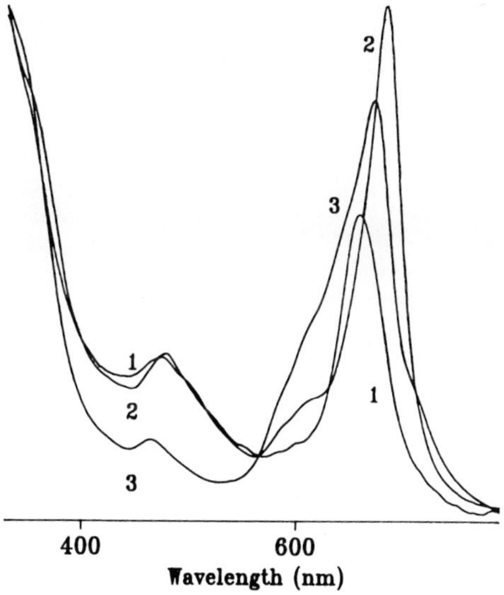

Fig. 2. Electronic spectra of $TbPc_2$: two LB monolayers (1), an LB multilayer (2) and a 200 nm evaporated film (3).

of a multilayer film, with five monolayers on each side of the substrate, differs from that of the monolayer. The maxima of the absorptions are at 682, 474 and 326 nm and the dichroism is no longer apparent. Hence, the partial ordering in the monolayer is not retained in the LB film.

The evaporated film shows a broadened Q-band centred at 672 nm, a Soret band with maxima at 280 and 326 nm, and a very weak band at 464 nm. When this spectrum is compared with those obtained for multilayers of the same compound or thin solid films of the late members of the $LnPc_2$ series [5, 13], it can be noticed that the relative intensity of the higher energy side of the Q-band has increased, and that the intensity of the 458 nm band has decreased. There is an intense absorption in the 640 nm region in the solid state spectrum of metal-free phthalocyanine (H_2Pc), but no bands appear around 450 nm. One possible rationalization for the observed spectral changes in the $TbPc_2$ spectra could be the partial decomposition of this compound to give H_2Pc. The explanation was indirectly supported by the IR results.

Raman spectra

Four laser lines were used for the Raman characterization of the samples: the 647.1 and 676 nm lines of the krypton laser, which were in resonance with the Q-band, and the 488 and 514.5 nm lines of the argon laser, which were off-resonance. The Raman scattering (RS) of the evaporated film obtained with the 488 nm line, the resonance Raman scattering (RRS) excited with the 647.1 nm line, and the surface enhanced resonance Raman scattering (SERRS) of a monolayer on a gold island film, excited with the 676 nm line are shown in Fig. 3. The observed fundamental frequencies, and the empirical

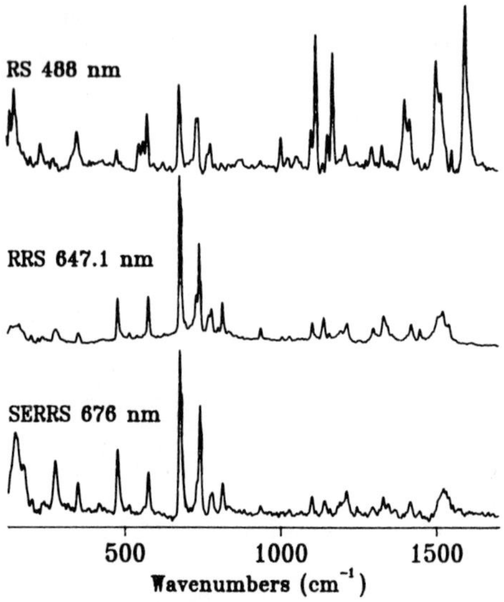

Fig. 3. RS and RRS of a $TbPc_2$ evaporated film on KBr, and SERRS of a monolayer on a 4 nm gold island film.

TABLE 1

Vibrational fundamental frequencies of an evaporated $TbPc_2$ film (numbers in parentheses are relative intensities)

RRS (647.1 nm)	Infrared	RS (488 nm)	Interpretation
140 (12)		140 (22)	
		153 (35)	
167 (13) p			
206 (7)		205 (19)	
240 (5)		238 (21)	metal–N stretch
283 (11) p			
354 (10) p		353 (6)	
479 (29)		480 (13)	
516 (6)			
		549 (20)	
576 (28) p		576 (37)	benzene radial
		627 (8)	
678 (100) p	678 (8)	678 (50)	Pc breathing
694 (9)	698 (6)		
731 (26) p	727 (100)	735 (35)	C–H wag
741 (57)	742 (14)	741 (36)	C–H wag
	757 (8)		Pc ring
772 (13)		771 (15)	
779 (14)	776 (11)	781 (20)	Pc ring
801 (7) p			
816 (24) p	811 (8)	815 (7)	
	884 (15)		C–H wag
938 (11)		942 (8)	benzene ring
1005 (6)	1003 (13)	1006 (23)	C–H bending
1029 (6)		1026 (12)	
		1032 (11)	
	1061 (10)	1054 (12)	C–H bending
	1073 (11)		C–H bending
1103 (12)		1103 (27)	C–H bending
	1114 (55)	1118 (83)	C–H bending
1138 (14)		1142 (10)	pyrrole breathing
1157 (6) p	1159 (6)	1156 (28)	C–H bending
		1173 (79)	C–H bending
1194 (9)			
1216 (14)		1215 (20)	
	1277 (5)		C–H bending
1299 (9)		1300 (18)	C–H bending
	1321 (63)		isoindole stretch
1333 (16) p	1329 (28)	1333 (19)	pyrrole stretch
1344 (11)			pyrrole stretch
	1366 (11)		isoindole stretch
	1407 (3)	1404 (46)	isoindole stretch
1419 (13)	1422 (5)	1420 (46)	isoindole stretch
	1438 (7)		
1447 (9)	1448 (21)	1447 (11)	isoindole stretch
		1470 (8)	
	1482 (7)		
1504 (13)	1502 (12)	1504 (66)	pyrrole stretch
1520 (18)	1525 (5)	1519 (47)	C=N aza group
		1555 (16)	
	1594 (4)	1595 (100)	benzene stretch
	1608 (7)		benzene stretch

vibrational assignment of the characteristic modes are listed in Table 1 [14]. The SERRS spectrum of the monolayer and the RRS of a multilayer correspond to the RRS spectrum of the evaporated film, with only slight differences in the relative intensity pattern.

Organization in the evaporated film and LB multilayers was probed by recording Raman spectra with different polarizations of the electric field vector of the incident and scattered light. A compressed pellet of $TbPc_2$ in a KBr matrix, where a random orientation of microcrystals is expected, was the reference. Depolarization ratios were measured by comparison between the SS (S-incident S-collected) and the SP (S-incident P-collected) spectra. Raman bands with a low SP/SS value in the pellet, are marked p in Table 1, and correspond to the totally symmetric modes. The spectra of the evaporated film and the LB film were very similar to those of the pellet, and the formation of microcrystals in both films is a possibility.

Infrared spectra

The middle IR spectrum of the evaporated film was studied to complete the vibrational characterization of $TbPc_2$. The observed fundamental frequencies are also listed in Table 1. The mutual exclusion rule for the Raman and IR bands is not observed, and thus the most likely molecular symmetry could be C_{4v}. A weak band was detected at about 3280 cm^{-1} in the spectrum of the film, but not in the pellet. This band has been attributed to the N–H stretching vibration in metal-free phthalocyanine [15]. The observation of the N–H vibration supports the assumption about the partial decomposition of $TbPc_2$ during evaporation in the vacuum chamber. The spectra of the KBr pellet and the evaporated film were otherwise very similar, in agreement with a random orientation of the microcrystallites in the film.

Conclusions

The preparation of Langmuir–Blodgett monolayers of $TbPc_2$ is reported here for the first time. Stable Langmuir monolayers were obtained by lowering the temperature of the subphase. Vibrational spectra of $TbPc_2$ in KBr pellet and evaporated films were recorded. The SERRS spectrum of a Langmuir–Blodgett layer on Au island films and RRS spectra of vacuum sublimed films of $TbPc_2$ were recorded. The assignment of the vibrational fundamentals has been attempted. It was concluded that LB films and evaporated films are most likely formed by randomly oriented microcrystallites.

Acknowledgements

Financial assistance from CICYT (MAT91–0303) and NATO scientific affairs division (Grant CGR 900582) is gratefully acknowledged.

References

1 P. Turek, P. Petit, J.-J. André, J. Simon, R. Even, B. Boudjema, G. Guillaud and M. Maitrot, *J. Am. Chem. Soc.*, *109* (1987) 5119.

2 M. Petty, D. R. Lovett, J. M. O'Connor and J. Silver, *Thin Solid Films*, *179* (1989) 387.

3 Y. Liu, K. Shigehara, M. Hara and A. Yamada, *J. Am. Chem. Soc.*, *113* (1991) 440.

4 W. R. Barger, H. Woltjen, A. W. Snow, J. Lint and N. L. Jarvis, in D. Schuetzle and R. Hammerle (eds.), *Fundamentals and Applications of Chemical Sensors*, American Chemical Society, Washington, DC, 1986, p. 155.

5 R. E. Clavijo, D. Battisti, R. Aroca, G. J. Kovacs and C. A. Jennings, *Langmuir*, *8* (1992) 113.

6 J. Souto, L. Tomilova, R. Aroca and J. A. de Saja, *Langmuir*, *8* (1992) 942.

7 L. S. Kirin, P. N. Moskalev and Yu. A. Makashev, *Zh. Neorg. Khim.*, *10* (1965) 1951.

8 H. Kuhn, D. Mobius and H. Bucher, in A. Weissberger and B. Rositer (eds.), *Physical Methods in Chemistry*, Wiley, New York, 1972, p. 577.

9 P. N. Moskalev and N. I. Alimova, *Russ. J. Inorg. Chem. (Engl. Transl.)*, *20* (1975) 1474.

10 D. Markovitsi, T. Tran-Thi, R. Even and J. Simon, *Chem. Phys. Lett.*, *137* (1987) 107.

11 T. Tran-Thi, D. Markovitsi, R. Even and J. Simon, *Chem. Phys. Lett.*, *139* (1987) 107.

12 D. W. Kalina and S. W. Crane, *Thin Solid Films*, *134* (1985) 109.

13 A. N. Sidorov and P. N. Moskalev, *Russ. J. Phys. Chem. (Engl. Transl.)*, *62* (1988) 1569.

14 R. Aroca, R. E. Clavijo, C. A. Jennings, G. J. Kovacs, J. M. Duff and R. O. Loutfy, *Spectrochim. Acta, Part A*, *45* (1989) 957.

15 M. P. Sammes, *J. Chem. Soc., Perkin Trans. II*, (1972) 160.

Synthetic Metals, 54 (1993) 237–241

Femtosecond dynamics, photoexcitation and ESR spectra of MX chain solids

J. Tinka Gammel

Materials Research Branch, Naval Command, Control and Ocean Surveillance Centre, RDT & E Division, San Diego, CA 92152-5000 (USA)

A. Saxena and A. R. Bishop

Theoretical Division and Centers for Nonlinear Studies and Materials Science, Los Alamos National Laboratory, Los Alamos, NM 87545 (USA)

I. Batistić

Institute of Physics of the University, 41001 Zagreb (Croatia)

Abstract

Non-linear adiabatic dynamics and ESR spectra associated with non-linear excitations in MX chain (M = Pt; X = Cl, Br, I) materials are numerically studied within a discrete, 3/4-filled, two-band, tight-binding extended Peierls-Hubbard model. Both Hartree-Fock (HF) adiabatic molecular relaxation and molecular dynamics techniques are employed to investigate the time evolution of solitons, polarons, bipolarons in charge-density-wave (CDW) ground state materials. The time evolution of excitons, defect pairs and/or breathers is studied subsequent to photoexcitation. The ESR spectra associated with the resulting spin carrying non-linear excitations such as neutral solitons, triplet excitons as well as electron and hole polarons are then calculated within the above model. The superhyperfine structure in the ESR spectra is attributed to a small spin density present on the halogen sublattice.

Introduction

Theoretical modelling, synthesis and experimental characterization of halogen-bridged mixed valence linear chain MX materials have been described recently in great detail [1, 2]. In particular, optical absorption, IR and resonance Raman spectra were computed and compared with experimental data. Furthermore, we found that the mixed-halide MX materials exhibit photoinduced and intrinsic electron–hole charge separation [3] which is potentially very useful for device applications. Here we focus on the theoretical predictions for the (i) dynamics of photoexcitation and (ii) electron spin resonance (ESR) spectra in pure MX materials. It is worth noting that certain defect states, e.g. excitons, are inaccessible via impurity doping, but may be obtained during photoexcitation. This is illustrated below in the case of a weak CDW (delocalized) material PtI. The ESR spectrum, however, is shown for a triplet exciton in the case of a strong CDW (localized) material PtCl. We present the results

0379-6779/93/$6.00

of theoretical calculations based on discrete, 3/4-filled, two-band, tight-binding, Peierls-Hubbard Hamiltonian for representative parameter sets [1].

Dynamics and photoexcitation

We have investigated the photodecay channel subsequent to photoexcitation in the ground state as well as in the presence of non-linear excitations and impurities using adiabatic molecular dynamics [1, 4]. Photoexcitation was simulated numerically by manually removing an electron from an occupied state and instantaneously placing it in an unoccupied state. The system was allowed to evolve adiabatically with no further changes in electronic occupations. Figure 1 shows the evolution on a PtI chain with 96 sites (48 Pt atoms and 48 I atoms) after a single electron is photoexcited across the Peierls gap, with the addition of the gap energy ($E_g \simeq 1.2$ eV) to the system. Periodic boundary conditions were employed. As is clear from Fig. 1(a), initially an exciton is formed within a phonon period. However, this exciton is unstable and, in principle, can decay into either (i) an electron polaron–hole polaron pair or (ii) a charged kink (i.e. electron and hole soliton) pair or (iii) a neutral kink–antikink pair. Based on the electronic occupancies of the excited state, the analysis of fully relaxed electronic states at a later time, and the energetics, we find that the initially formed exciton evolves into a slowly separating neutral kink–antikink pair. Since the creation energy of the kink–antikink pair is smaller than the gap energy, the remaining energy goes partly into the kinetic energy of the kinks, partly into the acoustic phonons and partly in the form of a small amplitude localized 'breather' (phonon bound state) between the kinks. Such a breather, a temporally and spatially coherent state of optical phonons [4], is clearly visible in Fig. 1(a). Alternatively, a breather is a charge-neutral, spatially localized, time-periodic, in general persistent, non-linear lattice excitation which has important implications for subpicosecond time-resolved absorption experiments. It has distinct signatures in the intragap absorption. Since PtI is a delocalized system the initially formed exciton is extended over about 40 sites. Similarly, the breather is also quite extended.

Figure 1(b) shows the time evolution of associated energy levels, in particular the characteristic gap states of a kink pair. Within $\sim$200 femtoseconds two continuum states are pulled into one (almost) degenerate midgap state indicating that the initial bound electron hole pair (exciton) quickly evolves into a kink–antikink pair. At the same time a breather level oscillates about the conduction band edge into the gap and persists with a time period larger than the phonon period. We have also studied dynamics of photoexcitation in PtCl and PtBr. Unlike PtI, at later times lattice discreteness effects hinder complete separation of the kink–antikink pair in PtCl which is a very localized CDW system.

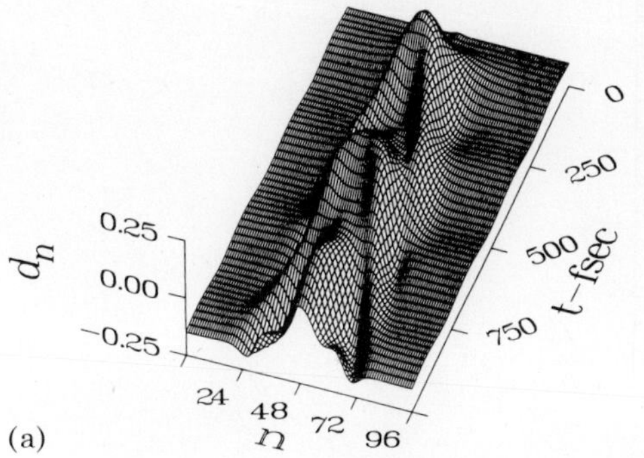

(a)

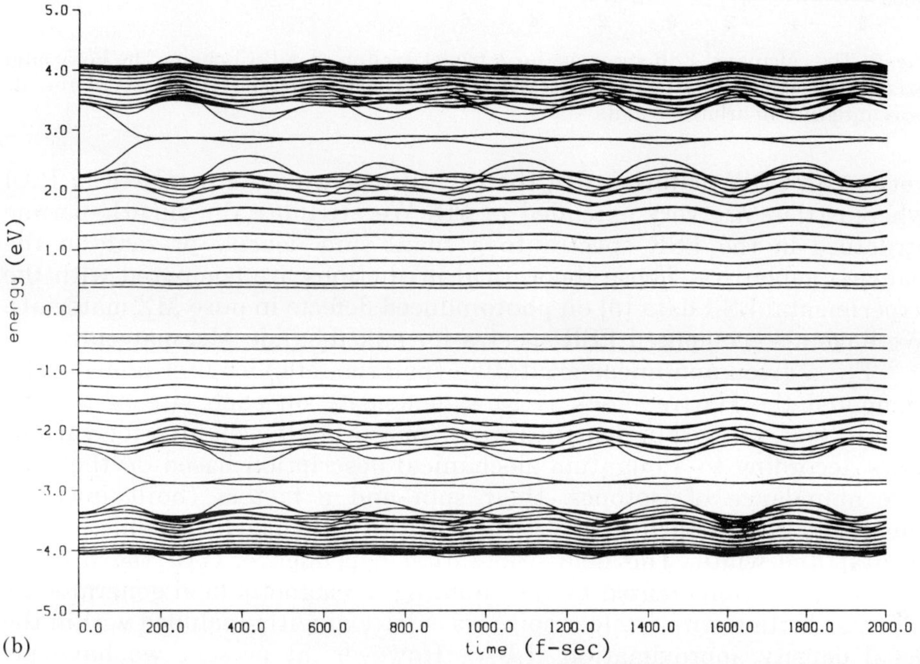

(b)

Fig. 1. Dynamics of photoexcitation in PtI. (a) CDW distortion as a function of time in units of 10^{-15} s; for clarity the negative of the CDW distortion is plotted. (b) Energy levels as a function of time for photoexcitation of the ground state.

ESR spectra

We calculated the electron spin resonance (ESR) spectra associated with spin carrying non-linear excitations such as neutral solitons, triplet excitons as well as electron and hole polarons for pure MX materials within the same two-band model used for the dynamics calculations. Our results contrast a strongly distorted CDW material, PtCl, with that of a

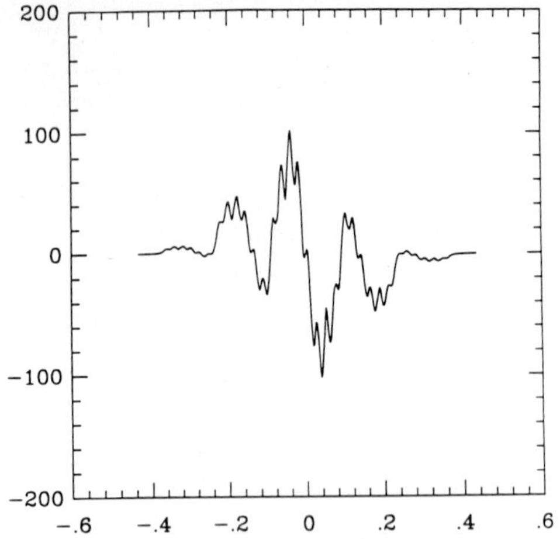

Fig. 2. The calculated ESR spectrum for a triplet exciton on a PtCl chain. The horizontal axis gives peak positions in terms of scaled magnetic field while the vertical axis represents peak intensity in arbitrary units.

very weak CDW material, PtI; the excitations are quite localized in PtCl whereas they are very extended in PtI. We attribute the superhyperfine structure in the ESR spectra to a small spin density present on the halogen sublattice. These theoretical predictions are compared with the experimental ESR data [5] on photoinduced defects in pure MX materials. We have also computed ESR spectra for mixed-halide MX materials.

For a given non-linear excitation (polaron, etc.) we first numerically computed the Hartree-Fock (HF) spin density on each site of an MX chain. Next, these spin densities were appropriately weighted for various lines according to a quantum mechanical description based on the relative abundance of isotopes, their spin and g factors (built into the numerical code). Each ESR was Lorentzian broadened to mimic experimental line width. The field orientation dependence, (i.e., the dipole–dipole interaction created by the coupling of magnetic field generated by electron motion with nuclear spin) was also explicitly included within the local density approximation (LDA). However, at present we have neglected dynamical effects on the line shape such as spin–flip and coherent soliton motion. We have also neglected the coupling between electron orbital and spin angular momenta as well as the dipole–dipole interaction between two unpaired electrons (S.D.S). The latter is unimportant except in the case of a triplet exciton where the zero field splitting parameters (D, E) may play a significant role. Akin to a linear molecule we expect $E = 0$ for a linear chain. The value of D depends on the distance between the two spins in a triplet exciton. Work is in progress to estimate this parameter for various MX chains.

For illustrative purposes, we show in Fig. 2 the calculated ESR spectrum for a triplet exciton in PtCl. Five main peaks (hyperfine structure) are clearly visible. Superimposed on these peaks is a small amplitude modulation (superhyperfine structure). A careful analysis of the spin densities reveals that the hyperfine structure arises from two parallel spins on two neighboring metal (Pt) ions in an exciton (a triplet exciton can be thought of as a bound state of an electron and a hole polaron with parallel spins). Most of the spin density ($\sim$95%) is on these two Pt atoms. However, the remaining spin density ($\sim$5%) is distributed over three Cl ions around and between the two Pt ions. This small spin density on the halogen ions is sufficient to produce the superhyperfine structure. We emphasize here that since the halogen sublattice is not explicitly included in a one-band model, the latter will not be able to reproduce the superhyperfine structure. The experimentally observed ESR spectra [5] (not shown here) from photoinduced defects in PtCl resembles rather well the calculated spectra (Fig. 2). Nevertheless, whether it arises from a triplet exciton, a kink–antikink pair or an electron and hole polaron pair remains an open question for the time being, although the evidence points more towards a polaron pair.

Conclusions

We have presented illustrative examples of theoretically calculated photodecay channel and ESR spectrum. Investigations of the influence of Coulomb interactions (both on-site and inter-site Hubbard terms), site and bond impurities on the photodecay channel are in progress in addition to the dynamics and ESR studies for mixed-halide MX materials. We believe that these studies will elucidate the processes of photoinduced charge transfer and self-doping in mixed-halide chains.

Acknowledgements

This work was supported by the US DOE. While this work was done J.T.G. held a National Research Council–NRaD Fellowship supported through a grant from the Office of Naval Research.

References

1 J. T. Gammel, A. Saxena, I. Batistić, A. R. Bishop and S. R. Phillpot, *Phys. Rev. B*, *45* (1992) 6408.

2 T. Mitani *et al.*, paper presented at *Symp. E. Synth. Mater. Nonlinear Opt. Electron., 1992 E-MRS Spring Conf., Strasbourg, France, June 2–5, 1992; Proc. ICSM '90, Tübingen, Germany*; *Synth. Met.*, 41–43 (1991).

3 A. Saxena, X. Z. Huang, A. R. Bishop, L. A. Worl, S. P. Love and B. I. Swanson, *Synth. Met.*, *54* (1993) 395.

4 S. R. Phillpot, A. R. Bishop and B. Horowitz, *Phys. Rev. B*, *40* (1989) 1839.

5 C. A. Arrington (Furman University, SC), personal communication.

Light induced conversion of chlorine containing polymers in the presence of pyrrole into conducting patterns

R. Baumann
Technische Hochschule Leipzig, Fachbereich Naturwissenschaften, Leipzig (Germany)

K. Lennarz and J. Bargon
Institute of Physical Chemistry, University of Bonn, Bonn (Germany)

Abstract

This paper outlines the generation of conducting patterns via light induced conversion of chlorine containing polymers in the presence of suitable monomers such as pyrrole. Pyrrole is introduced into the matrix by diffusion during the deposition of the polymer in a pyrrole vapor atmosphere. Subsequently, this precomposite is exposed to UV radiation. The photolysis of the host polymer is associated with the elimination of HCl in the irradiated areas only, yielding a composite material with enhanced electrical conductivity. The unexposed areas remain highly insulating. Via this procedure, a composite pattern with enhanced electrical conductivity can be obtained using a patterned mask for flood exposure. Thereafter, the pattern can be transferred into a three-dimensional structure by wet or dry development steps.

Introduction

Light sensitive polymers or composite systems are traditional materials in the patterning processes during the fabrication of microelectronic components. Thereby the physical and chemical properties of the starting organic materials are significantly changed by irradiation using UV light sources for flood exposure. Eventually, the demanded three-dimensional polymer patterns are obtained by wet, but preferably by dry processes like plasma etching or excimer laser ablation.

Among the large variety of polymers and monomers alike, heterocyclic compounds that give π-conjugated polymers are under intense investigation due to their semiconductor properties, their electrical conductivity in the partially oxidized state, and because of their non-linear optical activity in the neutral state. There are only a few reports on the light induced polymerization of pyrrole and of its derivatives, resulting in polymer patterns with dimensions required for microlithography. Photoelectrochemically induced polymerization processes of pyrrole have been outlined by Hikita *et al.* [1] and Okano *et al.* [2]. The light induced patterning of polymeric substrates in the presence of pyrrole vapor has been described by Bargon and co-workers [3–5], Ojio and Miyata [6], and Rodriguez and Gonzalez-Velasco [7], using $FeCl_3$ or organometallic com-

0379-6779/93/$6.00

plexes as the photosensitive agents. Additional procedures have been described by Nannini and Serra [8] using $CuCl_2$ precursor patterns, and by Abdou *et al.* [9] based upon photo-crosslinked π-conjugated prepolymers followed by wet development of the pattern.

In the following, we report on a new way to obtain patterns with modified electrical properties starting from insulating precursors consisting of chlorine containing polymers. This process is based upon the light induced generation of free radicals* as a result of photolyzing a matrix polymer in the presence of a suitable donor monomer, i.e. of pyrrole. Other monomers investigated were aniline, thiophene and substituted forms of both. Related phenomena were observed in all of these systems*. It, therefore, appears that the phenomenon observed here is neither peculiar for a particular matrix polymer nor for a specific monomer. Various combinations of monomers and polymers can lead to the same generic result, namely the formation of conjugated double bonds. The system pyrrole/PCAN works best, however.

Experimental

The chlorine containing polymers poly(vinylchloride) (PVC), poly(chloroprene) (PCP), poly(chlorostyrene) (PCS) and poly(chloroacrylonitrile) (PCAN) were obtained from Polyscience. Solutions of the polymers in tetrahydrofuran were spin coated onto sheets of glass or quartz in layers of 2 μm. To remove the solvent, the clear polymer films were dried at 45 °C in air for 2 h or under vacuum without any change in the polymer constitution, as measured by IR spectrosocpy. Subsequently, the dry films were transferred into an atmosphere of pyrrole vapor for 1–10 h. During this procedure, the heterocyclic monomer diffuses into the polymer matrix. At this point, the films are still clear, and show a linear superposition of both the polymer and the monomer spectrum in the UV–Vis as well as in the IR range. This light sensitive composite was exposed to the radiation of a 1000 W Hannovia mercury high pressure lamp under flood exposure conditions using a chromium mask on quartz. The UV–Vis and IR spectra were recorded with a Shimadzu double-beam spectrophotometer UV-210A and a Digilab FTS-40 FT-IR spectrometer, respectively.

Results

UV irradiation of gassed polymer films

When irradiating thin films of chlorine containing polymers after deposition in a pyrrole atmosphere with UV radiation, the polymer films

---

*NMR studies of a model system confirm the postulated reaction mechanism. Details will be published elsewhere.

turn from their virgin transparent form to black–brown within tens of seconds. The solubility of the resulting composite material in common organic solvents decreases with increasing irradiation time. The exposed black–brown areas represent a polymer composite consisting of a conjugated system of double bonds. These stem predominately from the starting matrix polymer, which, however, becomes interpenetrated by polypyrrole or oligomers of the pyrrole monomer. The light sensitivity of the polymer–monomer precomposites depends on the type of polymer and increases in the following order: PCS < PCP < PVC < PCAN. It has been found that the period between removal of the samples from the monomer vapor and subsequent irradiation can be up to 24 h in normal air without significant loss of sensitivity. Using a scanning electron microscope (SEM), we could show the difference between the electrical properties of PCAN and the exposed areas of composite that contain the conjugated double bonds afterwards.

Figure 1 shows an SEM micrograph of the pure surface (not sputtered with Au) of a polymer sample on a sheet of gold coated glass as a conducting and grounded substrate material. The polymer was exposed through a mask. The light areas received no light, and, consequently, no conjugated bonds were produced. Therefore, the polymer remains an insulator, resulting in a strong electrostatic charge build up due to the intense primary electron beam within the electron microscope. The dark cross consists of irradiated precomposite material, i.e., of modified poly(chloroacrylonitrile) interpenetrated by pyrrole. In these areas, the primary electrons do not build up an electrostatic charge, since the electrical conductivity of the UV-irradiated composite is sufficient to quench any space charge formation.

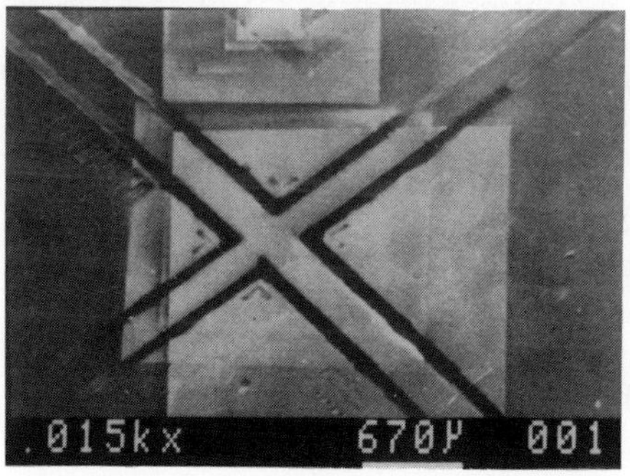

Fig. 1. SEM micrograph of patterned poly(chloroacrylonitrile)/pyrrole composite; the light areas represent the electrostatically charged, non-irradiated, and therefore insulating PCAN. The dark cross represents the conducting PCAN/pyrrole composite material produced by UV irradiation.

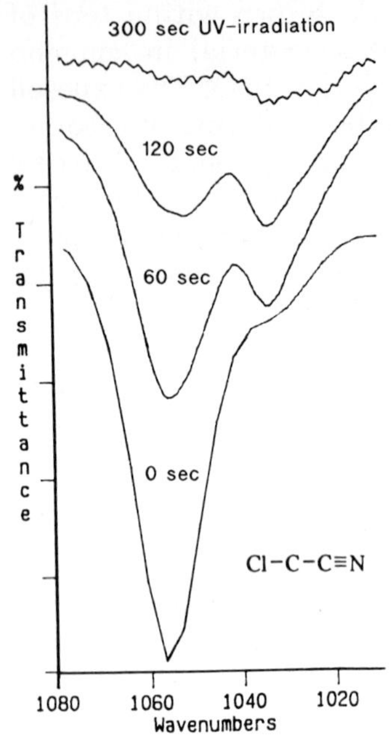

(a)

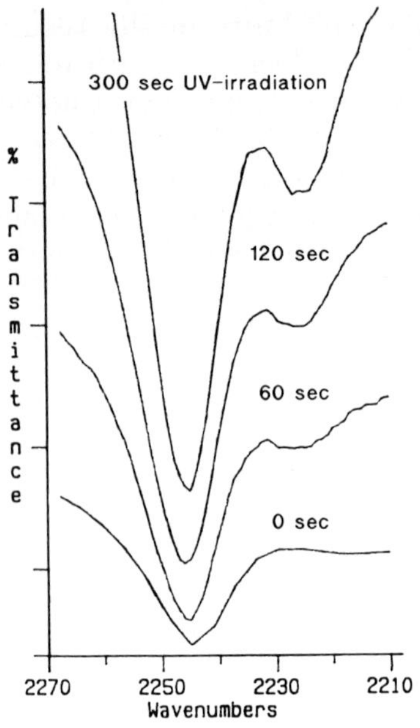

(b)

Fig. 2. UV photolysis of poly(chloroacrylonitrile): (a) decreasing intensity of the Cl–C–C≡N scissor vibration and (b) increasing intensity of the C≡N stretching vibration (2248 cm^{-1}) as a result of UV radiation induced dechlorination of the polymer in the absence of pyrrole.

Photolysis of poly(chloroacrylonitrile)

In order to obtain information about the initiation step of this photoinduced conversion, we investigated the photolysis of the most efficient matrix polymer, PCAN, by FT-IR spectroscopy in the absence of pyrrole. Whereas the UV-irradiated polymer does not change its color in the visible, Fig. 2(a) shows significant changes in the IR spectrum of PCAN as a result of advanced irradiation. The absorption band at 1056 cm^{-1} is assigned to the scissor vibration of the Cl–C–C≡N group. The intensity of this band decreases with increasing irradiation time, accompanied by a weak shift. This result suggests a loss of chlorine. On the other hand, the absorption band at 2248 cm^{-1} is assigned to the C≡N stretch vibration which is quenched by the first harmonic of the C–Cl valence vibration [10, 11] of the starting PCAN in the areas which received no exposure. An increase of this band with increasing irradiation time indicates a loss of chlorine likewise, see Fig. 2(b). Therefore, we conclude that the removal of chlorine from the polymer results in the formation of HCl. Since pyrrole and related heterocyclic monomers are very easily converted into autoxidized polymers by acid [12], we postulate an ensuing acid catalyzed oligomerization/polymerization mechanism of pyrrole, whereby the acid is generated photochemically. The observations that the UV-irradiated PCAN in the absence of pyrrole does not change its color, and the strong increase of the C≡N stretch vibration with increasing UV irradiation in the absence as well as in the presence of pyrrole speak against a C≡N bond opening and a subsequent internal cyclization.

Conversion of pyrrole in PCAN

In order to obtain more detailed information about the polymer generated by UV irradiation, we started to analyze the IR spectra of the composites. The irradiated samples were kept in vacuum for 10 h, to remove the unconverted pyrrole from the PCAN/polypyrrole composite. Figure 3 shows a comparison of the PCAN, pyrrole and the IR spectra of the composite. The annulment of the C≡N stretch quenching of PCAN ($2248\,cm^{-1}$) is significant again. Both the occurrence of typical pyrrole bands in the composite spectrum, and the resulting insolubility of the composite, point to the formation of polypyrrole, however, the exact structure of the resulting composite is still under investigation. From the systematical analysis of the C–H vibrations of the pyrrole product at 3130 and $3104\,cm^{-1}$, we hope to receive detailed information about the polypyrrole formation.

Polymer patterning

To prove the suitability of the system for producing three-dimensional polymer patterns, a 2 μm thick layer of PCAN on a glass substrate was put into a pyrrole vapor atmosphere for 5 h. Afterwards, the sensitized polymer layer was UV irradiated for 3 min through a chromium mask on quartz. The mask was in close contact with the polymer layer.

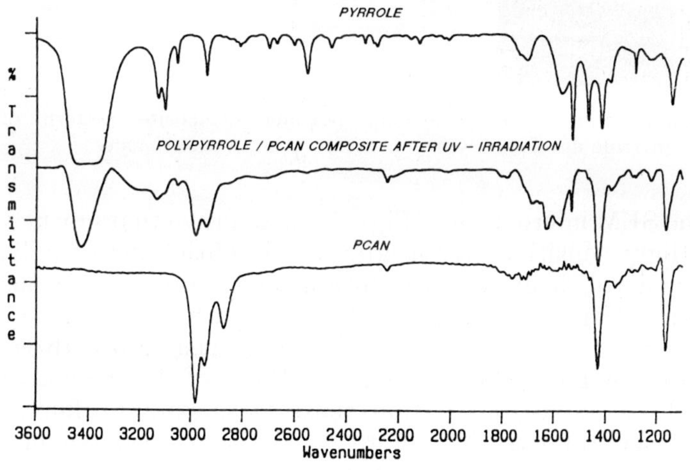

Fig. 3. Comparison of IR spectra of poly(chloroacrylonitrile), pyrrole and the polymer composite after subjecting the polymer to a pyrrole vapor atmosphere, irradiating it with UV light followed by treatment to remove the non-polymerized pyrrole.

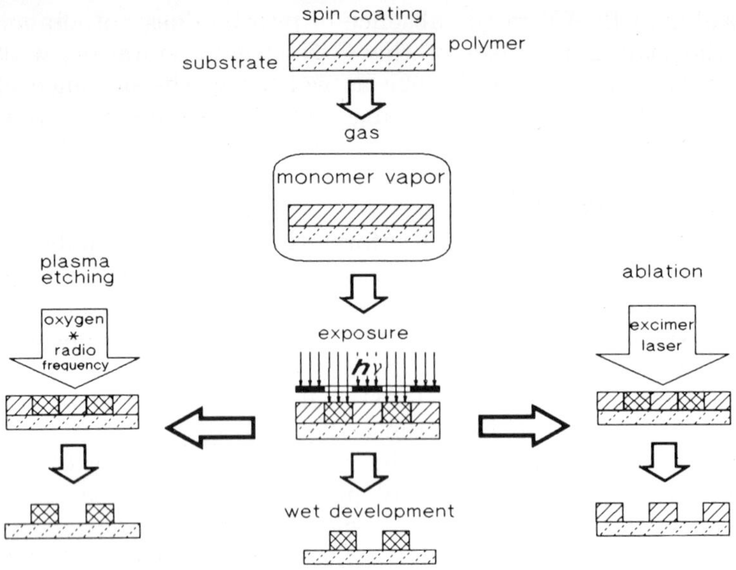

Fig. 4. Sequence of processing steps.

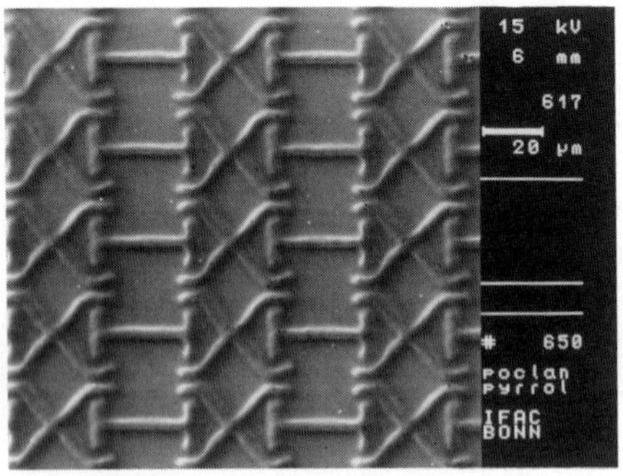

Fig. 5. SEM micrograph of a three-dimensional polymer composite pattern of poly(chloroacrylonitrile)/pyrrole developed with tetrahydrofuran in a wet mode.

Figure 4 and the SEM micrographs (Figs. 5 and 6) prove that our new PCAN/polypyrrole light sensitive system allows the fabrication of polymer patterns in the microlithographic resolution range applying various developing procedures. Using tetrahydrofuran, the clearly visible black–brown negative image of the mask can be wet developed into a three-dimensional structure owing to the different solubility of the irradiated versus the non-irradiated areas. In addition, these areas exhibit different rates for excimer laser ablation. Therefore, the UV exposed polymer composite can subsequently be dry developed by excimer laser radiation (248 nm KrF or 351 nm XeF), which removes the exposed polymer by ablation. Since the different regions (exposed/unexposed) show different

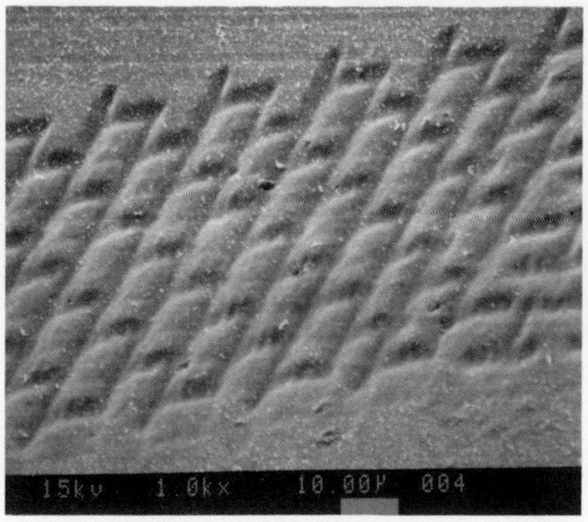

Fig. 6. SEM micrograph of a three-dimensional polymer composite pattern of poly(chloroacrylonitrile)/pyrrole developed by 248 nm KrF excimer laser ablation in a dry mode.

etching rates in an oxygen plasma as well, the image of the mask within the polymer film can also be dry developed by selective reactive ion etching in a third way. Using a post development baking procedure at 800 °C, the initial polymer pattern can be carbonized for even more processing options.

Acknowledgements

This work has been supported by the Fonds der Chemischen Industrie and by the Deutsche Forschungsgemeinschaft.

References

1 M. Hikita, O. Niwa, A. Sugita and T. Yamamura, *Jpn. J. Appl. Phys.*, *24* (1985) L79.
2 M. Okano, K. Itoh, A. Fujishima and K. Honda, *J. Electrochem. Soc.*, *134* (1987) 837.
3 T. Ueno, H. D.Arntz, S. Flesch and J. Bargon, *J. Macromol. Sci.*, *Chem.*, *A25* (1988) 1557.
4 J. Bargon, T. Weidenbrück and T. Ueno, *SPIE*, *1262* (1990) 564.
5 J. Bargon, W. Behnck, T. Weidenbrück and T. Ueno, *Synth. Met.*, *41* (1991) 1111.
6 T. Ojio and S. Miyata, *Polymer J.*, *18* (1986) 179.
7 I. Rodriguez and J. Gonzalez-Velasco, *J. Chem. Soc.*, *Chem. Commun.*, (1990) 387.
8 A. Nannini and G. Serra, *J. Mol. Electron.*, *6* (1990) 81.
9 M. S. A. Abdou, G. A. Diaz-Guijada, M. I. Arroya and S. Holdcroft, *Chem. Mater.*, *3* (1991) 1003.
10 H. Hiraoka and W. Y. Lee, *Macromolecules*, *11* (1978) 622.
11 N. Grassi and E. M. Grant, *J. Polym. Sci. C*, *16* (1967) 591.
12 G. F. Smith, *Adv. Heterocycl. Chem.*, *2* (1963) 287.

5. Ferroelectrictiy and Ferromagnetism in Organics

Synthetic Metals, *54* (1993) 253–261

Physical properties of achiral mesomorphic ferroelectrics

F. Tournilhac[a], L. M. Blinov[b], J. Simon[a], D. B. Subachius[b] and S. V. Yablonsky[b]
[a]*ESPCI, 10 rue Vauquelin, 75231 Paris Cédex 05 (France)*
[b]*Institute of Crystallography, Russian Academy of Sciences, 117333 Moscow (Russian Federation)*

Abstract

Physical properties of new mesomorphic organic ferroelectrics composed of achiral polyphilic molecules are described. Ferroelectricity is demonstrated by the measurements of the acoustically induced piezoelectric response, pyroelectric effect and repolarization currents. For the first time, a ferroelectric repolarizable mesophase is obtained with non-chiral molecular units. The mechanism of ferroelectricity is therefore different from that found in conventional chiral smectics C*.

Introduction

According to their symmetry all polar crystalline and mesomorphic phases are divided into two groups [1]. One group, which is composed of chiral molecules, has a point symmetry group ranging from C_1 to C_∞. Examples of this type are liquid crystalline chiral phases (smectic C*, F*, I*, etc.) differing in the character of the in-plane packing of their molecules within a smectic layer [2–5]. These phases manifest reversible switching of their spontaneous polarization, that is, they are ferroelectrics. The other group of polar materials comprises mirror-symmetric (achiral) phases with point symmetry groups of C_{2v} to $C_{\infty v}$ and among them one finds many classical solid crystalline ferroelectrics like $BaTiO_3$, triglycinesulfate, thiourea, etc. [6]. However, as far as we know, there has been no observation of ferroelectric properties in non-crystalline (mesomorphic) achiral molecular materials.

A chemical approach to construct new polar materials has been suggested [7] and the first observations of their ferroelectric properties reported [8]. The basic idea was to use the polyphilic effect to form 'building elements' of a polar phase. According to this concept, chemically different moieties of a molecule tend to segregate to form polar aggregates or lamellae which result in the formation of polar phases. The direction of the spontaneous polarization is allowed to be in the plane of the lamellae of layers (transverse ferro- or antiferroelectrics) or perpendicular to the layers (longitudinal ferro- or antiferroelectrics discussed theoretically in refs. 9 and 10). In this paper we present the experimental

0379-6779/93/$6.00

data on piezo- and pyroelectric effects and the switching behaviour of polyphilic compounds confirming their ferroelectric nature.

Experimental

Two polyphilic derivatives (Fig. 1, **I** and **II**) showing unusual smectic phases (later referred to as smectic X and X′) and their binary mixture have been studied. The two compounds are made up of three distinct parts: perfluoroalkyl and alkyl side chains and a biphenyl rigid core. They differ from each other by the direction of their longitudinal dipoles. Both pure compounds, apart from the well-known smectic A phase, have another, metastable and partially ordered smectic X phase which shows a tendency to crystallize. Their binary mixtures, in a certain concentration range, have a stable enantiotropic X′ phase with an optical texture slightly different from that of the X phase of the pure compounds. The X and X′ phases are miscible with each other and certainly distinct from a crystalline phase.

X-ray structure investigations have been carried out on unoriented samples of **I** (showing the following phase sequences: on heating, crystal—95 °C—smectic A—113 °C—isotropic; on cooling, isotropic—113 °C—smectic A—82 °C—smectic X stable down to room temperature and crystallized on the next heating cycle) and the mixture **M70** (70% **I**

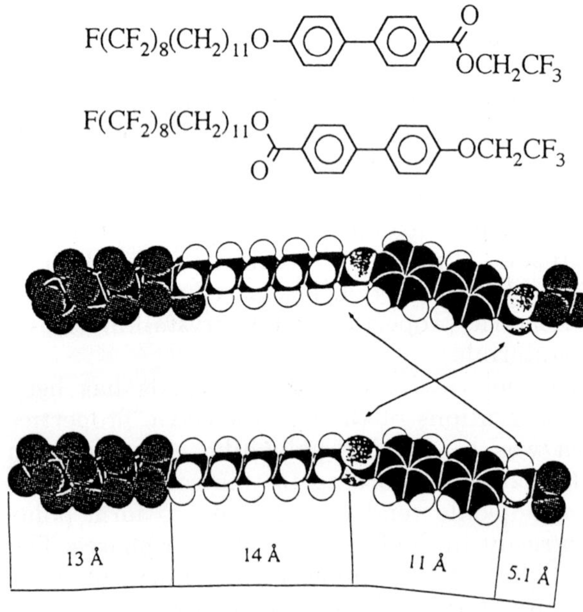

I

II

Fig. 1. Polyphilic compounds investigated. On molecular models different moieties having a tendency to segregation are easily seen and their characteristic dimensions are shown.

30% **II**) showing the phase sequence: smectic X′—82 °C—smectic A—105 °C—isotropic.

The X-ray diffraction pattern of the X phase (**I**) shows three incommensurate rather broad Bragg reflections at small angles (spacing 41.3, 27.5 and 17.5 Å with corresponding correlation lengths of 270, 60 and 280 Å) and a diffuse ring at wide angles (around 4.5 Å). The diffraction pattern for the X′ phase of the mixture also shows three broad reflections but, in contrast to the previous case, with commensurate spacings (34.5, 17.2 and 11.5 Å with a unique correlation length 100–110 Å). At wide angles the smectic X′ phase shows two distinct rings of comparable intensities at 4.8 and 4.4 Å against the diffuse background. In the smectic A phase of both samples two reflections commensurable with molecular length (42 and 21 Å for **I** and 43 and 21.5 Å for **M70**) are observed.

The short correlation length reveals a rather inhomogeneous structure of the X and X′ phases at the microscopic level indicative, probably, of a short-range ferro- or antiferroelectric packing of molecules. The diffraction patterns for the two phases are dramatically different from that observed for the crystalline phase of **I** (numerous sharp reflections for both small and wide angles). Thus, the physical properties to be discussed below are related to a mesomorphic state.

In our experiments three fundamental properties characteristic of ferroelectric substances were studied, namely, the field-induced switching of the piezoelectric response, the pyroelectric effect and the repolarization currents. The principles of the measurements of the piezo- and pyroelectric responses are shown in Figs. 2 and 3, respectively.

A drop of a substance of area 0.01 cm^2 was placed into a flat capillary of thickness 15–120 μm formed by two glasses covered with SnO_2 optically transparent electrodes. The electrodes play two parts: (i) when necessary

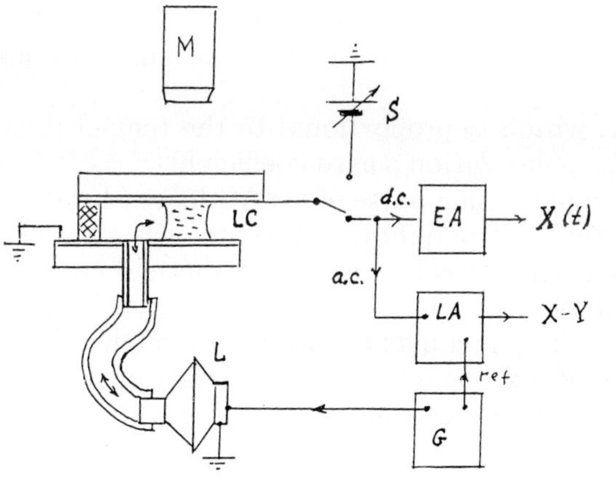

Fig. 2. Acoustoelectric cells: S, source of d.c. poling voltage; EA, electrometric amplifier; LA, lock-in a.c. amplifier; G, audio generator; L, loudspeaker; M, polarizing microscope; $X(t)$ and $X-Y$ are recorders.

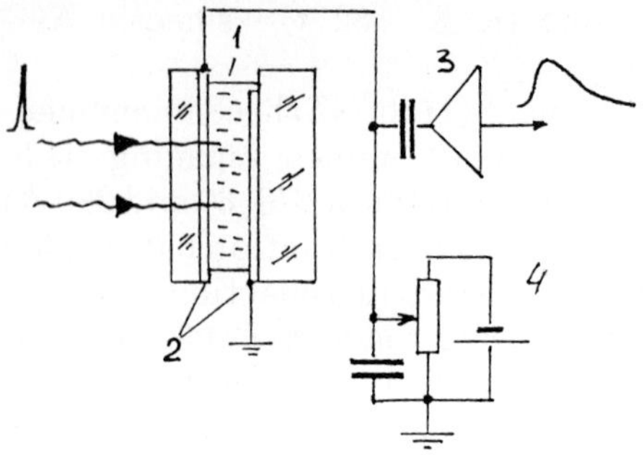

Fig. 3. Pyroelectric measurements: liquid crystal layer (1) between electrodes (2) transparent in the visible region and semi-transparent in the near-IR region; wide bandpass amplifier (3); d.c. voltage source (4).

the poling voltage from a d.c. source can be applied across them, (ii) an a.c. piezoelectric or pyroelectric signal or d.c. repolarization current can be measured using the same electrodes. The cells were placed in a thermostat and the optical texture of the drop observed under a polarizing microscope.

The piezoelectricity was induced by a mechanical distortion of the drop due to the a.c. air flow generated by a loudspeaker (Fig. 2). The acoustic wave of frequency 80 Hz was delivered to the capillary with an acoustic waveguide the length of which was chosen to have a resonance for the wave. The acoustic wave intensity at the end of the waveguide was about 10^4 Pa. The a.c. piezovoltage was observed at the fundamental frequency of the exciting acoustic wave (although, when necessary, measurements at the second-harmonic frequency, 160 Hz, were carried out) using a lock-in amplifier (input impedance 10 MΩ) synchronized from the same function generator which fed (through a 10 W audio amplifier) the loudspeaker. The a.c. piezo-response can be measured simultaneously with a d.c. current generated by the cell when the electric polarization of the substance is changed. This repolarization current was measured by an electrometric amplifier.

The pyroelectric response, which is proportional to the temperature derivative of the macroscopic polarization (pyro-coefficient $\gamma = \partial P/\partial T$) was observed by heating the sample with a pulse of a YAG laser (duration of the pulse was about 100 μs, Fig. 3). The light ($\lambda = 1.06\ \mu$m) was absorbed in the SnO_2 electrodes and the sample was heated due to thermal conductivity of the substance. The absolute measurements of the pyroelectric coefficient were carried out using a conventional ferroelectric liquid crystal with a known value of the spontaneous polarization (about 60 nC cm^{-2}) placed between the same electrodes in another experiment. The macroscopic polarization (either spontaneous or induced) was calculated by integrating the pyro-coefficient over temperature starting from the temperature T_o of the phase transition to the smectic A (or other paraelectric) phase:

$$P = \int_{T_0}^{T} \gamma(T)\, dT \qquad (1)$$

Reference 11 gives more details of the techniques.

Results

Pure compound I

In experiments with the pure compound **I**, the polar state can only be induced using a poling voltage. If the sample is poled during cooling from the isotropic phase, e.g. for 1 h, a piezoelectric response stable for days is observed in the metastable smectic X phase. Its amplitude ($U_p \approx 1$–10 mV) is comparable to the signal observed for conventional (chiral) ferroelectrics ($U_p \approx 15$–20 mV for a ferroelectric with spontaneous polarization of the order of 100 nC cm^{-2}. Nonferroelectric compounds show background piezoelectric signals at the level of a few microvolts. The phase of the a.c. piezo-response correlates with the sign of the poling voltage.

On heating, the piezo-voltage of **I** decreases even in the case when no crystallization is observed (Fig. 4, curve for +500 V). If, by chance, crystallization occurs, the signal drops dramatically and can even change its sign (Fig. 4, curve for −500 V). It is impossible to repolarize the sample

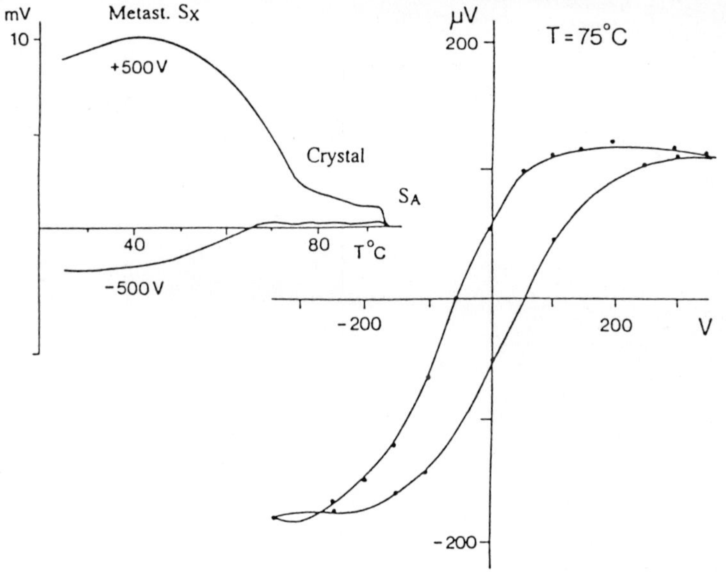

Fig. 4. Left: temperature dependences of the piezoelectric response for **I** poled either with +500 V or −500 V on cooling down from the isotropic phase. Right: hysteresis of the piezo-voltage taken at 75 °C. Each point of the loop is taken after applying the corresponding d.c. voltage for 2 min.

at room temperature but repolarization is easily attained with increasing temperature. The higher the temperature the faster is the process of repolarization. The corresponding hysteresis loop (in coordinates applied d.c. voltage – piezovoltage) is shown in the same Figure. The characteristic time of the repoling process at this temperature (75 °C) is of the order of a few minutes. With such a slow process it is impossible to observe repolarization currents for **I** as they are masked by parasitic ionic currents resulting from poling.

The results shown in Fig. 4 were obtained on rather thick ($d = 100\ \mu$m) unoriented samples of **I** and are quite reproducible. Thinner cells necessary for pyroelectric measurements have to be used with some precautions. The substance has a tendency to a spontaneous homeotropic orientation (easily seen in the smectic A phase). Homeotropically reoriented samples (poled as above) show a very weak piezoelectric signal and no pyroelectric response. Thus, the glasses for thin cells were treated with polyvinylketal solution and rubbed unidirectionally as is usually done for conventional nematic liquid crystals. The temperature behaviour of the piezoelectric response of thin cells is similar to that shown above.

The temperature behaviour of the pulsed pyroelectric response of a layer of **I** of thickness 11 μm placed between polymer-coated and rubbed SnO_2 electrodes is shown in Fig. 5. The sample was poled with a voltage of +30 V. The maximum value of the pyro-voltage corresponds to the pyro-coefficient of 0.07 nC cm^{-2} K^{-1}. The calculated value of the field-

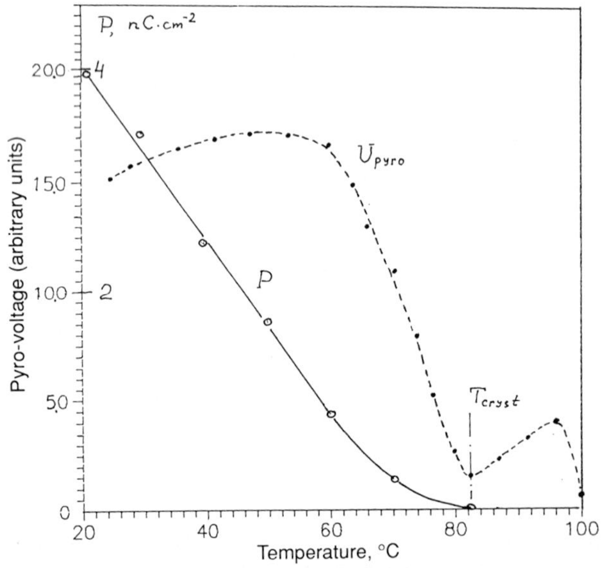

Fig. 5. Pyroelectric voltage and field-induced polarization of **I** as a function of temperature (poling d.c. voltage +30 V, cell thickness 11 μm).

induced macroscopic polarization (eqn. (1) with $T_o = 82$ °C being the crystallization point) is also shown in the same Figure. The magnitude of P at room temperature is about 4 nC cm^{-2} which is comparable (a little higher) with the spontaneous polarization of the first chiral liquid crystalline ferroelectric DOBAMBC.

The pyroelectric response depends on the spatial position of the focused laser beam on the texture. The sample seems to consist of domains having different magnitude and even sign of polarization although the response corresponding to the sign of the poling voltage prevails. Our setup allows a simultaneous measurement of the piezo- and pyro-response both in field-off and field-on regimes. In the field-off regime, the temperature behaviour of the piezo- and pyro-signals is very similar (see Figs. 4 and 5). In the field-on regime there is a considerable additional contribution to the piezo-response, but the pyro-response only slightly changes with d.c. field. Perhaps there is a certain contribution from the pressure-dependent dielectric permittivity $\varepsilon(p)$ to the field-on piezoelectric effect. For the pyroeffect, the corresponding contribution from the temperature dependence $\varepsilon(T)$ seems to be weaker.

Mixture ***M70***

Repolarization currents could not be measured on **I** because of a strong contribution of the parasitic ionic current due to the poling process. However, they have been measured on mixture **M70**. In the latter case, we have the stable smectic X′ phase over a wide temperature range including room temperature and the macroscopic polarization can also be induced by an external field with the same hysteresis behaviour (at least at higher temperatures). On cooling a sample from the smectic A phase, even without any applied field, spontaneous transition into a polar state occurs within the range $T = 55$–30 °C. It manifests itself in sharp steps (either one or several, up to five) of the piezoelectric voltage accompanied by sharp peaks of the repolarization current. In Fig. 6 the spontaneous transition to the polar state occurs in the smectic X′ phase at $T = 53$ °C. No change in the optical texture of the X′ phase is observed during the spontaneous transitions.

The sign of the current pulses correlates with the sign of the corresponding piezo-voltage steps. The spontaneous polarization appearing at the transition can be calculated by integrating the area under the $i(t)$ curve:

$$P_S = \frac{1}{S}\int i(t)\,\mathrm{d}t \tag{2}$$

where S is the area of the liquid crystal drop. For mixture **M70** the observed P_S values are three times less than the field-induced polarization of **I** (within the range 1–1.5 nC cm^{-2}).

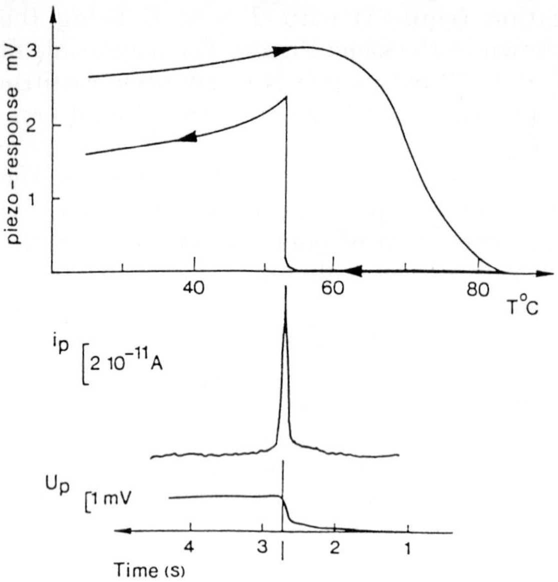

Fig. 6. Upper: temperature behaviour of the piezoelectric response during the cooling cycle. Lower: jump of the piezo-response $U_p(t)$ and the corresponding pulse of the repolarization current $i(t)$ are shown on the time scale.

Conclusions

The polyphilic compounds investigated therefore manifest ferroelectric properties. The smectic X′ phase consists of polar domains which can realign spontaneously. The direction of the macroscopic polarization can be switched, both in the X and X′ phases, by an external field with hysteresis loops typical of ferroelectrics. No electro-optical effect has been observed, which means that the polarization and director are decoupled as predicted for 'longitudinal' liquid crystal ferroelectrics [9]. On the other hand, the fact that homeotropic orientation prevents the observation of polar properties (at least on compound **I**) favours a model of a transverse ferroelectric (or antiferroelectric in the case of **I**). The pyroelectric effect and repolarization currents appearing during the spontaneous rearrangements of the polar domains allow the spontaneous polarization to be estimated: apparent value of the order of 1 nC cm^{-2} for spontaneously polarized domains in the X′ phase and about 4 nC cm^{-2} for uniformly poled X phase of samples **I**. A smaller magnitude of P_s for mixture **M70** could be explained by a mutual compensation of the longitudinal dipoles of compounds **I** and **II** if a model of longitudinal ferroelectric is true for mixture **M70**. This class of compounds will be the subject of further structural investigations.

Acknowledgements

We thank L. Beresnev for supplying conventional ferroelectrics as reference samples and for his help in orienting the samples, L. Bosio and T. Lobko for performing the X-ray measurements and J. Prost for many stimulating discussions.

References

1 Y. Sirotin and M. Chaskolskaia, *Fondements de la Physique des Cristaux*, Mir, Moscow, 1984.
2 R. B. Meyer, L. Liebert, L. Strzelecki and P. Keller, *J. Phys. (Paris) Lett.*, *36* (1975) L69.
3 H. R. Brand and P. E. Cladis, *J. Phys. (Paris) Lett.*, *45* (1984) L217–L222.
4 H. L. Brand and P. E. Cladis, *Mol. Cryst. Liq. Cryst.*, *114* (1984) 207–235.
5 L. A. Beresnev, L. M. Blinov, M. A. Osipov and S. A. Pikin., *Mol. Cryst. Liq. Cryst. (Suppl.) 158A* (1988) 1–150.
6 M. E. Lines and A. M. Glass, *Principles and Applications of Ferroelectrics and Related Materials*, Clarendon Press, Oxford, 1977.
7 F. Tournilhac F., L. Bosio, J. F. Nicoud and J. Simon, *Chem. Phys. Lett.*, *145* (1988) 452–453.
8 F. Tournilhac, L. M. Blinov, J. Simon and S. V. Yablonsky, *Nature (London)*, *359* (1992) 621.
9 J. Prost and P. Barois, *J. Chim. Phys.*, *80* (1983) 65.
10 R. G. Petschek and K. M. Wiefling, *Phys. Rev. Lett.*, *59* (1987) 343–346.
11 L. M. Blinov, V. A. Baikalov, M. I. Barnik, L. A. Beresnev, E. P. Pozhidayev and S. V. Yablonsky, *Liq. Cryst.*, *2* (1987) 121.

6. Poster Session-Nonlinear Optics: Materials and Devices

Photoluminescence of solid state fullerenes

H. J. Byrne, W. K. Maser, W. W. Rühle, A. Mittelbach, W. Hönle, H. G. von Schnering and S. Roth
Max Planck Institut für Festkörperforschung, Heisenbergstrasse 1, 7000 Stuttgart (Germany)

B. Movaghar
Institut für Theoretische Physik, Mainzer Gasse 33, 3550 Marburg (Germany)

Abstract

Time resolved photoluminescence studies have been carried out in films, powders and single crystals of C_{60}. The observed spectra for C_{60} in all forms are consistent with those reported for steady state conditions and may be associated with an intramolecular process. The spectrum is largely temperature independent with the exception of a broadening of the high energy edge with increasing temperature. At low temperatures, the observed C_{60} luminescence decays rapidly with a lifetime of 1.2 ns and can be interpreted as a low efficiency singlet decay or a decay from the molecules on the surface of the microcrystallites. No long lived decay from the triplet is observable at these intensities. In the case of the films, both the emission spectrum and lifetime are independent of excitation intensity. In the case of the powder and crystal a dramatic increase in the lifetime and broadening of the spectrum is observed at higher intensities. This broadened spectrum shows no fine structure and the emission maximum is red shifted by ~0.4 eV. Above a 'threshold' intensity, the luminescence intensity is seen to increase with the cube of the input intensity. This behaviour is interpreted as an abrupt onset of emission from the highly populated triplet state. The restrictions of the transition selection rules are relaxed by a banding of the intramolecular triplet states, a process which is dependent on a critical excited state population density. Thus a transition from intramolecular processes to an extended state, band like behaviour is observable at high excitation densities.

Introduction

To date, the photophysical properties of the fullerene C_{60} in solution have been well characterised. The symmetry forbidden HOMO-LUMO transitions become only weakly allowed through solvent environment fluctuations or vibrational interactions [1]. Excitation leads to population of the triplet state with an efficiency close to one [2]. This triplet state is extremely long lived [3], and no luminescent decay has been observed [3, 4]. In the solid state, C_{60} forms an f.c.c. crystal lattice bound by van der Waals' interactions with an interball spacing of 3 Å [5]. Using the orthogonalised linear combination of atomic orbitals method the electronic structure of f.c.c. C_{60} has been calculated [6], broad bands of width 0.5–1.0 eV with direct band gaps of 1.34 and 1.87 eV being predicted. Such a band picture leads to the conclusion that it is not adequate

0379-6779/93/$6.00

to approximate the electronic structure of an f.c.c. C_{60} with that of an isolated cluster. Studies of the dynamics of photoexcited states in C_{60} do little to support this, however, and no conclusive experimental evidence for electronic banding in solid state fullerenes has been reported, posing questions as to the degree of inter- versus intramolecular interactions in solid state fullerenes. In this study time resolved luminescence of C_{60} films, powders and crystals is carried out. A strong dependence of the luminescence on excitation intensity is observed in powders and crystals, but not in films. The results are discussed in terms of a transition between intramolecular and banded intermolecular behaviour at high excited state densities, which is dependent on microcrystallite size.

Sample preparation and experimental technique

C_{60} microcrystalline powders were generated from chromatographed solutions of starting materials prepared by the method of Krätschmer *et al.* [7]. Scanning electron microscopy shows these powders to be polycrystalline with crystallites of sizes ranging up to tens of microns. Film deposition was carried out by vacuum evaporation of these powders onto sapphire plates, under conditions described previously [8]. The resultant films were determined by depth profile meaurement to be 300 nm in thickness and are microcrystalline, with domain size of the order of 200 nm. Single crystallites were extracted by solvent evaporation from a supersaturated solution and had sizes in the order of $\sim 100\ \mu m$. Samples were fixed on a cold finger in a helium cryostat which could be cooled to ~ 10 K. Excitation was performed with a tunable Rhodamine 6G cavity dumped dye laser, operating at 585 nm with pulsewidth ~ 20 ps at a repetition rate of 400 kHz, pumped by a mode-locked Ar^{+}-ion laser with a repetition rate of 80 MHz. The luminescence from the samples was dispersed by a 0.32 m focal length monochromator and detected by a two dimensional synchroscan streak camera with an S-20 photocathode. The synchroscan is derived by an electrical 40 MHz trigger from the mode-locker of the source laser. The spectra were corrected for the spectral response of the system, and for any non-linearity of the temporal sweep.

Results and discussion

From all samples, a weak but readily observable luminescence was seen at moderate laser powers (~ 0.1 W). Figure 1 shows the spectrum of the C_{60} powder at room temperature. The observed spectra for C_{60} in all forms are consistent with those reported for steady state conditions [9, 10]. The onset of the emission overlaps that of the absorption band and higher resolution steady state measurements reveal a considerable degree of structure which is attributable to phonon replicas of electronic

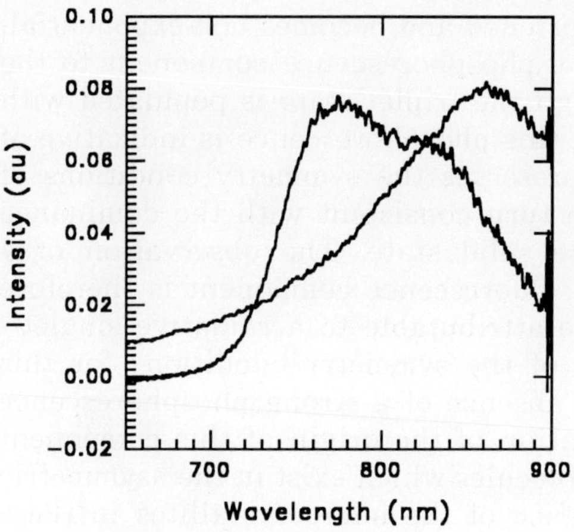

Fig. 1. Luminescence of C_{60} powder at room temperature under low illumination conditions, and the red shifted spectrum observed under illumination with 5×10^9 W cm^{-2}.

origin. The observed spectral width and vibronic structure can be correlated with that of the absorption spectrum of C_{60} solutions [9] and the luminescence may therefore be associated with an intramolecular process. The spectrum is largely temperature independent down to temperatures of ~10 K with the exception of a sharpening of the high energy edge with decreasing temperature, consistent with an energy relaxation to the lowest excited vibronic level followed by luminescence to the ground state vibronic manifold.

The temporal decay of the luminescence from all samples was seen to be spectrally uniform. At low temperatures, the decay fits well to a single exponential of lifetime 1.2 ns as shown in Fig. 2. At elevated tempera-

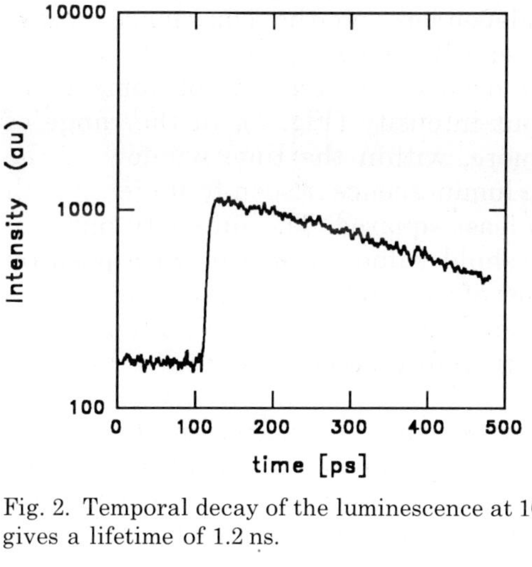

Fig. 2. Temporal decay of the luminescence at 10 K. A best fit to a single exponential decay gives a lifetime of 1.2 ns.

tures, the decay time is slightly shortened and becomes non-exponential. As in C_{60} solution, no long lifetime, phosphorescence component to the luminescence is observable, although the triplet state is populated with high efficiency [11]. The absence of this phosphorescence is indicative of a solid state environment which preserves the symmetry conditions of the isolated molecules, which is in turn consistent with the dominance of intramolecular processes in the solid state. The observation of a fast decaying, albeit low efficiency, fluorescence component is therefore surprising. This component may be attributable to a radiative singlet–singlet decay although relaxation of the symmetry conditions for this transition is inconsistent with the absence of a strong phosphorescence component. An alternative explanation of the origin of this component may lie in the high proportion of molecules which exist in the asymmetric environment presented by the surface of the microcrystallites in these polycrystalline materials. In these 'surface states' the restrictive selection rules are relaxed, leading to increased efficiency of the optical transitions.

In the case of the films, the magnitude of the luminescence increases linearly with input intensity over more than an order of magnitude (1×10^8 to 8×10^9 W cm^{-2}). In the case of the powder, however, a dramatic shift in the emission spectrum is observed with increasing intensity. Figure 1 shows the emission spectrum of the powdered sample at varying intensities. The emission undergoes a continuous shift towards the red and beyond the range of the streak camera photocathode. This shift in the emission spectrum is accompanied by a broadening and loss of vibrational structure as well as a dramatic increase in the luminescence efficiency, principally the result of the emergence of a long-lived component which persists during the reverse sweep of streak camera voltage sinusoid, which corresponds to a time 6 ns after excitation. The spectrally integrated temporal streak as a function of intensity is shown in Fig. 3. A notable feature is that the magnitude of the initially observed, fluorescence decay becomes intensity independent on the emergence of the long-lived, spectrally shifted component. In contrast, the magnitude of the long-lived component, as measured from the reverse swept component, increases with the cube of the input intensity (Fig. 4), in the range of 8×10^8 to 5×10^9 W cm^{-2}. Furthermore, within the time window of the reverse sweep, the decay time of the luminescence is seen to increase with intensity, with a dependence of at least squared. The intensity onset of this behaviour appears to be a threshold value and varies with position on the sample by a factor of the order of two and, within this variation, is independent of temperature. This phenomenon is also exhibited by the single crystallite samples and may therefore be considered a solid state phenomenon.

During the course of these measurements, a description of this phenomenon by Feldmann *et al.* [12] was communicated to one of the authors (W.W.R.). At low excitation levels, the luminescence spectrum is discussed in terms of a band picture and the occurrence of this luminescence

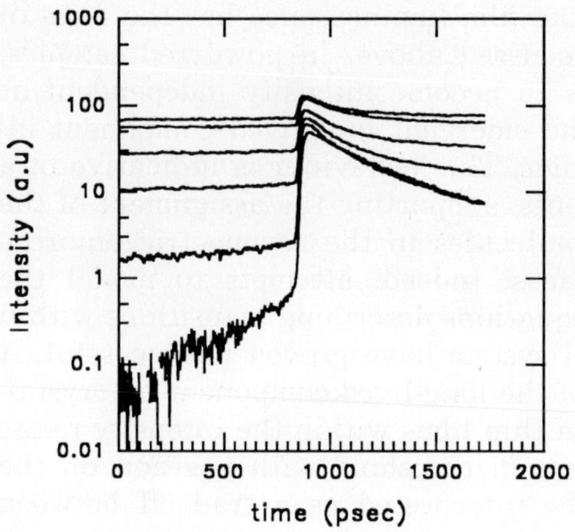

Fig. 3. Dependence of the temporal decay on increasing illumination from 8×10^8 to 5×10^9 W cm^{-2}, in steps of 8×10^8 W cm^{-2}.

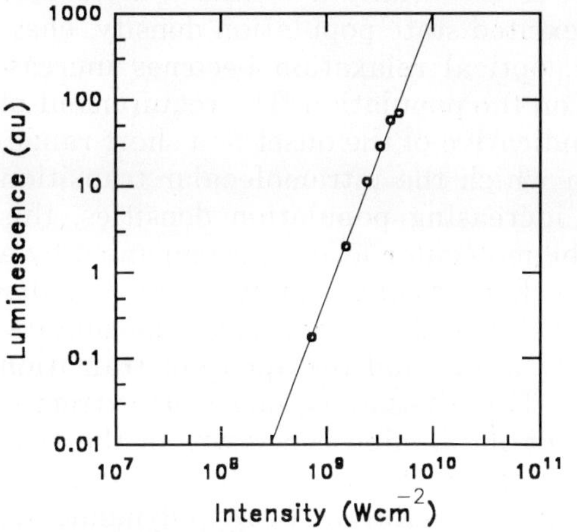

Fig. 4. Dependence of the long lived luminescence component on intensity. Solid line shows a slope of 3.

phenomenon is attributed to transitions between higher conduction bands which are populated via two step excitation processes. Although such two step excitations are well known in fullerenes [2], the second excited state is expected to thermalize rapidly. This interpretation does not account for a continuous shift in the emission spectrum. Furthermore, it is not compatible with the high order of non-linearity of the process, observed in the intensity dependence of both the luminescence efficiency and lifetime.

When considering this phenomenon, an important point is its absence in thin films, in which the microcrystallite size is of the order of 200 nm.

Over the intensity range studied, the film luminescence has the form of the fast fluorescence-like decay discussed above. In powdered samples, however, this component saturates to become intensity independent at the threshold intensity, whereas the emergent long-lived component increases with intensity to a high order. This behaviour is indicative of a separate origin of the two components, supporting the assignment of the fast component to transitions in molecules in the asymmetric environment of the microcrystallite surfaces. Indeed, attempts to model the observed behaviour based on rate equations describing transitions within a homogeneous, static energy level system have proved unsuccessful. A dependence of the intensity onset of the long-lived component on crystallite size, indicated by the absence in thin films within the intensity range covered, as well as by the variation in threshold with position on the powdered sample, may therefore be interpreted as a tradeoff between surface and bulk phenomena.

Within the bulk of the material, photoexcitation leads predominantly to population of the long-lived triplet manifold. The absence of optical relaxation from this manifold at low population densities, as is the case in solution, indicates that this triplet is essentially a localised molecular orbital. However, above a critical excited state population density, characterised by a threshold intensity, optical relaxation becomes increasingly favoured, strongly dependent on the population. The requirement of this critical population density is indicative of the onset of a short range, intermolecular interaction through which the intramolecular transition selection rules are relaxed. With increasing population densities, this interaction leads to a banding of the molecular levels, accompanied by a broadening of the emission profile with a loss of the individual molecular vibronic structure and a continuous shift to lower energies. The simultaneous increase of the transition probability and the apparent transition lifetime is difficult to conceptualise. This lifetime is, however, extracted from a fit to an exponential decay of the luminescence within the time window of the reverse sweep. The process must, however, be considered in terms of a level system which, after initial excitation, evolves dynamically as it decays, the probability of the optical transition decreasing with the cube of the population. The luminescence intensity therefore rapidly reduces as the population decreases towards the critical population density. This strong dependence of the luminescence intensity on the population is manifested as an apparent increase in the lifetime within a given time window (in this case 6.25–7 ns after excitation) with population density and therefore on incident intensity.

Conclusions

At low excitation densities, luminescence from C_{60} films, powders and single crystallites is attributable to intramolecular decay processes. The rapid decay of the luminescence is, however, in contradiction with the highly forbidden nature of intramolecular optical processes in these materials, indicating that the observed luminescence has origin in the symmetry breaking environment of the microcrystallite surfaces and that, similar to molecules in solution, bulk C_{60} exhibits no luminescent decay. The onset of a highly intensity dependent luminescence above a critical excited state population in large crystallite materials points towards the existence of significant intermolecular interactions only at highly excited state densities. Such interactions lead to a banding of the highly populated triplet levels, resulting in a broadening and spectral shift of the luminescence. Two important implications follow from such an interpretation of the observed phenomenon. Firstly, in the solid state, particularly with respect to optical measurements, the observed properties of C_{60} may be strongly dominated by those of the molecular states at the surfaces of the microcrystallites and therefore on the size of the crystallite. Secondly, the picture of bulk C_{60} which emerges is of a molecular solid which is finely balanced between an assembly of molecules governed by intramolecular processes, in the ground state, and a banded, semiconductor-like solid in which intramolecular processes are strongly suppressed, in a fully excited state. Investigations of the dependence of the threshold intensity on crystallite size are currently underway in order to clarify the competition between surface and bulk processes. Further investigations are also required to elucidate the nature of the banding interaction as well as the highly excited state properties.

Acknowledgements

W.W.R. would like to thank J. Feldmann and co-workers for communicating their results prior to publication.

References

1 I. Laszlo and L. Udvardi, *Chem. Phys. Lett.*, *144* (1988) 31.
2 J. W. Arbogast, C. S. Foote and M. Kao, *J. Am. Chem. Soc.*, *114* (1992) 2277.
3 M. R. Wasielewski, M. P. O'Neill, K. R. Lykke, M. J. Pellin and D. M. Gruen, *J. Am. Chem. Soc.*, *113* (1991) 1774.
4 J. W. Arbogast, A. P. Darmanyan, C. S. Foote, Y. Rubin, F. N. Diederich, M. M. Alvarez, S. J. Anz and R. L. Whetten, *J. Phys. Chem.*, *95* (1991) 11.
5 R. M. Fleming, T. Seigrist, P. M. March, B. Hessen, A. R. Kortan, D. W. Murphy, R. C. Haddon, R. Tycho, G. Dabbagh, A. M. Mujsce, M. L. Kaplan and S., M. Zahurak, *Mater. Res. Soc. Symp., Proc.*, *206* (1991).
6 W. Y. Chiang, M.-Z. Huang, Y.-N. Xu, W. G. Harter and F. T. Chan, *Phys. Rev. Lett.*, *67* (1991) 2045.
7 W. Krätschmer, L. Lamb, K. Fostiropolus and D. Huffman, *Nature (London)*, *347* (1990) 354.

8 M. Kaiser, J. Reichenbach, H. J. Byrne, J. Anders, W. Maser, S. Roth, A. Zahab and P. Bernier, *Synth. Met.*, *51* (1992) 251.

9 C. Reber, L. Yee, J. McKiernan, J. I. Zink, R. S. Williams, W. M. Tong, D. A. A. Ohlberg, R. L. Whetten and F. Diederich, *J. Phys. Chem.*, *95* (1991) 2127.

10 S. P. Sibley, S. M. Argentine and A. H. Francis, *Chem. Phys. Lett.*, *188* (1992) 187.

11 P. A. Lane, L. S. Swanson, Q.-X. Ni, J. Shinar, J. P. Engel, T. J. Barton and L. Jones, *Phys. Rev. Lett.*, *68* (1992) 887.

12 J. Feldman, R. Fischer, E. O. Göbel, S. Schmitt-Rink and W. Krätschmer, (1992) submitted for publication.

ODMR spectroscopy of photoexcitations in C_{60} and C_{70} polystyrene glasses

X. Wei and Z. V. Vardeny
Department of Physics, University of Utah, Salt Lake City, UT 84112 (USA)

D. Moses, V. I. Srdanov and F. Wudl
Institute of Polymers and Organic Solids, University of California, Santa Barbara, CA 93106 (USA)

Abstract

We have used a variety of optical characterization techniques to study photoexcitations in C_{60} and C_{70} fullerenes as dispersion in polystyrene glass. These techniques include photoluminescence (PL), photomodulation (PM) and their respective versions of optically detected magnetic resonance (ODMR). In both fullerenes at 4 K we found long-lived ($\sim$5 ms in C_{60} and 0.5 s in C_{70}) triplet excitons with two pronounced PA bands below the gap. The zero-field splitting parameters indicate that the triplet excitons are delocalized over the entire molecule: 6 Å in C_{60} and 7.8 Å in C_{70}.

The recent synthesis of macroscopic amounts of the all-carbon fullerenes C_{60} and C_{70} [1] has stimulated a great deal of experimental work. Most dramatically, C_{60} films show superconductivity at T_c as high as 34 K [2, 3] when doped with alkali metals. The icosahedral structure of C_{60} and its electronic energy bands have been measured with good agreement between experiment and theory [4]. Compared to C_{60}, substantially less extensive experimental studies have been carried out on C_{70}. The C_{70} molecule has a D_{5h} symmetry; its structure was recently determined [5] to be of prolate spheroid ('rugby ball'), where the long axis is 7.80 Å and the diameter of the equator is 6.94 Å. Recent optical studies of C_{60} and C_{70} include optical absorption of the ground [6–8] and excited states [9–12], photoluminescence [13–17] light induced ESR [16] and optically detected magnetic resonance of solutions and glasses [15], where the role of the triplet excitons has been emphasized.

In this work we study photoexcitations in C_{60} and C_{70} as a dispersion in polystyrene (PS) glass [18] using the following optical measurement techniques: photomodulation (PM), photoluminescene (PL), and their respective versions of optically detected magnetic resonance: the absorption detected magnetic resonance (ADMR) [19] and the PL detected magnetic resonance (PDMR) [15]. We measured in C_{60} and C_{70}:PS glasses metastable triplet excitons with two intense photoinduced-absorption

0379-6779/93/$6.00

(PA) bands in the triplet manifold. From the magnetic resonance spectra we obtained the exciton zero-field splitting (ZFS) parameters, D and E: $D = 0.0115\ \mathrm{cm}^{-1}$ and $E = 0.0008\ \mathrm{cm}^{-1}$ in C_{60}, whereas in C_{70} we measured $D = 0.0055\ \mathrm{cm}^{-1}$. We calculated from these ZFS parameters that the triplet excitons are delocalized over the entire respective molecule: 6 Å in C_{60} and 7.8 Å in C_{70}.

The PM and PL spectra have been excited by an Ar^+ laser beam at 458 nm (pump) with intensity I_L of 400 mWcm^{-2}, which was modulated at frequency f ranging from 20 Hz to 50 kHz, by an acousto-optical modulator. The PL from the sample and the transmission of a cw lamp (probe beam) were dispersed by a quarter-meter monochromator and measured by various solid state detectors in the spectral range 0.3–2.7 eV. For the ADMR measurements [19] the pump and probe beams constantly illuminate the sample mounted in a high Q microwave cavity at 3 GHz equipped with optical windows and a superconducting magnet producing a field H up to 3 Tesla. Microwave (μ-wave) resonant absorption, modulated at a frequency ϑ between 20 Hz and 40 kHz, leads to small changes δT in transmission T proportional to δn, the change in photoexcitations density n produced by the pump beam. δn is induced by transitions in the μ-wave range that change spin-dependent recombination rates. With suitable signal averaging, the system $\delta T/T$ sensitivity was about 10^{-7}. In PDMR [15] we measured changes δL in the PL intensity L with sensitivity $\delta L/L \simeq 10^{-5}$. Two types of ADMR (PDMR) spectra were obtained: the H-ADMR spectrum, in which δT (δL) is measured at a fixed probe wavelength λ while sweeping H, and the P-ADMR spectrum in which δT is measured at a constant H, in resonance, while λ (probe) is varied.

Purified C_{60} and C_{70} powders, with purity better than 99%, as established by Raman scattering spectroscopy [20], were mixed in a solution of degassed toluene/polystyrene, which was subsequently evaporated to form C_{60} and C_{70}:PS glasses.

The optical and magneto-optical measurements of C_{60}:PS glass at 4 K are shown in Fig. 1. The PM spectrum contains a strong PA band (T_1) which peaks at 1.65 eV [9, 10] and a shoulder (T_2) at 1.8 eV. Both PA bands increase linearly with I_L showing a monomolecular recombination kinetics for the photoexcitations. The H-ADMR spectrum at 1.65 eV (Fig. 1, lower inset) shows a 200 G wide triplet powder pattern at 'full-field' ($\Delta m_s = \pm 1$) around $H = 1071$ G with $\delta n/n = -6 \times 10^3$, and a narrow (5 G) signal at 'half field' ($\Delta m_s = \pm 2$) peaked at 531 G. To correlate the ADMR and PM spectra, we measured the P-ADMR spectra at 1010 (full field) and 531 (half field) G, respectively, as shown in Fig. 1. Since the P-ADMR and PM spectra in Fig. 1 are identical, we conclude that the two PA bands in the PM spectrum are due to photoinduced triplet transitions. An accurate measure of the triplet ZFS parameters can be obtained by taking the derivative of the H-ADMR spectrum [21]. From the six singularities, shown as arrows in Fig. 1 (lower inset), we calculate for the triplets in C_{60}:PS, $D = 0.0115\ \mathrm{cm}^{-1}$ and $E = 0.0008\ \mathrm{cm}^{-1}$, identical to the ZFS parame-

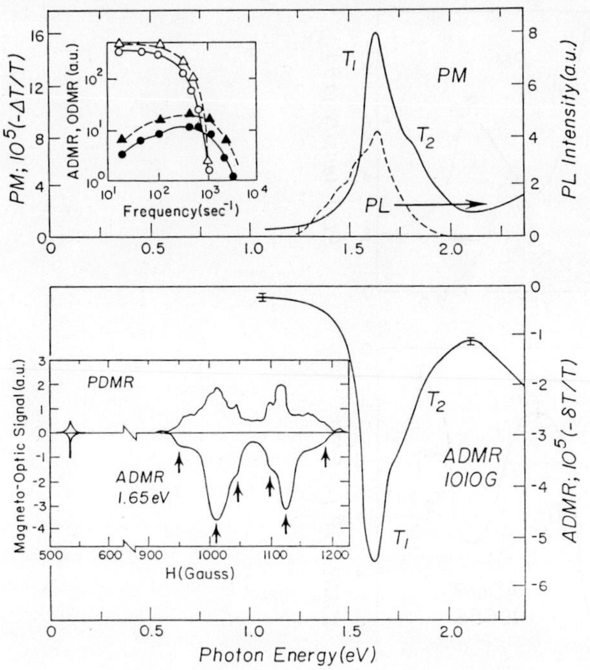

Fig. 1. PM, PL and P-ADMR spectra of C_{60}:PS glass at 4 K; the PA bands T_1 and T_2 are assigned. Lower inset shows the H-ADMR at 1.65 eV (T_1) and the H-PDMR spectra of the PL band at 'half' and 'full' fields, respectively; the six arrows point to the singularities associated with the triplet ZFS parameters. Upper inset shows the in-phase and out-of-phase ADMR and PDMR signals vs. the μ-wave modulation frequency.

ters determined previously by LESR [16]. The triplet state is thus nearly axially symmetric, of about 6 Å in size, i.e. delocalized over most of the C_{60} molecule (which is 7 Å in diameter).

The PL spectrum of C_{60}:PS is also shown in Fig. 1. It peaks at 1.65 eV [9, 13–15]. Excitation frequency dependence of the PL at 4 K shows that it contains fast and slow components. The fast PL is probably due to emission of singlet excitons, whereas the slow component is due to emission of triplet excitons, as determine by its H-PDMR spectrum [15]. As shown in Fig. 1 (lower inset) the H-PDMR and H-ADMR triplet powder patters are quite similar, however $\delta n < 0$, whereas $\Delta L > 0$. Also the in-phase and out-of-phase ADMR and PDMR μ-wave modulation frequency dependencies are identical (Fig. 1, upper inset). These two measurements show that the μ-wave induces an increase in the triplet recombination rate, which results in $\delta n < 0$ and $\delta L > 0$. Also from the frequency dependence of Fig. 1 we obtained a lifetime τ of ~5 ms [16] for the triplet excitons in C_{60}.

The optical and magneto-optical measurements of C_{70}:PS glass at 4 K are shown in Fig. 2. As in C_{60}:PS glass, the PM spectrum also contains

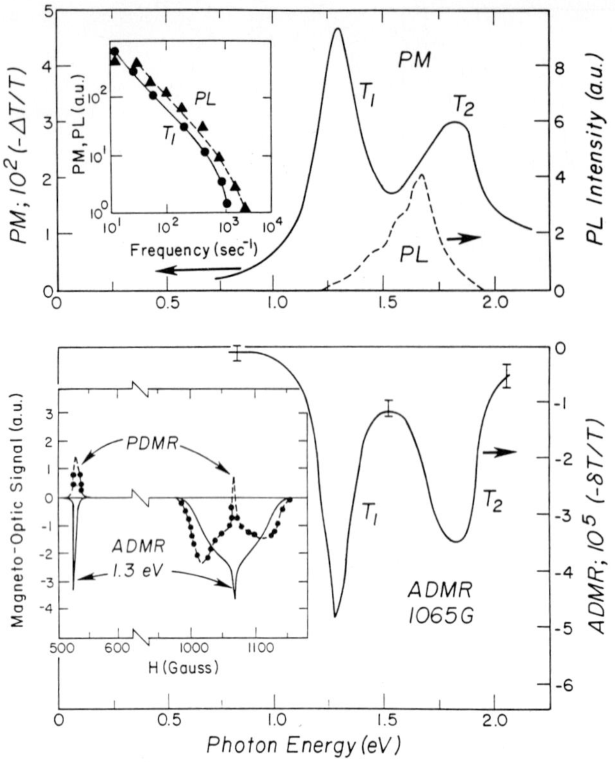

Fig. 2. PM, PL and P-ADMR spectra of C_{70}:PS glass at 4 K: the PA bands T_1 and T_2 are assigned. Lower inset shows the H-ADMR at 1.3 eV (T_1) (full line) and the H-PDMR spectra (dashed and circled line) of the PL band. Upper inset shows the laser modulation frequency dependencies of T_1 PA band (full line) and the slow component of the PL (dashed line).

two intense PA bands (now more separated): T_1 at 1.3 eV and T_2 at 1.8 eV. The H-ADMR spectrum at 1.3 eV (T_1) (Fig. 2, lower inset) shows a 100 G wide triplet powder pattern at 'full field' (FF) with $\delta n/n = -10^{-3}$, and a narrow (4 G) signal at 'half field' (HF) peaked at 532 G. To correlate the ADMR and PM spectra, we measured the P-ADMR spectra at 1050 G (FF) and 532 G (HF), respectively, as shown in Fig. 2. Since the P-ADMR and PM spectra in Fig. 1 are identical, we again conclude that the two PA bands in the PM spectrum are due to photoinduced optical transitions in the triplet manifold, as in C_{60}:PS glass.

The PL spectrum of C_{70}:PS glass is also shown in Fig. 2; it peaks at 1.7 eV [7, 8]. Excitation modulation frequency dependence of the PL at 4 K shows that it also contains fast and slow components [17]. The fast PL is probably due to emission of singlet excitons, whereas the slow component is due to emission of triplet excitons. As shown in Fig. 2 (upper inset), the slow PL component has the same f dependence as the T_1 band in the PM spectrum. Also the H-PDMR spectrum contains transitions at

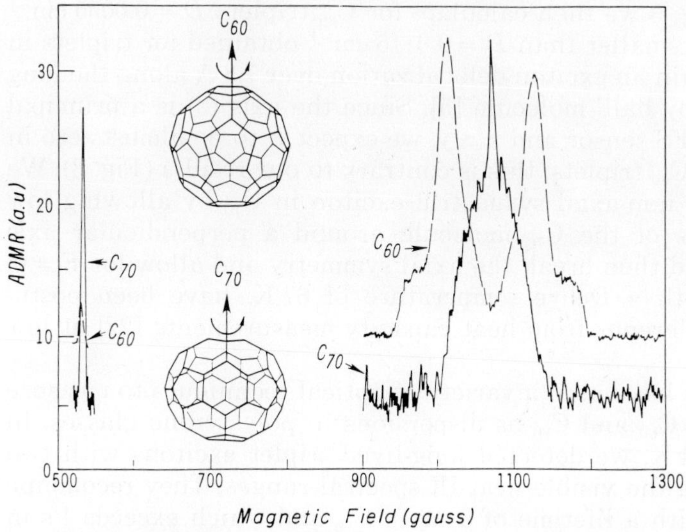

Fig. 3. The H-ADMR spectra of triplet excitons in PS glasses of C_{70} (full lines) and C_{60} (dashed lines). The molecular structure of C_{70} and C_{60} and a possible libration axis of the excitons, are also shown.

FF and HF (Fig. 2, lower inset). We note, however, that the H-PDMR triplet powder pattern in FF is not simply an integration display of a standard thermalized triplet ESR lineshape [7], since transition probabilities *as well as changes in the radiative recombination rates among the triplet sublevels*, are involved [21]; the H-PDMR spectrum contains $\Delta L < 0$ and $\Delta L > 0$ contributions which, however, belong to the same species (triplet excitons). From the frequency dependencies of T_1 and the triplet PL component shown in Fig. 2, we determine that some triplets in C_{70} are really long lived (metastable) with $\tau > 1$ s [16] at 4 K.

To more accurately determine the triplet ZFS parameters D and E for C_{70}, we plot in Fig. 3 the triplet H-ADMR spectrum of C_{70}:PS and compare it to that of C_{60}:PS. The difference between the C_{70} and C_{60} FF powder patterns is striking. The triplet FF powder pattern contains, in general, six singularities at $\pm D$, $\pm(D+3E)/2$ and $\pm(D-3E)/2$, respectively around H_0. Also simple theories involving bipolar interaction [12], relate the ZFS values with averages in the triplet excited state, where $D \sim <(r^2-3z^2)/r^5>$ and $E \sim <(y^2-x^2)/r^5>$. From Fig. 3 it seems that the FF powder patterns for triplets in both C_{60} and C_{70} contain fewer singularities, probably due to the highly symmetric excited states of these fullerenes. We can fit the powder pattern in C_{60} with nearly an axial symmetric exciton ($y \simeq x$), so that the FF spectrum contains four main singularities at $\pm D$ and $\pm D/2$, respectively, around H_0. In order to fit the powder pattern in C_{70}, however, we postulate $E \simeq D/3$, so that the FF spectrum shows three main singularities: two at $\pm D$, respectively, around H_0, and one at H_0. From Fig. 3 we then calculate for C_{70} triplets $D = 0.0055\ \mathrm{cm}^{-1}$ [16], which is much smaller than $D = 0.0115\ \mathrm{cm}^{-1}$ obtained for triplets in C_{60}. From D we obtain an exciton delocalization over 7.8 Å along the long axis of the C_{70} 'rugby ball' molecule [5]. Since the axis (z) is a principal axis of the triplet ZFS tensor and $x \simeq y$, we expect E to be almost zero in C_{70} similar to E for C_{60} triplets; this is contrary to our results (Fig. 3). We can understand the non-axial symmetric exciton in C_{70} by allowing low frequency librations of the C_{70} molecule around a perpendicular axis (Fig. 3). These would then break the axial symmetry and allow for $E \neq 0$. Such librations, with a Debye temperature of 67 K, have been postulated to exist in fullerenes from heat capacity measurements [22] at low temperatures.

In summary, we have used a variety of optical techniques to measure photoexcitations in C_{60} and C_{70} as dispersions in polystyrene glasses. In both fullerenes at 4 K, we detected long-lived triplet excitons with two intense PA bands in the visible near-IR spectral ranges. They recombine monomolecularly with a lifetime of 5 ms in C_{60} and which exceeds 1 s in C_{70}. The ZFS parameters that we measured from the FF triplet powder patterns in H indicate exciton delocalization over the entire fullerene molecules.

Acknowledgements

The work at the University of Utah was supported in part by the DOE grant no. DE-FG 02-89 ER 45409 and by ONR grant no. N00014-91-C-0104. The work at UCSB was supported by DMR 88-20933, NSF grant no. CHE 89-08323. V.I.S. acknowledges support by the NSF QUEST program at UCSB.

References

1 W. Krätschmer, L. D. Lamb, K. Fostiropoulos and D. R. Huffman, *Nature (London), 347* (1990) 354.

2 M. J. Rosseinsky, A. P. Ramirez, S. H. Glarum, D. W. Murphy, R. C. Haddon, A. F. Hebard, T. T. M. Palstra, A. R. Kortan, S. M. Zahurak and A. V. Makhija, *Phys. Rev. Lett., 66* (1991) 2830.

3. A. F. Hebard, M. J. Rosseinsky, R. C. Haddon, D. W. Murphy, S. H. Glarum, T. T. M. Palstra, A. P. Ramirez and A. R. Kortan, *Nature (London), 350* (1991) 600.

4 J. S. Miller, *Adv. Mater, 3* (1991) 362.

5 D. R. McKenzie, C. A. Davis, D. J. H. Cockayne, D. A. Muller and A. M. Vassallo, *Nature (London), 355* (1992) 622.

6 H. Ajie, M. Alvarez, S. Anz, R. Beck, F. Eiederich and R. Whetten, *J. Phys. Chem., 94* (1990) 8630.

7 J. P. Hare, H. W. Kroto and R. Tayl, *Chem. Phys. Lett., 177* (1991) 394.

8 C. Reber, L. Yee, J. McKiernan, J. I. Zink, R. S. Williams, W. M. Tong, D. A. A. Ohlberg, R. L. Whetten and F. Diederich, *J. Phys. Chem., 95* (1991) 2127.

9 Y. Kajii, T. Nakagawa, S. Suzuki, Y. Achiba, K. Obi and K. Shibuya, *Chem. Phys. Lett.*, *181* (1991) 100.
10 T. W. Ebbesen, K. Tanigaki and S. Kuroshima, *Chem. Phys. Lett.*, *181* (1991) 501.
11 J. W. Arbogast and C. S. Foote, *J. Am. Chem. Soc. 113* (1991) 8886.
12 K. Tanigaki, T. W. Ebbessen and S. Kuroshima, *Chem. Phys. Lett.*, *185* (1991) 189.
13 X. Wei, Z. V. Vardeny, D. Moses, V. I. Srdanov and F. Wudl, *Synth Met.*, *50* (1992) 549.
14 K. Pichler, S. Graham, O. M. Gelsen, R. H. Friend, W. J. Romanow, J. P. McCauley, Jr., N. Coustel, J. E. Fischer and A. B. Smith, *J. Phys.: Condensed Matter*, *3* (1991) 9259.
15 P. A. Lane, L. S. Swanson, Q. X. Ni, J. Shinar, J. P. Engel, T. J. Barton and L. Jones, *Phys. Rev. Lett.*, *68* (1992) 887.
16 M. R. Wasielewski, M. P. O'Neil, K. R. Lykke, M. J. Pellin and D. M. Gruen, *J. Am. Chem. Soc.*, *113* (1991) 2774.
17 S. P. Sibley, S. M. Argentine and A. H. Francis, *Chem. Phys. Lett.*, *188* (1992) 187.
18 X. Wei, Z. V. Vardeny, D. Moses, V. I. Srdanov and F. Wudl, *Solid State Commun.*, in press.
19 X. Wei, B. Hess, Z. V. Vardeny and F. Wudl, *Phys. Rev. Lett.*, *68* (1992) 666.
20 D. Dick, R. E. Benner, Z. V. Vardeny, D. Moses, V. I. Srdanov and F. Wudl, *Synth. Met.*, *50* (1992) 543.
21 H. C. Benner, in R. H. Clarke (ed.), *Triplet State ODMR Spectroscopy*, Wiley, New York, 1982.
22 W. P. Beyermann, M. F. Hundley, J. D. Thompson, F. N. Diederich and G. Gruner, *Phys. Rev. Lett.*, *68* (1992) 2046.

Time-resolved luminescence measurements in poly(*p*-phenylenevinylene)

I. D. W. Samuel[a], B. Crystall[b], G. Rumbles[b], P. L. Burn[c], A. B. Holmes[c] and R. H. Friend[a]

[a]*Cavendish Laboratory, Madingley Road, Cambridge CB3 0HE (UK)*

[b]*Department of Chemistry, Imperial College, South Kensington, London SW7 2AY (UK)*

[c]*University Chemical Laboratory, Lensfield Road, Cambridge CB2 1EW (UK)*

Abstract

We report time-resolved measurements of photoluminescence in poly(*p*-phenylenevinylene) (PPV) and related polymers, using the technique of time-correlated single photon counting. We find the photoluminescence is longer lived in less-conjugated samples, and that this accounts for the higher efficiency of light-emitting devices made from less-conjugated materials. At low temperatures, the luminescence is longer lived, and at 77 K there is an interesting red shift of the emission during the first nanoseconds after photoexcitation. We consider that this arises from migration of excitons to more-conjugated (and therefore lower energy) regions of the sample.

Introduction

Conjugated polymers possess π-electrons delocalised along the polymer chain. The associated electronic response is highly non-linear, with the formation of intra-gap states, and associated novel non-linear and electro-optical properties [1–4]. The intra-gap states form either following photoexcitation, or via charge injection in suitable device structures. An interesting example is luminescence, which can be either optically or electrically excited [2–4].

A key feature of conjugated polymers is that there is enormous flexibility of sample preparation [5]. Many of these materials, including the members of the poly(arylenevinylene) family, can be prepared via a precursor route, giving excellent processing properties. In this approach a non-conjugated solution-processable precursor polymer is fabricated into the desired sample shape, and then converted into the conjugated form. For example, light-emitting diodes of PPV can be prepared by spin coating a precursor polymer onto a suitable substrate, followed by thermal conversion and evaporation of contacts [3, 4].

Selection of the leaving group and conversion conditions gives control of the degree of conjugation (extent of π-electron delocalisation) [6, 7], as shown in Figs. 1–3. The THT leaving group precursor (I) may be converted to fully conjugated PPV (IV) by heating at 220–300 °C, and the

0379-6779/93/$6.00

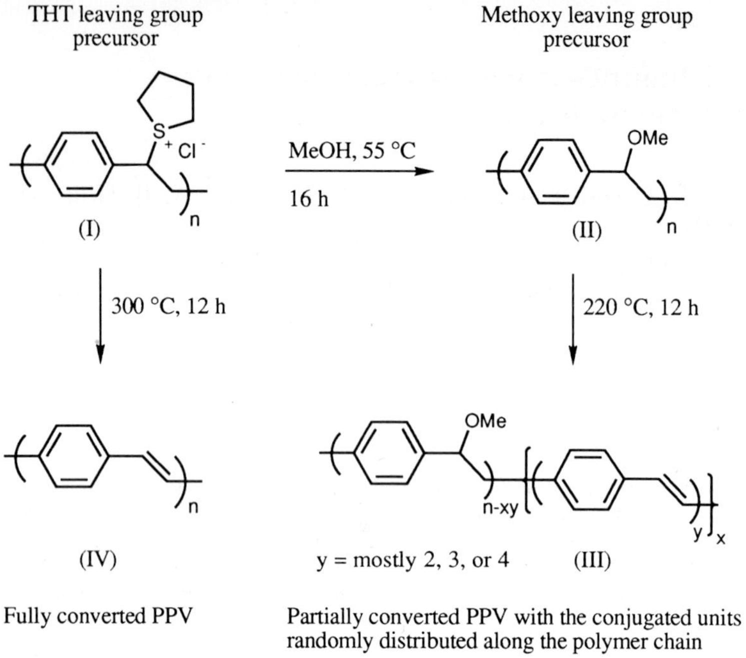

Fig. 1. Synthesis of PPV (simplified): the preparation conditions give control over the degree of conjugation.

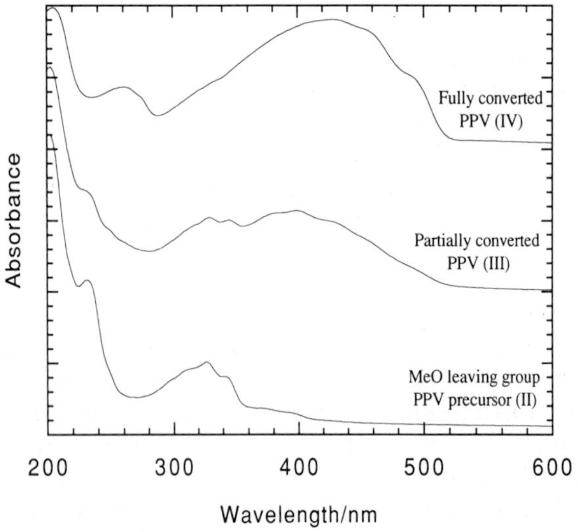

Fig. 2 . Absorption spectra at 295 K of PPV, partially conjugated PPV and PPV precursor.

reaction proceeds slowly even at room temperature. In contrast, the MeO leaving group precursor (II) does not convert significantly at room temperature, making it much more convenient for studies of a precursor. It requires both heat and acid catalysis for full conversion, and when heated

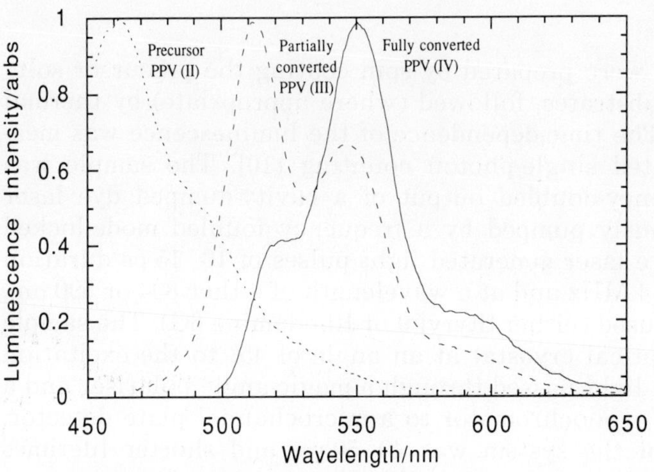

Fig. 3. Steady-state luminescence spectra (normalised) of PPV, partially conjugated PPV and PPV precursor. Measured at 295 K with excitation at 350 nm.

in the absence of acid, gives a random copolymer of conjugated and non-conjugated units (III). The precursor solutions contain some conjugated units, and this accounts for the absorption of the precursor (II) at 280–360 nm seen in Fig. 2. Figures 2 and 3 show that as the conjugation length increases, there is a shift of spectral features to longer wavelength. The band gap can also be tuned by synthesising copolymers, and further control can again be achieved by adjusting the conversion conditions: this has recently been used to pattern copolymers [4].

Time-resolved measurements of photoluminescence give important information about photoexcitations and their dynamics. Previous studies in PPV [8, 9] have used oriented films made from a precursor with a dimethyl sulfonium leaving group. Wong *et al.* [8] studied the decay of the total luminescence (without spectral resolution) as a function of conversion time, and found faster decay in more conjugated material. Furukawa *et al.* [9] made measurements with spectral resolution, but only studied fully conjugated material.

In the present work we have studied the time-dependence of photoluminescence in unoriented spin-coated films, as used for electroluminescent devices. Photoluminescence provides a much cleaner way than electroluminescence of studying radiative and (by inference) non-radiative decay processes in conjugated polymers. For example, it is known that the quantum efficiency of electroluminescent devices made from partially conjugated PPV (III) is approximately double that of devices made from fully conjugated PPV (IV) [7]. Transient photoluminescence measurements enable us to determine whether this is due to a real increase in the molecular luminescence efficiency, or some other device property such as carrier injection or quality of interfaces.

Experimental

Thin-film samples were prepared by spin coating the precursor solution onto spectrosil substrates, followed (where appropriate) by thermal conversion *in vacuo*. The time-dependence of the luminescence was measured by time-correlated single-photon counting [10]. The sample was excited by the frequency-doubled output of a cavity-dumped dye laser which was synchronously pumped by a frequency-doubled mode-locked Nd:YAG laser. The dye laser generated light pulses of 10–15 ps duration at a repetition rate of 4 MHz and at a wavelength of either 804 or 590 nm, depending on the dye used (either Styryl 9 or Rhodamine 6G). The sample was mounted in an optical cryostat at an angle of 45° to the excitation beam, and the emitted light passed through a magic-angle polariser and a subtractive-dispersion monochromator to a microchannel plate detector. The time-resolution of the system was 40–50 ps, and shorter lifetimes could be measured by deconvolving the impulse response from the measured response.

Results and discussion

Effect of conjugation

Figure 4 compares the decay of photoluminescence following excitation at 402 nm in precursor PPV (II), partially conjugated PPV (III) and fully conjugated PPV (IV). The results of exponential fits to the data are summarised in Table 1. The decay of luminescence in the precursor is approximately (single) exponential, with a time constant of 927 ± 90 ps.

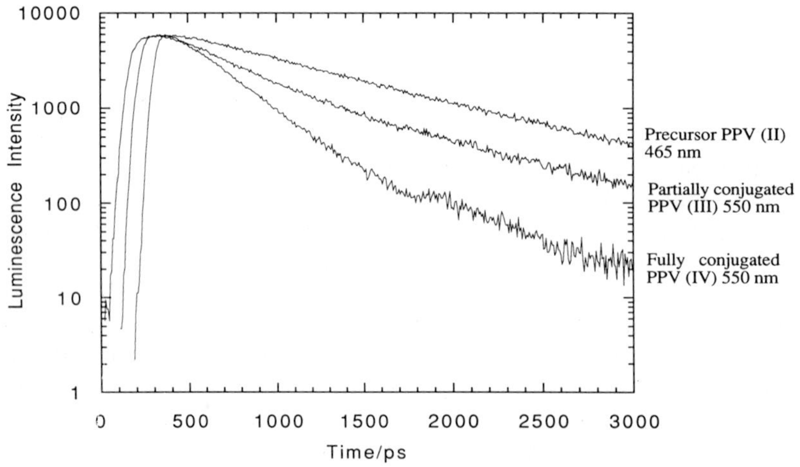

Fig. 4. Transient luminescence in PPV at 295 K, following excitation at 402 nm. Each curve is marked with the material, and the wavelength at which the emission was measured. The data have been rescaled to give the same peak size.

TABLE 1

Summary of exponential fits to luminescence decay

Material	λ_{exc} (nm)	λ_{em} (nm)	Temperature (K)	t_1 (ps)	A_1 (%)	t_2 (ps)	A_2 (%)	t_3 (ps)	A_3 (%)	χ^2
Precursor PPV (II)	402	465	295	927	100					1.35
Partially conjugated PPV (III)	402	550	295	314	45	936	55			0.97
Fully conjugated PPV (IV)	402	550	295	249	84	763	16			1.10
	295	550	295	253	70	538	27	2600	3	1.36
	295	550	150	397	29	1118	51	4580	19	1.14
	295	550	77	345	19	1104	59	4572	22	1.4

The Table shows the results of three-exponential fits to the luminescence decay after deconvolution from the instrument impulse response. t_1, t_2 and t_3 denote the time constants of the three exponentials, and A_1, A_2 and A_3 denote their relative (integrated) weights. Where there is no entry for t_2 and/or t_3, this means that an acceptable fit was achieved with one or two exponentials.

The decay is faster in partially conjugated PPV, and can be fitted by two exponentials with time constants of 314 ± 30 and 936 ± 90 ps. In fully converted PPV, the decay is faster still: the time constants are slightly shorter, and there is much greater weight in the fast component.

The dominant product of intra-chain photoexcitation is believed to be the formation of (singlet) polaron-excitons, which may decay by a combination of radiative and non-radiative pathways [2, 11]. In the model oligomer (V) the luminescence has a lifetime of 1100 ± 150 ps, and

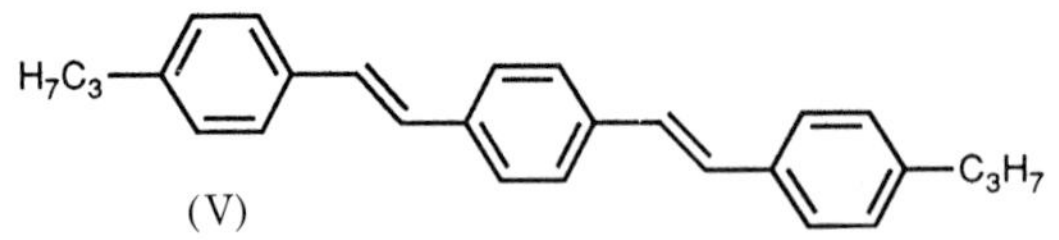

(V)

a quantum yield of 94% [12]. We expect the natural radiative lifetime in PPV to be similar, and consider that the faster decay of the luminescence arises from competing non-radiative decay channels. The time constant for decay of the luminescence, τ, is related to the rate constant for radiative decay ($1/\tau_R$) and the rate constant for non-radiative decay ($1/\tau_{NR}$) by $1/\tau = 1/\tau_R + 1/\tau_{NR}$, and the quantum yield $= \tau/\tau_R$.

Possible non-radiative decay processes include exciton migration to quenching sites, intersystem crossing to a triplet state, internal conversion to the ground state, and exciton–exciton collisional annihilation. We can rule out the last possibility as the excitation densities are modest (10^{10}–10^{11} photons/cm^2), and no intensity dependence is obverved. The migration of excitons to quenching sites will be faster in more-conjugated material, and provides a possible explanation for the more rapid decay of the luminescence in more-conjugated PPV. The slower decay of photoluminescence in partially conjugated PPV (III) suggests that the quantum yield for luminescence is 1.8 ± 0.2 times that for fully conjugated PPV (IV), and accounts for the doubling of the quantum yield for electroluminescence in devices made from the former material.

Wong *et al.* [8] also observed a faster decay of luminescence in more-conjugated material, although the decays were faster than we observed. The faster decay may be due to the greater order in oriented material, which may facilitate the migration of excitons to quenching sites.

Temperature dependence

The decay of the luminescence in fully conjugated PPV (IV) following excitation at 295 nm was studied as a function of temperature. At room temperature, the decay of the luminescence is approximately the same across the whole spectrum of the emission, although the bluer end decays slightly more quickly. At 150 K, the decay is slower and there is some change in the emission spectrum. At 77 K, there is a pronounced shift of the emission towards the red with increasing time. This effect is very clear, both in time-resolved emission spectra (Fig. 5), and in the time-dependence of the luminescence at several different wavelengths (Fig. 6). We consider that we initially photoexcite polymer chains with a range of conjugation lengths, and that the red shift arises from migration of excitons to more-conjugated (and therefore lower energy) regions of the sample.

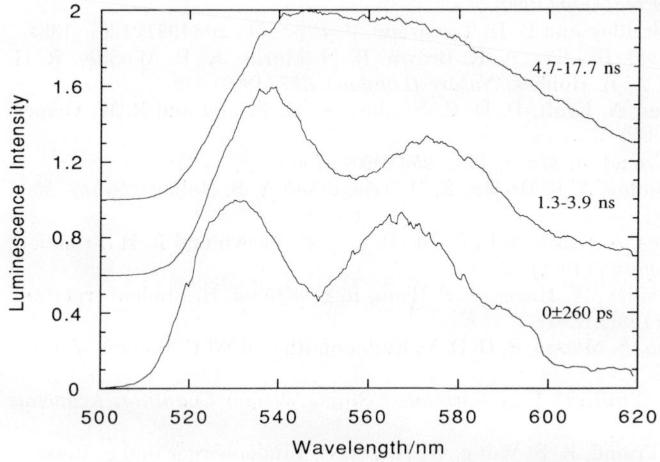

Fig. 5. Time-resolved emission spectra of fully conjugated PPV (IV) measured at 77 K following excitation at 295 nm. The data have been rescaled, but are not corrected for instrumental response functions. The maximum number of counts in a channel (before rescaling) was 2745 for 0 ± 260 ps, 1485 for 1.3–3.9 ns, and 3399 for 4.7–17.7 ns.

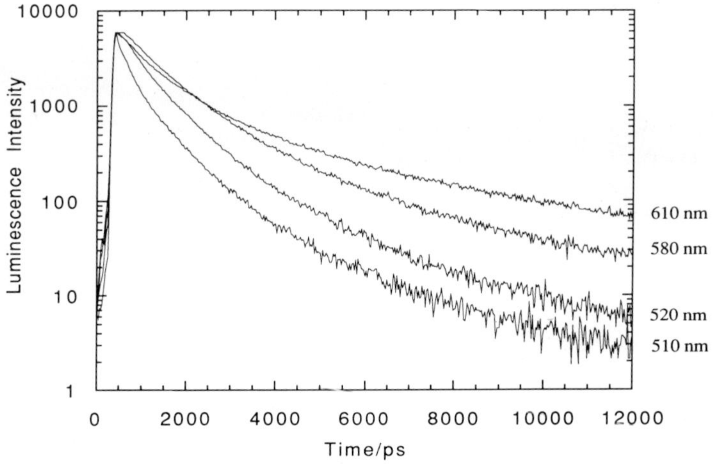

Fig. 6. Transient luminescence at various emission wavelengths in fully converted PPV (IV). Measured at 77 K, with excitation at 295 nm.

Acknowledgement

We are most grateful to Dr D. D. C. Bradley for many helpful discussions.

References

1 *Mater. Res. Soc. Symp. Proc.*, *173* (1990).
2 R. H. Friend, D. D. C. Bradley and P. D. Townsend, *J. Phys. D.*, *20* (1987) 1367–1384.
3 J. H. Burroughes, D. D. C. Bradley, A. R. Brown, R. N. Marks, K. D. Mackay, R. H. Friend, P. L. Burns and A. B. Holmes, *Nature (London)*, *347* (1990) 539.
4 P. L. Burn, A. B. Holmes, A. Kraft, D. D. C. Bradley, R. H. Friend and R. W. Gymer, *Nature (London)*, *356* (1992) 47.
5 W. J. Feast and R. H. Friend, *J. Mater. Sci.*, *25* (1990) 3796.
6 P. L. Burns, D. D. C. Bradley, A. R. Brown, R. H. Friend and A. B. Holmes, *Synth. Met.*, *41–43* (1991) 261.
7 P. L. Burns, A. B. Holmes, A. Kraft, D. D. C. Bradley, A. R. Brown and R. H. Friend, *J. Chem. Soc., Chem., Commun.*, (1992) 32.
8 K. S. Wong, D. D. C. Bradley, W. Hayes, J. F. Ryan, R. H. Friend, H. Lindenberger and S. Roth, *J. Phys. C*, *20* (1987) L187.
9 M. Furukawa, K. Mizuno, A. Matsui, S. D. D. V. Rughooputh and W. C. Walker, *J. Phys. Soc. Jpn.*, *58* (1989) 2976.
10 D. V. O'Connor and D. Phillips, *Time-Correlated Single Photon Counting*, Academic Press, London, 1984.
11 D. D. C. Bradley, R. H. Friend, K. S. Wong, W. Hayes, H. Lindenberger and S. Roth, in *Springer Series on Solid State Sciences 76, Electronic Properties of Conjugated Polymers*, Springer, Berlin, 1988.
12 I. B. Berlman, *Handbook of Fluorescence Spectra of Aromatic Molecules*, Academic Press, New York, 2nd edn., 1971.

Synthetic Metals, 54 (1993) 289–293

Non-linear optical properties modification in some substituted polydiacetylenes due to change of basicity constant of the side groups

N. V. Agrinskaya, E. G. Guk and L. A. Remizova

A. F. Ioffe Physico-Technical Institute, Polytechnicheskaya 26, St. Petersburg 194021 (Russian Federation)

Abstract

Four symmetrical derivatives of diacetylenes (1,6-bis-piperidine, 1,6-dimorpholine, 1,6-dimethylphenylamine and 1,6-bis-diphenylamine) of 2,4-hexadiene have been synthesized. The polymerization behavior and third-order non-linear optical susceptibilities ($\chi^{(3)}$) have been investigated as a function of the donor or acceptor properties of the side groups (side groups basicity constant, K). The values of $\chi^{(3)}$ appear to be rather large in resonance and near resonance spectral regions. In vacuum evaporated films we have obtained $\chi^{(3)}=1\times10^{-7}$ e.s.u. in resonance. The relaxation time for non-linear polarization has been evaluted as $<10^{-12}$ s. The observed dependencies of $\chi^{(3)}$ and the degree of polymerization from K allow the possibility to optimize the parameters that are necessary for application of these non-linear optical materials.

Organic materials like conjugated polymers with π-electrons delocalized along the polymer chain have attracted a great deal of attention due to the large values of third-order non-linear optical susceptibility $\chi^{(3)}$ found in these materials [1, 2]. In comparison with semiconductor quantum well structures they have advantages connected with the very small values of response time $\leqslant10^{-13}$ s and with small absorption losses. These features as well as the plasticity of the polymer materials provide prospects for their applications in non-linear optical devices [3].

Polydiacetylenes (PDA) are the materials that have been investigated most intensively. The PDA class is very large; their electronic properties and polarizability being dependent on the type of substituent R attached to the carbon backbone. According to theoretical calculations [4], the introduction of substituents containing phenyl groups (with acceptor properties) leads to a significant decrease of band gap E_g and to an increase of $\chi^{(3)}$. On the other hand substituents containing methyl groups (donors) lead to an increase of E_g. The electronic characteristics of the side groups enable the electronic structure of PDA to be finely tuned. The sensitivity of this tuning can be governed by the coupling strength between R and the polymer backbone. Unfortunately it has not been possible to verify the trends predicted because known diphenyldiacetylene derivatives cannot be polymerized.

We have studied the polymerizability and optical properties of PDA with an intermediate CH_2 group between the substituent and the di-

0379-6779/93/$6.00

acetylene fragment as a function of the chemical nature of the nitrogen-containing side groups (their chemical formulae are given in Fig. 1). It has been shown that the degree of polymerization and the $\chi^{(3)}$ parameter increase as a result of transition from a donor type substituent to an acceptor one.

Four symmetrical derivatives of diacetylenes (1,6-bis-piperidine, 1,6-dimorpholine, 1,6-dimethylphenylamine and 1,6-diphenylamine) of 2,4-hexadiene have been synthesized by the method described in ref. 5. Along this row the basicity constant K of the side groups varies in the following way: 11.1, 8.3, 5.04, 0.73. That is, substituent 1 is a strong donor, 2 is a weak donor, 3 is a weak acceptor, 4 is a strong acceptor. These monomers were dissolved in CCl_4 along with polymethylmethacrylate (PMMA), then the solution was placed on a glass substrate. Thin films of thickness 5–15 μm were formed after slow evaporation of CCl_4. Thin films of some monomers with a thickness of 0.05–0.1 μm were obtained by the vacuum evaporation technique [2]. Polymerization was produced by UV, gamma rays and heat treatment. It has been shown that the polymerization capabilities of these compounds are rather different. The monomers 1–3 are hardly polymerized even under gamma irradiation, while monomer 4 is polymerized completely in the presence of daylight. After polymerization the films 1–3 were yellow in color, while film 4 was red. The yellow films have a good optical quality while the red ones show strong optical scattering due to the presence of large non-oriented crystallites.

Absorption spectra for some polymerized films are shown in Fig. 1. The red film PDA-4 shows a pronounced maximum in the 0.56 μm region,

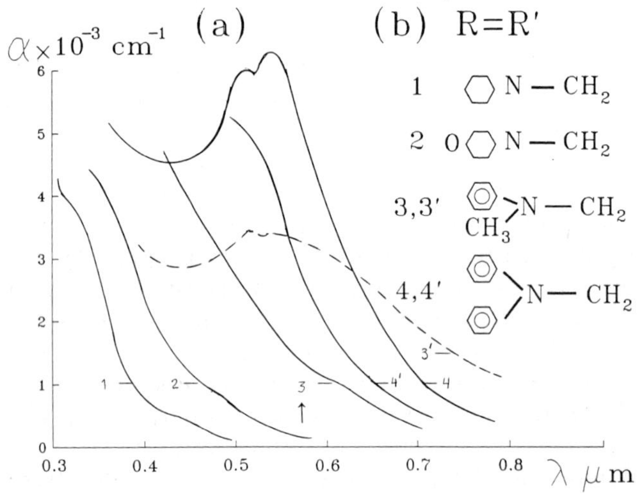

Fig. 1. (a) Optical absorption spectra. Curves 1–4′: PDA in PMMA matrix, curve 3′: vacuum evaporated PDA-3 film. The values of α for samples 1, 2, 3, 4′ are given in absolute units, for samples 3′, 4 in relative units. An arrow denotes excitation wavelength of 4-photon scattering. (b) Structural chemical formulae of the substituents R = R′.

connected with absorption of the polymer. The absorption coefficient (α) at the maximum of the band is equal to 2×10^4 cm^{-1}. Taking into account the monomer content of the matrix PMMA (15%) one estimates $\alpha \sim 10^5$ cm^{-1}. Such a value is characteristic for most PDA in exciton absorption maximum. Continuous absorption tails have been observed for the other samples (1–3) in the region 0.7–0.35 μm; the main maximum being situated in the UV spectral region (0.35 μm). It means that the polymer content is small for these samples, while the spectral behavior of α is governed by absorption of oligomers with different numbers of repeating fragments $N < 7$. Some structures have been observed for samples 1–3 in the regions 0.42, 0.45 and 0.59 μm, respectively, on the background of the absorption tails. If one prescribes this spectral structure to polymer absorption bands one can estimate from the α values the polymer content to be 0.5, 1 and 1% for samples 1, 2 and 3, respectively. From analysis of the absorption spectra one can conclude that the band gap decreases from 2.9 to 2.2 eV in the row 1–4. Such behavior is in accordance with the theory [4] that predicted an increasing E_g due to replacing of an acceptor type substituent by a donor type.

To determine $\chi^{(3)}$ we used the method of four-photon scattering under biharmonical pumping [6, 7]. The sample was irradiated by two light beams I_1 and I_2 (angle between them being 1°)with wave lengths λ_1 and λ_2, produced by a tunable dye laser, pumped by nanosecond pulses of second harmonic of a neodym laser ($\lambda_1 = 0.572$ μm). The regime close to degenerate $\omega_1 - \omega_2 = \Delta\omega = 1$–$10$ cm^{-1} was realized. Such a mismatch between wavelengths prevents scattering by the thermal diffraction grating, because the frequency of low-frequency polarization oscillations is much larger than the thermodiffusion rate. From the dependence of scattering signal intensity on frequency $\omega = 2\omega_1 - \omega_2$ from $\Delta\omega$ one can estimate the non-linear optical response relaxation time τ. For all of the samples studied $\tau \leqslant 10^{-12}$ s. Such a small response time is characteristic for PDA even for resonant excitation [2]. The values of $|\chi^{(3)}|$ estimated from dependencies of intensity of scattering signal on the pumping beam intensity are given in Table 1 for all samples investigated, except the red film PDA-4. To obtain optically transparent films PDA-4 we changed the polymerization procedure. It was carried out directly in solutions CCl_4 and PMMA. In this way the optically transparent yellow film PDA-4 was obtained. From the optical absorption spectra, the polymer content in these films is evaluated to be 1–2%. The values of $|\chi^{(3)}|$ are given Table 1.

It is known that the long polymer chains ($N > 7$) are responsible for the values of $\chi^{(3)}$ [1, 2]. This conclusion is supported by the fact that our films prepared from monomers or from light fraction oligomers (containing only chain fragments with $N = 2$–3) did not demonstrate the 4-photon scattering signal or it was very weak. Taking into account that the matrix and short oligomers do not contribute to $\chi^{(3)}$ and having information about the polymer content, one can estimate $\chi^{(3)}$ for completely polymerized samples 1–4: $< 2 \times 10^{-11}$, 1×10^{-9}, 6×10^{-9} and 1×10^{-8} e.s.u. One can

TABLE 1

The parameters of PDA with different substituents (R)

Sample no.	K	Polymer content (%)	$\lvert\chi^{(3)}\rvert$ e.s.u. $\lambda = 0.57\ \mu m$	α (cm^{-1}) $\lambda = 0.57\ \mu m$	n_2/α (cm^3/W)
1	11.1	0.5	$\leqslant 10^{-13}$	5×10^1	$\leqslant 10^{-16}$
2	8.3	1	$1-4 \times 10^{-11}$	2×10^2	$1.2-5 \times 10^{-15}$
3	5.04	2	1.2×10^{-10}	10^3	2.5×10^{-15}
4 (red)	0.79	15		2×10^4	
4′ (yellow)	0.79	2	2×10^{-10}	2×10^3	2.5×10^{-15}
3′ (vac. evap.)	5.04	10	1×10^{-7}	10^5	2.5×10^{-14}

see that the values of $\chi^{(3)}$ correlate with the basicity constant K of the substituent for samples 1–4. The smallest values of $\chi^{(3)}$ correspond to substituents exhibiting properties of a strong donor; the transition to an acceptor type substituent leads to a strong increase of $\chi^{(3)}$. It should be noted that the value of $\chi^{(3)}$ changes dramatically if the excitation wavelength approaches resonance. In particular for samples 3 and 4 we are apparently dealing with an excitation close to a resonance peak. A characteristic value of $\chi^{(3)}$ for a resonant excitation in a different PDA is 10^{-8} e.s.u. [2].

For the practical use of non-linear optical materials it is necessary to obtain an optimal relation between $\chi^{(3)}$ and α at excitation wavelength. Our samples with a small polymer content have some advantages because of the small value of α and the good optical quality. The parameter n_2/α calculated for our samples 1–4′ is given in Table 1.

$$n_2(cm^2/W) = \frac{|\chi^{(3)}|\ (\text{e.s.u.})}{19 n_0^{\,2}}$$

where n_0 is the linear refraction coefficient, $n_0 = 1.8$ for PDA.

The parameter n_2/α has a value $2.5–5 \times 10^{-15}\ cm^3/W$ for our samples. For instance typical PDA 4BCMU studied in detail has a maximum value of $n_2/\alpha = 4.7 \times 10^{-15}\ cm^2/W$ [8]. Therefore films with a small polymer content have advantages connected with small values of α and with possibilities to produce samples with small light scattering.

Thus our investigations have shown that electronic structure and non-linear optical properties of PDA depend on the chemical nature of the side groups, the values E_g decreasing and $\chi^{(3)}$ increasing as a result of transition from donor type substituent (containing saturated alkyl groups) to an acceptor one (containing phenyl groups). One can expect that completely polymerized oriented PDA films with phenyl groups will demonstrate much larger non-linear polarizability than has been observed in this study. In this respect the preparation of oriented PDA films by the vacuum evaporation method seems to have good prospects. Our preliminary result is $|\chi^{(3)}| = 1 \times 10^{-7}$ e.s.u. for vacuum evaporated PDA-3 film (excitation wavelength $\lambda = 0.572\ \mu m$ is near the resonance peak). As far as we know this result is a record for PDA.

Acknowledgements

The authors are grateful to A. G. Spiro and S. E. Kulya for the measurements of the samples' parameters by the 4-photon scattering method.

References

1 T. Kobayashi (ed.), *Nonlinear Optics of Organic and Semiconductors*, Springer, Berlin, 1989.
2 D. S. Chemla and J. Zyss, *Nonlinear Optical Properties of Organic Molecules and Crystals*, Academic Press, Orlando, FL, 1987.
3 G. I. Stegemon, R. Zanoni, K. Rochford and C. T. Seaton, *NATO ASI Series E*, Vol. 162, Kluwer, Dordrecht, 1989, p. 257.
4 B. J. Orchard and S. K. Triparthy, *Macromolecules, 19* (1986) 1844.
5 A. S. Hay, *J. Org. Chem., 27* (1962) 3320.
6 T. Yagima and H. Souma, *Phys. Rev. A, 17* (1978) 309.
7 N. V. Agrinskaya, E. G. Guk and L. A. Remisova, *Fiz. Tverd. Tela, 33* (1991) 2770.
8 K. S. Wong, S. G. Han and Z. V. Vardeny, *Synth. Met., 41–43* (1991) 3209.

Non-linear optical properties of conjugated polymers in the SSH model: third harmonic generation

Yuri I. Dakhnovskii and Kirill A. Pronin*
Institute of Chemical Physics, Russian Academy of Sciences, Kosygin Street 4, Moscow 117977 (Russian Federation)

Abstract

With the use of the Keldysh diagram technique we rigorously calculate the non-linear optical susceptibility of a conjugated polymer in the SSH model for third harmonic generation. The result is valid for an arbitrary band gap width and slow relaxation due to phonon scattering. A detailed comparison with the linearized TLM model is carried out. The latter neglects many processes incorporated in SSH but due to the compensation of contributions gives good results that differ only by correction terms proportional to the square of the ratio of the band gap to the band width.

Introduction

The non-linear optical (NLO) properties of conjugated polymers are of considerable interest because of the prospects of promising technical applications [1]. These materials exhibit an extremely short response time of the order of 10^{-13} s and extremely high hyperpolarizabilities [1–6]. For example for third harmonic generation (THG) the susceptibility $\chi^{(3)}$ is of the order of 10^{-11} e.s.u. for polydiacetylene [3], 10^{-10} e.s.u. for *cis*-$(CH)_x$ and polythiophene [4] and 10^{-9} for *trans*-$(CH)_x$ and up to 10^{-8} e.s.u. for oriented $(CH)_x$ samples [5, 6].

A large number of quantum chemical calculations of NLO properties of polymers have been performed [1]. Rigorous analytic results for the third order susceptibility of a Peierls dielectric having a dimerized degenerate ground state (*trans*-$(CH)_x$ without solitons) have been obtained by Wu [7] for the linearized TLM model using the Keldysh diagram technique. A non-degenerate conjugated polymer in the TLM model has been considered in refs. 8 and 9 taking into account weak damping of electrons. With the help of a Genkin–Mednis approach $\chi^{(3)}$ for THG has been calculated in refs. 10 and 11; the numerical curves presented agree with the results of ref. 7 and with refs. 8 and 9. Computer simulations of the soliton contribution to $\chi^{(3)}$ have been presented in ref. 12. In this paper we obtain rigorous analytic results for the susceptibility of a (non)degener-

*Author to whom correspondence should be addressed.

0379-6779/93/$6.00

ate conjugated polymer in the SSH model and compare the validity of SSH and TLM models in taking into account different diagrammatic processes contributing to $\chi^{(3)}$.

The model

We consider one-dimensional combined Peierls dielectrics [13] in the SSH model [14]. These substances have a non-degenerate ground state, i.e. they have a band gap due to their chemical structure even without considering the electron–lattice interaction. But the Peierls effect is of considerable importance and must be accounted for. *trans*-$(CH)_x$ having a degenerate ground state is also included as a limiting case.

One of our goals is to compare the discrete SSH model with its continuum linearized TLM version [15] for NLO calculations. For this reason considering SSH we use the representation of left- and right-going electrons that was utilized in TLM

$$H = 2t_0 \sum \sin q \Psi_q^+ \sigma_3 \Psi_q + \Delta \sum \cos q \Psi_q^+ \sigma_1 \Psi_q \tag{1}$$

Here $\Psi_q^+ = (\psi_1^+(q), \psi_2^+(q))$ is a two component wave function, σ_i = Pauli matrices, t_0 = the transfer integral for the undimerized chain and $\Delta = \Delta_s + \Delta_p$ = the complex gap. The component $\Delta_s = 2t_s$ corresponds to the difference in transfer integrals for unequivalent bonds in the initial rigid structure of the molecule while Δ_p originates from the Peierls distortion. The electron–electron interaction in the mean-field approximation renormalizes t_0, Δ producing no qualitative effects in the spectrum of $\chi^{(3)}$.

The external electric field we introduce in the following gauge

$$\Phi = 0, \quad A = -E_0 c\omega^{-1} \sin \omega t \tag{2}$$

Then the interaction Hamiltonian and the current operator in the SSH model take the form ($\hbar = 1$)

$$\Delta H = -\frac{1}{e} \sum \Psi_q^+ \left\{ \frac{eA}{c} j_0 + \frac{1}{2}\left(\frac{eA}{c}\right)^2 j_1 - \frac{1}{6}\left(\frac{eA}{c}\right)^3 j_0 \right\} \Psi_q \tag{3}$$

$$j = \sum \Psi_q^+ \left\{ j_0 + \frac{eA}{c} j_1 - \frac{1}{2}\left(\frac{eA}{c}\right)^2 j_0 - \frac{1}{6}\left(\frac{eA}{c}\right)^3 j_1 \right\} \Psi_q \tag{4}$$

where

$$j_0 = e(2t_0 \cos q\sigma_3 - \Delta \sin q\sigma_1), \quad j_1 = -dj_0/dq \tag{5}$$

In the TLM model the substitution $\sin q \Rightarrow q$, $\cos q \Rightarrow 1$ is made. Moreover, as the process of linearization precedes the introduction of the external field, only the first term, proportional to j_0 in figure brackets in eqns. (3) and (4) and in j_0 only the first term in square brackets proportional to σ_3 are left. Thus for TLM the summand of ΔH incorporates only one term linear in the applied field while five other terms that would also contribute to $\chi^{(3)}$ are omitted. The current operator incorporates only a part of the diffusion current and no terms depending on the electric field (in eqn. (4) there are seven of them). As known from the linear response theory, the omission of the diamagnetic current leads to unphysical low frequency divergences. All these arguments bring some doubts to the validity of the TLM model for the description of the NLO properties. On the other hand we note that the interchange of the sequence of linearization and the introduction of the external field will not make the model simpler than SSH.

Results and discussion

As we are considering a non-linear time-dependent problem with the field having a frequency comparable to the band gap width, we use the Keldysh non-equilibrium diagram technique [7, 16]. Then the time integration and the T-operator permutations are performed on the Keldysh contour having two branches. The matrix Green functions (GF) have lower indices labeling left and right electrons (matrices σ_1, σ_3) and the upper Keldysh indices.

$$G^{11}(q, v) = \frac{v + 2t_0 \sin q\sigma_3 + \Delta \cos q\sigma_1}{v^2 - (\omega_q - i\varepsilon)^2} \tag{6a}$$

$$G^{22}(q, v) = -\frac{v + 2t_0 \sin q\sigma_3 + \Delta \cos q\sigma_1}{v^2 - (\omega_q + i\varepsilon)^2} \tag{6b}$$

$$G^{12}(q, v) = -\frac{i\varepsilon(2w_q + v)}{\omega_q^2 + \varepsilon^2} \frac{-\omega_q + 2t_0 \sin q\sigma_3 + \Delta \cos q\sigma_1}{(v + \omega_q)^2 + \varepsilon^2} \tag{6c}$$

$$G^{21}(q, v) = -\frac{i\varepsilon(2\omega_q - v)}{\omega_q^2 + \varepsilon^2} \frac{\omega_q + 2t_0 \sin q\sigma_3 + \Delta \cos q\sigma_1}{(v - \omega_q)^2 + \varepsilon^2} \tag{6d}$$

where $\omega_q^2 = (2t_0 \sin q)^2 + (\Delta \cos q)^2$.

These GFs take account of the relaxation processes. The bare electron GFs correspond to the purely dimerized lattice and are given by eqn. (6) in the limit $\varepsilon \Rightarrow 0$, when the lorenzians in eqns. (6c) and (6d) are in fact δ functions. The imaginary part of the electron self energy arising from the interaction with the phonons, renormalized by Peierls distortion—'phasons' and 'amplitudons' [17], enters eqn. (6) as ε—for simplicity we take it independent of q.

The expression of the NLO susceptibility for THG in the SSH model is rather lengthy. It incorporates seven terms with the following correlators: $\langle T_c \Psi_q^+ j_1 \Psi_q \rangle$, $\langle T_c \Psi_q^+ j_0 \Psi_q \Psi_{q'}^+ j_0' \Psi_{q'} \rangle$, $\langle T_c \Psi_q^+ j_1 \Psi_q \Psi_{q'}^+ j_1' \Psi_{q'} \rangle$, $\langle T_c \Psi_q^+ j_1 \Psi_q \Psi_{q'}^+ j_0' \Psi_{q'} \Psi_{q''}^+ j_0'' \Psi_{q''} \rangle$ and $\langle T_c \Psi_q^+ j_0 \Psi_q \Psi_{q'}^+ j_0' \Psi_{q'} \Psi_{q''}^+ j_0'' \Psi_{q''} \Psi_{q'''}^+ j_0''' \Psi_{q'''} \rangle$. Here the matrix product over the indices of left and right electrons is implied and in the subsequent integration the sum over the branches of the Keldysh contour from $-\infty$ to $+\infty$ and back from $+\infty$ to $-\infty$ is taken. The TLM model generates only one term with the last correlator mentioned and with a simpler expression for j_0: $j_0 = 2t_0 e\sigma_3$. After cumbersome calculations we obtain for the SSH model the rigorous result for slow relaxation

$$\chi^{(3)}_{\mathrm{THG}} = Bz^{-8}\{[5 - 8z^2(1+\delta^2) + 20z^4\delta^2]f(z) - 8[1 - 4z^2(1+\delta^2) + 16z^4\delta^2]f(2z) + [3 - 24z^2(1+\delta^2) + 188z^4\delta^2]f(3z) - F\} \tag{7}$$

where

$$f(z) = \delta^2 n_1(1-\delta^2)^{-1/2}\Pi(\pi/2, n_1, (1-\delta^2)^{1/2}),$$
$$F = 8(1-\delta^2)^{1/2}\delta^2 z^4 E(\pi/2, (1-\delta^2)^{1/2}) \tag{8}$$

$$n_{\mathrm{m}} = \frac{1-\delta^2}{1-(mz\delta)^2 - i\varepsilon t_0^{-1} mz\delta}, \quad z = \frac{\omega}{2|\Delta|}, \quad \delta = \frac{\Delta}{2t_0}, \quad B = \frac{e^4}{3^2 2^9 \delta^6 t_0^3 (1-\delta^2)^{1/2}} \tag{9}$$

and E, Π are the complete elliptical integrals of first and third kind. Here for convenience we have used the notations of ref. 7. Equations (7) and (8) are accurate to the lowest order in ε and exact with respect to z and δ.

Next we analyze the low frequency domain. We note that the result for the TLM model without relaxation, presented by Wu (eqns. (11) and (12) of ref. 7) omits the terms of the order of δ^2 (or else it can be obtained by expanding the limits of integration over q to infinity). The rigorous expression for $f(z)$ in the TLM model is given by eqn. (10). Then in the low frequency limit $\chi^{(3)}$ diverges due to the reasons stated above. The exact eqn. (7) gives for $\chi^{(3)}(\omega = 0)$ a constant as it should—the SSH model has no drawbacks of linearization.

Equations (7) and (8) however are not convenient for qualitative analyses. To obtain a more transparent expression we perform the linearization in the last integral over q. We note that this approach goes beyond the TLM model that was linearized from the very start, as in our case all the terms entering SSH are preserved. We again obtain eqn. (7) with

$$f(z<1) = \frac{1}{z(1-z^2)^{1/2}} \arcsin \frac{z}{\left[1 + \dfrac{4\delta^2(1-z^2)}{\pi^2(1-\delta^2)}\right]^{1/2}} \tag{10a}$$

$$f(z>1) = \frac{1}{z(z^2-1)^{1/2}} \left\{ \ln\left| z - (z^2-1)^{1/2}\left[1 + \frac{4\delta^2}{\pi^2(1-\delta^2)}\right]^{1/2} \right| - \frac{\pi_i}{2} - \frac{1}{2}\ln\left[1 - \frac{4\delta^2(z^2-1)}{\pi^2(1-\delta^2)}\right] \right\} \tag{10b}$$

$$F = z^4\delta^2 \left\{ \pi^2(1-\delta^2)\left[1 + \frac{4\delta^2}{\pi^2(1-\delta^2)}\right]^{1/2} + 4\delta^2 \times \left[\ln\frac{\pi}{2} + \ln\left(1 + \left[1 + \frac{4\delta^2}{\pi^2(1-\delta^2)}\right]^{1/2}\right) - \ln\frac{\delta}{(1-\delta^2)^{1/2}} \right] \right\} \tag{10c}$$

Equations (10a) and (10b) are the correct forms that can be also used in the expression of $\chi^{(3)}$ of the TLM model. The results of ref. 7 without relaxation correspond to $\delta = 0$, $F = 0$.

To obtain tractable analytic expressions for the SSH model with non-zero damping we expand integrations over q to infinity and again get eqn. (7) with

$$f(mz) = g(mz + i\varepsilon\Delta^{-1}) - g(mz - i\varepsilon\Delta^{-1}) - i\frac{\pi}{2}(mz)^{-1}[mz + i\varepsilon\Delta^{-1})^2 - 1]^{-1/2}$$

$$g(z) = z^{-1}(z^2 - 1)^{-1/2}\ln\left[\left(\frac{z+1}{2}\right)^{1/2} - \left(\frac{z-1}{2}\right)^{1/2}\right] \tag{11}$$

Apart from the correction terms δ^2 this result coincides with our previous formula [9] for the TLM model. So in the main terms with respect to δ the analysis and all the asymptotics, presented in ref. 9, remain valid for this case also.

As it can be seen from eqns. (7), (10) and (11) despite the low frequency problems the TLM model gives a good description of the process, the difference lying in the terms proportional to δ^2. It happens because of the compensation of the numerous SSH terms additional to TLM that could not be expected beforehand. This conclusion for the THG process agrees well with the fact of qualitative similarity of numerical curves, presented in refs. 7 and 9 and 11 and 12.

In summary, we have rigorously calculated the NLO susceptibility for third harmonic generation in the SSH model for arbitrary gap width in the entire frequency range and in the lowest order in quasiparticle relaxation constants. Though the comparison shows that the linearized TLM model neglects many processes incorporated in SSH, the results of the two models are close because of the compensation of additional contributions.

Acknowledgements

The authors thank Professor A. D. Bandrauk for stimulating discussions and ELORMA for financial support.

References

1 J. Messier and F. Kajzar (eds.), *Organic Molecules for Nonlinear Optics and Photonics*, Kluwer, Dordrecht, 1991, 560 pp.
2 A. J. Heeger, D. Moses and M. Sinclair, *Synth. Met.*, *17* (1987) 343.
3 P. A. Chollet, F. Kajzar and J. Messier, *Synth. Met.*, *18* (1987) 459.
4 M. Sinclair, D. McBranch, D. Moses and A. J. Heeger, *Synth. Met.*, *28* (1989) D645.
5 E. Wintner, F. Krausz and G. Leising, *Synth. Met.*, *28* (1989) D155.
6 W. S. Fann, S. Benson, J. M. J. Madey, S. Etemad, G. L. Baker and F. Kajzar, *Phys. Rev. Lett.*, *62* (1989) 1492.

7 W. Wu, *Phys. Rev. Lett.*, *61* (1988) 1119.
8 Yu. I. Dakhnovskii and K. A. Pronin, in *Electron-Electron Correlation Effects in Low Dimensional Conductors and Superconductors (Proc., Kiev, May 1990)*, Springer, Berlin, 1991, p.73.
9 K. A. Pronin and Yu. I. Dakhnovskii, *Synth. Met.*, *43* (1991) 3759.
10 C. Wu and X. Sun, *Phys. Rev. B*, *41* (1990) 12845.
11 C. Q. Wu and X. Sun, *Synth. Met.*, *43* (1991) 3213.
12 D. M. Mackie, R. J. Cohen and A. J. Glick, *Phys. Rev. B*, *29* (1989) 3442.
13 S. A. Brazovskii and N. N. Kirova, *Sov. Sci. Rev. A*, *5* (1984) 99.
14 W. P. Su, J. R. Schriffer and A. J. Heeger, *Phys. Rev. B*, *22* (1980) 2099.
15 H. Takayama, Y. R. Lin-Liu and K. Maki, *Phys. Rev. B*, *21* (1980) 2388.
16 L. V. Keldysh, *Sov. Phys. JETP*, *20* (1965) 1018.
17 P. Lee, T. M. Rice and P. W. Anderson, *Solid State Commun.*, *14* (1974) 703.

Synthetic Metals, 54 (1993) 301–305

Characterization of optical third-order non-linearities by prism coupling and pulse shape analysis on a ps timescale

B. J. Offrein, J. B. P. Van Schoot, A. Driessen, H. J. W. M. Hoekstra and Th. J. A. Popma

MESA-Research Institute, Departments of Applied Physics and Electrical Engineering, University of Twente, P.O. Box 217, 7500 AE Enschede (Netherlands)

Abstract

Materials with an intensity dependent index of refraction and absorption coefficient—third-order optical non-linear (ONL) effects—offer the possibility of all-optical signal processing. Prism coupling is a well-known tool to investigate the intensity dependent refractive index, however, such experiments are often obscured by thermal effects. To avoid these we have studied the influence of the ONL effects on the shape of 70 ps pulses in non-linear prism coupling. The full width at half maximum (FWHM) of the in and out coupled pulses is compared simultaneously with the FWHM of the pulses of a reference beam. By measuring at various angles of incidence around that for optimal coupling it is possible to measure the sign and value of the change in both the refractive index and absorption coefficient. As a function of the coupling angle, the first leads to an asymmetric line shape for the ratio of the two FWHMs mentioned above, whereas the second leads to a symmetric one. From a comparison of simulations with experimental data, the values of the non-linear constants can be derived.

Introduction

There is a variety of methods to characterize the third-order optical non-linear coefficient, which describe the intensity dependent refractive index and absorption coefficient: the intensity dependent complex refractive index. We mention four wave mixing, pump and probe techniques and interferometric methods. The induced changes in the optical constants may have several physical origins depending on the material properties. Mostly various effects compete which can often be separated on their different response in time. Prism coupling is a well-known tool to investigate the non-linearity but is very sensitive to thermal effects. Not only does the thermo-optical effect influence the process, but also the coupling structure itself changes due to expansion of the layers. To be sure that only the fast electronic effect is measured we use 70 ps pulses in a prism coupling setup and look at the influence on the pulse shape in the time domain. Thermal effects can be ruled out as the time constants of the thermal effects are much larger than the pulse width [1].

0379-6779/93/$6.00

Theory

With the aid of simulations one is able to relate the measured change of the full width at half maximum (FWHM) of the pulses to values of the real and imaginary part of $\chi^{(3)}$. We have to distinguish between a change in the refractive index and in the absorption coefficient. The first influences the phase matching in the coupling process, which means that it has to be described with the theory of the non-linear prism coupler. The influence of an intensity induced absorption change depends on the propagation distance in the non-linear layer structure, which is essentially the distance between in and output coupler.

Influence of non-linear refraction

The theory of (non-linear) prism coupling of light into a planar structure has been described in refs. 2 and 3. Assuming that neither the modal field shape nor the coupling constants depend on the intensity of the modal field, the coupling process can be described by

$$\frac{\mathrm{d}A_{\mathrm{m}}}{\mathrm{d}z} = -k_0(K_{\mathrm{m}} + \beta_{\mathrm{i}})A_{\mathrm{m}} + t_{\mathrm{m}}K_{\mathrm{m}}k_0E_0 \exp\{\mathrm{i}(\Delta\beta z - \varphi_{\mathrm{r}})\} \tag{1}$$

The amplitude of the modal field, $A_{\mathrm{m},}$, is related to the electric modal field, E_{m}, by

$$E_{\mathrm{m}}(x, z) = A_{\mathrm{m}}(z)\xi_{\mathrm{m}}(x) \exp(-\mathrm{i}\beta_{\mathrm{m}}z) \tag{2}$$

The field shape ξ is defined such that $|A_{\mathrm{m}}|^2$ represents the power of the mode per unit length in the y direction. The modal absorption is given by β_{i}; $K_{\mathrm{m}}k_0$ and $t_{\mathrm{m}}K_{\mathrm{m}}k_0$ are the output and input coupling coefficients. The mismatch of the propagation constants of the incoming light parallel to the z axis, β_0, and the mode m, β_{m}, is given by $\Delta\beta$: $\Delta\beta = \beta_0 - \beta_{\mathrm{m}}$. The change in the real part of the refractive index is described with the non-linear phase shift φ_{r}, which is the integral over z of the change in the real part of the progation constant $\mathrm{Re}\{\Delta\beta_{\mathrm{NL}}\}$. This change is, for TE modes, related to $\chi^{(3)}$ and the electric field [3, 4] in the following way

$$\Delta\beta_{\mathrm{NL}} = \frac{\omega}{4P}\int 3\varepsilon_0\chi^{(3)}(-\omega; \omega, \omega - \omega\ |E_{\mathrm{y}}(\omega)\ |^4\ \mathrm{d}x \tag{3}$$

where P is the power of the mode. The coupling efficiency can now be calculated by numerical integration of eqn. (1). To obtain a large effect on the non-linear phase shift the interaction length and coupling efficiency need to be large. This means that the coupling should be weak with, for a good coupling efficiency, a large waist of the incoming beam. The non-linearity is assumed to be instantaneous and diffusion effects are not taken into account.

In the case of a negative non-linear coefficient an increase of power will decrease the effective index of the waveguide. If the incoming beam is tuned to an effective refractive index just a little below that of the waveguide, the coupling efficiency will increase with power. This results in a decrease of the FWHM of the pulses as the effect is the largest in the peak of the pulse. Tuning at an effective index a little bit above the waveguide index leads to an increase of the FWHM.

Influence of non-linear absorption

The influence of an intensity dependent absorption coefficient increases with the propagation distance of the mode. The attenuation can be written as

$$\frac{\mathrm{d}\,|A_{\mathrm{m}}|^2}{\mathrm{d}z} = -2(\beta_{\mathrm{i}}|A_m|^2 + n_2''|A_{\mathrm{m}}|^4) \tag{4}$$

We use this equation only in the limit where the absorption change is small compared to the linear absorption coefficient. The non-linear coefficient n_2'' is related to $\mathrm{Im}\{\chi^{(3)}\}$ with eqn. (3) by $n_2'' = \mathrm{Im}\{\Delta\beta_{\mathrm{NL}}\}/P$. Solving eqn. (5) we obtain

$$|A_{\mathrm{m}}(z)|^2 = \frac{\beta_{\mathrm{i}}\,|A_{\mathrm{m}}(0)|^2}{(\beta_{\mathrm{i}} + n_2''\,|A_{\mathrm{m}}(0)|^2)\,\exp(2\beta_{\mathrm{i}}z) - n_2''\,|A_{\mathrm{m}}(0)|^2} \tag{5}$$

As the pulse corresponds to an intensity distribution, the change in pulse width can be related directly to the value of the imaginary part of $\chi^{(3)}$ under the condition that the linear attenuation of the mode is known.

Experimental

In the experiment, pulses that are coupled in and out of the non-linear layer structure are simultaneously compared with undisturbed pulses from the same source. To do this part of the laser beam is split off while most is coupled into the non-linear layer structure. The excited mode is coupled out and both beams are combined and detected as shown in Fig. 1. We used a fast detector, 10 GHz bandwidth and a 50 GHz sampling scope. Because the path lengths of the beams differ the pulses can be detected separated in time. The measurements are performed with a mode-locked Nd:YLF-laser at 1313 nm having 70 ps pulses with a repetition rate of 76 MHz and peak power of 600 W. The layer structure consists of an optically non-linear substrate with a linear PECVD-deposited SiO_xN_y guiding layer. For the non-linear material we used semiconductor doped glass, Schott RG 850.

To characterize the linear structure, first the refractive index of the RG 850 substrate is determined by measuring the critical angle for refraction from the prism to the glass substrate. The SiON layer is characterized with the prism coupling method. It is important that the simulated shape of the coupling curve is well fitted to the measured one.

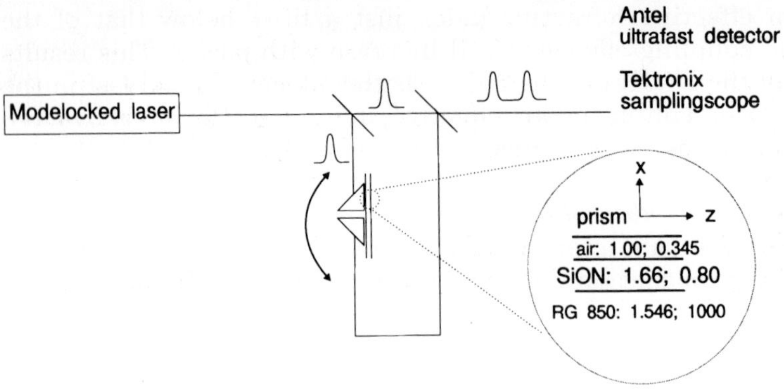

Fig. 1. Experimental setup. In the circle the refractive index and the thickness in μm of the layered structure are given, respectively.

Using a one parameter fit of the coupling strength with the gap thickness, a good agreement between experiment and simulations can be obtained. The structure is shown in the circle in Fig. 1. The laser beam is focused in the y direction by a cylindrical lens to 45 μm. The coupling efficiency of the TE_0 mode is determined to be 16% and the attenuation is 4 dB/cm. The distance between the prism is 14 mm. The peak power of the incoming pulses is 160 W and their FWHM about 70 ps. The experimental results are given in Fig. 2. With the described models we obtain

$$\chi^{(3)} = -4 \times 10^{-18} - i2 \times 10^{-20} \text{ (m/V)}^2 \qquad (6)$$

This represents a negative change of the refractive index and a decrease of the absorption coefficient.

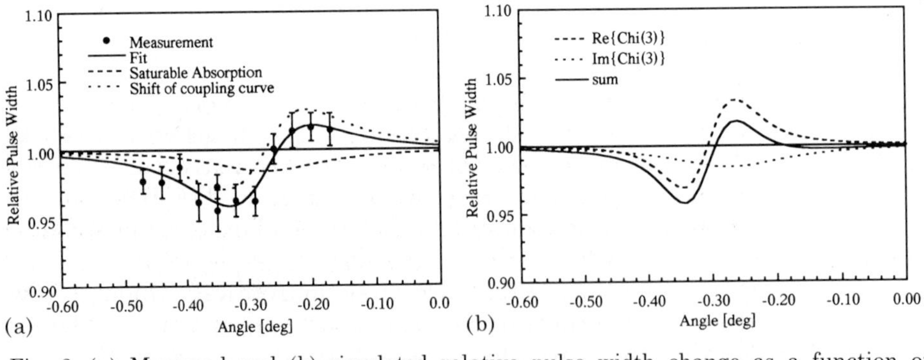

Fig. 2. (a) Measured and (b) simulated relative pulse width change as a function of the effective index of the incoming beam. The experimental result is decomposed into a symmetrical and asymmetrical part.

Discussion

We found a decrease of the refractive index and absorption coefficient with intensity, which is what may be expected for the electronic non-linearity of a semiconductor at wavelengths above the gap wavelength [5]. The non-linearity of semiconductor doped glasses for wavelengths around the gap is extensively investigated and found to be in the order of 10^{-18} $(m/V)^2$. Cotter and Manning [6] found, like ourselves, a large non-linearity at 1064 nm of $|\chi^{(3)}| = 10^{-20}$ $(m/V)^2$, with $Re\{\chi^{(3)}\}/Im\{\chi^{(3)}\} = 9$ and a relaxation much faster than 75 ps. For the real part of the non-linearity we find an even larger coefficient. However it is difficult to reproduce the results on the same sample, probably due to darkening of the sample [7]. The non-linearity of a darkened sample is reduced by about a factor of 20. On behalf of this technique we conclude that the experiment is simple because only one beam needs to be coupled into the guiding wave structure, however a reference beam is needed for reliable measurements. There will only be an effect on the pulse shape if the non-linearity is fast, so slow thermal effects are cancelled out. The time dependence of the non-linearity cannot be measured, only an upper limit can be given. For ONL organic polymers, which are promising for device applications [8], the time response of the electronic non-linearity is much faster than the pulse width of the mode-locked pulses. Both a change in refractive index and absorption coefficient can be measured this way.

Acknowledgements

We thank A. Hollink for technical support. This investigation was supported by the Dutch Foundation for Fundamental Material Research (FOM) and the Innovative Research Program of the Dutch Ministry of Economic Affairs (IOP).

References

1 R. Burzynski, B. P. Singh, P. N. Prasad, R. Zanoni and G. I. Stegeman, *Appl. Phys. Lett.*, *53* (1988) 2011.
2 R. Ulrich, *J. Opt. Soc. Am.*, *60* (1970) 1337.
3 G. M. Carter and Y. J. Chen, *Appl. Phys. Lett.*, *42* (1983) 643.
4 H. Kogelnik, in T. Tamir (ed.) *Topics in Applied Optics*, Springer, Berlin, 1990, Ch. 2.
5 G. R. Olbright, N. Peyghambarian, S. W. Koch, and L. Banyai, *Opt. Lett.*, *12* (1987) 413.
6 D. Cotter and R. J. Manning, Optical nonlinearity of semiconductor microcrystallites in glass: observation of large picosecond nonlinear refraction in the infrared, *Proc. Quantum Electronics Conf. 90, Washington, DC, USA*, QPDP33.
7 J. L. Coutaz and M. Kull, *J. Opt. Soc. Am.*, *B8* (1991) 95.

Surface-plasmon-resonance investigations of Langmuir–Blodgett films of donor–acceptor substituted polyenes: linear optical and electro-optic properties

Th. Wagner and S. Roth
Max-Planck-Institut für Festkörperforschung, Heisenbergstrasse 1, 7000 Stuttgart 80 (Germany)

Abstract

We present surface-plasmon-resonance (SPR) investigations on Langmuir–Blodgett films of two different types of donor–acceptor substituted polyenes. SPR spectroscopy in Kretschmann configuration enables determination of thickness and dielectric constants of films down to the monolayer range. The linear electro-optic susceptibilities $\chi^{(2)}(-\omega;\omega,0)$ of mono- and multi-layer films are determined as dielectric constant changes $\Delta\varepsilon$, which are seen as field-induced modulations in the mode-coupling condition in prism-coupled SPR. Knowing the magnitude and sign of $\chi^{(2)}(-\omega;\omega,0)$ and the orientation angle of the molecules in the film, obtained from FT-IR spectroscopy, we were able to determine the second hyperpolarisability $\beta(-\omega;\omega,0)$. In films of 7-*N*,*N*-dioctylaminophenyl-7,8′-diapocarotene-8′-al, β is found to have a very high value of 7.7×10^{-47} cm^3/V^2. The dependence of β on the film deposition type is discussed.

Introduction

Donor–acceptor substituted polyenes are molecules which exhibit significant second-order non-linear optical (NLO) behaviour because of the large difference in dipole moment between their ground and excited states and their non-centrosymmetric structure. To study the optical non-linearities of these molecules the Langmuir–Blodgett (LB) technique is a powerful tool to build up thin films with well-defined thickness and structure.

In this paper we report on the investigation of two different types of polyenes. Thickness measurements were performed by surface plasmon resonance (SPR) which is a very sensitive method to thickness changes on an angstrom scale. The linear electro-optic effect (Pockels' effect) was investigated with a slightly modified SPR configuration by applying an electric field.

Experimental

We prepared Y- and Z-type of films of 7-*N*,*N*-dioctylaminophenyl-7,8′-diapocarotene-8′-al (DAPC) and 3-methoxyphenyl-5,6-dihydro-3,8′-diapo-

0379-6779/93/$6.00

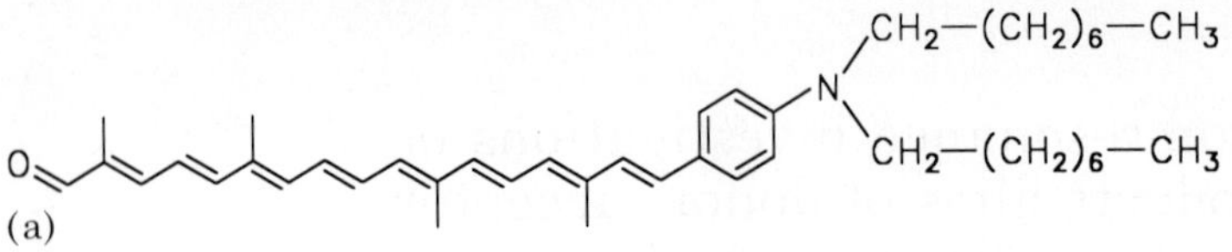

(a)

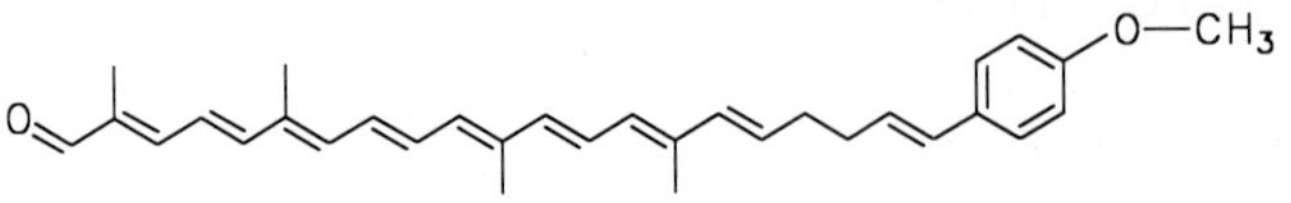

(b)

Fig. 1. (a) 7-*N,N*-Dioctylaminophenyl-7,8′-diapocarotene-8′-al (DAPC), (b) 3-methoxyphenyl-5,6-dihydro-3,8′-diapocarotene-8′ (MPDC).

carotene-8′ (MPDC), a long polyene with a spacer inside the conjugation chain, which leads to a disruption of the π-system delocalisation (Fig. 1). The most common, Y-type, deposition mode results in general in centrosymmetric packing of the molecules. For details of Y-type LB preparation see ref. 1. Z-type films were built up by spreading the monolayer film on the water surface after dipping in the substrate. Thus the monolayers were transferred only in the upstroke direction. To determine thickness and dielectric constants by SPR we used the Kretschmann ATR configuration [2]. A p-polarized HeNe laser beam was directed onto a half-cylindric prism which was in optical contact with a glass slide coated with a thin silver layer. The LB films were transferred in steps of 2, 4, 6 and 8 monolayers (Y-type films). We measured at two different wavelengths: 543 and 632.8 nm. The resonance curves were theoretically fiitted with the exact Fresnel equations.

The linear electro-optic effect of the LB films was studied by a configuration first described by Cross *et al.* [3]. An electric field of the order of 10^6 V/m is produced in the film by applying a low voltage (30 V) between the silver film and a second electrode which is placed parallel to the film at a distance of 6 μm with a Mylar spacer. Because the change of the refrective index is very small ($\sim 10^{-5}$) the electric field is modulated by a frequency of 1.8 kHz and the resulting modulation in the reflected light is measured.

Results and discussion

Figure 2 shows the experimental resonance curves (scatters) and theoretical analysis (full lines) for the LB films of DAPC recorded at $\lambda = 632.8$ nm. A linear shift of the resonance to higher angles of incidence without a change in the form is observed, so only the real part of the refractive index need be considered. At $\lambda = 543$ nm a slight decrease of the minimum depth results from an imaginary part of the refractive index

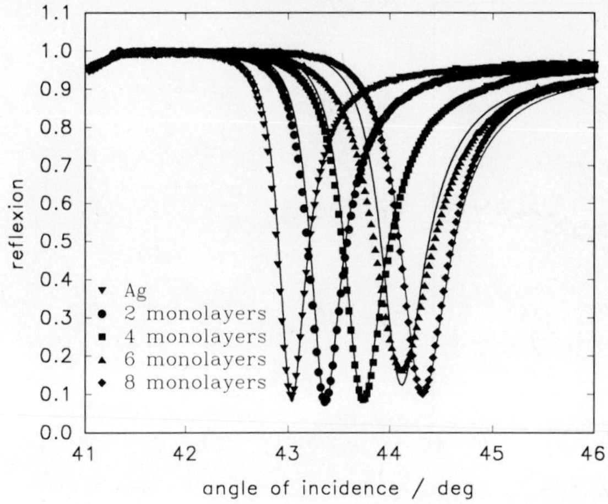

Fig. 2 Experimental resonance curves (scatters) and theoretically calculated (full lines) surface-plasmon-resonance curves of pure silver and silver covered with LB films of DAPC recorded at $\lambda = 632.8$ nm.

which is non-zero. The SPR curves for MPDC show at both wavelengths non-absorbing behaviour. Analysis of the resonsance curves gives an average monolayer thickness of DAPC of 13.9 Å and of MPDC of 14.9 Å. In Table 1 the fitted values of thickness and dielectric constants are listed.

Figure 3 shows the differential resonance curve $\Delta R(\theta)$ together with the corresponding resonance curve for a two monolayer thick Y-type film of DAPC. The solid line is the theoretical best fit. To calculate the second-order non-linear susceptibility $\chi^{(2)}$ and the surface susceptibility $\chi_S^{(2)\prime} = d\chi^{(2)}$, where d is the thickness of the LB film, we used an approximation explained by Loulergue *et al.* [4] to get the real part of the change in refractive index by a peak-to-peak fit. The imaginary part was calculated by fitting the difference of the exact calculated curve $\Delta R_{th} = R(n + \Delta n, \kappa + \Delta\kappa) - R(n, \kappa)$ to the measured ΔR curve.

TABLE 1

SPR results

No. of monolayers	$\lambda = 632.8$ nm				$\lambda = 543$ nm				
	DAPC		MPDC		DAPC			MPDC	
	n	d (Å)	n	d (Å)	n	κ	d (Å)	n	d (Å)
2	1.581	25	1.517	30	1.795	0.148	27.5	1.628	31
4	1.517	56	1.510	59	1.796	0.161	57	1.591	61
6	1.581	80	1.466	90	1.734	0.187	82.5	1.539	90
8	1.581	114	1.449	118	1.695	0.207	112	1.507	118

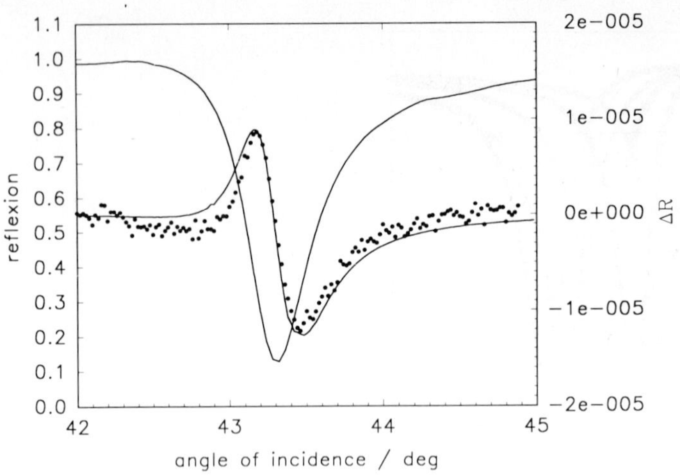

Fig. 3. The differential resonance curve $\Delta R(\theta)$ together with the corresponding resonance curve for a two monolayer thick Y-type film of DAPC.

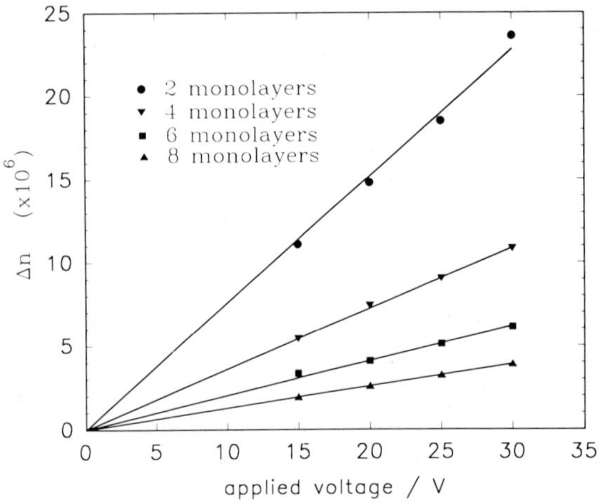

Fig. 4. Dependency of Δn on the applied field for Y-type DAPC films of different thickness.

Figure 4 shows the dependency of Δn on the applied field for Y-type DAPC films of different thickness. This linear dependence is characteristic of the linear electro-optic effect. The second order surface susceptibility is high for the two layer film but does not increase with the number of layers (Fig. 5). This behaviour would be expected for an odd number of layers. So we conclude that the first layer shows a random orientation with no contribution to the susceptibility. In the multilayer system, the contributions from individual layers cancel each other in pairs and only the upper layer gives a non-zero contribution. The small decrease of $\chi_S^{(2)}$ with the number of layers is effected by surface contributions. Z-type films

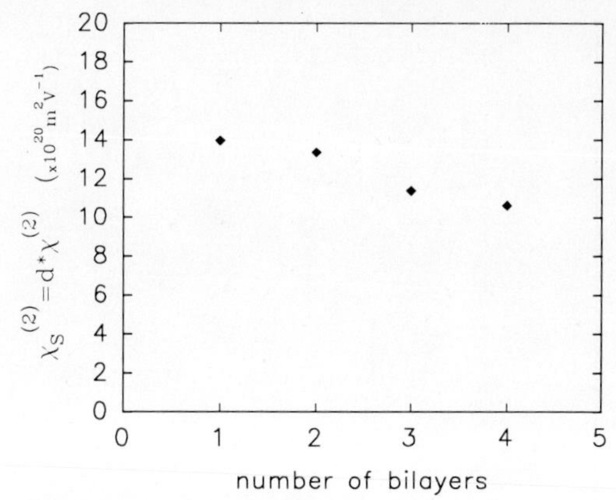

Fig. 5. Surface susceptibility of Y-type DAPC films vs. the number of layers.

show an increase of $\chi_S^{(2)}$ with increasing film thickness because each monolayer has the same orientation and so the $\chi_S^{(2)}$ effects will be additive. But the values are a factor of three lower than expected from the Y-type film measurements. If we look at the linear parameters (Table 2), we see a lower monolayer thickness and a lower refractive index. This behaviour is affected by a relaxation of some molecules from the forced Z-orientation back to the natural Y-orientation [5]. So a lower number of molecules per monolayer gives an additive $\chi_S^{(2)}$ effect.

With the tilt angle $\varphi = 83°$ of the conjugation chain from previous measurement of molecular orientation in Y-type films by FT-IR [6] and the relationship [7]

$$\chi_{zzz}^{(2)}(-\omega;\omega,0) = Nf(\omega)f(\omega)f(0)\beta_{zzz}(-\omega;\omega,0)\langle\cos^3\varphi\rangle$$

where N indicates the molecular density, known from the pressure-area-isotherms, $f(0)$ and $f(\omega)$ the local field factors, we may calculate the second-order hyperpolarisability $\beta_{zzz}(\omega;\omega,0)$. The values are listed in Table 3 for Y- and Z-type films. We found a very high value for the Y-type film of 7.7×10^{-47} cm^3/V^2.

Measurements on LB films of MPDC show the Y-type films form a good centrosymmetric system because the $\chi_S^{(2)}$ values are nearly one order lower than measured in Z-type films (Table 4). In Z-type films there are no differences in the linear optical parameters in comparison to Y-type films so we think of no molecular relaxation. Z-type films indicate the orientation of monolayers from the upstroke direction. From the negative sign of $\chi_S^{(2)}$ of Z-type and the positive sign of $\chi_S^{(2)}$ of the Y-type films we conclude that the amount of change in dipole moment due to the applied electric field in downstroke monolayers is a little bit larger than that of the upstroke monolayer. With a tilt angle of 62° meaured by FT-IR we calculate for the Z-type film: $\beta_{zzz}(-\omega;\omega,0) = 1.1 \times 10^{-49}$ cm^3/V^2.

TABLE 2

Measured values on LB films of DAPC

	N							
	Y-type				Z-type			
	2	4	6	8	1	2	3	4
n	1.58	1.517	1.58	1.58	1.449	1.414	1.510	1.529
d (Å)	25	56	80	114	11	21	34	45
$\Delta n \times 10^5$ [a]	2.36	1.08	0.611	0.389	0.708	0.667	0.472	0.556
$\Delta \varepsilon_r \times 10^5$ [a]	8.45	3.29	1.93	1.23	2.05	1.89	1.43	1.71
$\Delta \varepsilon_i \times 10^6$ [a]	15	7	2.5	1	1	1	0.7	0.1
$\chi^{(2)} \times 10^{11}$ (mV) [b]	5.14 + i1.1	2.38 + i0.5	1.42 + i0.18	0.885 + i0.07	1.48 + i0.079	1.35 + i0.07	0.89 + i0.05	1.10 + i0.07
$\chi_S^{(2)} \times 10^{20}$ (m^2/V) [b]	13.96	13.34	11.38	10.18	1.63	2.84	3.03	4.94

[a]Measured at $\Delta V = 30$ V peak–peak corresponding to $\Delta E = 1.389 \times 10^6$ V/m.
[b]Average of measurements at $\Delta V = 15$, 20, 25 and 30 V peak–peak.

TABLE 3

Values of $\beta_{zzz}(-\omega;\omega,0)$ (cm^3/V^2) of DAPC

N	Y-type	Z-type
1		1.13×10^{-47}
2	7.70×10^{-47}	2.11×10^{-47}
3		1.80×10^{-47}
4	7.35×10^{-47}	2.93×10^{-47}
6	6.28×10^{-47}	
8	5.86×10^{-47}	

TABLE 4

Measures values on LB films of MPDC

	N					
	Y-type				Z-type	
	2	4	6	8	1	2
n	1.517	1.510	1.466	1.449	1.549	1.549
d (Å)	31	61	90	118	16	33
$\Delta n \times 10^6$ [a]	0.722	0.222	0.222	0.097	−6.53	−1.95
$\Delta\varepsilon_r \times 10^5$ [a]	0.219	0.0670	0.0651	0.0281	−2.02	−0.603
$\chi^{(2)} \times 10^{11}$ (mV)[b]	0.158	0.0483	0.0469	0.0202	−1.38	−0.450
$\chi_S^{(2)} \times 10^{20}$ (m^2/V)	0.490	0.295	0.422	0.238	−2.21	−1.48

[a]Measured at $\Delta V = 30$ V peak–peak corresponding to $\Delta E = 1.389 \times 10^6$ V/m.
[b]Average of measurements at $\Delta V = 15$, 20, 25 and 30 V peak–peak.

Conclusions

We measured the thickness and refractive index of LB films of DAPC and MPDC by surface plasmon resonance. Measurements of the electro-optic effect on DAPC Y-type films showed a non-centrosymmetric system with a randomly oriented ground layer and gave a very high second-order hyperpolarisability. Investigations on MPDC showed the influence of the spacer part in the conjugation chain. The second-order hyperpolarisability is just two orders lower than in DAPC.

Acknowledgements

We thank Dr Sigurd Hagen for the synthesis of the molecules, and the DFG under SFB 329 and ESPRIT-Basic Research Action OLDS for financial support.

References

1 Th. Wagner, H. Schier, S. Hagen, S. Roth, S. Akari and K. Dransfeld, in H. Kuzmany, M. Mehring and S. Roth (eds.), *Electronic Properties of Polymers—Orientation and Dimensionality of Conjugated Systems (Kirchberg IV), Springer Series in Solid State Sciences*, Vol. 107, Springer, Berlin, 1992, p. 443.
2 E. Kretschmann, *Z. Phys.*, *241* (1971) 313–324.
3 G. H. Cross, I. R. Girling, I. R. Peterson and N. A. Cade, *Electron. Lett.*, *22* (1986) 1111.
4 J. C. Loulergue, M. Dumont, Y. Levy, P. Robin, J. P. Pochelle and M. Papuchon, *Thin Solid Films*, *160* (1988) 399.
5 A. Momose and Y. Hirai, *Thin Solid Films*, *204* (1991) 175.
6 Th. Wager, S. Hagen, S. Roth, S. Akari and K. Dransfeld, *Thin Solid Films*, *210/211* (1992) 652.
7 G. Khanarian, *Thin Solid Films*, *152* (1987) 265.

Dynamic light modulation in an electrochromic window consisting of polyaniline, tungsten oxide and a solid polymer electrolyte

B. P. Jelle and G. Hagen

Department of Applied Electrochemistry, NTH, N-7034 Trondheim (Norway)

S. Sunde and R. Ødegård

SINTEF Metallurgy, N-7034 Trondheim (Norway)

Abstract

The electrochemical and optical properties of an electrochromic window consisting of the two complementary electrochromic materials, polyaniline (PANI) and tungsten oxide (WO_3), have been investigated. Both PANI and WO_3 were electrochemically deposited on indium–tin oxide (ITO) glass substrates. Using the solid organic polymer electrolyte, poly(2-acrylamido-2-methyl-propane-sulfonic acid) (PAMPS), PANI and WO_3 were glued together. By applying a potential of ~1.5 V across the two external ITO contacts, we are able to modulate the light transmission, also in the near-infrared region (700–3000 nm), where about half of the solar energy lies, indicating that these 'smart windows' may significantly contribute to future energy savings in buildings. In order to study each of the electrochromic layers in the window, we have fabricated windows with holes in the electrochromic coatings, one window with a hole in the PANI film and another with a hole in WO_3. This enables us to study the optical properties of PANI and WO_3 separately.

Introduction

During recent years there has been a growing interest in environment protection and energy savings, and in this respect electrochromic windows offer an elegant solution for controlling the sun radiation through the window. Several electrochromic materials have been studied for this purpose [1–11].

The sun radiation at the earth surface is in the 300 to 3000 nm spectral region, of which about half of the energy lies in the near-infrared (NIR) region (700–3000 nm) [1]. Consequently, it is of primary interest to be able to regulate the light transmission through the window in the NIR region as well as in the visible part.

The main purpose of this work is to study the transmission spectra of the individual electrochromic layers in a window containing polyaniline and tungsten oxide.

0379-6779/93/$6.00

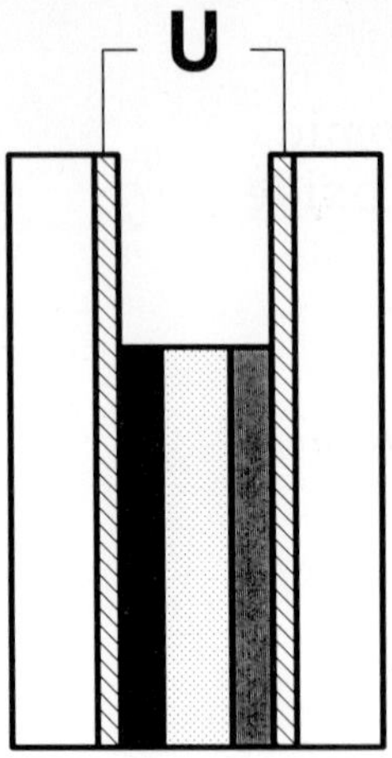

Fig. 1. A schematic drawing of the electrochromic window, Glass/ITO/PANI/PAMPS/WO_3/ITO/Glass, indicating the window configuration and the different layers used. U = applied voltage.

Experimental

We have studied an electrochromic window consisting of the two complementary electrochromic materials PANI and WO_3, the total glass sandwich being Glass/ITO/PANI/PAMPS/WO_3/ITO/Glass, as indicated in Fig. 1. PANI and WO_3 were deposited electrochemically on ITO coated glass plates (100 Ω/□) from aqueous solutions and glued together with PAMPS following a procedure described elsewhere [2, 3]. The size of the electrochromic window was 1 cm × 3 cm. The film thickness of PANI is typically 50 nm, the WO_3 coating 400 nm and the PAMPS layer thickness is *c.* 0.1 mm. Holes in the electrochromic coatings were made using an adhesive mask (diameter = 0.8 cm) during the film deposition.

A Cary 5 UV-VIS-NIR spectrophotometer was used to measure the transmission in the UV, Vis and near-IR region. The transmission spectra were recorded under potentiostatic control, and in order to stabilize the electrochromic PANI and WO_3 films, the window was held at a constant potential for about 10 min before recording the transmission spectra.

Results and discussion

From cyclic voltammograms (5 mV/s) of the electrochromic window (not shown here), in the potential region −1800 to +1600 mV, it was observed that only a low voltage (<2 V) and small currents (<50 μA/cm^2) are needed to drive the electrochemical processes. By applying a positive potential to the PANI electrode, both PANI and WO_3 turn to a blue

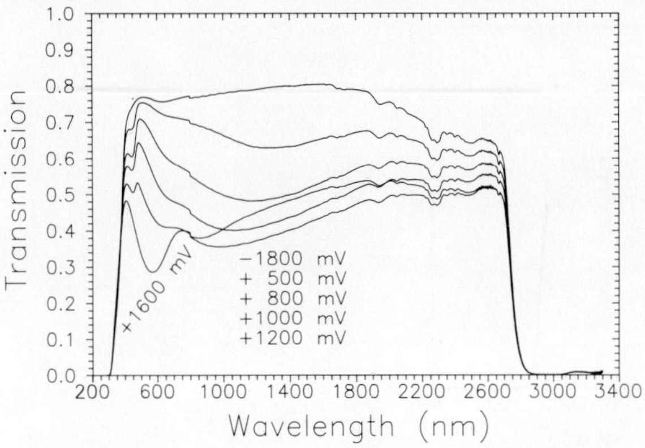

Fig. 2. Transmission vs. wavelength in the electrochromic window, Glass/ITO/PANI/PAMPS/WO_3/ITO/Glass, at different applied potentials.

colour, while the window is made almost transparent by reversing the polarity of the electrodes.

Transmission spectra in the 290–3300 nm wavelength region at different applied potentials (the sign refers to the PANI electrode) are shown in Figs. 2–4 for three windows; one without any hole (measuring transmission through the complete window), one with a hole in the WO_3 layer and one with a hole in the PANI film.

All transmission spectra are measured at a position 1.5 cm from the lower end of the window as shown in Fig. 1. Due to non-uniform current distribution during film formation the thickness of the electrochromic layers varies from top to bottom of the window. This inhomogeneity problem, which gives rise to diffraction patterns ('Newton rings') for the

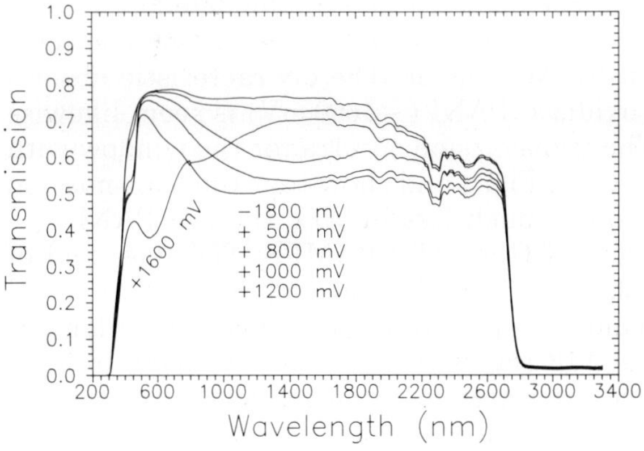

Fig. 3. Transmission vs. wavelength in an electrochromic window with a WO_3 hole at different applied potentials, thus measuring through Glass/ITO/PANI/PAMPS/ITO/Glass.

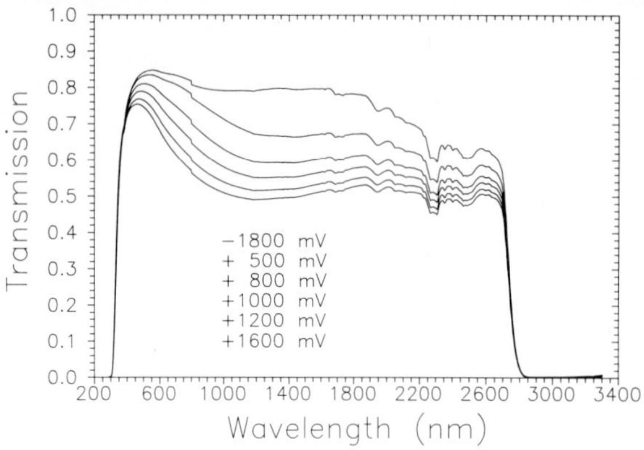

Fig. 4. Transmission vs. wavelength in an electrochromic window with a PANI hole at different applied potentials, thus measuring through Glass/ITO/PAMPS/WO_3/ITO/Glass.

WO_3 films, gets more pronounced for large surfaces. The variation in thickness of PANI and WO_3 over the film surface makes it difficult to study holes in the PANI and WO_3 layers on the same window sample. Consequently three individual windows had to be studied.

Figure 2 depicts the transmission through the entire window, while Figs. 3 and 4 lack the WO_3 and PANI component, respectively. At −1800 mV the window (Fig. 2) is in its transparent state, with a maximum transmission of 0.80 (1600 nm). Changing the potential to more positive values, the transmission decreases over the entire visible and NIR region down to 0.36 (1000 nm) at +1200 mV. Increasing the potential to +1600 mV causes the transmission in the NIR region to increase, while the transmission in the visible region is further decreased to 0.29 (600 nm). Reversing the potentials gives reproducible transmission spectra. (The hump in the spectra at 800 nm is caused by the detector shift in the CARY 5 spectrophotometer.) Figure 3 shows the transmission spectra for the window with a hole in the WO_3 layer. The characteristic absorption peak around 600 nm for oxidized PANI (+1600 mV) is seen, likewise the shoulder at 420 nm [12]. The transmission spectra for the window with a hole in the PANI film are given in Fig. 4 and show that the transmission modulation in the visible region is much less in WO_3 than in PANI.

Figure 5 gives the spectrum of Glass/ITO/PAMPS/ITO/Glass and of Glass/ITO, which shows that this solid polymer electrolyte has quite good transparency properties for window applications. The small irregularities in the PAMPS spectrum in the NIR region, which are also found in Figs. 2–4, are probably due to traces of water in the electrolyte, as we have previously made electrolytes with smooth transmission curves [3]. Large absorption in Glass/ITO/PAMPS causes sharp cutoffs in the transmission spectra located at 400 and 2700 nm.

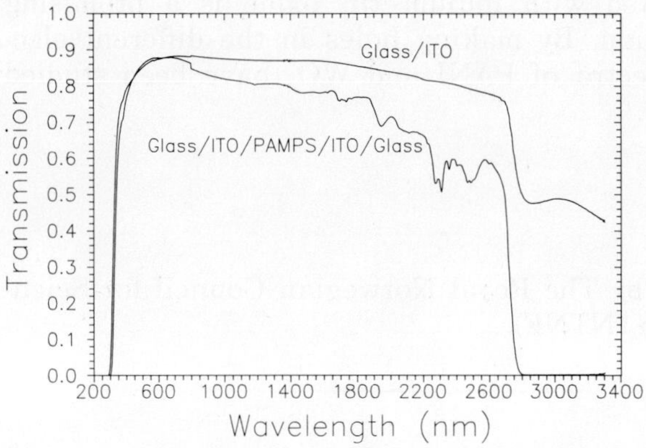

Fig. 5. Transmission vs. wavelength in Glass/ITO/PAMPS/ITO/Glass and in Glass/ITO.

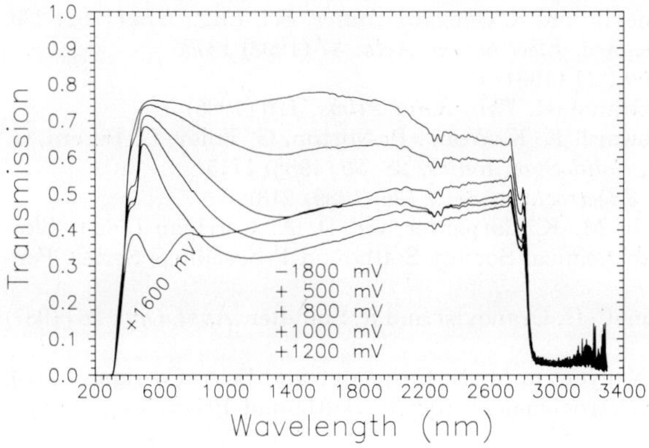

Fig. 6. Transmission vs. wavelength in a complete electrochromic window at different applied potentials, calculated from the transmission spectra in Figs. 3–5, thus using the windows with holes (in PANI and in WO_3) to modulate a complete window.

By combining the spectra in Fig. 3 (PANI) and Fig. 4 (WO_3), together with the spectrum in Fig. 5 (PAMPS), we can calculate the spectra of a complete window (without holes), which should, ideally, correspond to the spectra in Fig. 2. The combination spectra (from Figs. 3–5) are presented in Fig. 6 and show generally good agreement with the observed spectra (Fig. 2). This demonstrates that the spectral characteristics from the different electrochromic layers in the window can be studied individually as a function of applied potential by this ‘hole’ method.

Conclusions

The electrochromic window consisting of polyaniline, tungsten oxide, the solid polymer electrolyte poly(2-acrylamido-2-methyl-propane-sulfonic acid) and glass plates coated with indium–tin oxide is a promising candidate for light modulation. By making holes in the different electrochromic coatings, the spectra of PANI and WO_3 have been studied individually from each layer.

Acknowledgement

This work is supported by The Royal Norwegian Council for Scientific and Industrial Research (NTNF).

References

1 J. S. E. M. Svensson and C. G. Granqvist, in *Optical Materials Technology for Energy Efficiency and Solar Energy Conversion III, Proc. SPIE, 502* (1984) 30–37.
2 B. P. Jelle, G. Hagen, S. M. Hesjevik and R. Ødegård, *Mater. Sci. Eng., B13* (1992) 239.
3 B. P. Jelle, G. Hagen and R. Ødegård, *Electrochim. Acta, 37* (1992) 1377.
4 C. M. Lampert, *Sol. Energy Mater., 11* (1984) 1.
5 J. S. E. M. Svensson and C. G. Granqvist, *Thin Solid Films, 126* (1985) 31.
6 R. B. Goldner, T. E. Haas, G. Seward, K. K. Wong, P. Norton, G. Foley, G. Berera, G. Wei, S. Schulz and R. Chapman, *Solid State Ionics, 28–30* (1988) 1715.
7 M. T. Nguyen and L. H. Dao, *J. Electrochem. Soc., 136* (1989) 2131.
8 L. H. Dao and M. T. Nguyen, in M. K. Carpenter and D. A. Corrigan (eds.), *Electrochromic Materials*, The Electrochemical Society Softbound Proceedings Series, Pennington, NJ, 1990, pp. 246–260.
9 J. R. Stevens, J. S. E. M. Svensson, C. G. Granqvist and R. Spindler, *Appl. Opt., 26* (1987) 3489.
10 L. H. Dao, A. Guerfi and M. T. Nguyen, in M. K. Carpenter and D. A. Corrigan (eds.), *Electrochromic Materials*, The Electrochemical Society Softbound Proceedings Series, Pennington, NJ, 1990, pp. 30–44.
11 M. Akhtar and H. A. Weakliem, in M. K. Carpenter and D. A. Corrigan (eds.), *Electrochromic Materials*, The Electrochemical Society Softbound Proceedings Series, Pennington, NJ, 1990, pp. 232–236.
12 Y. H. Kim, C. Foster, J. Chiang and A. J. Heeger, *Synth. Met., 29* (1989) E285.

Magneto-optical characterization of excited states in short *trans* chains of partially isomerized polyacetylene

X. Wei and Z. V. Vardeny
Department of Physics, University of Utah, Salt Lake City, UT 84112 (USA)

E. Ehrenfreund*, D. Moses and Y. Cao
Institute of Polymers and Organic Solids, University of California, Santa Barbara, CA 93106 (USA)

Abstract

We have used photoluminescence (PL), photomodulation (PM) and their respective versions of optically detected magnetic resonance (ODMR); the PDMR for PL and the ADMR for PM, to characterize the long-lived photoexcitations in *trans* chains of thin films of partially isomerized $(CH)_x$ and $(CD)_x$. The PM spectra of ~80% *cis*/20% *trans* films contain two PA bands which belong to the *trans* chains; LE at 0.45 eV, which is due to charged solitons ($S^{\mp}$), and HE at 1.49 eV attributed to neutral solitons (S^0). The ADMR spectrum, however, reveals also a third PA band at 1.55 eV, which we tentatively identified as due to confined $S^0\bar{S}^0$ pairs in short *trans* chains. In addition we verified that the IR PL band observed at 1.3 eV is also associated with the confined $S^0\bar{S}^0$ pairs.

Introduction

The theoretical picture of electronic excitations in conducting polymers put forward about a decade ago [1] has created a great deal of interest both experimentally and theoretically. In this elegant picture, electronic excitations across the Peierls gap are entirely different from the electron–hole pairs of conventional semiconductors. Instead, the proper description of the quasiparticles in *trans*-$(CH)_x$ is 1D domain walls or solitons (S) which separate the two degenerate ground state structures. Since the soliton is a topological defect it can be either created in pairs of soliton–antisolitons (S–$\bar{S}$) or be created in polyacetylene chains with odd numbers of CH monomers upon isomerization from *cis*-$(CH)_x$. For the non-degenerate polymers, adding an extrinsic gap component to the electron–phonon Hamiltonian results in polarons and bipolarons as the proper description of their primary charged electronic excitations [1, 2].

To a great extent, experimental work on photoexcited $(CH)_x$ was stimulated by the prediction of soliton photogeneration [3]. Within the

*Permanent address: Physics Department, Technion, Haifa 32000, Israel.

0379-6779/93/$6.00

framework of the Su–Schrieffer–Heeger (SSH) model Hamiltonian [4], which contains only electron–phonon (e–p) interactions, it has been shown [3] that the photoexcited electron–hole (e–h) pair is unstable towards the formation of a soliton–antisoliton (S–$\bar{S}$) pair. Subsequently, it has been demonstrated [5] that as a consequence of the Pauli principle and charge conjugation symmetry in $(CH)_x$, the photogenerated soliton–antisoliton are oppositely charged ($S^+\bar{S}^-$).

The study of photoexcited $(CH)_x$, however, has revealed several unexpected phenomena, which were not predicted by the SSH model of the soliton. Most importantly, overall neutral states as well as charged photoexcitations have been observed [6]. It is now well established [1, 6, 7] that the lower energy (LE) photoinduced absorption (PA) band in *trans*-$(CH)_x$ at 0.45 eV (see Fig. 1) is due to photogenerated charged solitons ($S^{\mp}$), whereas the higher energy (HE) PA band at 1.35 eV is due to neutral solitons (S^0). The reason why S^0 are photogenerated in *trans*-$(CH)_x$ with high quantum efficiency, is still debated [7].

In this paper we extended our studies of photoexcitations in $(CH)_x$ to *cis*-rich films (~80% *cis*), where the *trans* segments contain a large fraction of short chains [8]. We used a variety of optical techniques such as photomodulation (PM), photoluminescence (PL) and their respective versions of optically detected magnetic resonance (ODMR), the ADMR for PM [7] and PDMR for PL [9]. We found a new PA band at 1.55 eV revealed only in ADMR spectra; we tentatively identify it as due to bound $S^0\bar{S}^0$ pairs in short *trans* chains. We also show that the IR PL band observed at 1.3 eV [10] is associated with these bound $S^0\bar{S}^0$ pairs.

Experimental

The PM and PL spectra were excited by an Ar^+ laser beam at 458 nm (pump) with intensity I_L of 400 mW cm^{-2}, which was modulated at frequency f ranging from 20 Hz to 50 kHz, by an acousto-optical modulator. The PL from the sample and the transmission of a cw lamp (probe beam) were dispersed by a $\frac{1}{4}$ met. monochromator and measured by various solid state detectors in the spectral range 0.3–2.7 eV. For the ADMR measurements, the pump and probe beams constantly illuminate the sample, mounted in a high Q microwave cavity at 3 GHz equipped with optical windows and a superconducting magnet producing a field H up to 3 Tesla. Microwave (μ-wave) resonant absorption, modulated at a frequency between 20 and 4 KHz, leads to small changes δT in the transmission T proportional to δn, the change in photoexcitation density n produced by the pump beam. δn is induced by transitions in the μ-wave range that change spin-dependent recombination rates. With suitable signal averaging, the system $\delta T/T$ sensitivity was about 10^{-7}. In PDMR we measured changes δL in the PL intensity L with sensitivity $\delta L/L \simeq 10^{-5}$. Two types of ADMR (PDMR) spectra were obtained: the H-ADMR spectrum, in which $\delta T(\delta L)$ is meaured at a fixed probe wavelength λ while sweeping H, and the P-ADMR spectrum in which δT is measured at a constant H, in resonance, while λ (probe) is varied.

The partially isomerized $(CH)_x$ and $(CD)_x$ samples were in the form of thin films ($d \simeq 1000$ Å) polymerized onto sapphire substrates by the Shirakawa technique. The films contain ~80% *cis* and 20% *trans*, as determined by resonant Raman scattering measurements [11].

Results and discussion

The PM spectrum of (80%) *cis*-rich $(CD)_x$ at 4 K is shown in Fig. 1 and compared to that of fully *trans*-$(CH)_x$. As in *trans*-$(CH)_x$ the PM spectrum in *cis*-rich $(CD)_x$ also contains two PA bands: a weak LE band at 0.5 eV and an intense HE band at 1.49 eV [8]. The relatively weak LE band shows either that long-lived $S^{\mp}$ in short chains are not generated with high quantum efficiency, or that their lifetime is long compared to the laser modulation frequency f [12]. The HE band, on the other hand, does not differ much from that in fully *trans*-$(CH)_x$, except for the blue shift to 1.49 eV, which is probably due to the relatively short *trans* chains in the film, with higher E_g. No long-lived PA bands associated with the *cis* segments in the film were detected [8].

The H-ADMR spectrum of the *cis*-rich $(CD)_x$ at 1.49 eV (HE band) at 4 K, is shown in Fig. 2. We observe two δn bands: an intense $\delta n < 0$ band with $\delta n/n \simeq 5\%$, which peaks at 1071 G, and a much weaker derivative-like band with maximum $\delta n/n$ of 10^{-5} and with zero crossing at 538 G. The intense δn band at 1071 G shows that spin $\frac{1}{2}$ excitations are correlated

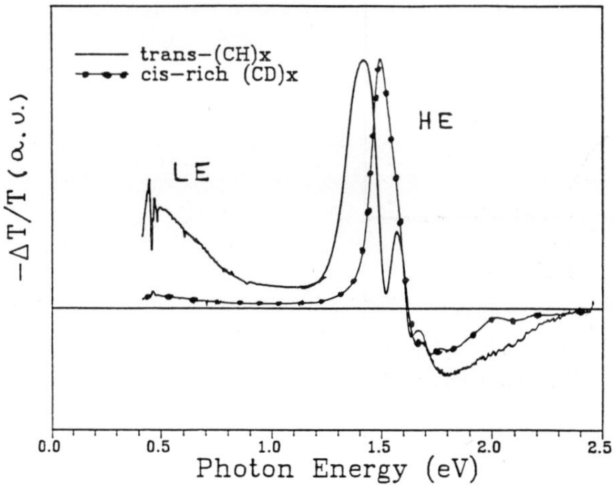

Fig. 1. PM spectrum of fully *trans*-$(CH)_x$ compared with that of 80% *cis*/20% *trans* $(CD)_x$, at 4 K. The LE and HE PA bands are assigned.

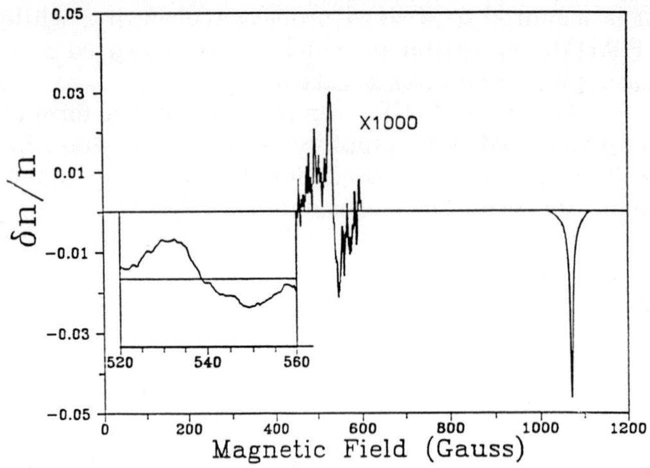

Fig. 2. H-ADMR of 80% *cis*-$(CD)_x$ at 1.49 eV. The inset shows in more detail the narrow, derivative-like feature at half field (537 G).

with the HE PA band and confirm, therefore, that it is indeed due to $S^0(\bar{S}^0)$ [7]. Careful studies of the ADMR band at 1071 G, however, reveal that it contains *two δn contributions*, with different dynamics as measured by their different response to the μ-wave modulation frequency ϑ: a narrow (N) component with FWHM of ~8 G, and a slower and broader (B) component with FWHM of 32 G. Figure 3 shows that the two δn components have the same P-ADMR spectrum. The P-ADMR associated with the N component peaks at 1.49 eV and contains an apparent sec-

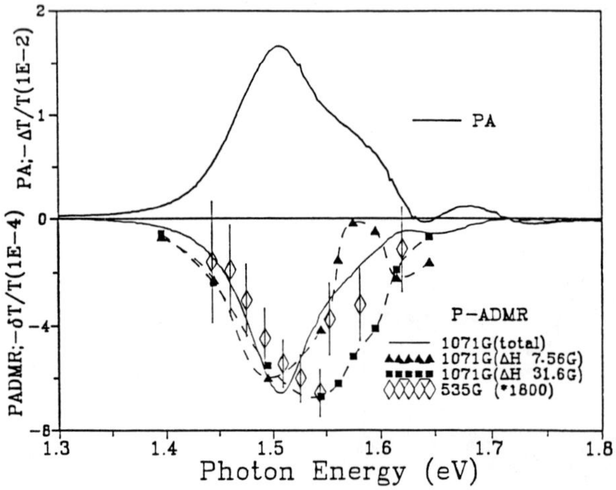

Fig. 3. The HE PA band compared to the P-ADMR spectra of the narrow (triangles) and broad (squares) ADMR resonances at 1071 G and ADMR resonance at 735 G (diamonds).

ondary oscillation at 1.61 eV, whereas the P-ADMR associated with the B component peaks at 1.55 eV and does not have any oscillation. It is obvious, therefore, that the two δn components do not share a common origin.

The much weaker ADMR signal at 538 G is close to the half field value of the stronger resonance at 1071 G and, therefore, we tentatively attribute it to a triplet powder pattern at half field (HF) associated with the HE PA band. We obtained δn dynamics by measuring the HF response at different ϑ. We found that its frequency response is identical to that of the B component of δn at 1071 G. Moreover, as shown in Fig. 3, the P-ADMR spectrum at 535 G (HF) is also peaked at 1.55 eV, as is the P-ADMR spectrum of the B component. We therefore identify the B component as the full-field (FF) powder pattern companion of triplets associated with the HE band. The HF and FF powder patterns in Fig. 2 are hence due to the same species, which we tentatively identified as due to a solitonic triplet [13] (bound $S^0\bar{S}^0$ pair) in short *trans* chains. From the FF powder pattern in Fig. 2 we calculate a triplet zero splitting parameter D [14] of 15 G ($D \simeq 0.0016\ \mathrm{cm}^{-1}$). If D is determined by bipolar interaction of $s = \frac{1}{2}$ e–h in the triplet state, then the extent of the solitonic triplet that we found is about 12 Å, which is an upper bound for the *trans* chain in which triplets are formed.

PL studies of partially isomerized $(CH)_x$ films revealed two PL bands [10, 15]. A visible PL band at 1.97 eV [16], which is due to the *cis* segments in the film, and an IR PL band at about 1.3 eV, which was attributed [15] to the *trans* segments. In Fig. 4 we show the H-PDMR spectrum at 1.3 eV; no PDMR signal has been measured at 1.97 eV, presumably due to a much shorter PL lifetime. As in ADMR, the PDMR contains two δL bands: a

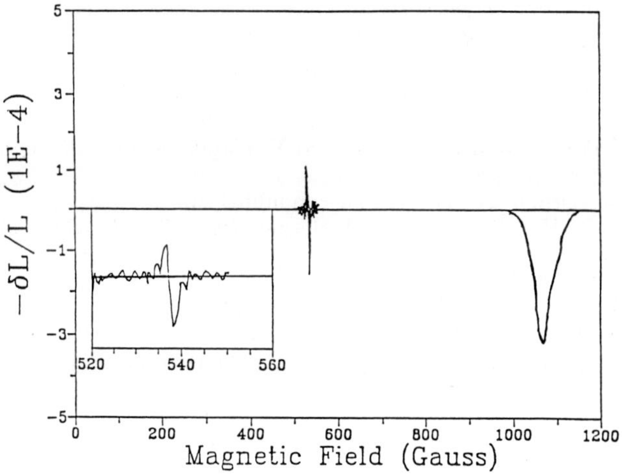

Fig. 4. The H-PDMR of the same *cis*-rich $(CD)_x$ film as in Figs. 1–3, for the PL at 1.3 eV. The inset shows in more detail the narrow derivative-like resonance at 737 G.

derivative-like band with zero crossing at 537 G and a much broader $\delta L > 0$ band at 1071 G with FWHM of 32 G. The two δL bands are correlated, due to identical dependencies with ϑ and I_L. Since the H-PDMR spectrum in Fig. 4 is identical to the triplet H-ADMR spectrum (Fig. 2), with narrow HF and broad FF powder patterns, we identify the PDMR as due to triplets and hence the IR PL is associated with bound $S^0\bar{S}^0$ in short *trans* chains. This explains its paradoxical dependence with % *trans* in $(CH)_x$ samples [15]: it decreases with increasing % *trans* and yet it originates from *trans* chains. As is well known, however, the % of short *trans* chains in the sample *decreases* with increasing % *trans*, in agreement with the IR PL band and our interpretation.

Acknowledgements

This work was supported in part by the DOE, grant DE-FG-02-89, ER 45409 and by ONR grant no. N00014-91-C-0104 at the Utah Laser Institute. At Santa Barbara the work was supported by NSF-PDMR89-15207. E.E. also acknowledges support from the US–Israel Binational Science Foundation (BSF. 89-00322), Jerusalem, Israel.

References

1 A. J. Heeger, S. Kivelson, J. R. Schrieffer and W. P. Su, *Rev. Mod. Phys.*, *60* (1988) 781, and refs. therein.
2 K. Fesser, A. R. Bishop and D. K. Campbell, *Phys. Rev. B*, *27* (1983) 4804.
3 W. P. Su and J. R. Schrieffer, *Proc. Natl. Acad. Sci. USA*, *77* (1980) 5626.
4 W. P. Su, J. R. Schrieffer and A. J. Heeger, *Phys. Rev. Lett.*, *42* (1979) 1698; *Phys. Rev. B*, *22* (1980) 2099.
5 R. Ball, W. P. Su and J. R. Schrieffer, *J. Phys. (Paris) Colloq.*, *44* (1983) C-3-429.
6 J. Orenstein and G. L. Baker, *Phys. Rev. Lett.*, *49* (1982) 1043.
7 X. Wei, B. C. Hess, Z. V. Vardeny and F. Wudl, *Phys. Rev. Lett.*, *68* (1992) 666.
8 Z. Vardeny, E. Ehrenfreund and O. Brafman, *Synth. Met.*, *17* (1987) 349.
9 L. S. Swanson, J. Shinar and K. Yoshino, *Phys. Rev. Lett.*, *65* (1990) 1140.
10 K. Yoshino, S. Hayashi, T. Sakai, Y. Inuishi, H. Kato and Y. Watanabe, *Jpn. J. Appl. Phys.*, *21* (1982) L653.
11 L. S. Lichtman, *Ph.D. Thesis*, Cornell University, 1981 (unpublished).
12 N. F. Colaneri, R. H. Friend, H. E. Schaffer and A. J. Heeger, *Phys. Rev. B*, *38* (1988) 3960.
13 W. P. Su, *Phys. Rev. B*, *34* (1986) 2988; *36* (1987) 6040.
14 L. Robins, J. Oreinstein and R. Superfine, *Phys. Rev. Lett.*, *56* (1986) 1850.
15 E. A. Imhoff, D. B. Fitchen and R. E. Stahlbush, *Solid State Commun.*, *44* (1982) 329.
16 L. Lauchman, S. Etemad, T. C. Chung, A. J. Heeger and A. G. MacDiarmid, *Phys. Rev. B*, *24* (1981) 3701.

Second optical harmonic generation in the vicinity of insulator-to-metal transition in polyaniline film

O. Aktsipetrov, A. Ermushev and A. Petukhov
Moscow State University, Moscow (Russian Federation)

L. Daikhin, M. Levi* and V. Vereta
A. N. Frumkin Institute of Electrochemistry, Leninski Prospect 31, Moscow 117071 (Russian Federation)

Abstract

In situ second harmonic generation reflected from the platinum electrode covered with polyaniline film has been observed. A drastic increase in the response in the vicinity of the insulator-to-metal transition is attributed to either non-linear electrochromic effect or variation in centrosymmetry of the polymer molecules in the course of electrochemical doping.

Introduction

Within the continuously growing class of conducting polymers, polyaniline (PANI) as well as its derivatives have many advantages such as the simplicity of the polymerization procedure (chemical or electrochemical oxidation of aniline, more often in aqueous solutions) and high environmental stability. Fascinating progress has been made in the last few years in the field of their processability [1].

Electrochemical doping (charge injection) or protonation of the partly oxidized PANI form (i.e. emeraldine) results in an increase of conductivity by more than ten orders of magnitude [2, 3]. The latter effect is accompanied by a rather narrow ESR peak [3, 4] as well as a sharp increase in the optical absorption [5, 6] and resonance Raman scattering [7]. Two possible reasons for this have hitherto been advanced; the first based on the thermopower and Pauli susceptibility measurements gives evidence for the formation of a polaron lattice [2] whereas the second one takes into consideration the conversion spin carrying species into spinless charged bipolarons [5–7].

Recently electronically conducting polymers have attracted attention because of their non-linear optical susceptibility; the prospective applications in optical processing etc, are reviewed in ref. 8. Studies of the non-linear optical processes can in principle provide a deeper insight into

*Present address: CENG, DRFMC/SESM/EM, 85X 38041 Grenoble Cédex, France.

0379-6779/93/$6.00

the electronic structure of π-conjugated polymers thus being complementary to the commonly used linear optical measurements.

It is the purpose of this work to present our recent results on the studies of the electrochemical doping of PANI using *in situ* second optical harmonic generation (SHG) [9]. It will be shown that a drastic increase in SHG response can be attributed to the insulator-to-metal transition whereas, more specifically, it seems to be connected with the existence of bipolarons in the emeraldine state of PANI equilibrated with high acidic solutions.

Experimental

PANI films were deposited onto the Pt electrode (visible surface area $s = 0.1\ cm^2$) in the cyclic voltammetric regime by oxidation of aniline.

SHG was observed in the reflected light of a pulsed YAC:Nd^{3+} laser with the following parameters: wavelength, $\lambda = 1064$ nm; pulse duration, $\tau = 10$–15 ns; frequency 12.5 Hz; energy density of a pulse *c.* 25 mJ/cm^2.

Results and discussion

Figure 1 shows the potential dependence of SHG intensity, $I_{2\omega}$, reflected from a Pt electrode covered with PANI film *c.* 500 Å thick (curve 1). For comparison the curve of *in situ* conductivity of PANI film obtained in approximately the same conditions is also plotted (the data are taken from ref. 10). It is clear that a drastic increase in SHG response occurs in the vicinity of the insulator-to-metal transition in this polymer. Other parameters such as low-frequency differential capacity and potential-modulated absorption [11], as well as integrated absorption [5, 6] display a very similar behaviour in this region.

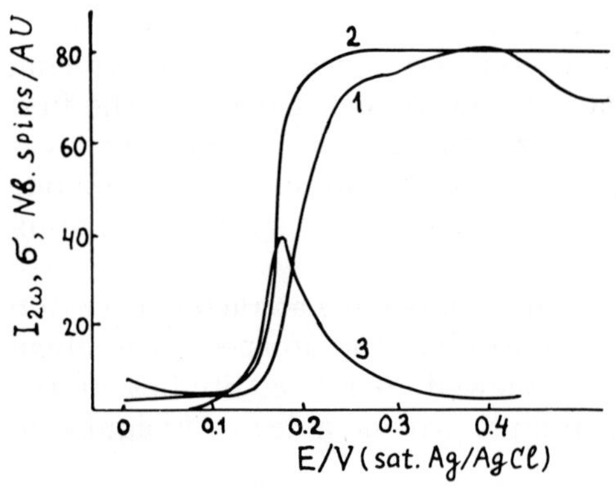

Fig. 1. Electrode potential dependence of SHG intensity, $I_{2\omega}$ (1) along with corresponding literature dependencies of conductivity, σ (2) [10] and number of spins (3) [4].

At present we are able to give at least two tentative explanations for the sharp increase in SHG response in the course of PANI film oxidation. The first one is connected with charge injection into the polymer bulk which results in new absorption bands in the electronic spectra. The latter can transform the whole process of SHG into the resonance regime which can be regarded as a non-linear electrochromism by analogy with the non-linear photochromic effect as reported in ref. 12. The second possibility takes into account the symmetry factor. We thus assume that the non-conducting leucoemeraldine molecule possesses a centrosymmetrical structure with SHG being induced by the quadrupole mechanism only. As a result its intensity should be rather small. A break in centrosymmetry of the protonated emeraldine molecules, if any, would certainly result in the dipole component SHG response which is much larger than the quadrupole one. A break in centrosymmetry of the molecules was assumed to be characteristic of bipolarons rather than polarons. To confirm this suggestion we reproduce the literature ESR absorption curve of PANI film (curve 3 in Fig. 1) [4] deposited under very similar conditions (the cyclic voltammogram observed in ref. 4 was almost completely identical to that observed in our work). We can thus establish that the drastic increase in SHG intensity occurs at the potential range where the ESR signal drops abruptly. This suggests that the SHG response enhancement is of bipolaron origin. A zero Pauli susceptibility would be expected for such an electronic configuration but this is not the case [2]. The electronic structure of PANI studied by *ex situ* magnetic measurements should not necessarily be identical with the *in situ* structure. In particular, one can imagine that the bipolaron structure is stabilized due to the solvent and (or) counterion penetration.

We have also studied the influence of the film thickness, d, on the SHG response. The cyclic voltammetric regime of PANI deposition is characterized by nearly quadratic dependence of d on the number of cycles, $d \approx N^2$ [13]. Figure 2 shows the dependence of SHG intensity for

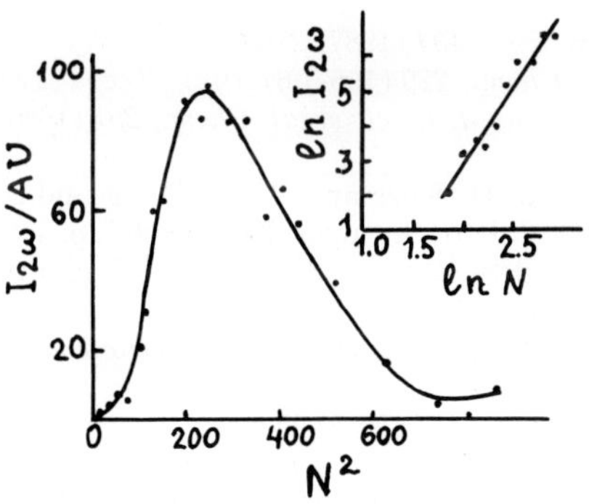

Fig. 2. Dependence of SHG intensity, $I_{2\omega}$, on the square of the number of polymerization cycles, N^2. The inset shows the $I_{2\omega}$ vs. N relationship in double-logarithmic coordinates.

the conducting state of PANI ($E = 0.4$ V). At small thicknesses $I_{2\omega}$ values should be proportional to d^2, i.e. $I_{2\omega} \approx N^4$. The inset in Fig. 2 shows the ln N dependence of ln $I_{2\omega}$ at $N < 15$, which can be approximated by linear function of the slope to be equal to 3.8. The latter seems to be in reasonable agreement with the above mentioned dependence of $I_{2\omega}$ on N, thus confirming the linear augmentation of the film growth rate on the increase of N. The SHG intensity starts to decrease when the maximum value at $N = 15$ is passed. The drop of $I_{2\omega}$ is likely to be a result of absorption inside the polymer bulk.

Conclusions

In situ SHG reflected from the platinum electrode coated with PANI film has been observed. The non-linear optical response increases considerably in the vicinity of the insulator-to-metal transition. The former can tentatively be attributed to either non-linear electrochromic effect (see above) or variation in centrosymmetry of the polymer molecules in the course of electrochemical doping. The observed relationship between the SHG intensity and the thickness of the film appears to be in reasonable agreement with the well-known linear augmentation of the film deposition rate on the increase of the number of polymerization cycles (for the relatively thin films, $N < 15$).

Further studies of SHG starting from the polyemeraldine base in the course of its protonation could in principle be very useful for discriminating between the two possible reasons for SHG mentioned above.

References

1 A. G. MacDiarmid and A. J. Epstein, *Faraday Discuss. Chem. Soc., 88* (1989) 317.
2 A. J. Epstein and A. G. MacDiarmid, *J. Mol. Electron., 4* (1988) 161.
3 M. Nechtschein, F. Genoud, C. Menardo, K. Mizoguchi, J. P. Travers and B. Villeret, *Synth. Met., 29* (1989) E211.
4 S. H. Glarum and J. H. Marshall, *J. Electrochem. Soc., 134* (1987) 2160.
5 E. M. Geniès and M. Łapkowski, *J. Electroanal. Chem., 220* (1987) 67.
6 R. J. Cushman, P. M. McManus and Sze Cheng Yang, *J. Electroanal. Chem., 291* (1986) 335.
7 M. Bartonek, N. S. Saricifitchi and H. Kuzmany, in H. Kuzmany, M. Mehring and S. Roth (eds.), *Springer Series in Solid State Sciences*, Vol. 91, Springer, Berlin, 1989, p. 330.
8 G. R. Mohlmann, in H. Kuzmany, M. Mehring and S. Roth (eds.), *Springer Series in Solid State Sciences*, Vol. 91, Springer, Berlin, 1989, p. 232.
9 O. Aktsipetrov, V. Bartenev, E. Mishina and A. Petukhov, *Quantum Electronics, 10* (1983) 1113 (in Russian).
10 S. H. Glarum and J. H. Marshall, *J. Electrochem. Soc., 134* (1987) 142.
11 R. S. Hutton, M. Kalaji and L. M. Peter, *J. Electroanal. Chem., 270* (1989) 429.
12 O. Aktsipetrov, N. Akhmediyev, W. Vsevolodov, E. Mishina and V. Bartenev, *DAN USSR, 293* (1987) 592 (in Russian).
13 G. Zotti, S. Cattarin and V. Comisso, *J. Electroanal. Chem., 235* (1987) 259.

Photoexcitations in polyacetylene with controlled conjugation length

G. Lanzani[a,b], A. Piaggi*[,c], R. Tubino[d], Z. V. Vardeny[b] and G. S. Kanner[b]

[a]*Istituto di Spettroscopia Molecolare del CNR, via dei Castagnoli 1, Bologna (Italy)*

[b]*Department of Physics, University of Utah, Salt Lake City, UT 84112 (USA)*

[c]*Dipartimento di Fisica 'A. Volta', Università di Pavia, via Bassi 6, Pavia (Italy)*

[d]*Istituto di Matematica e Fisica, Università di Sassari, via Vienna 2, Sassari (Italy)*

Abstract

Photoexcitation of solitons has been studied by steady state and transient picosecond photomodulation spectroscopy in a form of soluble polyacetylene with controlled conjugation length. A systematic blue shift of the charged soliton band is observed by decreasing the average conjugation length from a few hundred to 20 CH units, showing that the chain segmentation has strong effects when the conjugation length becomes comparable or shorter than the soliton spatial extent. The linear dependence of the charged soliton band on the inverse of the average conjugation length is discussed, and an evaluation of the electron correlation energy is given. The effects of the decreased conjugation length have been also investigated by time-resolved photoinduced absorption. The transient electronic response in the time domain from 5 ps to 1 ns is interpreted in terms of soliton recombination associated with diffusion on chains distributed in length.

Introduction

We report on our study of photoexcitations in samples of polyacetylene with controlled conjugation length. Steady state and transient picosecond photomodulation spectroscopy have been employed in order to clarify the effects of the segmentation when the average conjugation length becomes comparable or shorter than the soliton extent, and the finite chain length on the photoexcited solitons decay.

Samples of soluble polyacetylene (SCH) have been obtained by growing the polyenic chains onto activated sites of a flexible polybutadiene chain which acts as soluble carrier into common organic solvent [1]. The conjugation length of these samples, as detected by electronic absorption and Raman spectra, has been estimated to be of the order of a few hundred CH units. Electronic delocalization has been reduced by air exposure of the samples for a controlled amount of time. Thin films of SCH were obtained by evaporation in an inert atmosphere of soluble polyacetylene on SiO_2 substrate.

*Author to whom correspondence should be addressed.

0379-6779/93/$6.00

The steady state photoinduced changes in transmission have been measured in the energy range from 0.4 to 2.2 eV by phase-sensitive detection technique. The sample was excited by the 488 nm line of an argon laser chopped mechanically at 70 Hz. Measurements were carried out with the cast film placed on a cold finger assembly of a Cryo-tip system. The sample temperature could not be properly controlled due to the poor heat conduction of the substrate and the insulating nature of the polybutadiene matrix. Picosecond photoinduced absorption was measured with a synchronously pumped dye laser consisting of a frequency-doubled mode-locked Nd:YAG laser (with 76 MHz repetition rate) synchronously pumping two dye lasers, one for the pump beam (with 580 nm wavelength) and the other for the probe beam (with wavelength between 580 and 1020 nm). The room temperature photoinduced changes in transmission were measured using 10 MHz modulation. The time resolution of the system was about 5 ps.

Results and discussion

Steady state photoinduced absorption

Linear absorption spectra taken on samples with different average conjugation length show a blue shift of the absorption maximum upon shortening the conjugation length. According to the Kuhn model [2], the energy E of the absorption maximum in polyenes is related to the conjugation length 'n' (n = number of conjugated double bonds) by

$$E = A + \frac{B}{n} \qquad (1)$$

where A and B were evaluated from the observed position of the absorption maxima in a series of polyenes with known conjugation lengths [3]. The average conjugation length n of our segmented samples was then estimated by eqn. (1).

The photoinduced absorption spectra of four samples with different conjugation length are shown in Fig. 1. Strong oscillation is superimposed close to the isosbestic point spectral region; the oscillation disappears upon oxidation of the sample. A detailed analysis of this energy region, which exhibits significant differences from the Shirakawa polyacetylene, is reported in ref. 4. Here we discuss the low energy electronic transition correlated with the charged soliton ($S^{\pm}$) defect. We notice that, even for samples possessing the highest degree of delocalization, the $S^{\pm}$ peak position (0.61 ± 0.05 eV) is somewhat higher than the values reported for Shirakawa [5] (0.49 eV) and Durham [6] (0.55 eV) polyacetylenes. This discrepancy could be related to temperature effects due to the poor temperature control of the sample. The $S^{\pm}$ sub-gap absorption band shows a significant blue shift with decreasing electron delocalization. In particular, as reported in Fig. 2, the $S^{\pm}$-band position is linearly dependent on the inverse of the conjugation length.

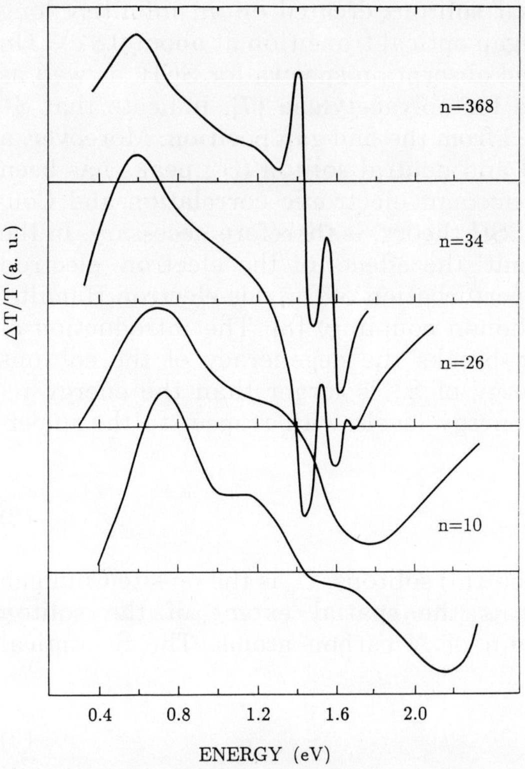

Fig. 1. Liquid nitrogen photoinduced absorption spectra of soluble polyacetylene with different conjugation lengths '*n*'.

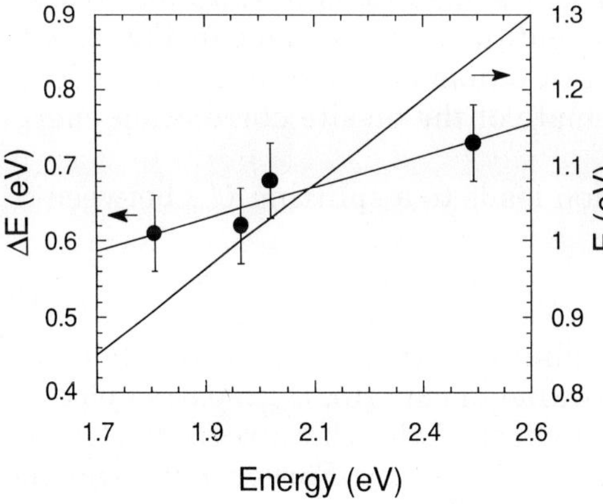

Fig. 2. Photoinduced absorption band maximum of $S^{\pm}$ as a function of the linear absorption band maximum. The straight line (referring to the right side axis) has been calculated using eqn. (4) for the case $U_0 = 0$.

The SSH model predicts that solitons created on an infinitely long chain are associated with a mid-gap optical transition at about 0.8 eV. On the other hand, the photoinduced absorption spectra for SCH, as well as those reported in the literature for polyacetylene [7], indicate that $S^{\pm}$ absorption is significantly shifted from the mid-gap position. Moreover, a large splitting between charged and neutral soliton (S^0) peaks has been observed. A model taking into account electronic correlation and Coulomb repulsion, not included in SSH theory, is therefore necessary. In the 'weak coupling perturbation limit' the effects of the electron–electron correlation can be treated as a perturbation of the one-electron Hamiltonian (including the electron–phonon coupling) [8]. The introduction of the electron–electron repulsion breaks the degeneracy of the solitons states because the creation energy of $S^{\pm}$ is larger than the energy required for S^0. The shift of the energy levels with respect to the unperturbed situation results in [9]

$$\delta E = \frac{U_0}{4}\left[(N-1) \pm \frac{2}{3\xi}\right] \tag{2}$$

where $+(-)$ refer to charged (neutral) solitons, U_0 is the on-site Coulomb repulsion parameter, and $2\xi a$ is the spatial extent of the soliton ($2a$ = lattice constant) in a chain of N carbon atoms. The $S^{\pm}$ optical transition is now located at

$$\Delta E_{S^{\pm}} = \Delta_0 - \frac{U_0}{3\xi} \tag{3}$$

where the soliton width is related to Δ_0, the energy gap parameter, by $\xi \sim W/(2\Delta_0)$ (W = bandwidth, estimated to be 10–12 eV). Finally one gets

$$\Delta E_{S^{\pm}} \simeq \Delta_0\left(1 - \frac{2U_0}{3W}\right) \tag{4}$$

Therefore, the transition energy for charged solitons is expected to scale linearly with the gap parameter, as confirmed by the data in Fig. 2, with a slope which depends on the on-site Coulomb repulsion. A linear least-squares fit of our data yields an estimate of the on-site correlation energy $U_0 \sim 9$–11 eV.

The electron–electron interaction leads to a splitting U_{eff} between $S^{\pm}$ and S^0 transition energies

$$U_{eff} = \Delta E_{S^0} - \Delta E_{S^{\pm}} = 2U_0/3\xi \tag{5}$$

The soliton spatial width 2ξ for an infinite chain has been estimated to be about 14 CH units. By inserting this value in eqn. (5), $U_{eff} \simeq 0.85$–1.05 eV is obtained. An interesting point is the effect of the chain segmentation on solitons. This problem has been investigated by Förner [10] who has shown that, within the framework of the SSH model, a soliton shrinkage occurs upon shortening of the chain length. This behaviour can possibly explain the reason why solitons can be supported also by very short conjugated segments, as we have observed in the present work. We have estimated the soliton width in samples of different conjugation length by the Förner approach, then we have determined the effective Coulomb repulsion U_{eff} from eqn. (5). As one could expect, U_{eff} increases upon chain shortening ($U_{eff} = 0.89$, 1, 1.17 and 2.56 eV for $n = 368$, 34, 26 and 10, respectively) due to a stronger confinement of excess electronic density associated with the conjugation defect.

Picosecond decay

A typical room temperature decay of the photoinduced absorption in SCH is shown in Fig. 3. The pump and probe wavelengths were 580 and 950 nm, respectively (no dependence of the transient signal on the probe wavelength was detected). The signal decays as $t^{-0.5}$ up to about 50 ps, then slows to $t^{-0.1}$. A similar transient behaviour has been observed in Shirakawa *trans*-polyacetylene [11, 12], where a longer crossover time t_c between the two power-law decays and an higher exponent α ($\sim$0.2–0.3) after t_c were found.

The initial $t^{-0.5}$ decay strongly suggests diffusion and geminate recombination of the soliton–antisoliton pairs along the chains. In fact, the survival probability of particles executing random walks on an infinite chain is [13]

$$S(u, t) = \mathrm{erf}\left(\frac{u - a}{2(2Dt)^{1/2}}\right) \tag{6}$$

where u is the distance of the two particles at $t = 0$ and D is the diffusion coefficient. In real samples chains are not infinite; for a finite chain, the $t^{-0.5}$ decay continues until the particles diffuse to the ends of the chain. In

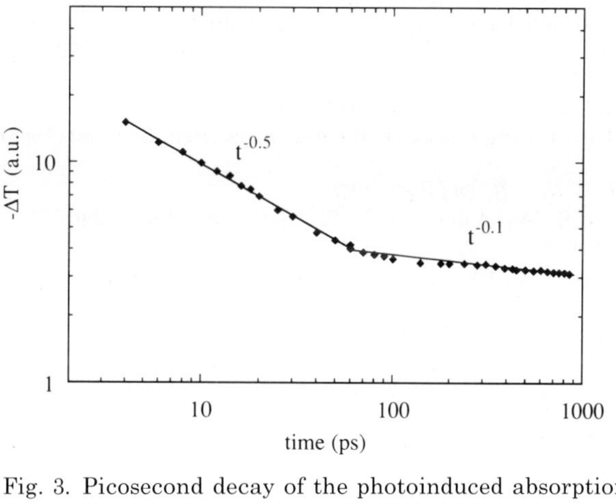

Fig. 3. Picosecond decay of the photoinduced absorption at room temperature. The pump and probe wavelengths are 580 and 950 nm, respectively. The lines are fits with power-law decays $t^{-\alpha}$ with $\alpha = 0.5$ and $= 0.1$, respectively.

this sense t_c is significantly related to the average chain length. It is then reasonable that in SCH t_c is lower than in solid *trans*-$(CH)_x$, due to the smaller chain lengths involved.

In principle, the long time survival probability for carrier diffusing along a chain of length L is [13, 14]

$$S(t) \propto \exp(-\pi^2 Dt/L^2) \tag{7}$$

Nevertheless, the measured decay $t^{-\alpha}$ calls for a model able to account for the slower and non-exponential behaviour. Weidman [15] proposed a model of diffusion-limited recombination in the presence of site disorder. The effect of a distribution of chain lengths $\Phi(L)$ is important in determining the long time exponent α. Slow decays in photoconductivity and photoluminescence of disordered materials have been explained in terms of a power-law distribution $G(\tau)$ of carrier lifetimes [16, 17]. Starting from a distribution of chain lengths in our samples, a distribution of decay times develops naturally. We take $G(\tau) \propto \tau^{-(1+\alpha)}$, and if the survival probability in eqn. (7) is averaged over $G(\tau)$, with τ varying from the minimum time constant τ_{min} that can be measured, i.e. the time resolution of our experimental set-up, and a maximum τ_{max} determined by the pump modulation frequency, the power-law $t^{-\alpha}$ is reproduced. The related distribution of chain lengths results in $\Phi(L) \propto L^{-(1+2\alpha)}$ (with L greater than the minimum length L_{min}). Smaller values of α are then connected to broader distributions of $\Phi(L)$, so that a broad distribution of lengths in SCH samples can be inferred.

References

1 S. Destri, M. Catellani and A. Bolognesi, *Makromol. Chem. Commun.*, *5* (1985) 353.
2 H. Kuhn, *Fortsch. Chem. Org. Naturst.*, *16* (1958) 169.
3 L. S. Lichtmann, *Ph.D. Thesis*, Cornell University, 1981, unpublished.
4 G. Lanzani, G. Kanner, S. Jeglinski and Z. V. Vardeny, *Synth. Met.*, (1992) *50* (1992) 461.
5 J. Orenstein, G. L. Baker and Z. V. Vardeny, *J. Phys. (Paris) Colloq. 44C3* (1983) 429.
6 P. D. Townsend and R. H. Friend, *Phys. Rev. B*, *40* (1989) 3112.
7 G. B. Blanchet, C. R. Fincher, T. C. Chung and A. J. Heeger, *Phys. Rev. Lett.*, *50* (1983) 1938.
8 S. Kivelson and D. E. Heim, *Phys. Rev. B*, *26* (1982) 4278.
9 D. Baeriswyl, D. K. Campbell and S. Mazumdar, *Phys. Rev. Lett.*, *56* (1986) 1509.
10 W. Förner, *Synth. Met.*, *30* (1989) 135.
11 Z. V. Vardeny, J. Strait, D. Moses, T. C. Chung and A. J. Heeger, *Phys. Rev. Lett.*, *22* (1982) 1657.
12 G. S. Kanner, *Ph.D. Thesis*, University of Utah, 1991, unpublished.
13 I. V. Zozulenko, *Solid State Commun.*, *76* (1990) 1035.
14 Y. Gaididei, A. I. Onipko and I. V. Zozulenko, *Phys. Lett. A*, *132* (1988) 329.
15 D. L. Weidman, *Ph.D. Thesis*, Cornell University, 1987, unpublished.
16 H. Scher and E. W. Montroll, *Phys. Rev. B*, *12* (1975) 2455.
17 C. Tsang and R. A. Street, *Phys. Rev. B*, *19* (1979) 3027.

7. Processing and Nanotechnologies

Scanning tunnelling microscopy of several alkylated molecular moieties in monolayers on graphite

J. P. Rabe, S. Buchholz* and L. Askadskaya
Max-Planck-Institut für Polymerforschung, Postfach 3148, D-6500 Mainz (Germany)

Abstract

Scanning tunnelling microscopy (STM) has been employed to characterize structure, dynamics and electronic properties at interfaces with molecular materials on the length and timescale of 10 pm and 10 μs, respectively. We report specifically on bias-dependent current images of monomolecular layers of several alkylated or alkoxylated molecular segments on the basal plane of graphite, including benzene, naphthalene, triphenylene, phenylbenzoate, disulfide and diacetylene. Derivatives with sufficiently long side chains tend to form lamellar two-dimensional crystals, which typically are incommensurate with the substrate along the lamellae, and commensurate perpendicular to them. The data indicate a pronounced dependence of conductivity on the relative position of adsorbate and substrate, as well as on the bias. In a bias range around 1 V, the conductivity of the aromatic segments, the disulfide and the diacetylene is generally larger than that of methylenes, methyls, hydroxyls and carboxyls. These results are in qualitative agreement with models in which the conductivity increases as the HOMO–LUMO gap of the free molecule decreases.

Introduction

Scanning tunnelling microscopy (STM) [1] has become a powerful tool for the investigation of molecules at surfaces and interfaces. Two classes of materials are directly accessible, i.e. molecular conductors and isolated molecules or ultrathin molecular layers on conductive substrates. Important contributions to this field have been reviewed recently [2]. Moreover, the capabilities and limitations of STM for a quantitative characterization have been analysed more generally for different classes of systems [3]. As far as conductive molecular cyrstals and polymers are concerned, we refer to the literature (e.g. refs. 4–8). In the following, results on ultrathin molecular layers are reported.

STM has been used successfully to determine various aspects of the molecular structure and dynamics in monomolecular layers on the length

*Present address: Harvard University, Department of Chemistry, Cambridge, MA 02138, USA.

0379-6779/93/$6.00

and timescale of 10 pm and 10 μs, respectively. It has been demonstrated that one can determine, e.g., (i) the molecular packing within crystalline monolayers [9–21], (ii) the orientation [12, 13] and commensurability [15] between adsorbate and substrate lattices, (iii) the molecular structure at domain boundaries within pure monolayers [14, 16] and mixtures [19, 22], as well as at steps in the substrate [23], and (iv) molecular dynamics, including reorientations of small domains [14, 15] and anisotropic molecular mobilities [22, 24, 28].

It must be noted that STM is based on the measurement of an electronic current through molecules, which means that an image contains information on molecular structure and dynamics, as well as on electronic properties, simultaneously. Fortunately, very precise information may be obtained on molecular structure and dynamics, without a detailed understanding of the electronic properties. This knowledge may then be used to determine the current–voltage characteristics through the individual molecules. Such an approach may lead towards the development of molecular electronics, i.e., electronics on the scale of individual molecules.

The present paper reports bias-dependent current images of monomolecular layers of various alkylated or alkoxylated molecular segments on graphite, including a series of aromatic systems, i.e. benzene, naphthalene and triphenylene, as well as phenylbenzoate, a disulfide and a diacetylene. The current–contrast between the alkyl chains and the various molecular segments is investigated as a function of tip bias and the relative positions of substrate and adsorbate. A study of one particular alkyl derivative, didodecylbenzene, on two other substrate materials, MoS_2 and $MoSe_2$, will be published elsewhere [25].

Experimental

The results reported below have been obtained with a home-built instrument, which together with some experimental procedures has been described elsewhere [26, 27, 28]. A schematic of the experimental setup is displayed in Fig. 1. The instrument was designed with an emphasis on low currents and high scanning speeds. A temperature-controlled sample stage allows imaging with atomic-scale resolution between room temperature and 120 °C with currents as low as 200 fA and scan speeds up to 1 ms per line [28]. The images were recorded on video tape in real time and digitized later for static image display. In the following, raw data without digital image processing are displayed. All images were recorded under ambient conditions using Pt/Ir tips, which were either electrochemically etched or mechanically formed. The images were recorded in the 'constant height mode' (a current profile is displayed) at an average current on the order of 1 nA. The tip bias was usually around +1 V, if not otherwise indicated.

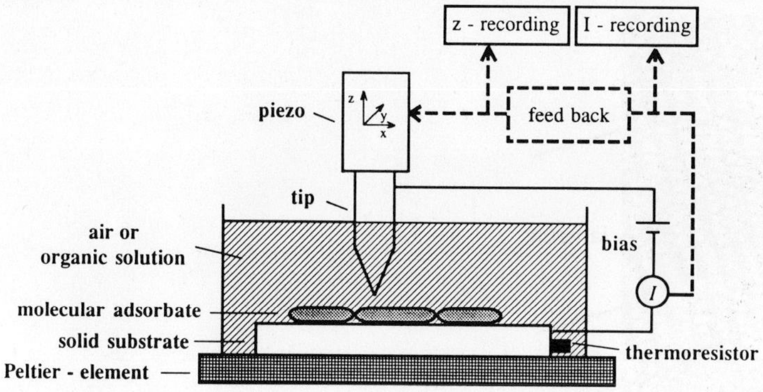

Fig. 1. Schematic of the *in situ* STM experiment [28].

The alkanes $C_{25}H_{52}$ and $C_{27}H_{56}$ (Aldrich, 99%), didodecylsulfide (Aldrich, 93%) and 10,12-octadecadiynoic acid (ABCR, Karlsruhe) were used as received. Monomolecular layers were prepared by adsorption at the interface between the basal plane of highly oriented pyrolytic graphite (HOPG, Union Carbide, quality ZYB) and a concentrated but not saturated organic solution in phenyloctane (Aldrich), as described before [15, 27].

Results and discussion

Long chain alkanes, as well as various alkyl derivatives, adsorb as monomolecular layers at the interface between HOPG and concentrated but not saturated organic solutions [29]. The packing within the monolayers of, e.g., long chain alkanes [13, 15, 16], fatty acids [15, 16], alkanols [15, 17, 19] or a dialkylbenzene [14] was determined *in situ* by STM.

In addition to the monolayer structure one may determine the relative orientation between adsorbate and substrate. One way to accomplish this is to image first the adsorbate layer, then to remove it by advancing the tip towards the substrate and finally to image the substrate [12]. A disadvantage of this procedure is that one needs to compare absolutely two images, and that between the recording of these two images the tip must be used to remove the monolayer. A more precise method to determine the relative orientation of monolayer and substrate lattices utilizes the fact that the tunnelling current is determined by both the substrate and adsorbate. It takes advantage of the crystal structure of the substrate surface being very precisely known and may be employed for an internal calibration.

Using this method it was found that the alkyl chains in alkane monolayers are packed commensurate with the graphite lattice, while in fatty acid monolayers they are incommensurate (or higher order commen-

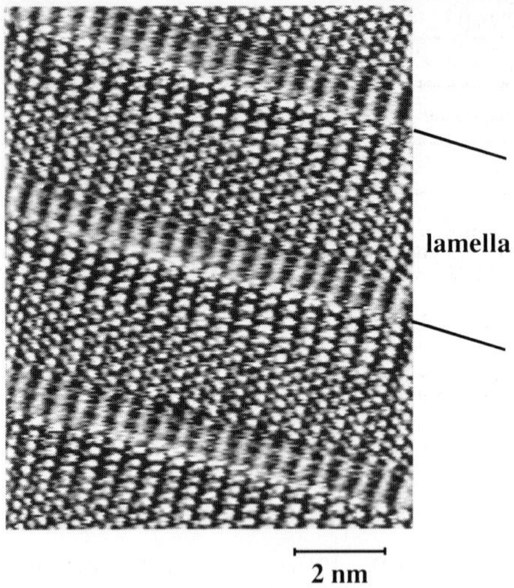

Fig. 2. STM image of a long chain alkane ($C_{27}H_{56}$) monolayer on HOPG. Under the employed tunnelling conditions the contrast is determined by both the substrate and adsorbate structures. The resulting moiré pattern indicates an incommensurability between the linear carbon chains within an individual alkane molecule and the basal plane of graphite. On the other hand, the molecular lattice is commensurate with the substrate both parallel and perpendicular to the lamellae.

surate) [15, 16]. In both cases the packing of the lamellae is commensurate with the substrate. The difference in the spacing of molecules within a lamella of less than 50 pm has been attributed to a different orientation of the carbon-backbone plane relative to the substrate. Contrary to an earlier interpretation [13], it was argued that the alkane-backbone plane is oriented predominantly perpendicular to the substrate surface [15, 24], with the main molecular axis parallel to the graphite surface. Moreover, it was shown that the interaction between the linear carbon chains in both the alkyls and the basal plane of graphite is not strong enough to force the chain lattices to become commensurate [15, 24]. As a consequence, at an appropriate bias the STM contrast along an alkyl chain varies from one end to the other, while the contrast perpendicular to the chain axes, i.e. along the alkane lamellae, is the same for every molecule (Fig. 2). This answers a question that has caused considerable controversy in the past [29, 30]. It also indicates that alkyl chains in alkyl derivatives will pack commensurate with HOPG only if the carbon planes are oriented perpendicular to the graphite surface, in contrast to earlier assumptions [12].

Noteworthy, the relative contributions of substrate and adsorbate depend strongly on the applied bias [3, 13, 15]. As demonstrated in Fig. 3 the transition is continuous, which bears the important implication that the monolayer is not destroyed, even though the image may only show the

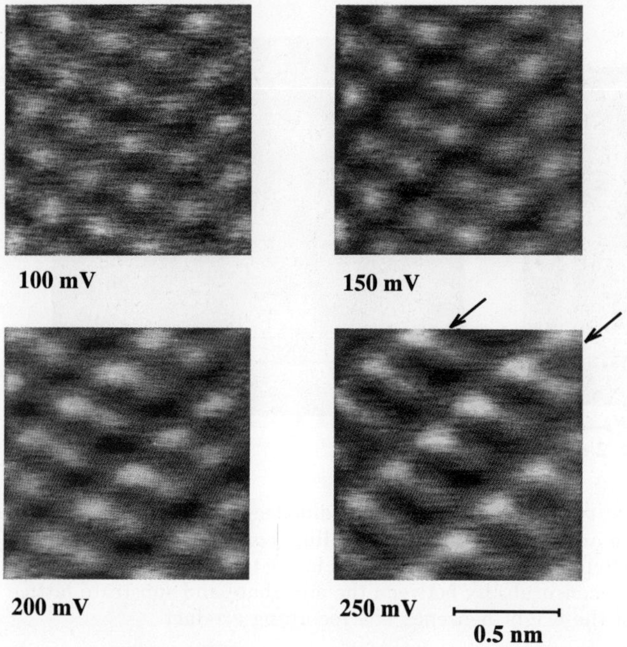

Fig. 3. Bias-dependent STM images of a long chain alkane ($C_{25}H_{52}$) monolayer on HOPG. The arrows indicate the orientation of the main molecular axes. At low bias the image is dominated by the substrate, while at higher bias the adsorbate dominates.

substrate structure. In addition there is some variation in the bias-dependent contrast on the tip condition, which makes it necessary to define operationally a 'standard' tip, e.g., with a test sample [3, 27]. Given the dependence of the STM contrast on both the relative orientation of adsorbate and substrate, as well as on the bias, the role of the chemical nature of an adsorbed species remains to be determined. Alkyl derivatives may be used to calibrate the conductivity of a particular molecular segment against the alkyl chains within one image. The important advantage of this method is that an identical tip is used for imaging both parts.

In order to determine whether the conductivity of aromatic moities in a certain bias range is in general larger than that of alkyl chains, we investigated a benzene, naphthalene, triphenylene and phenylbenzoate derivative. In general, it is found that in a bias range around +1 V or −1 V the current through the aromatic segments is significantly larger than through the alkyl parts. The same has been observed in a series of materials exhibiting liquid crystalline bulk phases [11, 31]. Since the difference exceeds the variation in contrast, which results from the different relative orientations of alkyls on HOPG [15], it is quite easy to identify the two parts.

Common to all the investigated molecular patterns is a close packing with the alkyl chains parallel to the graphite surface. Figure 4(a) shows

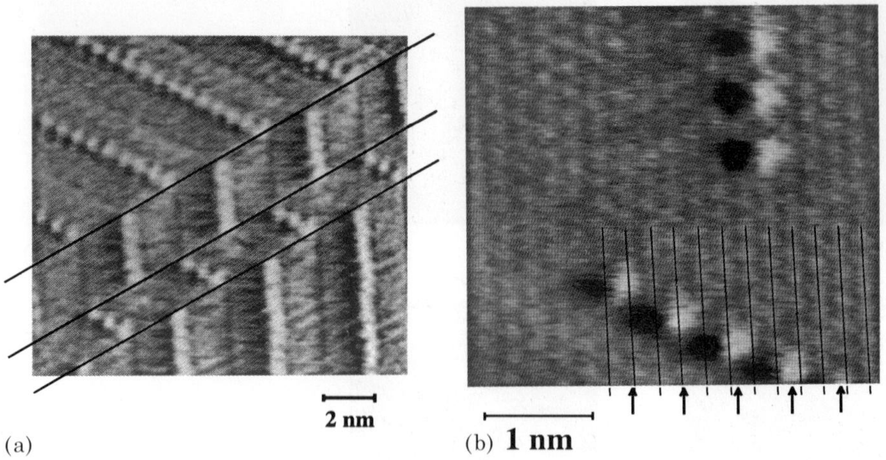

Fig. 4. (a) STM image of three twin boundaries in a 1,4-didodecylbenzene monolayer on HOPG. (b) At higher resolution and under the given tunnelling conditions one may clearly identify the positions of the individual benzenes (arrows) relative to graphite (lines), which is direct evidence for the incommensurability between the adsorbate and substrate lattice. The dark shadow to the left of the bright benzenes is a recording artefact.

a series of twin boundaries in a didodecylbenzene monolayer [15]. At higher resolution (Fig. 4(b)) it becomes clear that the tunnelling current is determined by one molecular segment only, since the domain boundary is molecularly sharp. Moreover, one can determine directly the incommensurability between adsorbate and substrate lattice. Figure 5 displays a

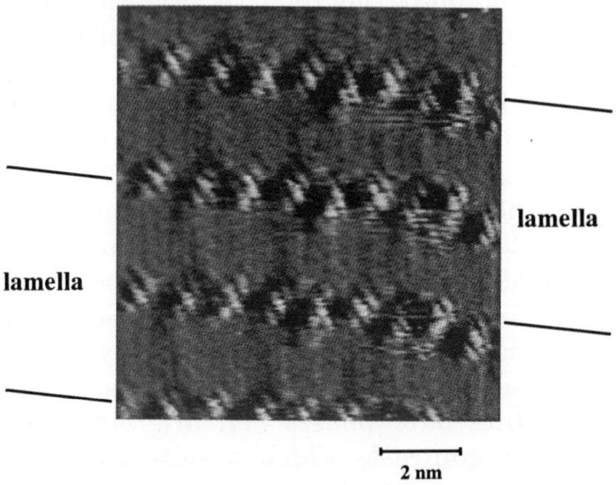

Fig. 5. STM image of a 1,5-didecyloxynaphthalene monolayer on HOPG. While the positions of the naphthalenes are resolved, the chains are not. The variable contrast of the naphthalenes indicates an incommensurate adsorbate lattice along the lamellae and a commensurate lattice perpendicular to them.

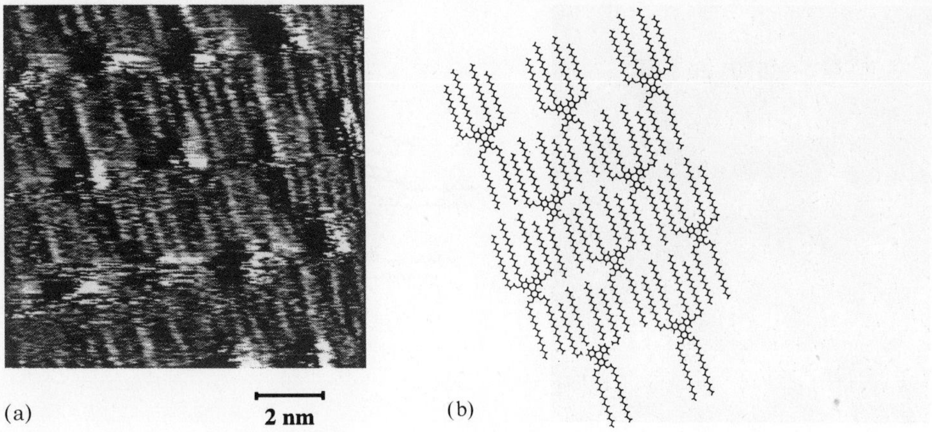

Fig. 6. STM image (a) and model (b) of a hexakis-hexadecyloxytriphenylene monolayer on HOPG.

didecyloxynaphthalene monolayer. From the size and shape of the unit cell one can conclude that the extended flat-lying molecules of adjacent lamellae interdigitate. Interestingly, the contrast along a lamella varies from molecule to molecule, while it repeats from lamella to lamella. The same is observed for the alkyl chains under the appropriate tunnelling conditions (image not shown). As in the case of fatty acids or didodecylbenzene [15, 16], this can be ascribed to an incommensurate packing within the lamellae and a commensurate packing perpendicular to them. It reflects the stronger intermolecular forces within a lamella as compared to between the lamellae. Figure 6 indicates that the packing of the hexakis-hexadecyltriphenylenes allows side-chain crystallization, while the triphenylene derivative with the shorter heptyl side chains does not [21, 32]. Worth mentioning also is the interdigitated packing of the phenylbenzoate derivative, which may be described as two alternating lamellae of pairs of molecules [8] (Fig. 7). Such a double-row structure, recently observed at the interface between HOPG and a mixture of phenylbenzoate derivatives, obviously does not necessarily indicate a mixture.

So far we have considered hydrocarbons with oxygen as the only heteroatom which, in various configurations, does not cause a significantly different contrast than the alkyl chains [15, 19, 20]. Another important heteroatom is sulfur, which has been imaged for the first time in alkylated sulfides. Dihexadecyldisulfide exhibits a gauche conformation of the central disulfide bridge, which in turn causes a V-shape of the individual molecule (Fig. 8). It is found that the conductivity of the sulfide is considerably larger than that of the alkyl chains. Similarly, the conductivity of the central sulfur in a didodecylsulfide monolayer has been found to be large compared to the alkyl part.

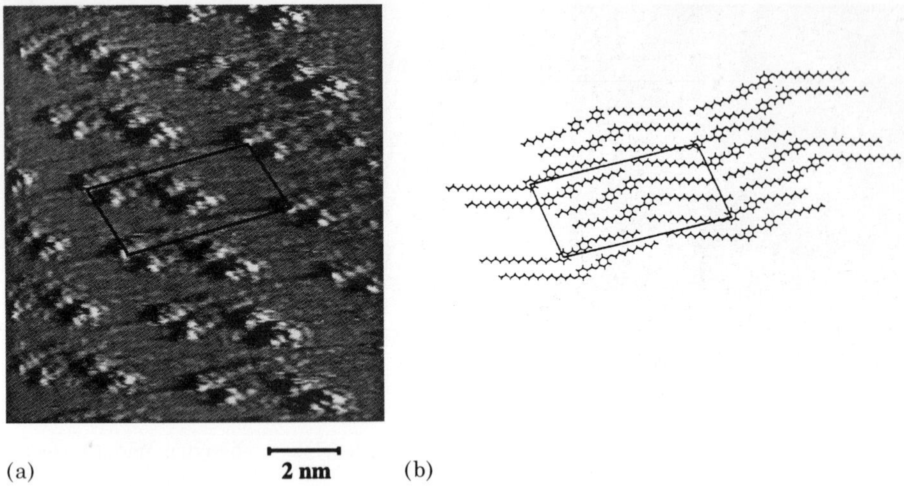

(a) (b)

Fig. 7. STM image with marked unit cell (a) and model (b) of a 4-decyloxy-4'-hexadecyloxyphenylbenzoate monolayer on HOPG, revealing a packing with two pairs of molecules per unit cell, providing close packing and compensation of dipole moments.

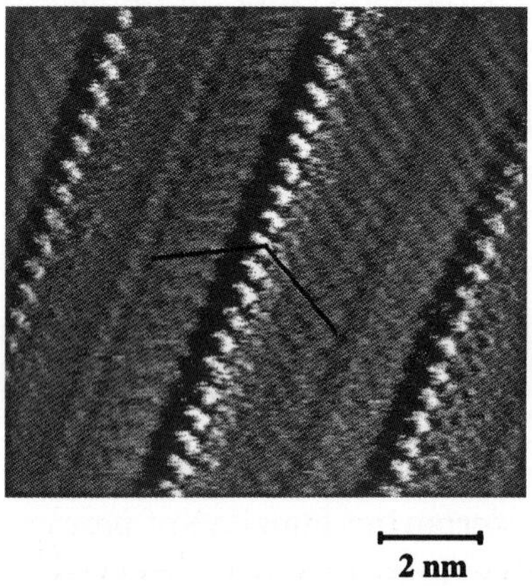

Fig. 8. STM image of a dihexadecyldisulfide monolayer on HOPG. The V-shape of the molecules is indicated for one molecule in the centre of the image.

Finally, we report on the first images of a diacetylene moiety, which allows us to study the current through carbon triple bonds. As evidenced in Fig. 9(a), at a bias around 1 V the current through the diacetylene moieties of 10,12-octadecadiynoic acid is larger than through the alkyl parts. Under tunnelling conditions, where the alkyl side chains are imaged simultaneously with the substrate, we observe a superstructure similar to the one observed for simple fatty acids [15, 16] (Fig. 9(b)). This indicates that the molecules form crystalline monolayers with the alkyl

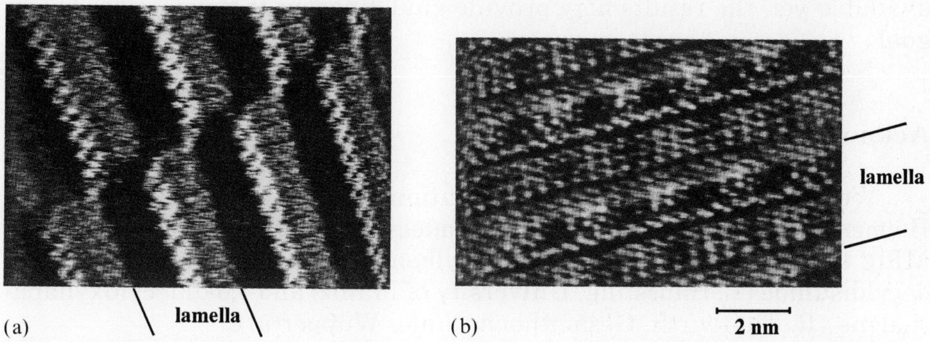

Fig. 9. STM images of a 10,12-octadecadiynoic acid monolayer on HOPG: (a) exhibiting a twin boundary and (b) showing a superstructure in the contrast along the lamellae due to the incommensurability between the adsorbate and substrate lattices in this direction.

chains perpendicular to the lamellae and the carbon-backbone planes parallel to the graphite. The resulting intermolecular distances are about 470 pm. The diacetylene units are tilted relative to the alkyl chains by about 45°. These data indicate that the packing is suitable for a solid state polymerization [34], which may eventually allow us to observe a polymerization along an individual chain.

Conclusions

Monomolecular layers of a series of alkyl derivatives have been imaged *in situ* at the interface between the basal plane of graphite and organic solutions at submolecular resolution. The observed structures indicate close packing with both intra- and inter-molecular distances as in the equivalent three-dimensional solids. The alkyl chains tend to orient parallel to the graphite surface and crystallize in lamellae if they are long enough. In such lamellar monolayers the adsorbate lattice is typically incommensurate with the substrate along the lamella direction and commensurate perpendicular to it. Fortuitous exceptions are the alkanes, which also exhibit a commensurate packing along the lamellae. The current through a particular molecular segment depends on the relative position of substrate and adsorbate segment, as well as on the bias. Moreover, it depends on the chemical nature of the molecule. At a bias around 1 V the current through various aromatic cores, as well as through sulfide and diacetylene segments, is considerably larger than through methylenes, methyls or various oxygen functions. The results are qualitatively consistent with models in which the conductivity increases as the gap between the highest occupied molecular orbital (HOMO) and the lowest unoccupied molecular orbital (LUMO) of the free molecule decreases [35]. Since a quantitative theory based on first principles is not available yet, the results may provide guidelines on the way towards this goal.

Acknowledgements

We gratefully acknowledge donations of 1,4-didodecylbenzene (F. Helmer-Metzman, MPI), hexakis-hexadecyloxytriphenylene (M. Werth, MPI), 4-decyloxy-4′-hexadecyloxyphenylbenzoate (C. Boeffel, MPI), dihexadecyldisulfide (L. Häussling, University of Mainz) and 1,5-didecyloxynaphthalene (R. Thorwirth, Gesamthochschule, Wuppertal).

References

1 G. Binnig and H. Rohrer, *Angew. Chem.*, *99* (1987) 622; G. Binnig and H. Rohrer, *Angew. Chem., Int. Ed. Engl.*, *26* (1987) 606.
2 J. P. Rabe, *Ultramicroscopy*, *42–44* (1992) 41.
3 H.-J. Butt, R. Guckenberger and J. P. Rabe, *Ultramicroscopy*, (1992) in press.
4 T. Sleator and R. Tycko, *Phys. Rev. Lett.*, *60* (1988) 1418.
5 J. P. Rabe and S. Buchholz, in J. L. Brédas and R. R. Chance (eds.), *Conjugated Polymeric Materials: Opportunities in Electronics, Optoelectronics, and Molecular Electronics, NATO-ARW Series E*, Kluwer, Dordrecht, 1990, pp. 483–493.
6 S. N. Magonov, J. Schuchard, S. Kempf, E. Keller and H.-J. Cantow, *Synth. Met.*, *40* (1991) 59.
7 S. N. Magonov, S. Kampf, H. Rotter and H.-J. Cantow, *Synth. Met.*, *40* (1991) 73.
8 J. P. Rabe, in W. Göpel and C. Ziegler (eds.), *Nanostructures Based on Molecular Materials*, VCH Publishers, Weinheim, 1992.
9 H. Ohtani, R. J. Wilson, S. Chiang and C. M. Mate, *Phys. Rev. Lett.*, *60* (1988) 2398.
10 J. S. Foster and J. E. Frommer, *Nature (London)*, *333* (1988) 542.
11 D. P. E. Smith, H. Hörber, Ch. Gerber and G. Binnig, *Science*, *245* (1989) 43.
12 D. P. E. Smith, J. K. H. Hörber, G. Binnig and H. Nejoh, *Nature (London)*, *344* (1990) 641.
13 G. C. McGonigal, R. H. Bernhardt and D. J. Thomson, *Appl. Phys. Lett.*, *57* (1990) 28.
14 J. P. Rabe and S. Buchholz, *Phys. Rev. Lett.*, *66* (1991) 2096.
15 J. P. Rabe and S. Buchholz, *Science*, *253* (1991) 424.
16 J. P. Rabe and S. Buchholz, *Makromol. Chem., Macromol. Symp.*, *50* (1991) 261.
17 G. C. McGonigal, R. H. Bernhardt, Y. H. Yeo and D. J. Thomas, *J. Vac. Sci. Technol. B*, *9* (1991) 1107.
18 R. Kuroda, E. Kishi, A. Yamano, K. Hatanaka, H. Matsuda, K. Egushi and T. Nakagiri, *J. Vac. Sci. Technol. B*, *9* (1991) 1180.
19 S. Buchholz and J. P. Rabe, *Angew. Chem.*, *104* (1992) 188; S. Buchholz and J. P. Rabe, *Angew. Chem., Int. Ed. Engl.*, *31* (1992) 189.
20 C. Ludwig, B. Gompf, W. Glatz, J. Petersen, W. Eisenmenger, M. Möbius, U. Zimmermann and N. Karl, *Z. Phys. B*, *86* (1992) 397.
21 J. P. Rabe, S. Buchholz and L. Askadskaya, *ACS Polym. Prepr.*, *33* (1992) 798.
22 R. Hentschke, L. Askadskaya and J. P. Rabe, *J. Chem. Phys.*, in press
23 L. M. Eng. H. Fuchs. S. Buchholz and J. P. Rabe, *Ultramicroscopy*, *42–44* (1992) 1059.
24 R. Hentschke, B. L. Schürmann and J. P. Rabe, *J. Chem. Phys.*, *96* (1992) 6213.
25 S. Cincotti and J. P. Rabe, in preparation.
26 J. P. Rabe, M. Sano, D. Batchelder and A. A. Kalatchev, *J. Microsc. (Oxford)*, *152* (1988) 573.

27 J. P. Rabe, *Proc. IMO Symp. Fall '90 Scanning Microscopy, ESPRIT Basic Research Series*, Springer, Berlin, 1992.
28 L. Askadskaya and J. P. Rabe, *Phys. Rev. Lett.*, *69* (1992) 1395.
29 G. H. Findenegg and M. Liphard, *Carbon*, *25* (1987) 119.
30 A. J. Groszek, *Proc. R. Soc.* (*London*), *Ser. A*, *314* (1970) 472.
31 J. K. Spong, H. A. Mizes, L. J. LaComb, Jr., M. M. Dovek, J. E. Frommer and J. S. Foster, *Nature* (*London*), *338* (1989) 137.
32 L. Askadskaya and J. P. Rabe, in preparation.
33 L. Sautière, S. Day and M. Miles, *Ultramicroscopy*, *42–44* (1992) 1054.
34 V. Enkelmann, *Adv. Polym. Sci.*, *63* (1984) 91.
35 P. H. Lippel, R. J. Wilson, M. D. Miller, Ch. Wöll and S. Chiang, *Phys. Rev. Lett.*, *62* (1989) 171.

Synthetic Metals, *54* (1993) 351–362

Epitaxial phthalocyanine thin films and phthalocyanine/C_{60} multilayers

G. E. Collins, V. S. Williams, L.-K. Chau, K. W. Nebesny, C. England, P. A. Lee, T. Lowe, Q. Fernando and N. R. Armstrong*
Department of Chemistry, University of Arizona, Tucson, AZ 85721 (USA)

Abstract

Ordered ultrathin films of various dye molecules can be created by vacuum deposition onto a variety of substrates where the interaction with the overlayer is weak and intermolecular forces dominate the ordering process. Examples are shown for the surface electron diffraction data obtained from ordered monolayers of phthalocyanines on metal dichalcogenide surfaces and the corresponding linear optical changes which result from the formation of thin multilayers of those dyes. Organic/organic′ multilayer assemblies are also presented, consisting of alternating layers of the fullerene C_{60} and InPc–Cl, which show promise for the formation of true organic superlattices, where excitonic interactions are confined in two dimensions.

Introduction

There is interest in the formation of highly ordered ultrathin films of organic molecules for purposes of the study and optimization of energy harvesting, photoconductivity, photography and electrophotography, electroluminescent display devices and optical logic materials [1–14]. Ordered thin films formed by self-assembly (SA) or by Langmuir–Blodgett (LB) techniques are well known [9]. Many large unsubstituted aromatic molecules, however, are also vacuum compatible and lend themselves to the deposition of monolayer to multilayer crystalline deposits with long range order and without the hydrocarbon spacers inherent in the LB or SA thin films [1, 3, 5, 7, 13, 14].

The ordering of organic dyes into two- or three-dimensional aggregates, oligomers or polymers (where the molecules are packed at van der Waals distances), perturbs both the ground and excited state energies of the individual molecules [1, 6, 8, 10, 12, 15]. The J aggregate and H aggregate cyanine dyes are among the best characterized of these systems [12], while aggregates of molecules with higher symmetry, such as the phthalocyanines and porphyrins, have also been studied [1, 8, 10, 15, 16]. The importance that such closest packing or aggregation might play in

*Author to whom correspondence should be addressed.

0379-6779/93/$6.00

transient linear and nonlinear optical processes has been recently noted (the concept of 'superradiance') [12a–c]. Fast optical switching devices based upon near-resonant or resonant excitation of dye aggregate thin films may also be feasible, provided that narrow excitonic bands can be produced in materials with high optical densities [11d].

We review here our development of highly ordered thin films of vacuum compatible phthalocyanine (Pc) dyes, where long range ordering is sustained from monolayers to 200–400 Å multilayers [1]. We also discuss the formation of multilayer assemblies consisting of two different dyes, deposited with thicknesses down to one molecular layer [5, 17]. There is interest in the formation of such dye/dye′ multilayers for optical logic applications and because exciton dissociation events leading to mobile charges can be confined to within a few molecular layers of an interface between two dissimilar thin dye films [5, 13, 14].

Substrates in our studies have consisted of (i) bulk crystals of MoS_2 and SnS_2, (ii) MBE-deposited ultrathin films of SnS_2, and (iii) single crystal sapphire, with all experimental procedures as described elsewhere [1]. SnS_2 and MoS_2 (freshly cleaved or freshly deposited by MBE) provide largely unreactive basal planes, consisting of hexagonally closest packed sulfur atoms, which interact only weakly with most organic overlayers, helping to relieve lattice strain brought about by the formation of thin films which do not form commensurate overlayers. In addition, SnS_2 has an optical bandgap of 2.2 eV, making it transparent over a large fraction of the visible wavelength region [2, 18]. We routinely use surface electron diffraction techniques, such as reflection high energy electron diffraction (RHEED) and low energy electron diffraction (LEED), to monitor the type and extent of ordering of the organic overlayer (see below). Absorbance spectra or *in situ* reflection/absorbance spectra complement the surface electron diffraction techniques, since spectral lineshapes are sensitive to packing geometry of these dyes. With conductive samples of SnS_2 we also photoelectrochemically characterize the ultrathin organic film, generating photocurrent action spectra for monolayer to multilayer deposits [1a, b, f, g].

Ordered ultrathin films of the chlorometallated phthalocyanines

Figure 1 shows a schematic diagram proposed for the vacuum deposition of the chlorometallated phthalocyanines and copper phthalocyanine (CuPc) near the completion of the first monolayer (ML), on substrates such as MoS_2 and SnS_2 [1d]. Under suitable deposition conditions it is possible to produce highly ordered structures which RHEED and LEED data suggest consist of closest packed square lattices of Pc, lying parallel to the substrate surface, aligned along the principal axes of the chalcogenide basal plane [1, 3a, f, g]. The proposed geometry of each square lattice is a structure known to exist in one face of certain single crystals of AlPc–Cl and GaPc–Cl [19]. As shown in the LEED diagrams (Fig. 2),

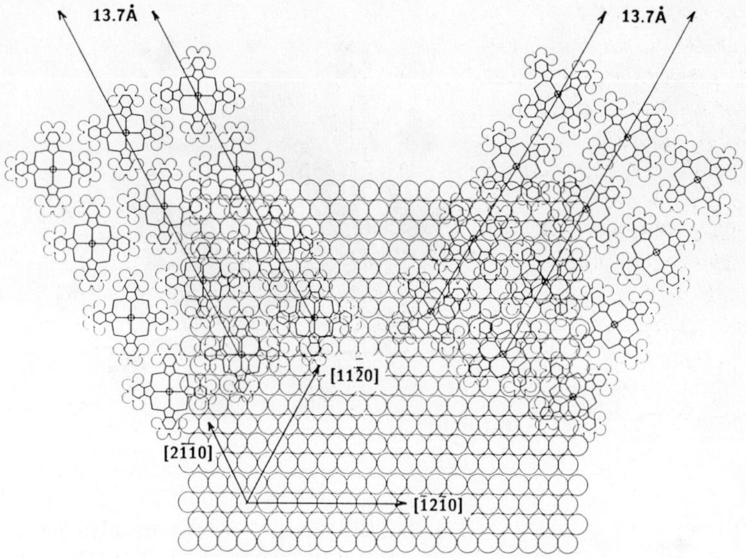

Fig. 1. Schematic diagram of the proposed packing of the trivalent metal phthalocyanines (e.g. InPc–Cl) and CuPc on substrates such as MoS_2 and SnS_2 [1].

three equivalent domains are produced at monolayer coverage, rotated by 60° with respect to each other, corresponding to the orientation of the three principal axes of the substrate [1]. The four-fold symmetric Pc is not capable of forming a truly commensurate layer with such substrates. The weak interactions between the chalcogenide basal plane and the Pc overlayer, however, allow for coincident 3×2 $R = 0°$ (MoS_2) or 3×2 $R = \pm 4°$ (SnS_2) lattices to form [1e]. Defects in the surface of MoS_2 or SnS_2 also play a role in the nucleation of these deposits and it may be either terrace sites and/or basal plane defects which initiate oriented growth of these Pc square lattices [1a, d, 17]. The epitaxial layer growth of these four-fold symmetric phthalocyanines on the six-fold symmetric chalcogenide surface is probably best described as 'seeded epitaxy' — the diameter of each Pc nucleus varies from a few hundred angstroms up to several microns.

Figure 3 shows the linear absorbance spectra in the Q-band region for ultrathin films of InPc–Cl/SnS_2/mica, compared with the spectrum for the monomer in solution [1h, g]. Compared with less-ordered films of InPc–Cl, these spectra are quite narrow (FWHM = 40–60 nm), and show a red shift of the main absorbance peak relative to the position for the isolated monomer. The red-shifting at low coverages of the Pc (also seen for comparable coverages of CuPc, VOPc, GaPc–Cl and AlPc–Cl) [1g, h] is believed to be due to the staggered, 'head-to-tail' alignment of the transition dipoles in the surface layer [1g, 6]. Addition of a second layer of InPc–Cl to the thin film causes a further red-shifting in the visible

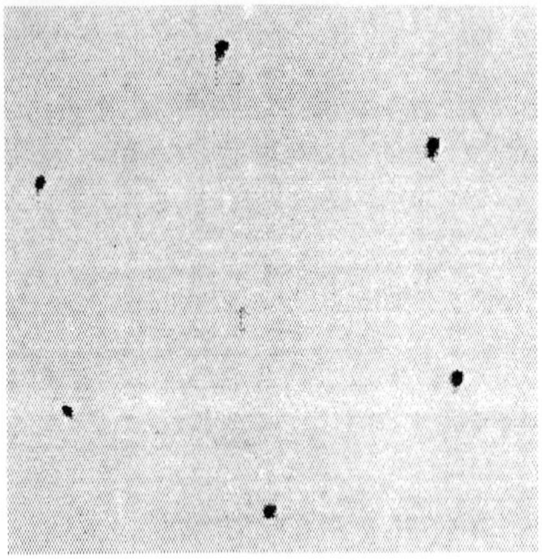

(a)

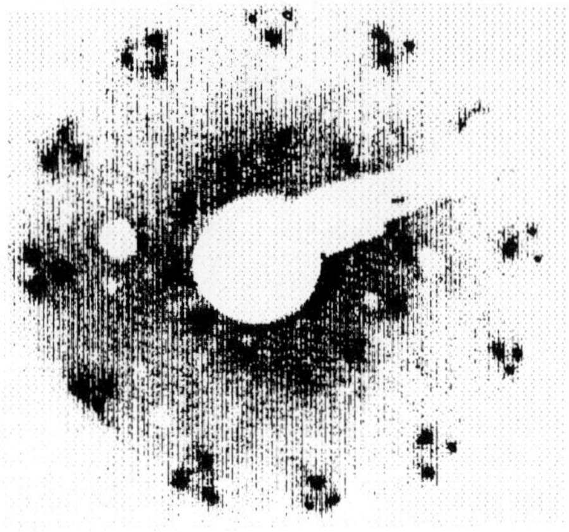

(b)

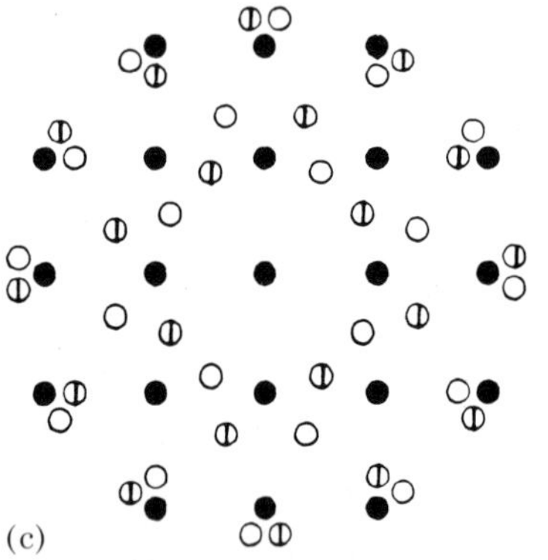

(c)

Fig. 2. LEED images of (a) a freshly cleaved MoS_2 crystal surface ($E_0 = 50$ V), (b) the same crystal with one monolayer of CuPc or InPc–Cl deposited ($E_0 = 15$ V) and (c) a schematic diagram of the reciprocal lattice images expected in the first four Laue zones, where the phthalocyanines are packed as shown in Fig. 1, with three equivalent square lattice domains aligned along the principal axes of the metal dichalcogenide substrate. The LEED patterns are shown as the output from CCD camera images taken of the phosphor screen. Only the first, and parts of the third and fourth Laue zones are observable.

spectrum, but no change in the surface electron diffraction data, provided that a 'layer-by-layer' growth mode is sustained [1d, g, h]. Structures which are consistent with such an observation place the Pc rings in the second layer parallel to the first layer and either (i) cofacially aligned with rings eclipsed, or (ii) staggered in both the x and y directions by half a molecular diameter [3, 4, 18]. Both configurations are known for the trivalent metal phthalocyanines, but only the staggered configuration produces the red-shifting of the Q-band spectrum [1g, 6, 8, 10, 16]. In contrast, the linear cofacial Pc stacks show a strong blue shift in the Q-band [6, 8a, c]. In this orientation the transition dipoles are aligned parallel and only the higher energy optical transition in the exciton manifold of the aggregate is allowed.

The shifting of the Q-band spectra for such aggregates is predicted from the theories for interactions of lamellar arrays of chromophores, where the magnitude of these spectral shifts is dependent upon the intermolecular spacing and the precise orientation of transition dipoles

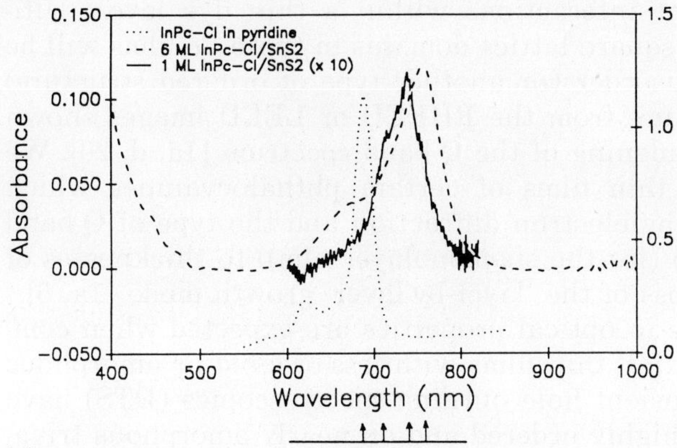

(a)

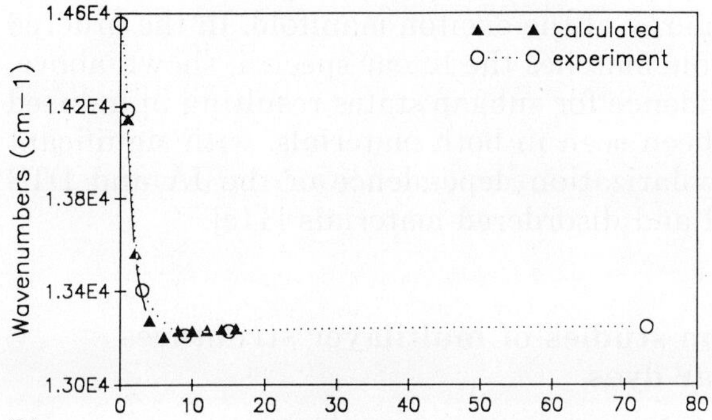

(b)

Fig. 3. (a) Visible absorbance spectra of the InPc–Cl monomer in solution, 1 ML InPc–Cl/SnS_2 and 6 ML InPc–Cl/SnS_2, where RHEED images suggested layer-by-layer growth of the Pc films, with rings parallel to the substrate. (b) Comparison of the predicted and observed shifts of the Q-band maximum for InPc–Cl (vs. the monomer in solution) for epitaxially grown layers like those shown in (a) [1f, g, 6].

coupled in all three dimensions [6b]. We have recently extended this treatment to include the ultrathin films of the trivalent metallated phthalocyanines, where the interactions between chromophores are considered in all dimensions as the thickness of the epitaxial thin film increases [1g]. Figure 3(b) shows the calculated and observed shifts in the Q-band transition for InPc–Cl as a function of the number of monolayers in the ordered thin film, adopting the staggered geometry discussed above. In addition to the good agreement between expected and observed peak shifts, it is interesting to note that the shift in the Q-band ceases after the ordered thin film reaches a thickness of about six monolayers. The interactions between adjacent chromophores have an R^{-3} dependence (R = chromophore separation distance) [6]. Considering such interactions in three dimensions, such a limit in the shift of the Q-band is expected and establishes an upper limit to the film thickness needed to optimize optically relevant interactions within a thin dye layer. Ultimately the intersection of square lattice domains in these Pc films will be sites for propagation of disorder (or another type of ordered structure) which is seen as a departure from the RHEED or LEED images shown above and significant broadening of the Q-band spectrum [1a, d, 20]. We have been able to grow thin films of certain phthalocyanines which exhibit ordering (as seen by electron diffraction and the type of Q-band spectra shown in Fig. 3(a) (for the six-monolayer film)) to thicknesses of about 200–400 Å, before loss of the 'layer-by-layer' growth mode [1a, 5].

Significant differences in optical properties are expected when comparing the ordered, 'epitaxial' thin films with less-ordered or amorphous systems. Femtosecond transient hole burning spectroscopies (DTS) have been carried out on both highly ordered and on nearly amorphous trivalent metal Pc thin films [11]. Decay of the excitation energy occurs in the disordered material in the subpicosecond regime to produce a narrow spectral hole at the lowest energy of the exciton manifold. In the ordered film, however, the spectral hole matches the linear spectra, shown above, for most of the band [12]. Evidence for subgap states resulting in induced absorption (IA) signals has been seen in both materials, with significant differences in IA, and the polarization dependence of the IA and DTS signals, between the ordered and disordered materials [11c].

Surface electron diffraction studies of multilayer structures created from two dissimilar dyes

Some of our recent experiments suggest that monolayer control over layer thickness in bilayer and multilayer dye assemblies will be achievable with phthalocyanines such as InPc–Cl, annulene spacer molecules such as perylenes, coronene, and fullerenes such as C_{60} [17]. Such control will have significance in fundamental studies of excitonic dissociation or recombination processes at dye/dye′ interfaces, as well as in the formation of new optical materials [5, 13, 14]. Figure 4 shows representative LEED data for (a) the freshly cleaved MoS_2 substrate, (b) an ordered monolayer of C_{60}/MoS_2, and (c) an ordered monolayer of InPc–Cl/C_{60}/MoS_2. The C_{60} layer shows h.c.p. packing, as expected from the wide variety of real space and reciprocal lattice images obtained for ultrathin films of this molecule [21]. This packing is sustained for multilayers of C_{60} on a substrate such as MoS_2 with no rotation of the h.c.p. C_{60} lattice with respect to the principal axes of the substrate [16]. There are likely to be multiple domains of C_{60} formed on the dichalcogenide substrate, as has been seen for adsorption of C_{60} on other substrates [21]. These domains would logically have the same principal axes as the basal plane of the metal dichalcogenide and the six-fold symmetry of the h.c.p. C_{60} lattice is identical to the symmetry of the substrate. The appearance of the LEED pattern is therefore only altered in size (correcting for differences in

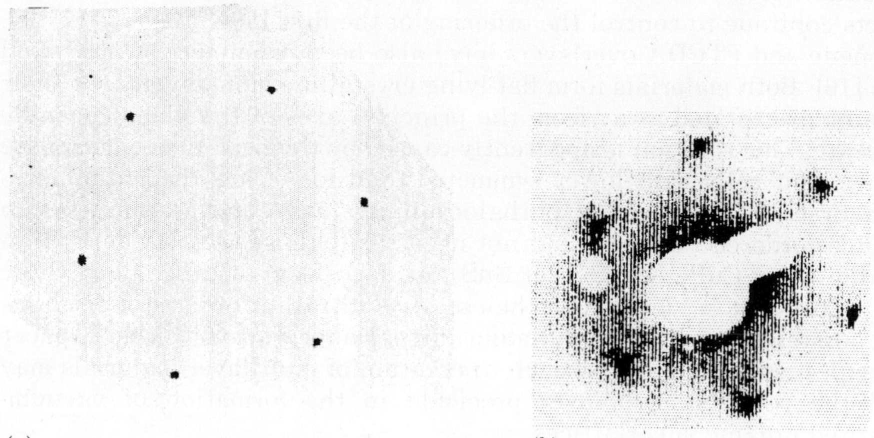

(a) (b)

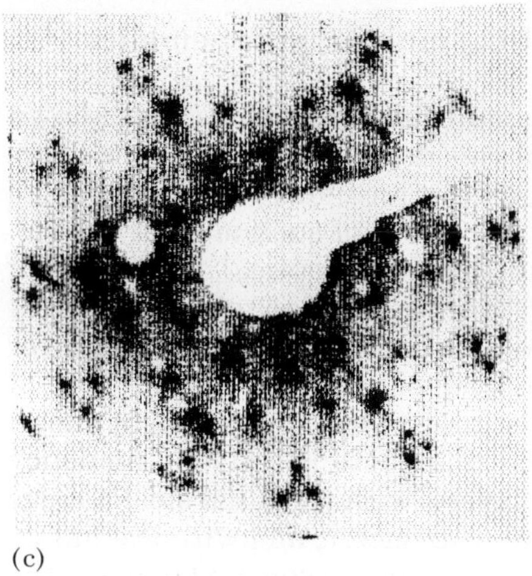

(c)

Fig. 4. LEED images of (a) the freshly cleaved MoS_2 surface ($E_0 = 50$ V), (b) the same surface with one monolayer of C_{60} ($E_0 = 20$ V) and (c) the surface in (b) with one monolayer of InPc–Cl added ($E_0 = 16$ V). In (c) the diffraction image for the C_{60} layer is still visible, note the difference in excitation voltage making the spacing in the C_{60} slightly different from that in (b). Parts of the second Laue zone for the Pc overlayer are now visible in these diffraction data.

electron beam excitation energy) to reflect the larger spacing (*c.* 10 Å) of the scattering centers. The low electron beam energies used to characterize these layers provides an inelastic mean free path for the diffracted electrons which allows us to monitor the slow disappearance of the MoS_2 diffraction pattern as the C_{60} coverage increases and then the slow disappearance of the C_{60} pattern as the Pc coverage increases to monolayer levels and beyond.

It is interesting to note that the Pc overlayer has adopted the same scattering pattern seen when this molecule is deposited directly on a bare MoS_2 or SnS_2 substrate. The formation of three equivalent square lattice domains is once again indicated, aligned along the principal axes of the MoS_2 substrate. This result suggests the C_{60}/MoS_2 monolayers may propagate defects in the MoS_2 surface, such as terrace sites, which gave rise to nucleation of the first of the C_{60} overlayers. It would appear that these defects continue to control the ordering of the first Pc/C_{60} layers [17, 21]. Coronene and PTCDA overlayers have also been examined on MoS_2 and SnS_2 [16]. Both materials form flat-lying crystalline monolayers, but their domains are rotated away from the principal axes of the dichalcogenide substrate. This rotation is apparently caused by the need to accommodate surface unit cells with lower symmetry than for either the phthalocyanines or C_{60}. Deposition of phthalocyanines over such assemblies also produces ordered Pc layers, but not necessarily those seen for deposition of InPc–Cl or CuPc on MoS_2 or SnS_2, or deposition of InPc–Cl or CuPc on C_{60}/MoS_2. It is clear, nevertheless, that ultrathin ordered assemblies are possible using the phthalocyanines and various other spacer molecules and that this approach to creation of multilayer materials may yield the required monolayer precision in the formation of vacuum-deposited organic superlattices.

Linear optical properties of InPc–Cl/C_{60} multilayer assemblies

Previous studies of VOPc/PTCDA multilayer materials had shown that, as the Pc and PTCDA film thicknesses decreased, the Q-band spectrum of the phthalocyanine narrowed, giving an indication that Pc–Pc interactions could be controlled by the formation of such multilayers [5]. Better control over layer thickness and the linear optical properties of such an assembly has been achieved by the creation of C_{60}/InPc–Cl multilayers on single crystal sapphire substrates (Fig. 5). Multilayer assemblies created on MBE-grown SnS_2, etc., will be the subject of forthcoming publications. Our earlier studies with the trivalent metal chloride phthalocyanines have suggested that there is a strong tendency for these molecules to form 'dimer-like' layers, with the adjacent Pc rings in each layer staggered by half a molecular diameter, and with the halide in one layer pointed toward the adjacent layer, accommodated by the space created at the intersection of any four Pc rings (see Fig. 1) [1d, g, 19]. For an ordered layer of Pc, separated by even one layer of C_{60}, Pc–Pc interactions between layers should be already minimized [1g, 6]. Similar confinements of excitonic interactions within an ordered cyanine or phthalocyanine layer are achievable through the formation of LB layers of those dyes, where the chromophore aggregates are separated by saturated, rather than aromatic, hydrocarbon spacers [8c, 12e].

Such confinement is confirmed by the spectra in Fig. 5(a), where multilayer assemblies are compared possessing the same thicknesses of Pc (two monolayers (2 ML) in each layer) and either one or two monolayers of C_{60} in each layer. The Pc Q-band spectrum is narrowed in $(C_{60,1ML}/InPc{-}Cl_{2ML})_{20}$ relative to less-ordered material, with a peak position of *c.* 760 nm and an HWHM on the low energy side of the band of *c.* 60 nm. Extending the layer thicknesses of the fullerene to ($C_{60,2ML}$/InPc–

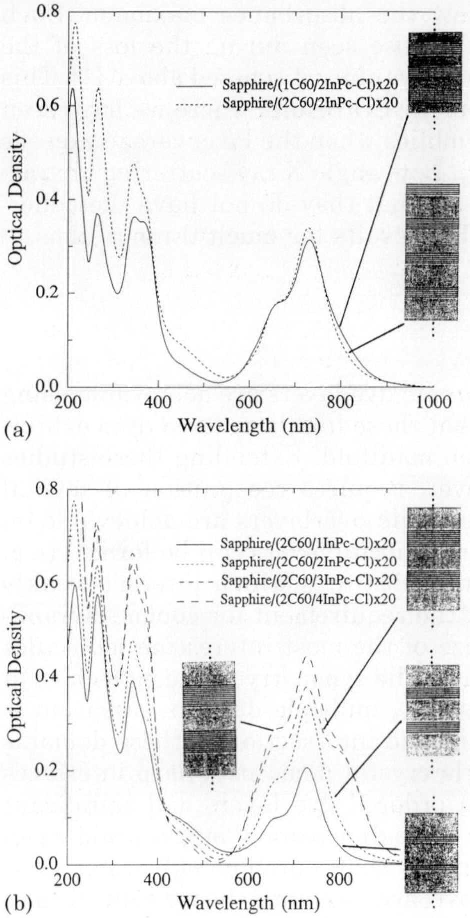

Fig. 5. Linear absorbance spectra of multilayers of InPc–Cl and C_{60}: (a) two different combinations are shown using thicknesses of the C_{60} spacer of one and two equivalent monolayers, and the same two-monolayer thickness of InPc–Cl in each lattice period (×20 layers); (b) the same thickness of C_{60} is used (2 equivalent monolayers) and thicknesses of 1, 2, 3 and 4 equivalent monolayers of InPc–Cl are used, ×20 layers of each molecule.

$Cl_{2ML})_{20}$ narrows the Q-band by only *c.* 10 nm versus the 1 ML C_{60} assemblies and does not further shift the Q-band spectrum, suggesting that the principal Pc–Pc interactions are within each layer.

Figure 5(b) shows the effect of maintaining a constant C_{60} layer thickness and increasing the Pc layer thickness from about one to four monolayers per period. For both the 1 ML and 2 ML Pc film thicknesses the spectra have a comparable position and width, and the optical density scales with the total amount of chromophore in the Pc layers. The comparable spectral shapes suggest that aggregates have already formed at one equivalent monolayer. With thicknesses of three or more monolayers of Pc in each layer the Q-band spectra quickly broaden, with addition of spectral intensity above and below the absorbance maximum. Such spectral changes are consistent with those seen during the loss of the 'layer-by-layer' growth mode for phthalocyanines discussed above [1]. This observation is also consistent with our LEED results where we have seen the loss of ordering in the Pc/C_{60} assemblies when the Pc coverage exceeds about two monolayers on ordered C_{60}. Low angle X-ray scattering characterization of these multilayers suggests that they do not have the coherence in structure implied by the LEED results for much thinner films.

Conclusions

It is clear that highly ordered organic dye layers are achievable using vacuum deposition approaches and that these highly ordered dyes exhibit the expected narrowing of the exciton manifold. Extending these studies to true organic superlattices, however, requires recognition of several important issues. The most-ordered organic overlayers are achievable on crystalline substrates for which commensurate layers can be formed (e.g. CuPc/Cu(100) [3f, g] or on crystalline substrates which interact weakly with the organic overlayer and relax the requirement for commensurability [1, 3a–e]. Because of the large size of the most interesting molecules and because of the difficulty in matching the symmetry of the molecules in the overlayer with those of the substrate, multiple domain formation is inevitable in most assemblies [17], and the intersection of these domains limits the size of the individual single crystal domains. Sharp interfaces are nevertheless achievable between ordered dye layers and significant control can be exercised over two- or three-dimensional aggregation, even in multilayer assemblies produced on less than optimum substrates. Several of the assemblies described above are currently being refined, and their linear and nonlinear optical characteristics determined, focusing on comparisons between ordered and disordered ultrathin dye layers.

Acknowledgements

We gratefully acknowledge support for this research from the National Science Foundation (Small Grants for Exploratory Research Program and Division of Chemical Sciences), Motorola, 3M and the Optical Data Storage Center, University of Arizona, and the Materials Characterization Program, State of Arizona.

References

1 (a) K. W. Nebesny, G. E. Collins, P. A. Lee, L.-K. Chau, J. Danziger, E. Osburn and N. R. Armstrong, *Chem. Mater*, *3* (1991) 829; (b) N. R. Armstrong, K. W. Nebesny, G. E.

Collins, P. A. Lee, L.-K. Chau, C. Arbour and B. A. Parkinson, in R. A. Lessard (ed.), *Photopolymer Device Physics and Applications, Proc. SPIE, 1559* (1991) 18; (c) N. R. Armstrong, K. W. Nebesny, G. E. Collins, L.-K. Chau, P. A. Lee, C. England, D. Diehl, M. Douskey and B. A. Parkinson, in R. P. H. Chang (ed.), *Science and Technology of Thin Films for the 21st Century, Thin Solid Films, 216* (1992) 90; (d) G. E. Collins, K. W. Nebesny, C. D. England, L.-K. Chau, P. A. Lee, B. A. Parkinson and N. R. Armstrong, *J. Vac. Sci. Technol. A, 10* (1992) 2902; (e) G. E. Collins, C. England and N. R. Armstrong, manuscript in preparation; G. E. Collins, *Ph.D. Dissertation*, University of Arizona, 1992; (f) C. Arbour, L.-K. Chau, G. E. Collins, K. W. Nebesny, P. A. Lee, C. D. England, N. R. Armstrong and B. A. Parkinson, *J. Phys. Chem.*, in press; (g) L.-K. Chau, C. D. England, S.-Y. Chen and N. R. Armstrong, *J. Phys. Chem.*, in press; (h) N. R. Armstrong, L.-K. Chau, G. E. Collins, S.-Y. Chen, K. W. Nebesny, V. A. Williams, P. A. Lee, C. Arbour, J. Danziger, E. Osburn, D. F. O'Brien and B. A. Parkinson, *Optical Materials Technology for Energy Efficiency and Solar Energy Conversion, Proc. SPIE, 1729* (1992) in press.

2 (a) B. A. Parkinson, *Langmuir, 4* (1988) 967; (b) B. A. Parkinson, in D. S. Ginley, A. Nozik, N. R. Armstrong, K. Honda, A. Fujishima and T. Sakata (eds.), *Photoelectrochemistry and Electrosynthesis on Semiconducting Materials*, Electrochemical Society Publications, Princeton, NJ, 1988, p. 160.

3 (a) M. Hara, H. Sasabe, A. Yamada and A. F. Garito, *Jpn. J. Appl. Phys., 28* (1989) L306; (b) H. Tada, K. Saiki and A. Koma, *Jpn. J. Appl. Phys., 30* (1991) L306; (c) A. J. Dann, H. Hoshi and Y. Maruyama, *J. Appl. Phys., 67* (1990) 1371 and 1845; (d) M. R. Fahey, H. Fujimoto, A. J. Dann, H. Hoshi, H. Inokuchi, Y. Maruyama and M. R. Willis, *Phys. Scr., 41* (1990) 550; (e) H. Yanagi, M. Ashida, J. Elbe and D. Wörhle, *J. Phys. Chem., 94* (1990) 7056; (f) J. C. Buchholz and G. A. Somorjai, *J. Chem. Phys., 66* (1977) 573; (g) P. H. Lippel, R. J. Wilson, M. D. Miller, Ch. Wöll and S. Chiang, *Phys. Rev. Lett., 62* (1989) 171.

4 (a) J. R. Fryer and M. E. Kenney, *Macromolecules, 21* (1988) 259; (b) J. R. Fryer, *Mol. Cryst. Liq. Cryst., 137* (1986) 49; (c) M. Ashida, in J. R. Fryer and D. L. Dorset (eds.), *Electron Crystallography of Organic Molecules*, Kluwer, Dordrecht, 1990, pp. 227–240.

5 (a) J. Danziger, J.-P. Dodelet, P. A. Lee, K. W. Nebesny and N. R. Armstrong, *Chem. Mater., 3* (1991) 821; (b) P. E. Burrows, M. Hara and H. Sasabe, *Mol. Cryst. Liq. Cryst., Sect. B: Nonlinear Opt., 2* (1992) 193.

6 (a) M. Kasha, in B. DiBartole (ed.), *Spectroscopy of the Excited State*, Plenum, New York, 1976; (b) R. M. Hochstrasser and M. Kasha, *Photochem. Photobiol., 3* (1964) 317; (c) V. Czikkely, H. D. Försterling and H. Kuhn, *Chem. Phys. Lett., 6* (1970) 11; (d) H. Büchner and H. Kuhn, *Chem. Phys. Lett., 6* (1970) 183.

7 H. Yanagi, S. Douko, Y. Ueda, M. Ashida and D. Wörhle, *J. Phys. Chem., 96* (1992) 1366.

8 (a) T. Sauer, W. Caser and G. Wegner, *Mol. Cryst. Liq. Cryst., 183* (1990) 387; (b) A. Kaltbeitzel, D. Neher, C. Bubeck, T. Sauer, G. Wegner and W. Caseri, in H. Kuzmany, M. Mehring and S. Roth (eds.), *Electronic Properties of Conjugated Polymers*, Springer, Berlin, 1989, pp. 220–224; (c) T. Sauer, T. Arndt, D. N. Batchelder, A. A. Kalachev and G. Wegner, *Thin Solid Films, 187* (1990) 357.

9 A. Ullman, *Ultrathin Organic Films*, Academic Press, Orlando, FL, 1991, pp. 159–176 and refs. therein.

10 C. H. Griffiths, M. S. Walker and P. Goldstein, *Mol. Cryst. Liq. Cryst., 33* (1970) 149.

11 (a) V. S. Williams, J. P. Sokoloff, Z. Z. Ho, C. Arbour, N. R. Armstrong and N. Peyghambarian, *Chem. Phys. Lett., 193* (1992) 317; (b) V. S. Williams, S. Mazumdar, N. R. Armstrong, Z. Z. Ho and N. Peyghambarian, *J. Phys. Chem., 96* (1992) 4500; (c) V. S. Williams, Sandolphon, N. R. Armstrong, S. Mazumdar and N. Peyghambarian, *Proc. Int. Quantum Electronics Conf. (IQEC), Vienna, Austria, June, 1992*; (d) V. S. Williams, Z. Z. Ho, N. Peyghambarian, W. M. Gibbons, R. P. Grasso, M. K. O'Brien, P. J. Shannon and S. T. Sun, *Appl. Phys. Lett., 57* (1990) 2399.

12 (a) A. A. Muenter, D. V. Brunbaugh, J. Apolito, L. A. Horn, F. C. Spano and S. Mukamel, *J. Phys. Chem., 96* (1992) 2783; (b) T. Tani, T. Suzumoto, K. Kemnitz and K. Yoshihara, *J. Phys. Chem., 96* (1992) 2778; (c) F. C. Spano and S. Mukamel, *Phys. Rev.*

A, *40* (1989) 5783; (d) F. C. Spano and S. Mukamel, *J. Chem. Phys.*, *91* (1989) 683; (e) T. L. Penner, *Thin Solid Films*, *160* (1988) 241; T. L. Penner, *J. Chim. Phys.*, *85* (1988) 1081; (f) I. Yamazaki, N. Tamai, T. Yamazaki, A. Murakami, M. Mimuro and Y. Fujita, *J. Phys. Chem.*, *92* (1988) 5035.

13 (a) C. W. Tang, *Appl. Phys. Lett.*, *48* (1986) 183; (b) C. W. Tang and S. A. VanSlyke, *Appl. Phys. Lett.*, *51* (1987) 913; (c) P. Panayotatos, G. Bird, R. Sauers, A. Piechowski and S. Husain, *Sol. Cells*, *21* (1987) 301; (d) P. Panayotatos, D. Parikh, R. Sauers, G. Bird, A. Piechowski and S. Husain, *Sol. Cells*, *18* (1986) 71.

14 (a) S. R. Forrest, L. Y. Leu, F. F. So and W. Y. Yoon, *J. Appl. Phys.*, *66* (1989) 5908; (b) J. F. Lam, S. R. Forrest and G. L. Tangonan, *Phys. Rev. Lett.*, *66* (1991) 1614; (c) D.-Y. Zang, Y. Q. Shi, F. F. So, S. R. Forrest and W. H. Steir, *Appl. Phys. Lett.*, *58* (1991) 562; (d) F. F. So and S. F. Forrest, *Phys. Rev. Lett.*, *66* (1991) 2649; (e) F. F. So, S. F. Forrest, Y. Q. Shi and W. H. Steier, *Appl. Phys. Lett.*, *56* (1992) 674; (f) U. Simmermann and N. Karl, *Surf. Sci.*, *268* (1992) 296; (g) C. Ludwig, B. Gompf, W. Glatz, J. Petersen, W. Wisenmenger, M. Möbius, U. Zimmermann and N. Karl, *Z. Phys. B, Condensed Matter*, *86* (1992) 397.

15 (a) T. Mezza, N. R. Armstrong and M. Kenney, *J. Electroanal. Chem.*, *137* (1982) 227; (b) T. Mezza, N. R. Armstrong and M. Kenney, *J. Electroanal. Chem.*, *176* (1984) 259.

16 (a) M. Mimuro, T. Nozawa, N. Tamai, K. Shimada, I. Yamazaki, S. Lin, R. S. Knox, B. P. Wittmershaus, D. C. Brune and R. E. Blankenship, *J. Phys. Chem.*, *92* (1989) 7503; (b) R. G. Alden, S. H. Lin and R. E. Blankenship, preprint kindly supplied.

17 G. E. Collins, C. D. England, V. S. Williams and N. R. Armstrong, manuscript in preparation.

18 (a) E. Mooser (ed.), *Physics and Chemistry of Materials with Layered Structures*, Vols. 1–6, Reidel, Dordrecht, 1976–1979; (b) V. Grasso (ed.), *Electronic Structures and Electronic Transitions in Layered Materials*, Reidel, Dordrecht, 1986.

19 K. J. Wynne, *Inorg. Chem.*, *23* (1984) 4658.

20 T. Sims, J. E. Pemberton, P. A. Lee and N. R. Armstrong, *Chem. Mater.*, *1* (1989) 26; T. Sims, *M.S. Thesis*, University of Arizona, 1987.

21 J. H. Weaver, *Acc. Chem. Res.*, *25* (1992) 143 and refs. therein.

Processing of conducting DCNQI radical ion salt films by solid state reactions*

H. Wachtel**
Institut Charles Sadron, 6 rue Boussingault, 67083 Strasbourg Cédex (France)

Abstract

Self-organization of small molecular units in stack-forming radical ion salts can be used to prepare anisotropic thin films. The alkali-2,5-dimethyl-*N*,*N'*-dicyanoquinonediimine (DMe-DCNQI) radical ion salt films represent model systems to study solid state redox reactions when their components are deposited on cooled substrates. While heating the films this reaction is monitored by *in situ* measurements of the conductivity, the UV–Vis–IR absorption, the electron spin resonance and the photoelectron emission of the valence band and shallow alkali core levels. Initial experiments demonstrate the use of the alkali DCNQI films as wireless sensors for spin-carrying gases, thus pointing at future device applications.

1. Introduction

Thin films of radical ion salts are receiving increasing interest because of their possible applications as anisotropic conductors and tailored sensor devices. The use of these two-component film systems opens the way to further organic materials. The thin films enable the study of the reaction of their constituents because they are well suited for a wide range of spectroscopic investigations. We deal with alkaline radical anion salts of 2,5-dimethyl-*N*,*N'*-dicyanoquinonediimine (DMe-DCNQI, Fig. 1(a)) [1]. The DMe-DCNQI-based films represent model systems which undergo diffusion-controlled solid state redox reactions after the alkali metal and the DCNQI molecules were deposited in separate layers at 77 K [2]. The quasi-one-dimensional properties of the charge and spin carriers are discussed on the basis of the angle-resolved photoemission (ARUPS) and ESR investigations on the thin films and are related to the bulk properties [3]. First applications of the DCNQI films are mechanical pressure sensors and wireless sensors for spin-carrying gases [4].

*This presentation gives a survey on a research project of the 3. Physikalisches Institut in Stuttgart (Professor Wolf).

**On leave from 3. Physikalisches Institut der Universität Stuttgart, Pfaffenwaldring 57, 7000 Stuttgart 80, Germany.

0379-6779/93/$6.00

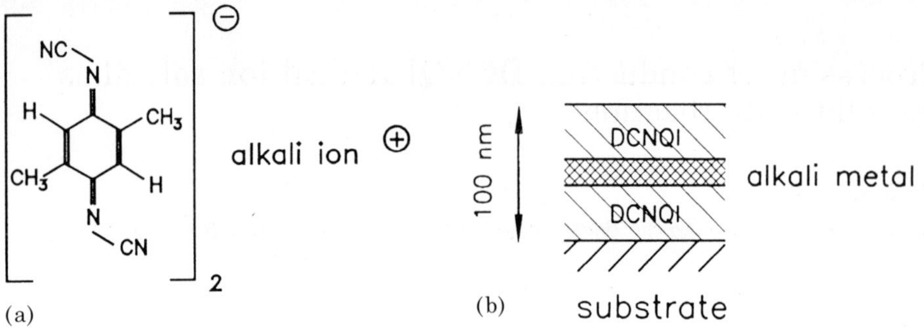

(a) (b)

Fig. 1. (a) Structure of the DCNQI salts used for film preparation (alkali ion = Li^+, Na^+, K^+, Rb^+, Cs^+). (b) Structure of the three-layer films on an arbitrary substrate. The accessible thickness ranges between 30 and 200 nm.

2. Experimental

Subsequent layers of DMe-DCNQI and alkaline metals were deposited onto cooled ($T = 77$ K) quartz substrates by evaporation. For the experiments described below, a three-layer structure was chosen (Fig. 1(b)). The vacuum was below 10^{-7} mbar. The thicknesses of the individual layers were measured by home-made quartz balances working at the substrate temperature. This ensured identical sticking coefficients on the balance and substrate. The thicknesses of the layers were controlled in order to achieve the stoichiometry of the salt. Three-layer structures have proven to be useful in the range of 30 to 200 nm. These limits are imposed by the need for sufficient accuracy of the composition and acceptable diffusion times. By heating the multi-layer films, the salt-forming reactions were started. These reactions were monitored by *in situ* measurements applying different techniques which are discussed here.

3. Results

Optical spectra in the UV–Vis range were acquired by transmission measurements. During the annealing of the sandwich structure, the absorption around 350 nm decreased while the absorption band of the DMe-DCNQI anions around 700 nm increased (Fig. 2). As the UV–Vis spectra are rather broad, they are of limited use for a detailed analysis of the oxidation state of the molecules. However, infrared reflection absorption spectroscopy (IRRAS) enables the observation of the change of vibronic states during the reduction of the DMe-DCNQI molecules. Thin film investigations eliminate matrix effects which are induced by conventional IR techniques using KBr pellets. Figure 3 gives a comparison between the known spectra of DMe-DCNQI and its potassium salt in KBr pellets and as thin films.

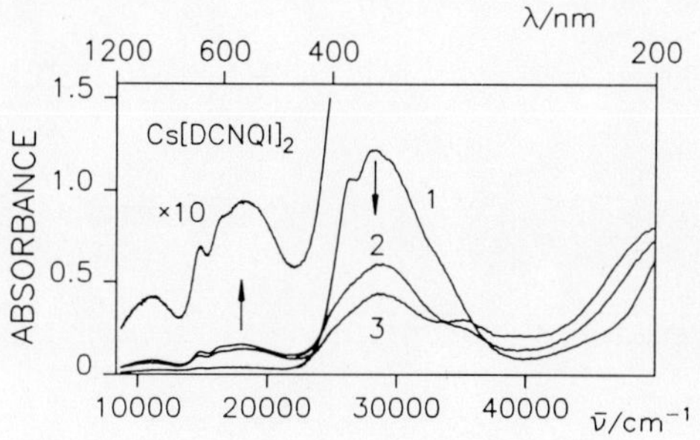

Fig. 2. Change of the optical absorbance during the solid state reaction of DMe-DCNQI and Cs: 1, immediately after film deposition ($T = 77$ K); 2, after 10 h ($T = 250$ K); 3, after 30 h ($T = 300$ K). Curve 3 is magnified by a factor of 10.

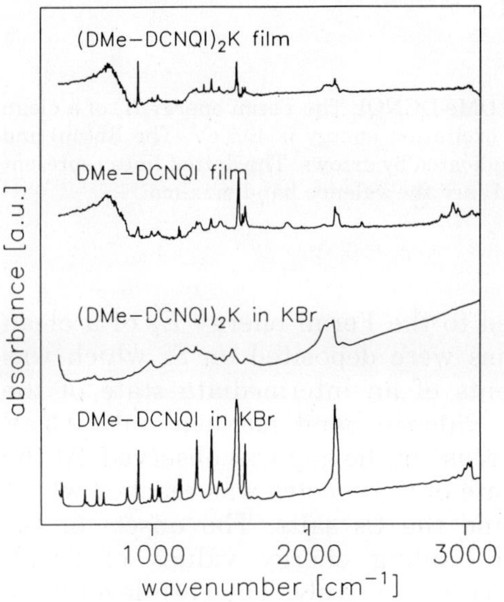

Fig. 3. Comparison of the infrared spectra of DMe-DCNQI and its potassium salt as films (100 nm, IRRAS, angle of incidence 20°) and as KBr pellets in transmission at $T = 300$ K.

Conductivity measurements were performed in a four-probe configuration. This enabled an in-plane measurement. Starting with a three-layer configuration as described above, the conductivity increased by more than two orders of magnitude in the temperature range between $T = 240$ K and $T = 250$ K. The heating rate was 15 K/h. During annealing, the conductivity reached its maximum at temperatures of 300 to 350 K. At the end of the reaction, room temperature conductivities of 10^{-2} S/cm (Li, Na, K salts) to 5×10^{-4} S/cm (Rb, Cs salts) were reached for the alkaline salt films. The conductivity was thermally activated with activation energies of $E_A = 0.06$ eV ((DMe-DCNQI)$_2$Li), $E_A = 0.07$ eV ((DMe-DCNQI)$_2$Na, (DMe-DCNQI)$_2$K) and $E_A = 0.24$ eV ((DMe-DCNQI)$_2$Cs).

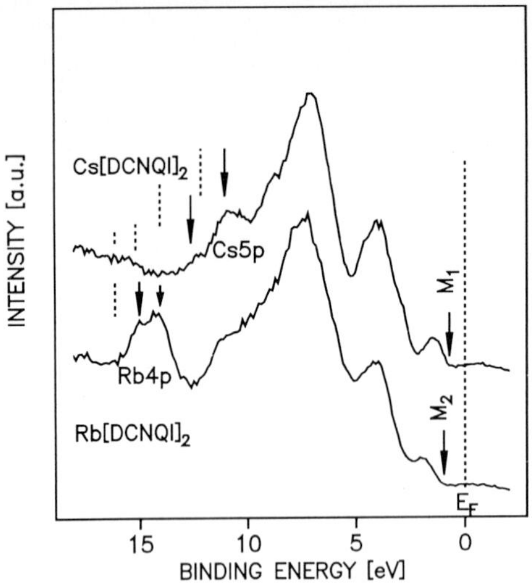

Fig. 4. UPS spectra of the Rb and Cs salts of DMe-DCNQI. The Fermi energy E_F of a clean Cu single crystal is given as reference. The excitation energy is 40.8 eV. The Rb(4p) and Cs(5p) core levels of the ions in the salt are indicated by arrows. The dotted lines represent the energies of the metallic states. M_1 and M_2 are the valence band maxima.

UPS investigations were referred to the Fermi energy E_F of a clean Cu single crystal reference. The films were deposited on Si which was coated by a gold layer. Measurements of an intermediate state of the reaction yielded spectra with their valence band maxima only 0.3 eV below E_F. During the reaction an increase in the gap was observed. At the end of the redox reaction a steady state of the spectra was reached which is displayed in Fig. 4 for the Rb and the Cs salts. The onsets of the photoemission are extrapolated to binding energy values of 0.9 eV (M_2, Rb salt) and 0.65 eV (M_1, Cs salt), respectively. The core level binding energies of Rb(4p) and Cs(5p) states (arrows) are shifted from the positions of the neutral states (dotted lines) towards the Fermi level. Angular resolved UPS measurements [5] showed a dispersion-like behaviour of the two bands which are closest to the Fermi level. A dispersion in the order of 0.5 eV was observed in normal emission by varying the excitation energy between 15 and 45 eV.

ESR data were obtained using a specially designed evaporation chamber which allows the sample transfer into the ESR cryostat in the cold state immediately after the deposition of the multi-layer structures. The onset of the ESR signal shows a similar behaviour as the conductivity. Figure 5 gives an example of the development of the ESR ($g = 2$) signal during the reaction. Starting at $T = 77$ K, the ESR signals of a $(DMe\text{-}DCNQI)_2K$ film were recorded while the sample was heated at a rate of 1 K/min. The strong increase of the ESR signal around $T = 250$ K

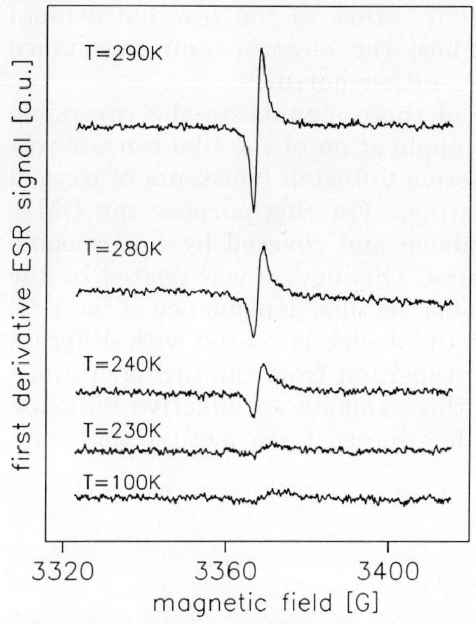

Fig. 5. Change of the ESR spectra during the annealing of a three-layer structure of DMe-DCNQI and potassium. Total thickness 200 nm.

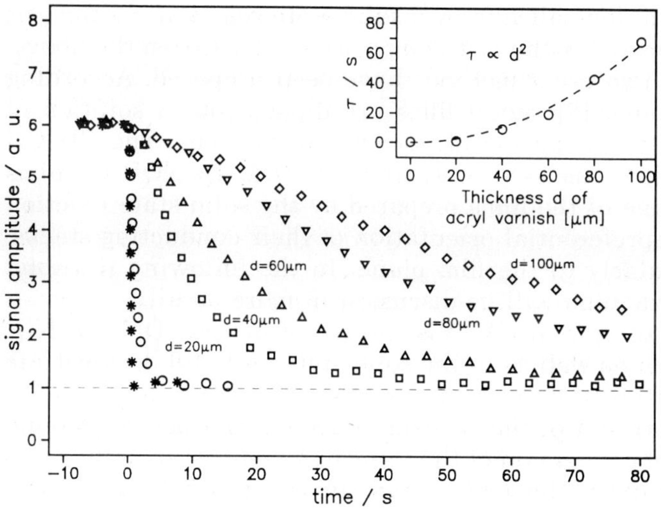

Fig. 6. Time dependence of the ESR first-derivative signal amplitude of a thin film sensor after the resonator is flushed with oxygen. The polymethacrylate under test was applied in the range of thickness between 20 and 100 μm.

coincides with the onset of conductivity. Due to the low-dimensional electronic motion in fully reacted films, the electron spin resonance signals of the conduction electrons are rather narrow.

We investigated the dependence of these signals on the reversible physisorption of gaseous oxygen. The application of the film sensors was demonstrated by determining the effective diffusion constants of oxygen and nitrogen in polymethacrylate coatings. For this purpose the DME-CDNQI salt films were prepared as above and covered by a permeable protective LiF layer of 300 nm thickness. This device was coated by the varnish under test. Figure 6 demonstrates the time dependence of the ESR first-derivative signal amplitude when the device is coated with different amounts of varnish and undergoes a transition from the nitrogen atmosphere to the oxygen atmosphere. In this example an effective diffusion constant of $D = 3.7 \times 10^{-7}$ cm^2/s was determined for a methyl methacrylate polymer.

4. Discussion

The interesting physical properties of the DCNQI family raise the question for possible applications. This is connected with the search for thin films which can be deposited on substrates that do not possess epitaxial properties and yield nevertheless a sufficient degree of order. First steps in this direction represent the single source evaporations of TCNQ charge transfer salts [6] which were claimed to be evaporated as complex [7]. The small and relatively flat DMe-DCNQI molecules, in addition to the favourable crystal structure of the salt, encourage the synthesis of a radical anion salt film by a solid state reaction. As the salt itself cannot be evaporated without decomposition, apart from the above-mentioned technique, two other methods have been proposed. According to the solution technique [8], metal films are dipped into a solution of DCNQI which starts the formation of polycrystalline films. The CH_3CN vapour-assisted reaction was used to synthesize $(DMe\text{-}DCNQI)_2Cu$ films [9]. The main advantage of the films prepared by the solid state reaction is that they possess a preferential orientation of their conducting stacks, which run perpendicularly to the film plane. In the following the solid state reaction under vacuum will be discussed in more detail.

The comparisons of optical UV–Vis spectra of neutral DMe-DCNQI and the salt in solution as well as in the solid state [2, 9, 10] demonstrate the formation of the radical anion with its absorption band between 600 and 700 nm. The absorption of the neutral DMe-DCNQI species (around 350 nm) decreases. The conversion of the former three-layer structure into the salt film also changes the background of the absorption spectra, therefore isosbestic points referring to the two oxidation states of DMe-DCNQI are not expected. As these bands reflect molecular properties, the optical spectra can be considered as molecular probes. During the annealing process electrical measurements show an increase of conductivity which covers several orders of magnitude. This proves a macroscopic charge transport by the salt films. The d.c. conductivity of the salt films yields a low activation energy which demonstrates that the influence of limited order and grain boundaries does not affect too much the overall performance of macroscopic DMe-DCNQI salt films.

UPS measurements have shown the oxidation of the alkaline metals by the shifts of their core levels with respect to the neutral states. However, the dominant features of the photoemission spectra are due to DMe-DCNQI, which is shown by the exchange of the counter ion [11]. The evaluation of extended-Hückel energy band structures [12] also hints at the dominant role of C- and N-derived states in the electronic structure of the DCNQI salts. The discrepancy between the position of the valence band maxima and the electrical activation energies is tentatively ascribed to a relaxation process which is slower than the escape of the electron from the low-dimensional conductor. The interpretation of the $E(k)$ dispersion yields a periodicity of the charge density distribution which is twice the lattice constant in the stacking direction [5]. Therefore, already at room temperature, the UPS investigations yield a dynamically dimerized electronic structure of the salt films which is not observed in the crystals. This charge distribution strongly hints at a pinned charge density wave.

IR measurements on charge transfer systems have proven to be an excellent tool in order to determine the degree of charge transfer [13]. In the case of DMe-DNCQI salts, the C=N stretch vibration of the imine positions shows a linear variation of its frequency as a function of the charge transfer ρ. The observation of this shift during the reaction is more sensitive to local changes than the conductivity or the ESR data. Concentrating on the C=N stretch mode around 1550 cm^{-1}, first measurements on $(DMe\text{-}DCNQI)_2K$ films indicate a charge transfer by an emerging new vibrational mode which is located 10 cm^{-1} below the decreasing line of the neutral species. However, the absolute positions of the peaks in the neutral film as well as in the salt film are at variance with the data obtained from KBr pellets. It seems possible that a solvent shift of 13 cm^{-1} is induced by the KBr surrounding in the pellets. The question of film-specific influences will be solved by further experiments using different substrates and films of other salt families.

Electron spin resonance experiments have shown the preferential orientation of conducting stacks in the alkaline DMe-DCNQI radical ion salt films. A preferential orientation of the stacking direction perpendicular to the film plane was deduced [2]. The development of the ESR signal shows the run of the reaction. Immediately after film deposition, a broad line of low intensity is present which exhibits a typical Dysonian shape (not clearly visible in the scale of Fig. 5). It is ascribed to the neutral layer of the alkali metal. The narrow line, which is attributed to the mobile electrons in the DMe-DCNQI salt, emerges around 250 K. The final state is reached after a considerable delay with respect to the optical and the conductivity data. Apparently the motion of the spin carriers in the salt is highly sensitive to structural imperfections causing line broadening. These defects are partly cured during the annealing process. This is in line with all ESR data of the DCNQI salt films which are generally slightly broadened with respect to the single crystal data.

5. Applications

The use of DMe-DCNQI radical ion salt films as electrical sensors for mechanical uniaxial pressure has been demonstrated previously. Here a wireless sensor for spin-carrying gases is presented. The motionally narrowed ESR lines of the alkaline DMe-DCNQI salts are broadened by the presence of O_2. The application of $(DMe\text{-}DCNQI)_2$ salt films is based on the crystal structure of the salt and the preferential orientation within the films. As known from X-ray analysis of single crystals of the salts [14], the segregated stacks leave enough space which can act as pores for the diffusion of gases along the stacking direction. Thus, a fast penetration is enabled. The broadening mechanism is thought to be a weak spin exchange between the electron spins on the DMe-DCNQI stacks and the two unpaired spins of the oxygen molecules. For practical applications, the ESR first-derivative amplitude is recorded. The relation between the amplitude and the O_2 concentration is calibrated by known mixtures of O_2 and inert gases, e.g. N_2 or He. The main field of practical application is seen in testing paint and coating materials, which can be directly spread on the device as well as oxygen detection in the field of clinical research [15].

Acknowledgements

The experiments presented above were obtained in team work with A. Edelmann, S. Hamm, J. U. von Schütz and H. C. Wolf at the 3. Physikalisches Institut der Universität Stuttgart (Germany). The collaboration with D. Schmeisser (Tübingen, Germany) enabled the UPS experiments to be undertaken. S. Hünig and his team (Würzburg, Germany) are acknowledged for the synthesis of DCNQI. This work has been supported by the Deutsche Forschungsgemeinschaft via SFB 329. A research grant by the French government is also acknowledged.

References

1 A. Aumüller, P. Erk, G. Klebe, S. Hünig, J. U. von Schütz and H.-P. Werner, *Angew. Chem.*, *98* (1986) 759.

2 H. Wachtel, J. U. von Schütz and H. C. Wolf, *Synth. Met.*, *42* (1991) 1789.
3 J. U. von Schütz, M. Bair, H. J. Gross, U. Langohr, H.-P. Werner, H. C. Wolf, D. Schmeisser, K. Graf, W. Göpel, P. Erk, H. Meixner and S. Hünig, *Synth. Met.*, *27* (1988) B249.
4 S. Hamm, J. U. von Schütz, H. Wachtel and H. C. Wolf, *Appl. Surf. Sci.*, *55* (1992) 239.
5 D. Schmeisser, W. Jaegermann, Ch. Pettenkofer, H. Wachtel, A. Jimenez-Gonzales, J. U. von Schütz, H. C. Wolf, P. Erk, H. Meixner and S. Hünig, *Solid State Commun.*, *81* (1992) 827.
6 P. Delhaes, personal communication.
7 C. Reinhard, W. Vollmann, C. Hamann, L. Libera and S. Trompler, *Krist. Techn. 15* (1980) 243.
8 P. Erk, *Thesis*, University of Würzburg, 1989.
9 S. Yamaguchi, C. A. Viands, H. S.-W. Hu and R. S. Potember, *Synth. Met.*, *45* (1991) 71.
10 K. Yakushi, A. Ugawa, G. Ojima, T. Ida, H. Tajima, H. Kuroda, A. Kobayashi, R. Kato and H. Kobajashi, *Mol. Cryst. Liq. Cryst.*, *181* (1988) 217.
11 D. Schmeisser, J. U. von Schütz, H. Wachtel and H. C. Wolf, *J. Phys. I*, *1* (1991) 1347.
12 W. Koch and F. Seelig, *Z. Naturforsch.*, *Teil A*, *42* (1987) 875.
13 G. Lunardi and C. Pecile, *J. Chem. Phys.*, *95* (1991) 6911.
14 R. Kato, H. Kobayashi, A. Kobayashi, T. Mori and H. Inokuchi, *Chem. Lett.*, (1987) 1579.
15 H. M. Swartz, *Int. Symp. Recent Advances in ESR Spectroscopies, University of Padua, Italy, Sept. 8–12, 1991.*

Synthetic Metals, *54* (1993) 373–381

Photoinduced orientation of azo dyes in polymeric films. Characterization of molecular angular mobility

Z. Sekkat and M. Dumont
Institut d'Optique Théorique et Appliquée, Unité Associée au CNRS No. 14, Bâtiment 503, B.P. 147 F-91403 Orsay Cédex (France)

Abstract

The reversible *cis–trans* photoisomerization of azobenzene derivatives is very sensitive to the polarization of the pumping beam which induces birefringence and dichroism. Recently, we have demonstrated a new method for poling azo dye-containing polymers, at room temperature, by applying a d.c. electric field during the optical pumping process. Here, we present a model for the dynamics of the photoinduced modifications of the angular distribution of chromophores and we compare a guest–host system (Disperse Red 1 (DR1) in PMMA) and a PMMA–DR1 copolymer.

1. Introduction

The Weigert effect, which is the ability of some materials to become dichroic and birefringent when they are illuminated by a polarized light, is known from the beginning of this century in photographic materials. In the sixties, it was widely studied in azo dye in viscous solutions [1] and recognized as a consequence of the *cis–trans* photoisomerization. Later it was used in polymer films for holographic applications [2]. Recently the interest for this process has increased considerably as it gives a method for designing optoelectronic devices by acting on the anisotropic indices of polymeric waveguides [3–5]. In two recent letters [6, 7], we have shown the possibility of breaking the centrosymmetry of DR1-containing PMMA films by pumping them with a resonant circularly polarized pump beam in the presence of a d.c. electric field. In DR1-doped PMMA (guest–host system) the mobility of DR1 is large and the photoinduced $\chi^{(2)}$ has a poor remanence. On the contrary, large and stable second-order nonlinearities have been obtained with a PMMA–DR1 copolymer.

In Section 2 we rapidly recall the experimental methods for measuring the time variations of the linear and nonlinear optical properties of polymer films and we give some results. In Section 3 we present a theoretical model for the evolution of the orientation of uniaxial molecules submitted to an optical pumping process. This model takes into account the reversible *cis–trans* photoisomerization and the angular relaxation of molecules.

0379-6779/93/$6.00

2. Experimental methods and results

2.1. The samples

The doped polymer is obtained by dissolving commercially available DR1 in a PMMA solution (chlorobenzene, chloroform or MIBK solution). The PMMA–DR1 copolymer is synthesized by Thomson CSF-LCR. Solutions are filtered and deposited by spin coating onto glass slides. Films are then backed (a few minutes at $\approx 120\,^{\circ}\mathrm{C}$) in order to remove the remaining solvent. Film thicknesses range from 0.6 to 1.8 μm.

2.2. Dichroism measurement

The sample is pumped by a linearly polarized 488 nm Ar^+ laser beam, with a normal incidence. A 514 nm probe beam (from the same Ar^+ laser) crosses the pump beam with a small angle; its polarization is at 45° from that of the pump. A Wollaston prism, located after the sample, splits the probe into two orthogonally polarized beams, parallel and perpendicular to the pump polarization. The simultaneous measurement of the transmitted intensity on both polarizations gives the time variations of absorbances:

$$A_{\parallel} = -\log(I_{\parallel}/I_0) \propto \mathrm{Im}(\chi_{33}^{(1)})$$

and

$$A_{\perp} = -\log(I_{\perp}/I_0) \propto \mathrm{Im}(\chi_{11}^{(1)})$$

The symmetry axis, labelled 3, is the polarization of the pump beam. Figure 1(a) shows the pump-induced variations of the transmission of a 10% DR1-doped PMMA film and Figs. 1(b) and (d) show the time transients of the absorbance of a 2.5% sample for two different pump intensities. Fig. 1(c) gives the variations of $A_{\parallel}$ and $A_{\perp}$, measured three minutes after the pump is switched on as a function of pump intensity.

2.3. Birefringence and electro-optic measurements by attenuated total reflection (ATR)

For ATR experiments [8], a thin semi-transparent silver layer is evaporated onto the glass slide prior to film deposition and a second thick silver layer is evaporated above the polymer. The glass slide is put in optical contact with a half-sphere through which an infrared laser beam (829 nm) is reflected onto the sample. The laser diode is polarized at 45° from the plane of incidence and a Wolaston prism divides the reflected beam into a TM and TE beam (parallel and perpendicular to the plane of incidence, respectively) which are detected by two photodiodes. The laser and the detector assembly can be symmetrically rotated in order to record TE and TM reflectivities as a function of incidence angle. On each detector a set of reflectivity dips appears corresponding to the successive TE or TM guided modes in the polymer film. The angular positions of these two sets of modes permits the measurement of the thickness of the sample (h) and

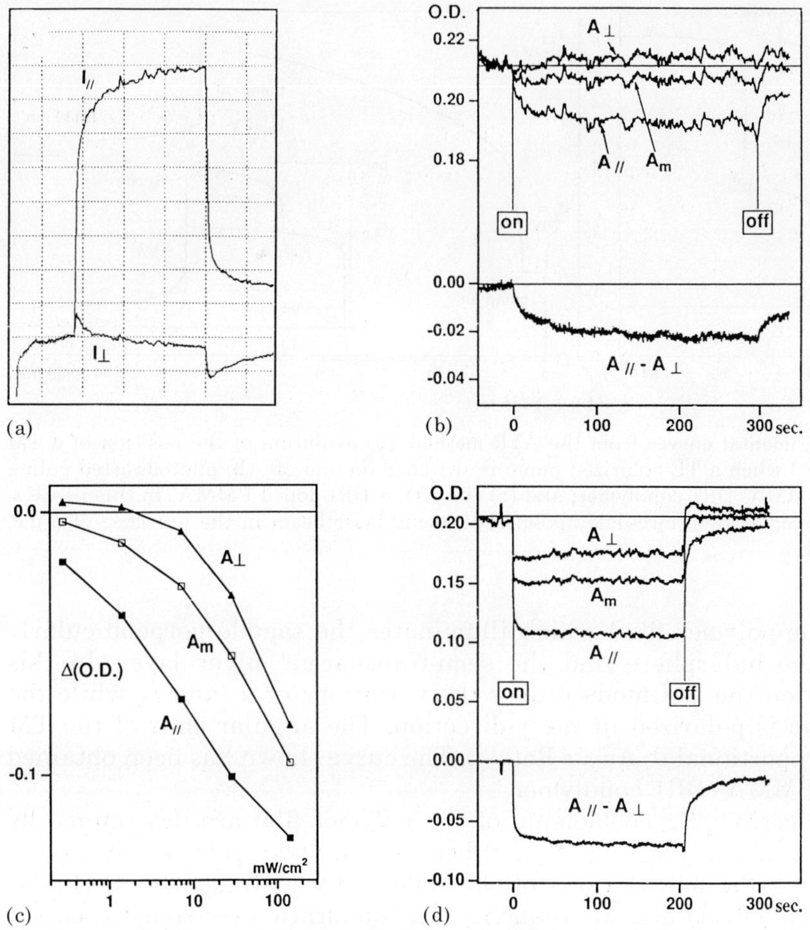

Fig. 1. (a) Variations of the transmission of the probe beams, $I_\parallel$ and $I_\perp$, through a 10% DR1-doped PMMA film when the pump beam is switched on and off. This curve is shown as it is characteristic of the two mechanisms described in Section 3, but the optical density of this sample is too large for precise measurements. Low optical density samples (concentration 2.5%) with (b) $I_p = 0.28\ \mathrm{mW/cm^2}$ and (d) $I_p = 28\ \mathrm{mW/cm^2}$ permit a quantitative determination of absorbances $A_\parallel = -\log(I_\parallel/I_0)$ and $A_\perp = -\log(I_\perp/I_0)$ and of $A_m = (A_\parallel + 2A_\perp)/3 \propto N_T$ and $(A_\parallel - A_\perp) \propto N_T A_2$. (c) Variations of $A_\parallel$ and $A_\perp$ measured 3 min after the pump is switched on.

the three principal indices n_x, n_y and n_z, (z is the normal to the sample and $\{x, z\}$ is the plane of incidence: in our experiments, these axes are always the principal axes of the susceptibility tensor, but z is not always the symmetry axis which is labelled 3).

Figure 2(a) shows the variations of the angular position of a TM mode obtained by recording the reflectivity of the sample with a fixed angle of incidence, on the side of this mode. The angular shift of the mode is induced by a pumping beam (488 nm; TE polarized; $30\ \mathrm{mW/cm^2}$ at the

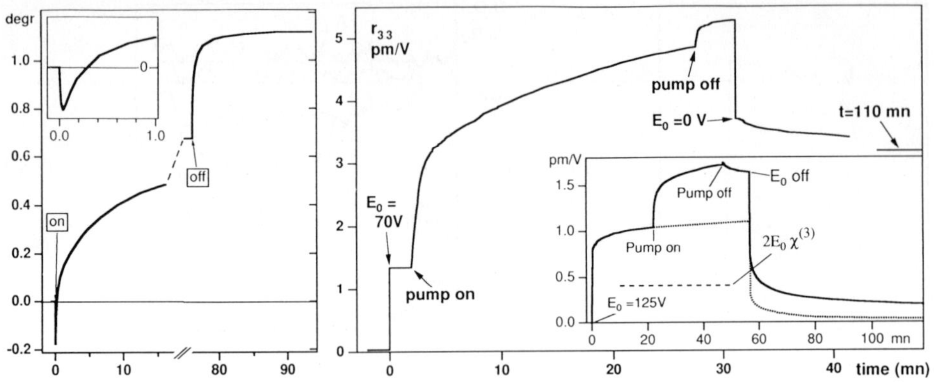

(a) (b), (c) (insert)

Fig. 2. Experimental curves from the ATR method: (a) evolution of the position of a TM mode ($\propto \Delta n_\perp$) when a TE-polarized pump is switched on and off; (b) photoassisted poling process in PMMA–DR1 copolymer; and (c) (insert) in DR1-doped PMMA. In this process, the film is pumped by a circularly polarized 488 nm laser beam in the presence of a d.c. electric field.

level of the polymer film) which illuminates the sample perpendicularly through the half-sphere and the semi-transparent silver layer. In this configuration the TM mode probes the variations of n_x and n_z while the pump beam is polarized in the y direction. The angular shift of the TM mode is proportional to $\Delta n_\perp \propto \mathrm{Re}(\chi_{11}^{(1)})$. The curve shown has been obtained with the PMMA–DR1 copolymer.

The electro-optic coefficients of the polymer film are determined by applying an a.c. voltage ($V_1 \cos \Omega t$) between the two metallic layers and by recording the modulations of the reflectivity at frequency Ω (linear electro-optic effect) and at frequency 2Ω (quadratic electro-optic effect) near each TM and TE mode. A simple computer fit evaluates the amplitudes of modulations of the thickness and of the three indices, from which Pockels and Kerr coefficients are obtained ($i = x, y, z$):

$$2\chi_{iiz}^{(2)} = -r_{iz} n_i^4 = 2n_i \Delta n_i(\Omega)(h/V_1) \tag{1}$$

$$3\chi_{iizz}^{(3)} = -s_{iz} n_i^4 = 4n_i \Delta n_i(2\Omega)(h/V_1)^2 \tag{2}$$

The variations of $\chi^{(2)}$ can be followed continuously by recording the reflectivity modulation (Ω), while the goniometer is maintained at a fixed angular distance from a chosen mode. Figure 2(b) (from ref. 7) shows the variation of the electro-optic coefficient (r_{33}) of a DR1-doped PMMA film and of a copolymer film submitted to a poling cycle: at time $t = 0$ a d.c. field (E_0) is applied which produces an electric field-induced Pockels effect (EFIPE) which is solely due to a third-order effect ($\chi^{(3)}E_0$) in the case of the copolymer, but which contains a part due to the rotation of free molecules in the guest–host case (induced $\chi^{(2)}$). The $\chi^{(3)}$ value is known from the modulations of ATR modes at twice the a.c. field frequency (2Ω).

When the circularly polarized pump beam is applied the photo-induced rotation of chromophores builds progressively the Pockels coefficient. After the successive removal of the pump and of the d.c. field, the $\chi^{(3)}$ part immediately falls while r_{33} rapidly relaxes in the case of DR1-doped PMMA, but is highly remanent in PMMA–DR1 copolymer. In this last material we obtained $r_{33} = 5.7$ pm/V after two days. This poling technique has also been demonstrated by second-harmonic generation [6, 9].

3. Theory of photoinduced orientation of molecules

The photochromism of azo dyes is due to the *cis–trans* isomerization of molecules with respect to the N=N double bond. The *trans* stable form presents a strong visible absorption band (peak at 490 nm for DR1). When a molecule absorbs a photon, it comes back to the ground state by a nonradiative decay, either in the *trans* form or in the *cis* form. At room temperature, the *cis* form spontaneously relaxes towards the *trans* form, with a time constant which depends on its environment (i.e. the nature of the physical or chemical links with the polymer). In the visible range, the absorption by *cis* molecules is small: for the sake of simplicity, we will neglect the *cis* contribution to the linear and the nonlinear susceptibilities of the sample.

The *trans* form is almost a uniaxial molecule: with a good approximation, the polarizability tensor and the second- and third-order nonlinear hyperpolarizabilities of the *trans* form have only one component (α_Z, β_Z and γ_Z, respectively) parallel to the dipole moment, μ, of the molecule. All the anisotropy and the nonlinear optical properties of the sample are given by the angular distribution of the dipole moment of *trans* molecules. As we will consider systems with an axial symmetry, the angular distribution depends only on the angle θ between the dipole moment and the symmetry axis and can be expressed in the standard basis of Legendre polynomials [10]:

$$n_{\rm T}(\theta, \varphi) = (N_{\rm T}/2\pi)G(\cos\theta) = (N_{\rm T}/4\pi)\Sigma(2l+1)A_l P_l(\cos\theta) \qquad (3)$$

$$\chi^{(1)}_{33} = \tfrac{1}{3}N_{\rm T}\alpha_z(1+2A_2); \qquad \chi^{(1)}_{11} = \tfrac{1}{3}N_{\rm T}\alpha_z(1-A_2) \qquad (4)$$

$$\chi^{(2)}_{333} = \tfrac{1}{5}N_{\rm T}\beta_z(3A_1+2A_3); \qquad \chi^{(2)}_{311} = \tfrac{1}{5}N_{\rm T}\beta_Z(A_1-A_3) \qquad (5)$$

where $N_{\rm T}$ is the density of dye molecules in the *trans* state. When an initially isotopic sample is pumped by a linearly polarized beam (e.g. experiments of Fig. 1), the symmetry axis, 3, is the polarization of the pump beam.

The probability for each molecule to undergo a *trans–cis* photisomerization is $\sigma_{\rm T} I_{\rm p} \Phi_{\rm TC} \cos^2\theta$, where $\sigma_{\rm T} \propto \alpha_z$ is the absorption cross section, $I_{\rm p}$ the pump intensity and $\Phi_{\rm TC}$ the quantum yield of the photoisomerization. This optical pumping produces a selective depletion of *trans* molecules parallel to the pump ('angular hole burning') which reduces $N_{\rm T}$ and induces a negative A_2 ('alignment'). A dichroism and an anisotropy appear as expressed by eqn. (4) and as shown by experimental curves in Figs. 1(a) and (b). In these Figures, $A_m = (A_\parallel + 2A_\perp)/3 \propto N_T$ and $(A_\parallel - A_\perp) \propto N_T A_2$ have been drawn. The variations of N_T express the population exchanges between *trans* and *cis* states: when the pump is switched off, the thermal *cis* to *trans* relaxation is complete, but the dichroism lasts much longer than the *cis* population. This proves that the orientation of molecules changes during the photoisomerization process. This 'angular redistribution' process is not surprising since the shape of the molecule completely changes in the excited states and in the *cis* form. After several optical pumping cycles the alignment of *trans* molecules can be large: it depends on the competition with the third mechanism, the 'angular relaxation' inside the *trans* form. The evolution of the angular distribution in *trans* and *cis* states has the general form:

$$\frac{\mathrm{d}}{\mathrm{d}t} n_T(\Omega) = -I \cos^2\theta n_T(\Omega) + \frac{1}{\tau_c} \int R_{CT}(\Omega' \longrightarrow \Omega) n_C(\Omega') \,\mathrm{d}\Omega'$$

$$- n_T(\Omega) \int D_T(\Omega \longrightarrow \Omega') \,\mathrm{d}\Omega' + \int D_T(\Omega' \longrightarrow \Omega) n_T(\Omega') \,\mathrm{d}\Omega' \quad (7)$$

$$\frac{\mathrm{d}}{\mathrm{d}t} n_C(\Omega) = -\frac{1}{\tau_c} n_C(\Omega) + I \int P_{TC}(\Omega' \longrightarrow \Omega) n_T(\Omega') \cos^2\theta' \,\mathrm{d}\Omega'$$

$$- n_C(\Omega) \int D_C(\Omega \longrightarrow \Omega') \,\mathrm{d}\Omega' + \int D_C(\Omega' \longrightarrow \Omega) n_C(\Omega') \,\mathrm{d}\Omega' \quad (8)$$

where $I = \sigma_T I_p \Phi_{TC}$; $n_T(\Omega)(n_C(\Omega))$ is the number of *trans* (*cis*) molecules oriented in the solid angle $\mathrm{d}\Omega$ around the direction $\Omega = \{\theta, \varphi\}$; τ_c is the *cis* lifetime; $P_{TC}(\Omega' \longrightarrow \Omega)$ and $R_{CT}(\Omega' \longrightarrow \Omega)$ are the probabilities for the molecule to rotate from Ω' to Ω in the *trans*–*cis* or in the *cis*–*trans* transition, respectively (angular redistribution); $D_T(\Omega' \longrightarrow \Omega)$ and $D_C(\Omega' \longrightarrow \Omega)$ are the probabilities per unit of time that the molecule rotates spontaneously from Ω' to Ω (angular relaxation). This general formalism can be simplified when relaxation and redistribution processes are isotropic, which means that they depend only on the rotation angle α between directions Ω and Ω' (they neither depend on the initial direction nor on the azimuth of the rotation). Then, eqns. (7) and (8) can be expressed on the basis of Legendre polynomials:

$$n_T = (1/4\pi)\Sigma(2l+1)T_l P_l(\cos\theta); \qquad n_C = (1/4\pi)\Sigma(2l+1)C_l P_l(\cos\theta) \quad (8)$$

$$R_{CT} = (1/4\pi)\Sigma(2l+1)R_{CT,l} P_l(\cos\alpha); \qquad P_{TC} = (1/4\pi)\Sigma(2l+1)P_{TC,l} P_l(\cos\alpha) \quad (9)$$

$$D_T = (d_T/4\pi)\Sigma(2l+1)D_{T,l} P_l(\cos\alpha); \qquad D_C = (d_C/4\pi)\Sigma(2l+1)D_{C,l} P_l(\cos\alpha) \quad (10)$$

With these definitions: $T_l = N_T A_l$; $R_{CT,0} = P_{TC,0} = D_{T,0} = D_{C,0} = 1$. In eqns. (7) and (8), terms with $\cos^2\theta$ couple Legendre polynomials with $\Delta l = \pm 2$, but isotropic relaxations and angular redistributions do not couple different l (this is a particular case of a general result in the Fano theory of irreducible tensors). On obtains two infinite sets of coupled equations (for odd and even l):

$$\frac{dC_l}{dt} = IP_{TC,l}[K_{l-}T_{l-2} + K_{l0}T_l + K_{l+}T_{l+2}] - C_l[\gamma_0 + d_C(1 - D_{C,l})] \quad (11)$$

$$\frac{dT_l}{dt} = -I[K_{l-}T_{l-2} + K_{l0}T_l + K_{l+}T_{l+2}] + \gamma_0 R_{CT,l}C_l - d_T(1 - D_{T,l})T_l \quad (12)$$

with

$\gamma_0 = \tau_c^{-1}$

$K_{l-} = l(l-1)(2l-1)^{-1}(2l+1)^{-1}$

$K_{l0} = (2l^2 + 2l - 1)(2l-1)^{-1}(2l+3)^{-1}$

$K_{l+} = (l+1)(l+2)(2l+1)^{-1}(2l+3)^{-1}$

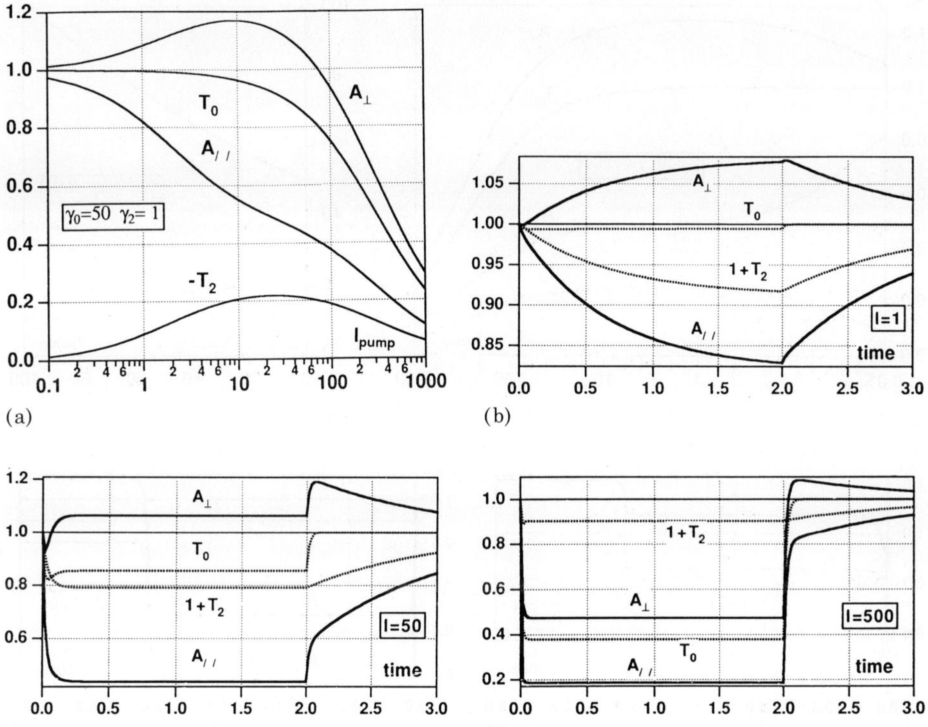

Fig. 3. Results of the simplified theory for $\gamma_0 = 50$ and $\gamma_2 = 1$: (a) asymptotic values ($t = \infty$) of *trans* state population (T_0) and alignment (T_2) and of absorbances $A_\parallel$ and $A_\perp$; (b), (c) and (d) time transients of T_0, T_2, $A_\parallel$ and $A_\perp$. The pump is switched on at $t = 0$ and off at the normalized time $t = 2$.

These sets of equations can be solved by truncating them above any arbitrary l order (it is reasonable to assume that C_l and T_l decrease when l increases). As a first approximation, we neglect all $l > 2$ and we assume that the redistribution completely loses the initial orientation ($R_{\mathrm{CT},l} = 0$ for $l \neq 0$): then, we have only two equations for $T_0 = N_\mathrm{T}$ and $T_2 = N_\mathrm{T} A_2$. The orientation T_1 will be considered in another publication. Figures 3 and 4 show the time evolution of T_0 and T_2, and of the normalized absorbances $A_\parallel$ and $A_\perp$ when the pump is switched on and off. The values of T_0, T_2, $A_\parallel$ and $A_\perp$ for $t = \infty$ are also shown as a function of normalized pump intensity, I. Figure 4(b) represents the two time constants, λ_0 and λ_2 as a function of I. When I tends towards zero, λ_0 tends towards γ_0 and λ_2 towards $\gamma_2 = d_\mathrm{T}(1 - D_{\mathrm{T},2})$. Figure 3 represents a behaviour similar to that of the DR1-doped polymer (Fig. 1), with a rapid relaxation of molecular alignment. In Fig. 4, a very long relaxation time γ_2^{-1} has been chosen: the behaviour is similar to that of the PMMA–DR1 copolymer as shown by Fig. 2(a).

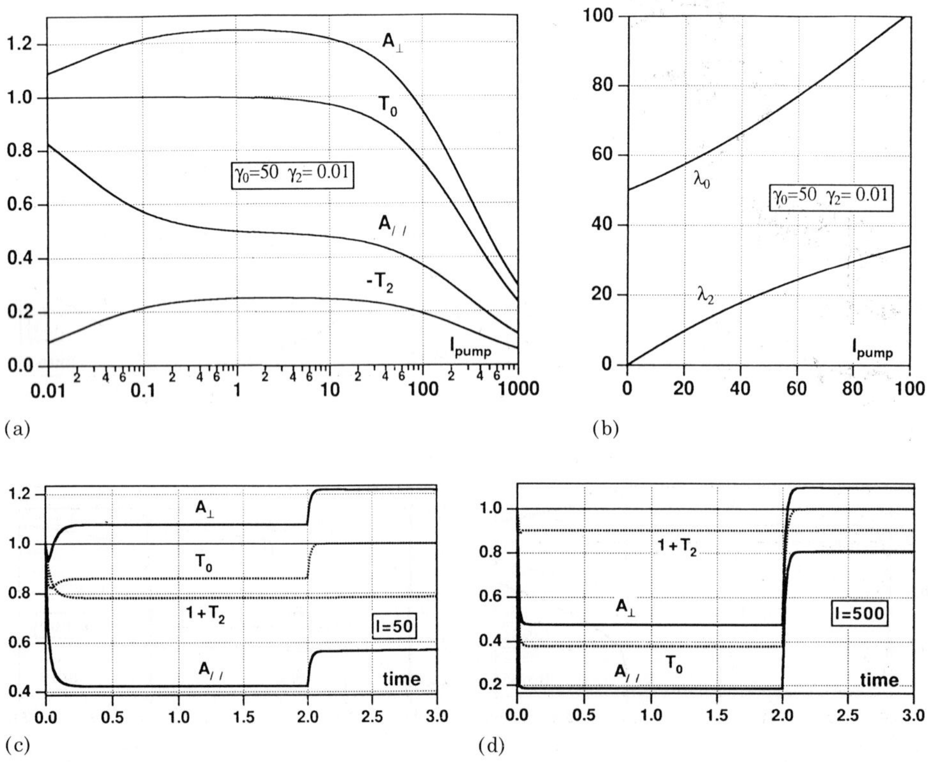

Fig. 4. Results of the simplified theory for $\gamma_0 = 50$ and $\gamma_2 = 0.01$: (a) asymptotic values ($t = \infty$) of *trans* state population (T_0) and alignment (T_2) and of absorbances $A_\parallel$ and $A_\perp$; (b) intensity dependence of the two time constants; (c) and (d) time transients of T_0, T_2, $A_\parallel$ and $A_\perp$. The pump is switched on at $t = 0$ and off at the normalized time $t = 2$.

4. Discussion

We have introduced a description of the polarization-selective photoisomerization in azo dyes which qualitatively explains our experiments. It also explains the results of Todorov *et al.* [2] who obtained $\Delta A_{\perp} = -\Delta A_{\parallel}/2$ in a dichroism experiment; in their case the pump intensity was very small compared to the fast relaxation of the *cis* state, as in Fig. 3(b). The opposite borderline case of saturated angular hole burning, which predicts $\Delta A_{\perp} \geqslant \Delta A_{\parallel}/3$ [3, 6], corresponds to large saturations (Figs. 3(d) and 4(d)). The detailed observation of experimental results shows that the assumption of the complete loss of orientation is too restrictive: this will be the subject of further studies with our general equations. Our relaxation model for the orientation of molecules is more general than the Brownian motion considered by several authors [11], but it can include it. Finally, let us notice that a limitation to our model comes from the fact that the population relaxation and the alignment relaxation, as observed after the pump is switched off, are not monoexponential, probably because of the different possible sites for DR1 molecules inside the polymer.

Acknowledgements

The copolymer used in the above experiments has been synthesized and the films have been deposited at Thomson CSF-LCR. We are particularly indebted to E. Chastaing, P. Le Barny and Ph. Robin for this contribution. J. Delaire, R. Loucif-Saïbi and K. Nakatani have confirmed our results by SHG measurements in their laboratory (Laboratoire de Physico-chimie des Rayonnements, in Orsay). We thank all of them for this efficient collaboration.

References

1 B. S. Neporent and O. V. Stolbova, *Opt. Spektrosk.*, *10* (1961) 287 (*Opt. Spectrosc.*, *10* (1961) 146); *14* (1963) 624 (*Opt. Spectrosc.*, *14* (1963) 331); A. M. Makushenko, B. S. Neporent and O. V., Stolbova, *Opt. Spektrosk.*, *31* (1971) 557 (*Opt. Spectrosc.*, *31* (1971) 295); *31* (1971) 741 (*Opt. Spectrosc.*, *31* (1971) 397).
2 T. Todorov, L. Nikolova and N. Tomova, *Appl. Opt.*, *23* (1984) 4309.
3 Yongqiang Shi, W. H. Steier, Lupin Yu, Mai Chen and L. R. Dalton, *Appl. Phys. Lett.*, *58* (1991) 1131; *59* (1991) 2935.
4 Z. Sekkat and M. Dumont, *Appl. Phys. B*, *53* (1991) 121.
5 P. Rochon, J. Gosselin, A. Natansohn and S. Xie, *Appl. Phys. Lett.*, *60* (1992) 4.
6 Z. Sekkat and M. Dumont, *Appl. Phys. B*, *54* (1992) 486.
7 Z. Sekkat and M. Dumont, *Mol. Cryst. Liq. Cryst., Sect. B: Nonlinear Opt.*, *2* (1992) in press.
8 M. Dumont, D. Morichère, Z. Sekkat and Y. Levy, in R. Lessard (ed.), *Photopolymer Device Physics, Chemistry, and Applications II*, *SPIE Proc.*, Vol. 1559, 1991, p. 127.
9 M. Dumont, Z. Sekkat, R. Loucif-Saïbi, K. Nakatani and J. Delaire, *Mol. Cryst. Liq. Cryst., Sect. B: Nonlinear Opt.*, accepted for publication.
10 M. G. Kuzyk, J. E. Sohn and C. W. Dirk, *J. Opt. Soc. Am., Part B*, *7* (1990) 842.
11 H. Benoit, *Ann. Phys. (Paris)*, *6* (1951) 561; J. W. Wu, *J. Opt. Soc. Am., Part B*, *8* (1991) 142.

4. Discussion

We have introduced a description of the polarisation selective photoisomerization in azo dyes which qualitatively explains our experiments. It also explains the results of Todorov et al. [illegible] who obtained $\Delta A_{\perp} = -\Delta A_{\parallel}/2$ in a dichroism experiment; in their case the pump intensity was very small compared to the fast relaxation of the cis state (see [illegible] Fig. 2(b)). The opposite borderline case of saturated angular hole burning, which predicts $\Delta A_{\perp} \approx \Delta A_{\parallel}$ [illegible], corresponds to large saturation (Figs. 4(d) and 4(e)). [illegible]

[illegible]

[illegible]

8. Non-Carbon Based Polymers

Synthetic Metals, *54* (1993) 385–390

Photoinduced electron–hole charge separation, intrinsic self-doping and novel device applications of Q1D heterojunction materials: mixed-halide MX solids

A. Saxena, X. Z. Huang, A. R. Bishop, L. A. Worl[a], S. P. Love[a] and B. I. Swanson[a]

Theoretical Division and Centers for Nonlinear Studies and Materials Science,
[a]Isotope and Structural Chemistry Group, Los Alamos National Laboratory, Los Alamos, NM 87545 (USA)

J. Tinka Gammel

Materials Research Branch, Naval Ocean Systems Center, San Diego, CA 92152 (USA)

Abstract

We have extensively studied, both experimentally and theoretically, the following three mixed-halide systems: (i) $PtCl_xBr_{1-x}$, (ii) $PtCl_xI_{1-x}$ and (iii) $PtBr_xI_{1-x}$ for a variety of concentrations x. We have obtained direct spectroscopic evidence (vibrational modes) for charge separation from resonance Raman (RR) experiments on doped and photoexcited single crystals of $PtCl_xBr_{1-x}$. In particular, electron polarons preferentially locate on the PtBr segment while hole polarons are trapped within PtCl segments. This polaron charge selectivity, which is potentially very useful for photovoltaic device applications, is demonstrated within theoretical calculations based on a discrete, 3/4-filled, two-band, tight-binding, extended Peierls-Hubbard model, showing strong hybridization of PtCl and PtBr electronic bands as the driving force for separation. In contrast to $PtCl_xBr_{1-x}$, we find that both the $PtCl_xI_{1-x}$ and the $PtBr_xI_{1-x}$ systems exhibit qualitatively different behavior, such as intrinsic 'self-doping'. In other words, our simulations predict that electronic charge is spontaneously removed from the PtI segment and injected into the PtCl (or PtBR) segment, thus creating hole defects in PtI and electron defects (polarons/bipolarons) in PtCl (or PtBr).

Introduction

Halogen-bridged transition metal complexes (or MX chains), as well as being important in their own right, provide an important test case for theoretical techniques and issues in low-dimensional materials with strong electron–electron and electron–phonon interactions [1–3]. This is particularly true because of the extreme range of broken-symmetry ground states that are achieved by varying M (Pt, Pd, Ni) and X (Cl, Br, I, etc.) as well as ligands, pressure, magnetic field—ground states ranging from a strongly disproportional CDW (e.g. PtCl) to a weak CDW (e.g. PtI) to SDW or spin-Peierls (and other mixed CDW/SDW) [4]) phases (e.g. NiBr, NiCl), in addition to long-period 'superlattice' structures [5]. Here we focus on the important consequences of mixing two CDW materi-

Elsevier Sequoia

als ($PtX_xX'_{1-x}$) with different strength of CDW amplitudes. We have modeled effective isolated single chains of MX materials in terms of a 3/4-filled, 2-band Peierls extended-Hubbard model. In an isolated mixed-halide chain [6] we replace a segment, containing m X atoms, by X′ atoms where X,X′ = Cl, Br, I.

Photoinduced charge separation

In this section we briefly describe photoinduced charge separation in $PtCl_xBr_{1-x}$ both as observed in resonance Raman (RR) experiments and simulations within the Peierls-Hubbard model.

In the RR experiment, upon photolysis of mixed-halide materials, photoinduced charged defects were produced in high concentration. Figure 1 depicts the RR spectrum probed at 1.34 eV of a mixed $PtCl_{0.75}Br_{0.25}$ crystal before and after photolysis at 2.54 eV (beyond the band edge of pure PtCl). This is an excitation energy (1.34 eV) where resonance enhancement is known to occur for the P^+ and P^- in PtCl, and a region of pre-resonance for the electron bipolaron and P^- in PtBr. Prior to photolysis the $Cl-Pt^{IV}-Cl$ symmetric stretch chain mode at 308 cm^{-1} dominates the spectra. Weak features at ~285 and 325 cm^{-1} are attributed to P^- defects and edge state modes in PtCl segments, respectively. In the spectral region for PtBr, modes at 213 and 181 cm^{-1} are observed which are vibrational signatures for PtBr edge states. The broad feature at 150 cm^{-1} is ascribed to a P^- in

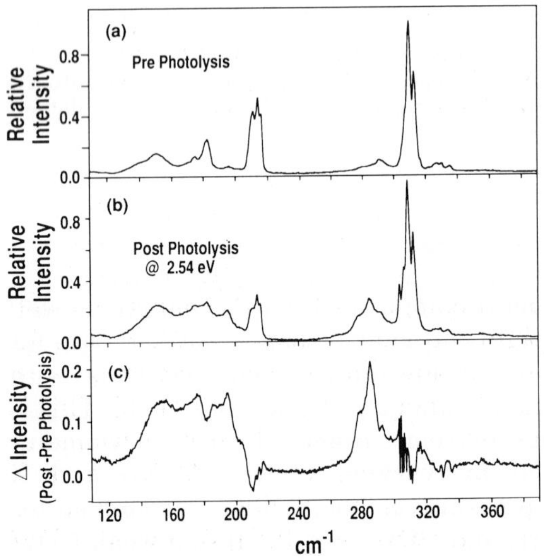

Fig. 1. Resonance Raman (RR) spectra at 1.34 eV of $PtCl_{0.75}Br_{0.25}$ mixed-halide crystal (a) before and (b) after photolysis with 2.54 eV excitation at 2 mW for 1.5 h at 13 K. (c) Difference spectra: post-photolysis intensity minus pre-photolysis intensity.

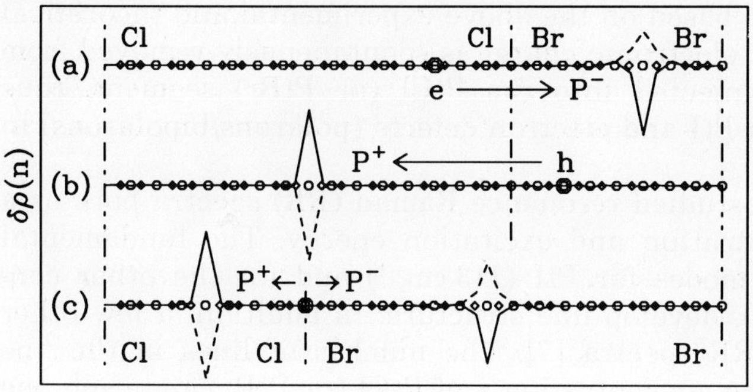

Fig. 2. Excess charge (——) and spin (– – –) densities with respect to the mixed-chain ground state showing polaron migration for (a) $P_{\uparrow}^{-}$ and (b) $P_{\downarrow}^{+}$ on a 16 PtCl and 8 PtBr unit chain, and (c) subsequent to photoexcitation of an up spin electron between levels 36 and 37 on an 8 PtCl and 16 PtBr unit chain. The large circles indicate initial defect location, though the result is insensitive to initial condition.

a PtBr segment. Photolysis within the band gap of PtCl and in the ultragap region of PtBr (2.54 eV) causes an increase in P^+ defects localized within the PtCl segments at $\sim 285\ cm^{-1}$. These results are in contrast to the photolysis of pure materials: in pure crystals of PtCl, RR studies show an increase in *both* P^- ($263\ cm^{-1}$) and P^+ ($287\ cm^{-1}$) local modes. Significantly, in the mixed materials, *no* P^- features appear in the PtCl segments with photolysis. In the PtBr segments, the P^- defect at $150\ cm^{-1}$ also increases in intensity. In a second experiment, photolysis on a mixed crystal of $PtCl_{0.95}Br_{0.05}$ exhibited similar results [6].

To illustrate charge separation within our model we systematically studied the lattice relaxation. Upon doping, if we initially placed a P^- in the PtCl segment (i.e. the 'wrong' segment) and allowed the system to self-consistently evolve to the minimum energy configuration, we found the P^- migrated to a Pt^{IV} site in the PtBr segment (Fig. 2). A P^+ placed in the PtBr segment similarly migrated to a Pt^{II} site in the PtCl segment. Analogously, the photogenerated exciton was invariably unstable and broke into a P^+ and a P^-, the P^+ migrating to the PtCl segment and its companion P^- to the PtBr segment. Both doping and photoexcitation lead to charge separation with a specific polaron charge selectivity consistent with experimental observations. The analysis of electronic wavefunctions revealed strong hybridization of PtCl and PtBr electronic bands as the driving force for charge separation.

Intrinsic self-doping

Apart from the $PtCl_xBr_{1-x}$ mixed-halide system discussed above we also explored $PtCl_xI_{1-x}$ and $PtBr_xI_{1-x}$ systems. Quite interestingly, these systems exhibit qualitatively different behavior such as intrinsic 'self-doping'. In other words, based on the above experimental and theoretical techniques we find that electronic charge is spontaneously removed from the PtI segment and injected into the PtCl (or PtBr) segment, thus creating hole defects in PtI and electron defects (polarons/bipolarons) in PtCl (or PtBr).

Experimentally we studied resonance Raman (RR) spectra both as a function of PtI concentration and excitation energy. The fundamental symmetric stretch (ν_1) modes for PtI (113 cm^{-1}) and for the other constituent, PtCl (or PtBr), develop fine structure. In addition, a few other lines are seen in the RR spectra [7]. The number of lines in the fine structure varies with the concentration x of PtCl (or PtBr) and each one of these lines comes into resonance at a different excitation energy. We are in the process of assigning specific vibrational modes to these lines based on the above theoretical model. Further, X-ray diffraction experiments in conjunction with thermal ellipsometric studies reveal that, subsequent to the formation of a mixed-halide crystal, the PtI segment contracts while the PtCl (or PtBr) segment expands [8].

Figure 3 depicts excess charge density with respect to the corresponding charge-density-wave (CDW) ground state in each segment as predicted by the two-band Peierls-Hubbard model for $PtCl_xI_{1-x}$ (Fig. 3(a)) and $PtBr_xI_{1-x}$ (Fig. 3(b)). Clearly, one observes formation of electron bipolarons in the PtCl (or PtBr) segment (as well as at the interface) with a concomitant formation of hole bipolarons in the PtI segment. However, since PtI is a very delocalized system (the correlation length of a bipolaron is very long) the hole bipolarons are squeezed into a small segment of PtI. Therefore, the shape of a bipolaron in a PtI segment is not as clearly visible as in the PtCl (or PtBr) segment. The driving force for the ejection of electrons from PtI into PtCl (or PtBr) can be understood from the calculated electronic band structure (not shown here). The highest unoccupied PtCl (or PtBr) electronic state is found to be lower in energy than the lowest occupied Pt band of PtI. The system lowers its electronic energy by pumping electrons into the PtCl (or PtBr) bands from the PtI (self-doping). From this perspective the PtI based mixed-halide systems behave like an intrinsic p–n junction. Further analysis of theoretical results shows that the PtI segment contracts while PtCl (or PtBr) expands consistent with the X-ray diffraction experiments [8]. We are in the process of exploring dielectric, IR, RR, ESR, d.c./a.c. conductivity, nonlinear optical (NLO) and other interesting properties to further characterize these novel electronic materials.

Device applications and conclusions

The above discussion suggests the possibility of fabricating Q1D devices that utilize polaronic charge separation/injection. We emphasize that, from a band structure point of view, mixed-halide chains represent a

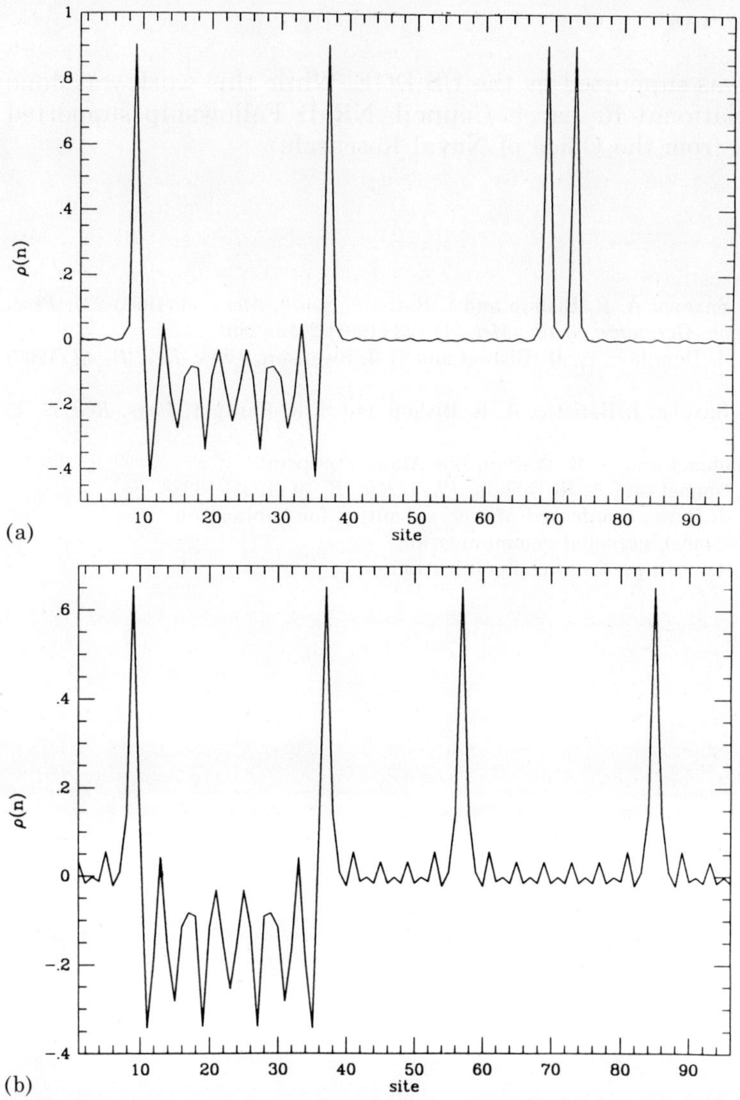

Fig. 3. Excess charge density with respect to the corresponding CDW ground state in each segment for (a) $PtCl_xI_{1-x}$ and (b) $PtBr_xI_{1-x}$.

1D analog of heterojunctions in semiconductors. When light is shone on such a device the photogenerated electrons migrate to the PtBr side of the junction while the holes move to the PtCl side. Thus, an open circuit voltage develops across the device and essentially one has a mixed-halide MX crystal based solar cell. Similarly, as indicated above, PtX_xI_{1-x} chain materials are one-dimensional analogs of p–n junctions. We are also in the process of modeling a compositional superlattice based on the PtCl–PtBr system.

Acknowledgements

This work was supported by the US DOE. While this work was done J.T.G. held a National Research Council–NRaD Fellowship supported through a grant from the Office of Naval Research.

References

1 J. T. Gammel, A. Saxena, A. R. Bishop and I. Batistić, *Synth. Met.*, *54* (1993) 237; *Proc. ICSM '90, Tübingen, Germany; Synth. Met.*, *41–43* (1991) 2715–2801.
2 J. T. Gammel, R. J. Donohoe, A. R. Bishop and B. I. Swanson, *Phys. Rev. B*, *42* (1990) 10566.
3 J. T. Gammel, A. Saxena, I. Batistić, A. R. Bishop and S. R. Phillpot, *Phys. Rev. B*, *45* (1992) 6408.
4 H. Röder, J. T. Gammel and A. R. Bishop, Los Alamos preprint.
5 I. Batistić, J. T. Gammel and A. R. Bishop, *Phys. Rev. B*, *44* (1991) 13228.
6 L. A. Worl *et al.*, *J. Phys.: Condensed Matter*, submitted for publication.
7 S. P. Love (Los Alamos), personal communication.
8 B. L. Scott (Los Alamos), personal communication.

Synthetic Metals, *54* (1993) 391–397

Exciton kinetics in poly(di-n-hexylsilane)

R. G. Kepler
Sandia National Laboratories, Albuquerque, NM (USA)

Z. G. Soos
Princeton University, Princeton, NJ (USA)

J. R. G. Thorne*
University of Pennsylvania, Philadelphia, PA (USA)

Abstract

We have previously shown that charge carriers are generated by exciton–exciton annihilation in solid films of poly(di-n-hexylsilane). Using this phenomenon we show that the exciton–exciton annihilation rate constant γ at ambient temperature is not a function of the photon energy used to create the excitons even at energies well out into the long wavelength tail of the exciton absorption band. We also show that the excitons remain highly mobile throughout their 600 ps lifetime and that they diffuse distances comparable to the crystallite size in the film. The value of γ obtained in these studies is in excellent agreement with that obtained previously in fluorescence intensity studies.

Introduction

Polysilanes are σ-conjugated polymers which have recently been shown to display many interesting electronic properties similar to those exhibited by some of the more familiar π-conjugated polymers. The first allowed electronic transition is a σ–σ^* transition in the near UV, at far lower energy than that exhibited by their σ-bonded carbon counterparts. The polysilanes exhibit reasonably large non-linear optical coefficients [1], are excellent photoconductors [2], and fluoresce in the near UV with a very high quantum efficiency [3]. The absorption spectra of the long, typically several thousand silicon atoms, chains are described in terms of many short, approximately 20 silicon atom, segments of varying length [4]. The electronic states including two-photon allowed transitions [5, 6] are well described within the Hückel approximation, plus electron–electron interactions, with alternating transfer integrals along the Si backbone, $t(1 \pm \delta)$, where $\delta \sim 0.33$, and the two integrals are one between σ orbitals on different Si atoms and the other between σ orbitals on the same Si atom [7].

*Present address: University of Manchester, Manchester, UK.

0379-6779/93/$6.00

The fluorescence emission band is on the low energy edge of the absorption band and is narrower. A number of experiments appear to show that fluorescence is emitted from longer segments which have been populated by energy transfer from the initially excited, shorter, higher energy segments. Studies of the time dependence of fluorescence anisotropy in dilute solutions as well as the crystalline phase have shown that the anisotropy decays only a small amount within the lifetime of the excitation and the results were interpreted as showing that the energy is transferred only from segment to segment along the chain for a relatively short distance [8, 9]. Studies of laser induced photochemical processes in solid films of poly(di-n-hexylsilane) have indicated that energy is transferred between polymer segments in the highly ordered crystallites but not between the crystallites at 1.4 K [10]. Similarly, site selective fluorescence studies at cryogenic temperatures (<10 K) have indicated that as the excitation energy is scanned toward the red edge of the first allowed transition, the exciton transition, energy migration becomes blocked [11].

In earlier work, we have reported that excitons appear to be highly mobile in solid polysilane films [12]. This conclusion was based on studies of fluorescence intensity versus excitation intensity in which a saturation effect was interpreted as evidence for exciton–exciton annihilation. In PDHS the interaction rate constant, γ, was found to be about 1×10^{-7} cm^3 s^{-1} and, in poly(n-propylmethylsilane), independent of temperature down to about 20 K [13]. This value for γ is suprisingly large since it is ten times larger than that found for γ in anthracene single crystals [14].

In this paper we present some results obtained from studies of carrier generation by exciton–exciton annihilation, a process which we recently discovered [15], and discuss their implications for exciton kinetics.

Experimental

Polymer samples for these studies were cast from PDHS toluene solutions onto a semitransparent aluminum electrode which had been vacuum deposited onto a UV quartz substrate. A second, thick aluminum, electrode was vacuum deposited on top of the cast film. An electric field of 1×10^5 V cm^{-1} was applied to the polymer film and the number of carriers generated in the film by a pulse of light was measured by integrating the amount of charge displaced from the electrodes by the pulse with an electrometer or a charge sensitive preamplifier. Two light sources were used, a Nd:YAG pumped dye laser and a tripled Nd:YAG laser with a 30 ps pulse. The pulse length of the dye laser was 5 ns.

Two types of experiment were performed. One in which the number of carriers generated was measured as a function of photon energy and of light intensity at each photon energy. The other involved the illumination of the sample with two superimposed 30 ps, 355 nm pulses and the number of carriers generated was measured as a function of the difference in time of arrival of the two pulses.

In order to compare our photon energy dependence results with theoretical predictions it was important that we obtain a reasonably accurate measure of the absorption coefficient versus photon energy. We used a standard spectrophotometer to measure the optical density of several thin films and measured the film thickness with a Dektak. This procedure worked well until we reached photon energies well out on the low energy side of the exciton absorption band where scattering becomes a significant factor. We attempted to use photothermal-deflection spectroscopy [16] to extend our range but found that the quantum efficiency for fluorescence is too high. For absorption coefficients, α, less than $10^4\ cm^{-1}$ on the low energy side of the absorption band we assumed that the number of carriers generated at a given light intensity was proportional to α^2. We had shown that the number of carriers generated by light pulses of a given photon energy in this region varied as the square of the intensity, so we believe that this is a reasonable assumption.

Results

All the experimental results, both the carrier generation results presented here and the results of the fluorescence intensity experiments published earlier, have been analyzed using the following rate equation for the exciton concentration

$$\frac{dn}{dt} = G(t,z) - \beta n - \gamma n^2 \tag{1}$$

where n is the exciton concentration, G an exciton generation term which depends on time and position in the sample, β is the reciprocal of the exciton lifetime and γ is the exciton–exciton annihilation rate constant. We assume that exciton annihilation occurs when two excitons interact to create one highly excited state, most of which rapidly decay radiationlessly down to the original exciton state but a small fraction results in a free carrier pair of which only the hole is mobile.

In the dye laser experiments the laser pulse length was 5 ns. Since the exciton lifetime is 600 ps, we assumed that an equilibrium concentration was reached and set $dn/dt = 0$ in eqn. (1). With this assumption, it can be shown that the total number of exciton–exciton annihilation events per absorbed photon F is

$$F(A, d) = \frac{H-4}{H} - \frac{4}{A(1-\exp(-\alpha d))} \ln \frac{1}{H} \times [1-(1-\exp(-\alpha d))(\{A+1\}^{1/2}-1)] \tag{2}$$

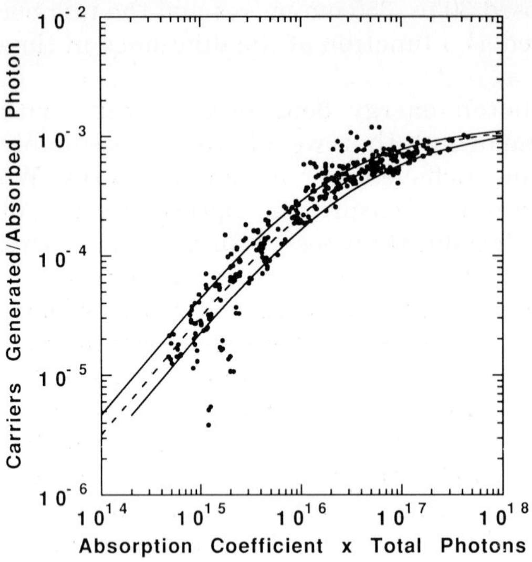

Fig. 1. The number of carriers generated per absorbed photon vs. the absorption coefficient times the incident light intensity in total photons per pulse. The upper solid line is the prediction of eqn. (2) for $\alpha d \ll 1$, the lower solid line for $\alpha d \gg 1$ and the dashed line for $\alpha d = 1$.

where d is the sample thickness, $A = 4\alpha\gamma I/\beta^2$ with I being the intensity of the incident light in photons $\mathrm{cm}^{-2}\,\mathrm{s}^{-1}$ and H is given by

$$H(A, d) = (A + 1)^{1/2} + (A \exp(-\alpha d) + 1)^{1/2}$$

In fitting the data we have assumed that the beam area is 0.4 cm^2 and that the intensity is constant over that area and zero elsewhere.

Equation (2) predicts that within a factor of two, all the dye laser experimental data should be a function of only αI. In Fig. 1 we have plotted all the data for photon energies below 4.5 eV. Above 4.5 eV a linear process, presumably a band-to-band transition, becomes important. The predictions of eqn. (2) with $\beta = 1.67 \times 10^9\ \mathrm{s}^{-1}$, $\gamma = 2 \times 10^7\ \mathrm{cm}^3\,\mathrm{s}^{-1}$ and a quantum efficiency for carrier generation by exciton–exciton annihilation of 1.3×10^{-3} are shown as the two solid and one dashed lines. The upper solid line is for $\alpha d \ll 1$, the lower $\alpha d \gg 1$, and the dashed line is for $\alpha d = 1$. These results are clearly in excellent agreement with the earlier estimate [12] of $1 \times 10^{-7}\ \mathrm{cm}^3\,\mathrm{s}^{-1}$ for γ.

For the double 30 ps pulse experiments we assumed in eqn. (1) that G is a delta function for each pulse and it can be shown that eqn. (1) then predicts that the ratio of the total number of carriers produced by the two pulses, Q, separated in time Δt, divided by the sum of the number of carriers produced by the two, equal intensity pulses individually, $2q$, is approximately

$$\frac{Q(\Delta t)}{2q} = 1 + \frac{\exp(-\beta t)}{(1 + B)^2} \tag{3}$$

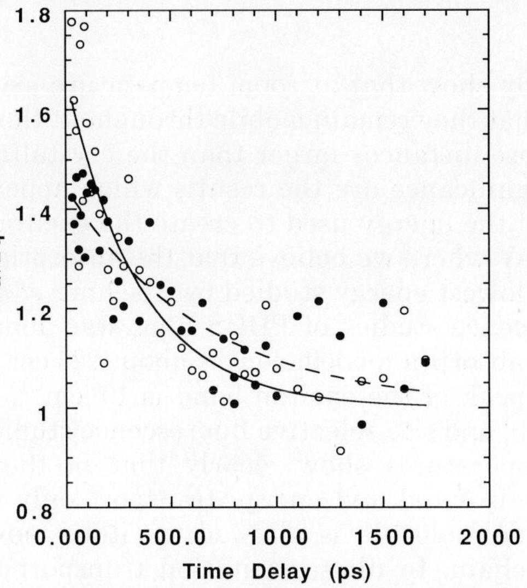

Fig. 2. The total number of carriers generated by two 30 ps pulses divided by the sum of the number of carriers generated by each pulse independently vs. the difference in the time of arrival of the pulses. The open circles were data taken with the polarization of the light in the two pulses parallel and the solid circles with polarizations perpendicular. The solid line is a least-squares fit of eqn. (3) to the open circles and the dashed line a fit to the solid circles.

if $B < 1$ at the sample surface where $B = n\gamma/\beta$. For our experiments B was about 0.25.

In Fig. 2 we have plotted some results from the double 30 ps pulse experiments. The open circles are the data obtained when the polarization of the light in the two pulses is parallel and the solid circles with the polarizations perpendicular. The solid line is a least-squares fit of eqn. (3) to the parallel polarization data and the dashed line a fit to the perpendicular polarization data. The solid line gives an exciton lifetime of 350 ps while the dashed line is a lifetime of 650 ps. Fluorescence lifetime measurements on films cast from the same batch of PDHS exhibited a lifetime of 600 ps which was essentially independent of the exciting and fluorescing wavelength. We believe that the difference between the two sets of data arises because the polymer film is crystalline and with parallel polarizations the excitation is highly inhomogeneous while with perpendicular polarizations the excitation is more homogeneous. With inhomogeneous excitation the excitation concentration decays more rapidly because the excitons diffuse out of the excited region as well as decay. If this interpretation is right, the excitons are rapidly diffusing out of the crystallites in which they are created. We do not know how large the crystals are in our film but studies of the morphology of PDHS show that in carefully prepared films the lamella are about 0.1 μm [17], ten times thicker than in most polymers.

Discussion

These results appear to clearly show that at room temperature excitons in PDHS are highly mobile, that they remain mobile throughout their 600 ps lifetime and that they diffuse distances larger than the crystallite size in our films. Of particular significance are the results which appear to show that γ is not a function of the energy used to create the excitons even for photon energies of 3.165 eV where we believe that the absorption coefficient is about 100 cm^{-1}. The lowest energy studied by Elschner *et al.* [11] in their site-selective fluorescence studies of PDHS films was about 3.18 eV where we believe that the absorption coefficient is about 225 cm^{-1}. The absorption coefficient at the peak of the exciton band is 10^5 cm^{-1}.

The fluorescence anisotropy [9] and site-selective fluorescence studies [11] on glasses at very low temperatures show clearly that in these systems, where the molecules are isolated and energy transport only to neighboring segments on the same molecule is allowed, excitons move only a short distance along the chain. In films, excitation transport to neighboring segments on different chains by the Förster mechanism is possible and perhaps much more probable than to neighboring segments on the same chain. Fluorescence anisotropy measurements were not made on films and the site-selective fluorescence studies of PDHS films appear to be less conclusive.

Our studies were conducted at ambient temperature. However in other studies we found little or no temperature dependence for γ in poly(n-propylmethylsilane) films [13], but we have not made such measurements in PDHS. In this work, we assumed an α to match our data. Perhaps absorption by chains at defect sites becomes dominant far into the red edge and these states produce the fluorescence while excitons created in the crystalline region have a high γ but do not interact with the defect sites.

Fluorescence intensity measurements of γ in PDHS at low temperature with excitation in the red edge are needed to resolve these issues.

Acknowledgements

The technical assistance of P. M. Beeson is gratefully acknowledged. The thermal deflection spectroscopy measurements were made by C. H. Seager. The work at Sandia National Laboratories was supported by the US Department of Energy under Contract No. DE-AC04-76DP00789 and the work at Princeton University was partially supported by NSF Grant No. NSF-DMR-8921072.

References

1 F. Kajzar, J. Messier and C. Rosilio, *J. Appl. Phys.*, *60* (1986) 3040.

2 R. G. Kepler, J. M. Zeigler, L. A. Harrah and S. R. Kurtz, *Phys. Rev. B*, *35* (1987) 2818.
3 L. A. Harrah and J. M. Zeigler, *Macromolecules*, *20* (1987) 601.
4 K. S. Schweizer, *Synth. Met.*, *28* (1989) C565.
5 J. R. G. Thorne, Y. Ohsako, J. M. Zeigler and R. M. Hochstrasser, *Chem. Phys. Lett.*, *162* (1989) 455.
6 Z. G. Soos and R. G. Kepler, *Phys. Rev. B*, *43* (1991) 11908.
7 Z. G. Soos and G. W. Hayden, *Phys. Rev. B*, *40* (1989) 3081.
8 Y. R. Kim, M. Lee, J. R. G. Thorne, R. M. Hochstrasser and J. M. Zeigler, *Chem. Phys. Lett.*, *145* (1988) 75.
9 J. R. G. Thorne, R. M. Hochstrasser and J. M. Zeigler, *J. Phys. Chem.*, *92* (1988) 4275.
10 H. P. Trommsdorff, J. M. Zeigler and R. M. Hochstrasser, *Chem. Phys. Lett.*, *154* (1989) 463.
11 A. Elschner, R. F. Mahrt, L. Pautmeier, H. Bässler and M. Stolka, *Chem. Phys.*, *150* (1991) 81.
12 R. G. Kepler and J. M. Zeigler, *Mol. Cryst. Liq. Cryst.*, *175* (1989) 85.
13 R. G. Kepler, *Synth. Met.*, *28* (1989) C573.
14 R. G. Kepler, in N. B. Hannay (ed.), *Treatise on Solid State Chemistry*, Vol. 3, Plenum, New York, 1976, p. 615.
15 R. G. Kepler and Z. G. Soos, in L. Y. Chiang, A. F. Garito and D. J. Sandman (eds.), *Electrical, Optical and Magnetic Properties of Organic Solid State Materials, Symposium Proceedings Series*, Vol. 247, Materials Research Society, Pittsburgh, 1992, p. 655.
16 C. H. Seager and C. E. Land, *Appl. Phys. Lett.*, *45* (1984) 395.
17 F. C. Schilling, F. A. Bovey, A. J. Lovinger and J. M. Zeigler, in J. M. Zeigler and F. W. G. Fearon (eds.), *Silicon Based Polymer Science*, Vol. 224, Advances in Chemistry Series, American Chemical Society, Washington, DC, 1990, p. 341.

9. Devices

9. Devices

Conjugated polymer electroluminescence

D. D. C. Bradley
Cavendish Laboratory, Madingley Road, Cambridge CB3 0HE (UK)

Abstract

In this review paper we present some research results on the use of conjugated polymers for electroluminescence (EL). In the introduction a comparison is made between the relative potential for conjugated polymers and the various other classes of EL material. We then discuss the structure and operation of conjugated polymer based devices and the progress that has been made in enhancing their efficiency. The importance of obtaining balanced carrier injection and transport is demonstrated by the effectiveness of multilayer heterostructure devices. A discussion of the likely non-radiative decay processes competing with EL is possible on the basis of earlier studies of the photophysics of the polymer emissive layers. Specific examples of molecular level polymer engineering in copolymer systems are presented and their use to allow lateral patterning of a film is considered. The range of colours available through tuning the polymer $\pi-\pi^*$ energy gap is reported and a discussion of possible future directions for research is given.

Introduction

Conjugated polymer electroluminescence is a rapidly expanding field of interest [1–15], and there is a strong possibility that these materials will find commercial use in a variety of display applications. Since the first literature report on poly(*p*-phenylene vinylene) (PPV) (see Fig. 1) polymer devices [1], following an earlier patent application [2], there has been a rapidly increasing number of publications demonstrating the ubiquity of the phenomenon amongst fluorescent conjugated polymers, with many of the materials anticipated in the patent application now experimentally confirmed as suitable for use in EL devices [4, 6, 7]. Important and continuing improvements in all features of device performance are being reported, and the colour range accessible already includes the three colours (red, green and blue) required for full colour displays [4, 6, 7, 8, 10, 14]. The combination of polymer processing techniques for low cost film deposition and an emissive light output operating at low d.c. drive voltages with high brightness and good efficiency is very attractive for large area displays. Moreover, unlike many other sectors of the electronics industry, large area EL panels do not at present have a firmly entrenched material of choice.

The competition faced by conjugated polymers for EL applications falls into four main categories [16], namely (i) inorganic semiconductor

0379-6779/93/$6.00

light emitting diodes, (ii) fluorescent ion doped inorganic semiconductor thin film devices, (iii) inorganic semiconductor powder devices, and (iv) sublimed molecular dye devices [17, 18]. Commercial products are available based on (i), (ii) and (iii) but there are drawbacks with all three, especially for large area applications.

Light emitting diodes based on forward biased inorganic semiconductor p–n junctions find widespread use in the display panels of many electrical machines, but are limited to applications where small area devices can be used. Growth of uniform samples with sufficiently low concentrations of non-radiative centres for use in large area devices is difficult and turns out to be uneconomic. Colours for commercial devices are limited to the green and longer wavelengths. There are also problems in finding materials which combine good p–n junction formation with high radiative recombination yields. GaP diodes utilize a zinc–oxygen defect complex to provide red emission with a relatively high efficiency of $\approx 1\%$. Quantum well devices give higher efficiencies but at a high fabrication cost.

Fluorescent ion doped inorganic semiconductor thin film devices operate via a different mechanism. Here, electrons or holes are accelerated to high enough energies to ionize or impact excite the fluorescent ion centres, e.g. Mn, Eu, etc. High fields are required for this process and difficulties in making uniform thin films of the inorganic semiconductors, e.g. ZnS, SrS, etc., lead to a requirement for high operating voltages, which increases the cost of driver circuitry for displays. Efficiencies of $\approx 2\%$ are achieved but higher efficiencies do not look feasible. Deposition by sublimation allows fabrication of large area devices, at a cost, and changing the fluorescent ion allows some colour tuning. There remains a problem over instability of the semiconductor film in part due to the high fields required for operation.

Inorganic semiconductor powder devices, as exemplified by the ZnS:Cu system dispersed in various polymer matrices, are the most efficient ($\approx 3\%$) of the commercially available visible EL materials, and the simplest to deposit. They suffer, however, from requiring a high voltage a.c. drive (increasing the cost of drive circuitry), from poor stability, and from little colour tunability. The mechanism of operation in these devices is poorly understood.

Sublimed molecular dye based devices triggered the recent resurgence of interest in organic materials for EL [17, 18]. Devices fabricated with an electron transporting emissive layer and a hole transport layer [17], or with both electron and hole transport layers sandwiching the emissive layer [18], have been shown to provide high brightness outputs under low voltage d.c. biases. The efficiencies of these devices are $\approx 1\%$ and red, green and blue outputs have all been demonstrated [17, 18]. Stability was initially a serious concern for these devices but recent reports [19] suggest that many of the problems are being ironed out as the efficiency improves, and the drive voltages are lowered.

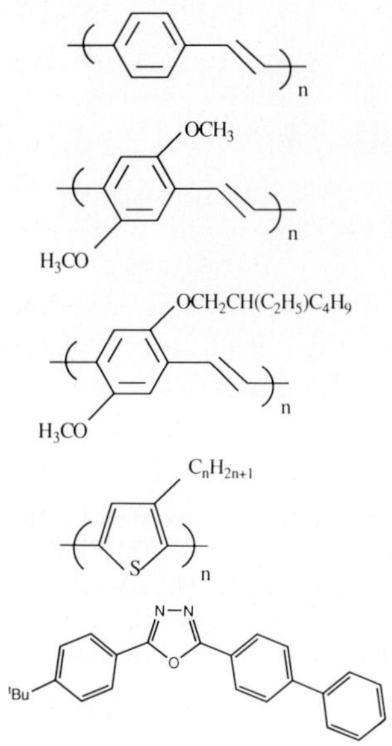

PPV

PDMeOPV

MEH-PPV

P3AT

PBD

Fig. 1. Chemical structures of selected conjugated polymer EL materials: poly(*p*-phenylene vinylene) (PPV); poly(2,5-dimethoxy-*p*-phenylene vinylene (PDMeOPV); poly(2-methoxy-5-(2′-ethyl-hexoxy)-*p*-phenylene vinylene) (MEH-PPV); poly(3-alkyl thiophene) (P3AT); and the electron transport molecule 2-(4-biphenylyl)-5-(4-tert-butylphenyl)-1,3,4-oxadiazole (PBD).

Conjugated polymers are clearly related to the latter category of material, and indeed some of those recently used are oligomeric analogues of PPV [20]. The polymers are distinguished by not requiring the presence of more than one organic layer in the device structure, though this may be advantageous under certain circumstances, and by the greater intrinsic stability of the emissive layer against morphological rearrangements during device operation.

Figure 1 shows the chemical structures of the fluorescent conjugated polymers and the transport molecule that are discussed in this article. We will not discuss here the details of polymer synthesis or film deposition and processing. References are given to earlier papers where these details may be found.

Conjugated polymer electroluminescent devices

The first conjugated polymer device to be reported [1] consisted of a very simple sandwich structure in which a PPV film was deposited on top of an aluminium electrode on which a thin oxide layer had been allowed to form. A second aluminium layer was then evaporated on top of the PPV film. In forward bias (top aluminium contact at negative potential, bottom

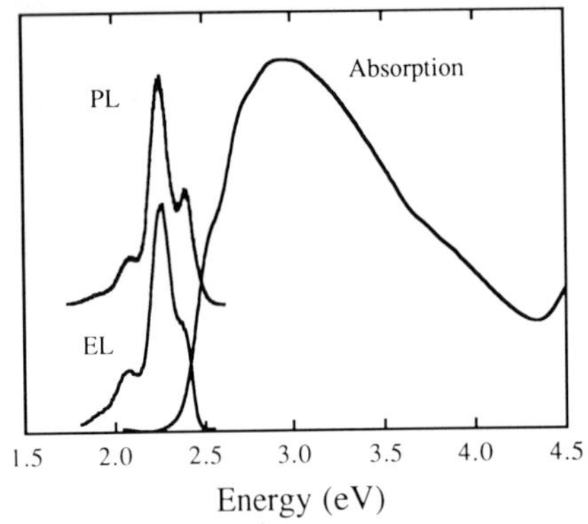

Fig. 2. PPV absorption and emission spectra. Photoluminescence (PL) was excited by the 457.9 nm output from a CW Ar^+ ion laser. Electroluminescence (EL) is for an ITO:PPV:Ca device. Very similar results are seen for other electrode combinations, e.g. ITO:PPV:Al. The different levels of self-absorption are clearly seen in the different lineshapes of the highest energy (0–0) vibronic peak. In the EL spectrum this peak shows up only as a shoulder whereas in the PL spectrum it is clearly resolved.

contact at positive potential) the device showed a turn on at about 25 V (threshold field 2×10^6 V/cm) with a yellow green emission seen under room lighting for currents of the order of 100 mA/cm^2. The efficiency calculated in terms of photons emitted per electronic charge transported was less than or of the order of 0.01%, a value too small to be of interest for applications. The EL emission spectrum, shown in Fig. 2, closely resembles the photoluminescence (PL) spectrum of PPV [21], a fact which implies that the same excited state, namely a singlet exciton, gives rise to both emissions. This in turn strongly suggests that the EL results from double charge injection with positive and negative carriers travelling though the polymer film and combining to form an exciton. An estimate of the theoretical limit on the efficiency achievable with PPV devices can be made from consideration of spin statistics. For an exciton formed by non-geminate pair combination (as in EL) there are four possible ways to combine the half-integer spins of the two charge carriers. Of these three give a resultant spin of one, i.e. a triplet state, and only one gives a zero-spin singlet state. The triplet state is not linked by a dipole allowed transition to the ground state and thus does not contribute to EL emission. In the case of geminate pair combination (which dominates for PL), the resultant state is always a singlet. We thus expect that the EL yield will be less than or equal to one quarter that observed in PL. This treatment is over simplistic, ignoring as it does both the influence of the energy difference between singlet and triplet excited states on the combination process and also the possibility for interconversion between them. It does however provide a useful rough guide to what can be expected. For PPV the PL efficiency is estimated, from the dominant component in the decay which has time constant ≈ 250 ps [22], to be as much as 25%. This assumes that the radiative lifetime is ≈ 1 ns, a reasonable figure given the known decay times of related oligomeric molecules [23], and the observation of components in the decay with characteristic times of up to this value but not in excess of it [22]. These data suggest therefore that an EL efficiency of up to 6% is a realistic target, a figure that would make PPV highly attractive for commercial applications. The question to be answered is how can such an enhanced value be realized? Other important questions that must be considered are: (i) how to control the device characteristics to allow bright emission at low drive voltages and currents?, (ii) how to tune the colour of emission?, and (iii) how to engineer multi-colour, multi-pixel elements and more complex device structures desired for application purposes? The rest of this article illustrates the lines that have been followed in Cambridge to achieve some of these goals. Reference is also made where appropriate to the work of other research groups active in this field.

The approaches to modifying the device characteristics fall into two main categories, namely (i) electrode engineering and (ii) polymer engineering. The former concerns optimization of the electrodes to match, as far as possible, and within the limits of the available materials, the electronic structure of the polymer. The latter offers scope for molecular level engineering of quantum efficiency, colour, and transport and injection properties, and for supramolecular level engineering both of device structure through lateral patterning, and of transport and injection properties through multilayer device fabrication.

Device physics and characterization

We consider two processes in the operation of our devices, firstly formation of the exciton and secondly its decay. The former encompasses (i) carrier injection, (ii) carrier transport and (iii) carrier combination, and the latter involves a competition between radiative and non-radiative decay channels.

Formation of excitons

Balanced injection and transport are prerequisites for highly efficient operation. If either of these is not achieved then an unproductive (without emission) current is likely to flow and recombination will become localized in a region at, or near to, one of the polymer/electrode interfaces where quenching effects may be important. To ensure balanced injection it is necessary to equalize the barriers that arise at the two polymer/electrode interfaces due to the difference in the workfunction of the positive (negative) electrode and the ionization potential (electron affinity) of the polymer. Moreover, to ensure that injection is readily achieved at low drive voltages we want these barriers to be small. There is unfortunately

TABLE 1

Examples of suitable electrode materials for conjugated polymer electroluminescence devices

Hole injection (high ϕ)	Electron injection (low ϕ)
Indium–tin oxide (ITO)	Al
$Al{:}Al_2O_3$	Mg/Ag
Au	Ca
Pt	In
p-type Si	n-type Si
etc.	etc.

very little experimental data in the literature on the ionisation potentials and electron affinities of conjugated polymers and the values reported for theoretical calculations show a large spread. This makes it rather difficult to select electrode materials with workfunctions that match the polymer electronic structure. There is in fact also a large uncertainty in the work function values for the electrode materials, with significant variations due to different deposition conditions and due to composition differences in, for example, the chemical structure of indium–tin oxide coatings.

For the combination of electrode materials used in the first PPV device discussed above the aluminium acts as an electron injecting contact and the aluminium with oxide surface layer as a hole injecting contact. Many other possible combinations of electrode materials exist and several examples for both electron and hole injection materials are listed in Table 1.

The use of a transparent ITO film as the hole injection electrode allows the EL emission to readily exit from the device, an obvious requirement for applications. Varying the electrode materials used in PPV devices produces interesting results [11]. The negative electrode material has been varied in devices with ITO coated glass as the positive electrode material. For ITO:PPV:Al devices the efficiency was 0.002%. Changing Al to Ca*, a metal with a significantly lower workfunction, results in an enhancement in efficiency to 0.1%. Changing to gold which has a larger workfunction leads to an efficiency of only 0.000005%! and, indeed, reversing the bias for ITO:PPV:Au devices such that Au becomes the hole injecting electrode gives better results, with an efficiency of 0.0001%.

These variations are consistent with estimates for the expected barriers taking electrode workfunctions $\phi_{Au} \approx 5.3$ eV, $\phi_{ITO} \approx 4.8$ eV, $\phi_{Al} \approx$

*The use of Ca as an electron injecting electrode was first reported by Braun and Heeger [4] who were able to achieve efficiencies of 1% for the highly fluorescent dialkoxy substituted PPV, poly(2-methoxy-5-(2′-ethyl-hexoxy-*p*-phenylene vinylene) (MEH-PPV) (see Fig. 1).

4.2 eV, $\phi_{Ca} \approx 2.87$ eV, polymer ionization potential I_p(PPV) ≈ 5.1 eV, and polymer electron affinity E_a(PPV) ≈ 2.6 eV. Using these values we get a hole injection barrier between ITO and PPV of 0.3 eV, and electron injection barriers of 2.7 eV for PPV:Au, 1.6 eV for PPV:Al and 0.27 eV for PPV:Ca. For Au:PPV:ITO, with Au as the positive electrode, the hole and electron injection barriers are estimated to be −0.2 and 2.2 eV. It must however, be re-emphasized that these values are not known accurately. Moreover, little is currently known about the nature and extent of any interfacial layers formed between the polymer film and the electrode material. For PPV prepared via the precursor route, HCl is a by-product of the thermal conversion, a substance widely used to etch ITO. Consequently, even though most of the HCl should escape from the film under conversion *in vacuo*, there is still likely to be an interfacial layer at the ITO electrode surface. Similarly, it has recently been shown that, for example, evaporation of Al onto PPV results in part in a chemical reaction of the Al with the polymer, which again may lead to an interfacial layer [24].

The use of a polymer emissive layer lends itself to fabrication of devices on a variety of substrates and these need not be planar or rigid. ITO can readily be deposited on many different substrates, including flexible plastic sheet and curved preforms. Other electrode materials, including, for example, doped conjugated polymer films [25], can also be used to allow fabrication of flexible diode structures. This diversity in device construction is an additional attractive feature of conjugated polymer EL devices.

To achieve balanced carrier transport requires that the injected positive and negative charge carriers have similar mobilities. The evidence to date is, however, that positive carrier transport is strongly favoured. Thus, for instance, in time-of-flight mobility measurements no negative carrier transport signals are observed [26]. Very low negative carrier mobilities are seen in many organic solids due in part to the strong trapping capability of oxgyen [27]. Additional evidence for the imbalance comes from EL spectra for which the amount of self-absorption, observed as an asymmetry in the 0–0 emission line, is significantly stronger than seen in PL spectra (see Fig. 2). For PL, recombination occurs in regions that extend through the sample but with regions near to the illuminated surface weighted more strongly in accordance with the concept of an absorption depth. The greater self-absorption of EL is consistent with that emission arising from recombination which, on average, takes place deeper within the film, i.e. closer to the negative electrode. There is then a long self-absorption path for the emission viewed through the ITO coated glass substrate.

In order to compensate for the imbalance in carrier transport it is possible to fabricate devices that incorporate carrier transport layers whose properties complement those of the emissive polymer [12]. For PPV we require an electron transporting material, of which there are rather few organic examples. Guided by earlier work on heterostructure devices utilising sublimed molecular films [18], we have chosen the molecule 2-(4-biphenylyl)-5-(4-tert-butylphenyl)-1,3,4-oxadiazole (PBD) (see Fig. 1). A device was constructed starting with ITO coated glass onto which a 3500 Å PPV layer was formed; a 300 Å PBD layer was then deposited on top of the PPV in the form of a solid state dispersion in poly(methylmethacrylate) (PMMA). Finally a Ca negative electrode was evaporated on top of the PBD/PMMA layer. The latter layer has the dual role of transporting negative charges away from the negative electrode and also of inhibiting the transport of positive carriers straight through the device. This inhibition of positive carrier leakage currents is expected to have the added benefit of causing the build up of a positive space charge at the PPV:PBD/PMMA interface that will increase the field across the PBD and should give enhanced electron injection (see Fig. 3). With this electron-conducting, hole-blocking layer the efficiency of the device increased from 0.1 to 0.8% at brightnesses of the order of 500 Cd/m^2. At lower brightness the efficiency was 1.1%, slightly higher, with the decrease in efficiency with increasing brightness possibly attributable to onset of hole leakage currents, alteration of the PBD/PMMA through sample heating, or exciton quenching in the presence of a high space charge density.

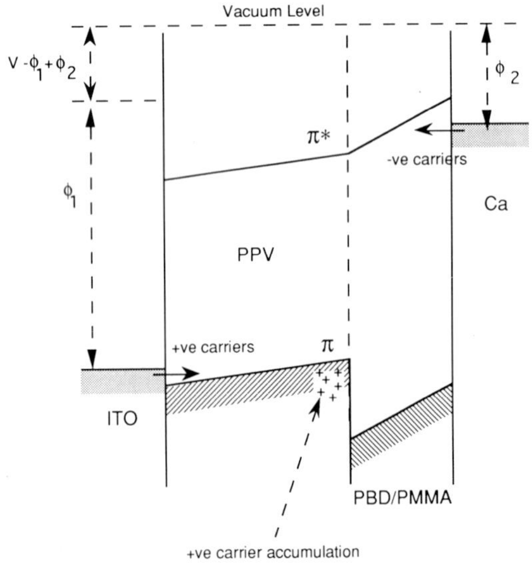

Fig. 3. Schematic electron level structure for ITO:PPV:PBD/PMMA:Ca multilayer heterostructure device under forward bias of V volts, ϕ_1 and ϕ_2 are the workfunctions of ITO and Ca, respectively. In the absence of experimental information, the full potential offset between PPV and PBD has for simplicity been drawn at the π electron level. An accumulation of +ve charge at the PPV:PBD/PMMA interface is indicated, and the resulting expected imbalance in the field distribution is illustrated.

The third step in exciton formation involves the combination of oppositely charged carriers. In conjugated polymers these carriers are expected to be polaron-like self-localized excitations [28]. For non-degenerate ground state materials, such as PPV, there are expected to be two carrier types, namely singly charged polarons and doubly charged bipolarons, with the latter favoured at high charge density. The energy levels of polarons and bipolarons lie within the $\pi-\pi^*$ energy gap, with bipolaron levels more strongly displaced into the gap. This has consequences for exciton formation since the energy of the excition level is essentially the $\pi-\pi^*$ gap energy. The question is then whether it is energetically favourable for either two polarons or two bipolarons to combine to form an exciton. No theoretical studies have yet been reported addressing this point. In the case of two polarons the answer is likely to be yes, but the situation of two bipolarons is much less clear. Consideration should, however, be given to the effects of the high electric fields ($\approx 10^6$ V/cm), present during device operation, upon the extent of lattice relaxation around the injected charges. It is conceivable that lattice relaxation is reduced, making exciton formation more likely.

Also of consequence to the process of exciton formation are the spin states of the charge carriers that combine. We have already used arguments based on spin statistics to discuss the theoretical limit on EL efficiency. The importance of spin state to the combination processes has been directly observed in optically detected magnetic resonance experiments that detect microwave induced transitions between the Zeeman split levels of the various excitations generated following charge injection [13]. Electroluminescence, photoluminescence, and conductivity detected magnetic resonance signals have been recorded that support the assignment of a double charge injection process as the mechanism of operation in PPV devices [13].

Exciton decay

As a consequence of the fact that the same singlet exciton gives rise to both EL and PL we can make use of the many earlier studies of photoluminescence [21] to gain an insight into the processes competing with EL emission. The reason why PL yields in fluorescent conjugated polymers are not as high as in their small molecule organic dye analogues is that the non-radiative decay processes are more efficient. As already commented above the quantum yield for PPV films is estimated to be of the order of 25% at room temperature. This compares with yields approaching 100% for small molecule analogues in solution, e.g. bis-isopropyldistyrl-benzene, effectively a dimer of PPV, has a quantum yield of 94% [23]. What distinguishes fluorescent conjugated polymers like PPV from non-fluorescent ones like poly(2,5-thienylene vinylene) [29], is currently a topic of great interest both from the viewpoint of scientific curiosity and now also from the viewpoint of selecting materials for potential use in EL devices. In general there are many possible non-radiative decay mechanisms that are well known in organic solids. Amongst these are included: (i) multiphonon emission, (ii) intersystem crossing from singlet to triplet manifold, (iii) exciton–exciton fusion at high excitation densities, (iv) migration to quenching sites, (v) 1^1B_u to 2^1A_g excited state energy transfer, (vi) self-trapped exciton formation and tunelling to ground state configuration coordinate surface, etc. Mechanisms (ii), (iii) and (iv) are relevant to PPV and its derivatives as evidenced by photoinduced absorption experiments in which there are clear triplet–triplet absorption signals [21], from the intensity dependence of PL decay curves [21], and from a variety of experiments [30] showing quenching of PL in the presence of dopant induced, photoexcited and field effect device injected, charge carriers. For mechanism (v) to be important correlation effects must be sufficiently large that the first ^1A_g excited state lies lower in energy than the first ^1B_u excited state. This is expected to be the case in *trans*-polyacetylene [31] and calculations are currently being undertaken on other polymer systems to assess whether this particular excited state ordering is likely to occur [32]. Mechanism (vi) has been discussed as a model to reconcile the variations in fluorescence yield seen for different polydiacetylene polymers and alkyl-substituted polythiophenes but there is little direct evidence to support the existence of the self-trapped exciton state [33].

Spin dependent magnetic resonance measurements can provide extremely useful information on the details of excitation dynamics and the interconversion of singlet and triplet states. Photoinduced absorption detected magnetic resonance measurements have given important confirmation of the nature of excited states in conjugated polymers [34] and a combination of electroluminescence and conductivity detected magnetic resonance measurements on EL diode structures provides a very useful means to study in detail the processes involved in device operation and how they may be controlled. Preliminary results have already revealed several interesting features, and show the unexpected presence of two distinct triplet states [13].

Colour tuning and film patterning

Control of the emission colour is clearly a requirement for the achievement of full-colour displays. Emission from fluorescent conjugated polymers is generally observed as a broad band, with superposed vibronic structure, that emerges at energies just below the onset of the $\pi-\pi^*$ absorption spectrum (see Fig. 2). The approach to colour tuning is thus to select known examples of fluorescent conjugated polymers or synthesize new ones that have $\pi-\pi^*$ energy gaps at energies just above those desired for the emission. PPV has a $\pi-\pi^*$ energy gap of ≈ 2.5 eV and its EL emission is yellow. To tune to the red one selects polymers with lower $\pi-\pi^*$ energy gaps. Examples of materials satisfying this criterion include dialkoxy substituted PPV polymers such as poly(2,5-dimethoxy-*p*-phenylene vinylene) (PDMeOPV) [35] (see Fig. 1) and MEH-PPV ($\pi-\pi^*$ energy gaps $\approx$2.1 eV) [4], poly(3-alkyl thiophenes) (see Fig. 1) such as poly(3-dodecyl-thiophene) ($\pi-\pi^*$ energy gaps $\approx$2.1 eV) [6, 36], and various copoloymers, e.g. of PPV and PDMeOPV ($\pi-\pi^*$ energy gaps in the range $\approx$2.1 to 2.5 eV) [9, 11]. Tuning towards the blue involves choosing polymers with larger $\pi-\pi^*$ energy gaps. Examples of these include poly(*p*-phenylene) ($\pi-\pi^*$ energy gap $\approx$2.7 eV) [7], poly(9,9-dihexylfluorene) ($\pi-\pi^*$ energy gap $\approx$2.9 eV) [8], and copolymers containing interruptions in the conjugated sequences, e.g. thermally converted methoxy leaving group PPV [10, 14].

Copolymers offer considerable flexibility for fine-tuning the emission colour. We have studied copolymers of PPV and PDMeOPV and have seen a linear shift in the emission spectra with varying composition [9]. These copolymers in fact allow both red and blue shifts in emission from that seen for PPV. The red shift is as expected following the introduction of segments of PDMeOPV, which as a homopolymer has a lower $\pi-\pi^*$ energy gap than PPV [35]. The blue shift is not expected and is achieved by careful control of the *in vacuo* elimination reaction that converts the precursor polymer into its conjugated form. The synthesis (see Fig. 4) of the copolymer precursors introduces two different leaving groups with one of these selectively introduced on the PDMeOPV precursor units (polymer **I** in Fig. 4). The elimination of this methoxy group requires more severe conditions than for elimination of the standard tetrahydrothiophenium (THT) moiety. In particular, acid is required as a catalyst to remove the methoxy groups (yielding polymer **III**) whereas heat alone suffices to eliminate the THT groups. As a consequence, if films are subject only to *in vacuo* thermal conversion (from polymer **I** to **II**), the PPV precursor units are preferentially converted with the PDMeOPV units remaining largely unconverted. The result is a polymer in which limited sequences of PPV units are separated from each other by non-conjugated segments (polymer **II**). This conjugated/non-conjugated polymer has a larger $\pi-\pi^*$ energy gap on account of the limited extent of conjugation, and is thus blue shifted relative to PPV.

These copolymers offer an additional attractive feature in that the control of the conversion reaction allows patterning of a film [9]. To selectively control, within a single film, the areas that undergo full conversion and those that undergo conversion of only the PPV sequences is simply a matter of laying down a patterned capping layer of, for example, aluminium. This can be done by standard photolithographic techniques using a lift-off process. The sample is then thermally treated *in vacuo* under the standard conditions for elimination of THT leaving groups. The difference here is that the capping layer acts to trap HCl, a by-product of the elimination of THT leaving groups, and this acid then catalyses the elimination of the methoxy leaving groups. Consequently in the capped regions full conversion (to polymer **III**) occurs whilst in the uncapped regions the conversion is limited to PPV sequences (yielding

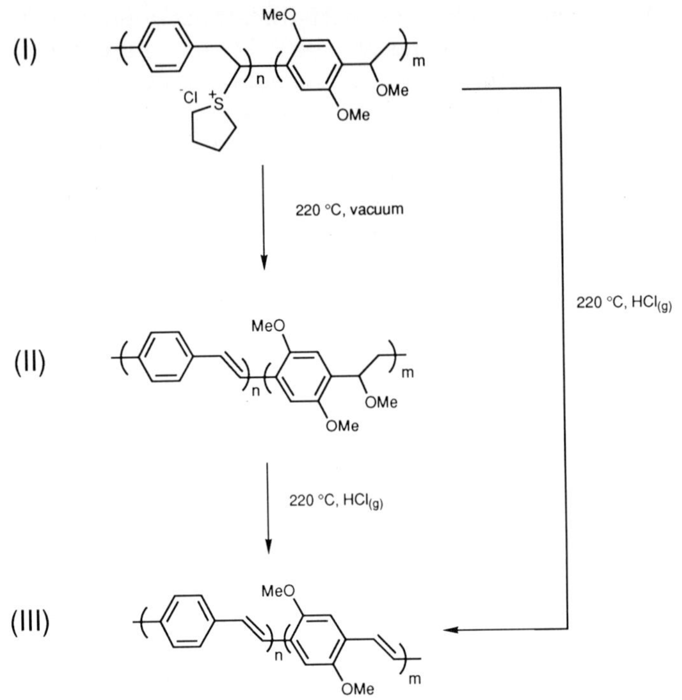

Fig. 4. Precursor route synthesis of PPV-PDMeOPV copolymers. The precursor **I** contains two different leaving groups with methoxy moieties preferentially present next to methoxy substituted phenylene rings. Thermal conversion *in vacuo* yields the conjugated/non-conjugated copolymer **II**, which requires subsequent heating under HCl gas to catalyse the elimination of the methoxy moieties in order to generate the fully conjugated polymer **III**. Polymer **III** may also be reached direct from precursor **I** by thermal conversion in the presence of HCl gas. N.B. HCl is a by-product of the elimination of the tetrahydrothiophenium moiety, and judicious control of its presence allows control of the compositional balance between polymer structures **II** and **III**.

polymer **II**). As a result, patterns can be made in the copolymer film, with the different regions possessing different $\pi-\pi^*$ energy gaps and different refractive indices. This approach offers the possibility of creating multicolour elements within a single film and of defining waveguide structures.

Assessment and conclusions

The extremely rapid progress to data in the development of conjugated polymer electroluminescence bodes well for the future application of these materials. Red, green and blue emission colours have all been demonstrated. For PPV, efficiences have risen by two orders of magnitude from a paltry 0.01% to a new 'best' figure of 2% (albeit at low brightness operation) for a device incorporating an electron transport layer [37].

This figure is already competitive with commercial EL materials. Brightnesses in excess of 1500 Cd/m^2 are now achievable, sufficient for interior display applications, and the lower brightnesses of the order of 100 Cd/m^2 needed for computer screen displays are readily achieved at moderate current levels. Threshold voltages for light emission of 4 V and less can also be demonstrated. Moreover, these figures are not expected to be the ultimate performance figures for these materials. Improvements in efficiency to values of 5% are anticipated and concomitant improvements in power consumption should follow. The aromatic based precursor-route polymers such as PPV and derivatives appear well suited to fabrication of robust devices and their processing procedures allow large area device fabrication on a range of different substrates and patterning to form multi-colour elements within a single film. It is important to emphasise that lifetime tests performed on the devices that we currently fabricate in a standard laboratory environment are unlikely to give good indications of the lifetime achievable for devices fabricated under rigorous manufacturing (e.g. clean room) conditions. Nevertheless, devices have been run under experimental conditions for periods of up to 3 days without significant degradation in performance. Further purification of the polymers and improved deposition techniques should also assist. The stability is expected to increase steadily as the device performance figures improve since, for example, improved injection and transport should lead to a reduction in the electric fields required for device operation and increased efficiency will result in less heat dissipation.

What needs to happen now is the simultaneous optimization of all the device parameters. In addition more information is required on the nature of the interfaces between the inorganic electrodes and the organic layers and how these affect device operation, and also how they may be controlled. There is also considerable scope for chemical optimisation of the emissive polymer layer to enhance its fluorescence quantum yield and of the transport layers to provide efficient carrier transport with high carrier mobilities.

Finally, there remain many interesting aspects of the science of these devices to be investigated including the nature of the charge carrier (polarons versus bipolarons and the extent of lattic relaxation), the different processes that allow changes in spin state between the singlet and triplet manifolds, and the reasons behind the very different fluorescent yields seen for various conjugated polymers.

Acknowledgements

The work described above is that of a team of researchers involving collaboration between the Cavendish and University Chemical Laboratories of Cambridge University. I gratefully acknowledge the contributions of colleagues A. R. Brown, P. L. Burn, J. H. Burroughes, R. H. Friend,

N. C. Greenham, R. W. Gymer, A. B. Holmes, A. M. Kraft and R. N. Marks to this programme of research. I thank the UK Science and Engineering Research Council and Cambridge Research and Innovation Ltd. for their support of this work.

References

1 J. H. Burroughes, D. D. C. Bradley, A. R. Brown, R. N. Marks, K. MacKay, R. H. Friend, P. L. Burn and A. B. Holmes, *Nature (London)*, *347* (1990) 539.
2 J. H. Burroughes, R. H. Friend and D. D. C. Bradley, *Br. Patent Applic. No. 8909011.2* (1989); *Int. Patent Applic. No. PCT/GB90/00584* (1990).
3 D. D. C. Bradley, A. R. Brown, P. L. Burn, J. H. Burroughes, R. H. Friend, A. B. Holmes, K. D. Mackay and R. N. Marks, *Synth. Met.*, *43* (1991) 3135.
4 D. Braun and A. J. Heeger, *Appl. Phys. Lett.*, *58* (1991) 1982.
5 D. D. C. Bradley, A. R. Brown, P. L. Burn, R. H. Friend, A. B. Holmes and A. Kraft, *Electronic Properties of Polymers, Springer Solid State Sciences 107*, Springer, Berlin, 1992, p. 304.
6 Y. Ohmori, M, Uchida, K. Muro and K. Yoshino, *Solid State Commun.*, *80* (1991) 605; *Jpn. J. Appl. Phys.*, *30* (1991) L1938.
7 G. Grem, G. Leditzky, B. Ullrich and G. Leising, *Adv. Mater.*, *4* (1992) 36.
8 Y. Ohmori, M. Uchida, K. Muro and K. Yoshino, *Jpn. J. Appl. Phys.*, *30* (1991) L1941.
9 P. L. Burn, A. B. Holmes, A. Kraft, D. D. C. Bradley, A. R. Brown, R. H. Friend and R. W. Gymer, *Nature (London)*, *356* (1992) 47.
10 P. L. Burn, A. B. Holmes, A. Kraft, D. D. C. Bradley, A. R. Brown and R. H. Friend, *J. Chem. Soc., Chem. Commun.*, (1992) 32.
11 P. L. Burn, A. Kraft, A. B. Holmes, A. R. Brown, D. D. C. Bradley and R. H. Friend, *Proc. MRS Meet., Boston, MA, USA, Dec. 1991.*
12 A. R. Brown, D. D. C. Bradley, P. L. Burn, J. H. Burroughes, R. H. Friend, N. C. Greenham, A. B. Holmes and A. M. Kraft, *Appl. Phys. Lett.*, (1992) in press.
13 L. S. Swanson, J. Shinar, A. R. Brown, D. D. C. Bradley, R. H. Friend, P. L. Burn, A. Kraft and A. B. Holmes, *Phys. Rev. B*, (1992) in press.
14 A. R. Brown, P. L. Burn, D. D. C. Bradley, R. H. Friend, A. M. Kraft and A. B. Holmes, *Proc. UPS'91, Okazaki, Japan*; *Mol. Cryst. Liq. Cryst.*, *216* (1992) 111.
15 S. Karg, W. Riess, V. Dyakonov and M. Schwoerer, *Synth. Met.*, *54* (1993) 427.
16 J. W. Allen, *Proc. Int. Conf. Electroluminescence, 1992.*
17 C. W. Tang and S. A. VanSlyke, *Appl. Phys. Lett.*, *51* (1987) 913; *J. Appl. Phys.*, *65* (1989) 3610.
18 C. Adachi, T. Tsutsui and S. Saito, *Appl. Phys. Lett.*, *55* (1989) 1489; *56* (1990) 799; *57* (1990) 531.
19 T. Tsutsui, personal communication.
20 C. Hosokawa, H. Tokailin, H. Higashi and T. Kusumoto, *Acta Polytech. Scand.*, *170* (1990) 219.
21 N. F. Colaneri, D. D. C. Bradley, R. H. Friend, P. L. Burn, A. B. Holmes and C. W. Spangler, *Phys. Rev. B*, *42* (1990) 11670, and refs. therein.
22 I. D. W. Samuel, B. Crystall, G. Rumbles, P. L. Burn, A. B. Holmes and R. H. Friend, *Synth. Met.*, *54* (1993) 281.
23 I. B. Berlman, *Handbook of Fluorescence Spectra of Aromatic Molecules*, Academic Press, New York, 1971.
24 W. R. Salaneck, personal communication.
25 G. Gustafsson, Y. Cao, G. M. Treacy, F. Klavetter, N. Colaneri and A. J. Heeger, *Nature (London)*, *357* (1992) 477.

26 T. Takiguchi, D. H. Park, H. Ueno and K. Yoshino, *Synth. Met.*, *17* (1987) 657; J. Obrzut, M. J. Obrzut and F. E. Karasz, *Synth. Met.*, *29* (1989) E103; P. Strohriegl and D. Haarer, *Makromol. Chem. Macromol. Symp.*, *44* (1991) 85.
27 M. Stolka and M. A. Abkowitz, *Synth. Met.*, *54* (1993) 417.
28 A. J. Heeger, S. Kivelson, J. R. Schrieffer and W.-P. Su, *Rev. Mod. Phys.*, *60* (1988) 781.
29 A. J. Brassett, N. F. Colaneri, D. D. C. Bradley, R. A. Lawrence, R. H. Friend, H. Murata, S. Tokito. T. Tsutsui and S. Saito, *Phys. Rev. B*, *41* (1990) 10586.
30 D. D. C. Bradley and R. H. Friend, *J. Phys. Condensed Matter*, *1* (1989) 3671; H. B. Gu, T. Takiguchi, S. Hayashi, K. Kaneto and K. Yoshino, *J. Phys. Soc. Jpn.*, *54* (1987) 3997; K. E. Ziemelis, A. T. Hussain, D. D. C. Bradley, R. H. Friend, J. Rühe and G. Wegner, *Phys. Rev. Lett.*, *66* (1991) 2231.
31 B. S. Hudson, B. E. Kohler and K. Schulten, in E. Lim (ed.), *Excited States 6*, Academic Press, New York, 1982, p. 1.
32 Z. Soos, *Synth. Met.*, *54* (1993) (these proceedings).
33 T. Kobayashi, *Synth. Met.*, *54* (1993) 75.
34 Z. V. Vardeny, *Synth. Met.*, *54* (1993) (these proceedings).
35 H. S. Woo, S. C. Graham, D. A. Halliday, D. D. C. Bradley, R. H. Friend, P. L. Burn and A. B. Holmes, *Phys. Rev. B*, *46* (1992) 7379.
36 J. Rühe, N. F. Colaneri, D. D. C. Bradley, R. H. Friend and G. Wegner, *J. Phys. Condensed Matter*, *2* (1990) 5465.
37 N. C. Greenham, personal communication.

Trap-free charge transport in polymers, with tunable mobilities

Milan Stolka and Martin A. Abkowitz
Xerox Corporation, Webster Research Center, Webster, NY 14580 (USA)

Abstract

Some insulating polymers with the ability to transport injected or photogenerated charge carriers also display useful features which until recently were believed to be unobtainable in organic compounds and are difficult or sometimes even impossible to achieve with classical solid state materials. Among these features are: (i) the existence of perfectly ohmic contacts with adjacent electrodes, (ii) trap-free non-dispersive transport with high carrier range ($\mu\tau$) far exceeding 10^{-5} cm^2/V, (iii) the ability to control drift mobilities from less than 10^{-10} to about 10^{-1} cm^2/V s without introducing deep traps and without increasing conductivity, and (iv) no detectable degradation of transport characteristics when space-charge-limited currents (SCLC) are passed through the specimens for long periods of time. These features are briefly reviewed and discussed for representative materials.

Introduction

Organic electronic materials are gaining recognition as viable device materials and are being rapidly developed for practical applications. Polymers capable of transporting photoinjected charge carriers are already used commercially as components of organic photoreceptors and thus support a $>\$50 \times 10^9$ per year copying and laser printing business. The purpose of this communication is to demonstrate that these polymers not only transport charge carriers with reasonably high drift mobilities in the range of 10^{-6} to 10^{-4} cm^2/V s, which are sufficient for application in electrophotographic receptors, but also have certain other characteristics which are often very difficult to duplicate even with well understood electronic materials [1, 2]. Among these characteristics are the ability to form perfectly ohmic contacts with metallic or carbon electrodes, trap-free non-dispersive transport and, above all, the opportunity to reduce or increase the drift mobility over many orders of magnitude without changing the carrier range, and without increasing the conductivity. These characteristics and the methods of manipulation of drift mobility are best illustrated in two charge transport materials, a silicon-chain polymer, poly(methylphenylsilylene) (PMPS) and *N,N′*-diphenyl-*N,N′*-bis(3-methylphenyl)-[1,1′-biphenyl]-4,4′-diamine (TPD) and its derivative. The former is an intrinsically transporting polymer with backbone-derived transport states and the latter is a transport-active

0379-6779/93/$6.00

small molecule molecularly dispersed in a transport-inactive polycarbonate binder.

Ohmic contacts

Ohmic contact between an electrode and a charge transporting insulator is operationally defined as a contact which is capable of supplying enough carriers to sustain space-charge-limited current (SCLC). For a layer of a unipolar system (most organic polymers can transport only one sign carriers and all practical polymers transport holes), and provided the carrier transit time ($t_{Tr} = d^2/\mu V_0$) is less than the dielectric relaxation time, the ohmic contact will, in the absence of deep traps, supply a steady state SCLC [3]

$$I_{(SCLC)} = (9/8)\,\varepsilon\mu E^2/d \qquad (1)$$

where d is the sample thickness, ε is permittivity, μ is the carrier drift mobility and E is the electric field. If the measured current is in a trap free system, the existence of SCLC and *not* a linear I versus E characteristic is an indication of ohmicity of the contact. (Linear I versus E characteristic can be observed even when the contacts are emission limited, i.e. non-ohmic.) The measured current would be less than the expected SCLC if the material has hole traps, and free carriers would provide additional ohmic conductivity at low electric fields.

Figure 1(a) demonstrates that hole injection from a carbon black electrode and transport in TPD/polycarbonate take place under trap-free space charge limited conditions (TFSCLC) [4, 5]. Observed steady state

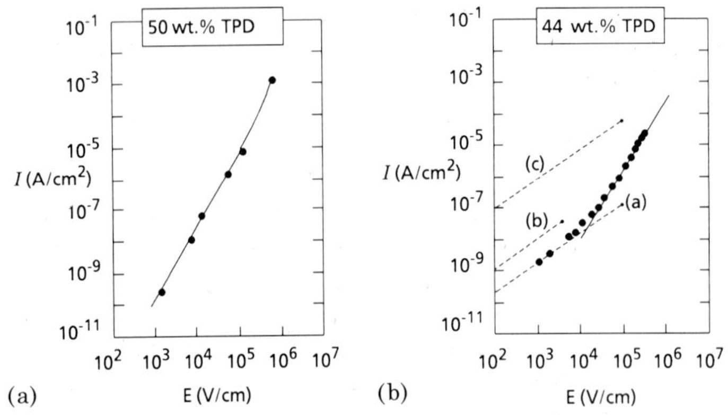

Fig. 1. (a) I vs. E. plot for 50% TPD/polycarbonate. Solid line: calculated TFSCLC from eqn. (1) using μ from independent TOF measurements [4]. (b) I vs. E. plot for 44% TPD/polycarbonate. (a), (b) and (c) indicate increasing levels of free carriers introduced by doping with $SbCl_5$. For (c) the free carrier density is estimated to be $2 \times 10^{15}/cm^3$ [5].

currents (data points) perfectly coincide with currents calculated using independently determined mobility values (solid line). Figure 1(a) also shows that there is no apparent contribution of conductivity to the current. On the other hand, when free carriers are deliberately introduced to the system by doping with traces of oxidizing agents, the low-field portion of the I versus E curve assumes a linear dependence on E [5] as shown in Fig. 1(b). Figure 1 thus provides self-consistent evidence of trap-free transport, the absence of free carriers and the existence of an ohmic contact.

In the case of TPD/polycarbonate, ohmic contacts exist not only with carbon black and Au electrodes but also with Ag.

Trap-free non-dispersive transport

Other indicators of trap-free non-dispersive transport are the quality of TOF transients obtained in relatively thick films and the independence of mobility on the sample thickness. Figure 2(a) shows the TOF transients with well developed plateaux of steady motion of the carrier packets through the bulk of about 25 μm thick films of TPD and PMPS. Well developed TOF transients were seen even at low fields of 10^2 V/cm [4] and with 300 μm thick films. Figure 2(b) shows that μ is independent of thickness [4, 6]. Charges transit through thick films with no detectable trapping. One can therefore assume that the carrier range in TPD is much greater than the minimum value of 10^{-5} cm^2/V.

Throughout the history of insulator physics, the existence of trap-free conditions has be viewed as an idealization. However, many organic polymeric charge transport media now display transport which is essentially free of traps with the carrier range (mobility–lifetime product $\mu\tau$) larger than 10^{-5} cm^2/V [5].

Control of drift mobility

The carrier mobilities μ (drift velocities per unit electric field) in most known hole transport materials are in the range of 10^{-6} to 10^{-4} cm^2/V s

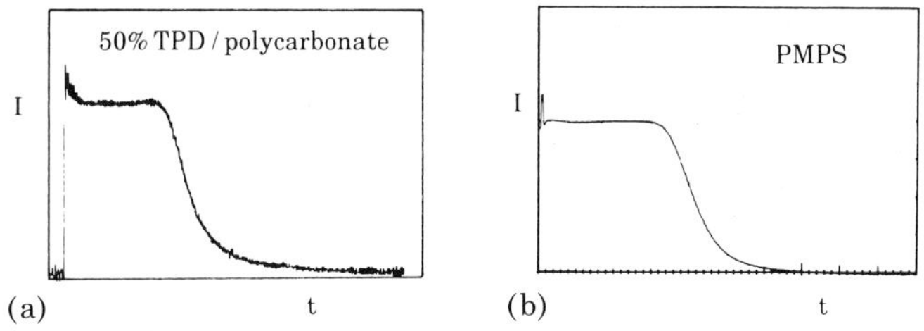

Fig. 2. TOF transients for TPD/polycarbonate (a) and PMPS (b).

at room temperature and an electric field of 10^5 V/cm. Under certain circumstances, however, it may be desirable to tune the carrier mobility μ to lower, or eventually higher values, as needed. Organic transport polymers provide this unique opportunity. The methods of controlling the drift mobilities are as follows.

Variation of the intersite hopping distance

It has been established that charge transport in a broad class of amorphous solids including a-Se and organic solids is an electric-field enhanced, thermally assisted sequence of electron transfers among localized states [2]. The energies of these localized states are most likely organized in a Gaussian distribution the width of which is controlled by disorder. In organic systems, these states are associated either with the transport-active molecules such as TPD or transport-active groups in polymers such as carbazole in poly(*N*-vinycarbazole) (PVK), or segments of σ-bonded silicon chain in polysilylenes, etc. The charge transfer rates among these discrete sites depend on the average separation distance between the transport-active moieties, the electric field, temperature and the chemical nature of the polymer. The dependence of μ on these variables can be expressed by an empirical formula [7]

$$\mu = a_0 \rho^2 \exp(-2\rho/\rho_0) \exp(\Delta/kT) \exp[\beta E^{1/2} (1/kT - 1/kT_0)] \qquad (2)$$

where ρ is the mean calculated distance between the hopping sites, ρ_0 is the wavefunction decay length, E is the electric field, Δ is the zero-field transport activation energy, T is temperature and T_0 is the empirically determined critical temperature. At a constant electric field and temperature, a linear relationship between log μ and the separation distance ρ is predicted

$$\ln[\mu(E, T = \text{const.})/\rho^2] \approx -2\rho/\rho_0 \qquad (3)$$

Figure 3(a) illustrates that the hole mobility in TPD/polycarbonate can be varied by about seven orders of magnitude just by changing the average hopping distance, or the concentration of the transport-active molecule. Figure 3(b) illustrates an agreement with eqn. (3).

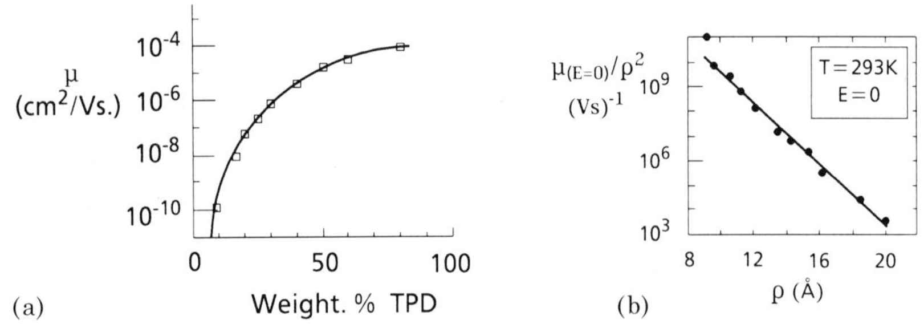

Fig. 3. (a) Hole mobility μ as a function of concentration of TPD in PC at 2.5×10^5 V/cm and 295 K. (b) Hole mobility μ as a function of the intersite distance.

It has to be noted that the desired change in the hole drift mobility by varying the average intersite distance can be achieved without decreasing the range ($\mu\tau$) and without increasing the conductivity σ.

It is obvious that this method of controlling the charge carrier mobilities is applicable only for 'molecularly doped' polymers. For polymers with intrinsic charge transporting capability such as poly(*N*-vinylcarbazole) (PVK), where the position of transport-active groups is fixed by covalent bonding to the chain, or in polymers with a transport-active backbone such as PMPS or poly(phenylene vinylene), this technique is not useful. However in all these polymers, as well as in so called molecularly doped polymers, drift mobilities can be controlled by chemical doping.

Single-trap controlled transport

The primary transport states lay in a narrow Gaussian distribution of energies. The introduction of energetically distinct localized states lying *above* the primary manifold, i.e. states with lower oxidation potential, as illustrated in Fig. 4, will create a situation where the hole localized at the dopant site can only be filled by an electron withdrawn from the neighboring molecule at a lower energy level [8]. This process requires an additional energy above the activation normally associated with the hopping process thus slowing down the carrier migration rate. It is also apparent that the greater the difference in the respective energy levels, the greater the activation energy is needed to de-trap the holes. This is illustrated in an example of PMPS doped with charge transporting molecules with three different oxidation potentials. At an arbitrarily chosen concentration of the dopant molecules (1 wt.%), μ scales qualitatively with the half-way oxidation potential $\varepsilon^{ox}_{1/2}$ of the dopant (Fig. 4(a)). The strongest effect is seen with PPR (1-phenyl-3-(4′-diethylaminostyryl)-5-(4″-diethylaminophenyl)pyrazoline ($\varepsilon^{ox}_{1/2} = 0.43$ V). Doping of PMPS with 1 wt.%

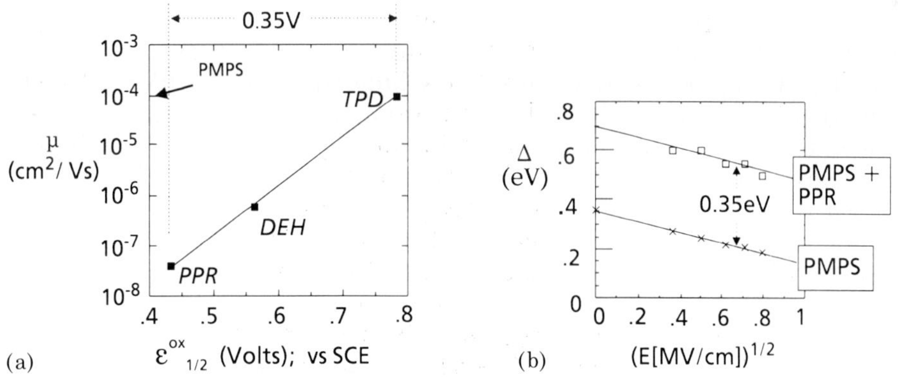

Fig. 4. (a) The effect of $\varepsilon^{ox}_{1/2}$ of the dopant at 1 wt.% on μ in PMPS at 293 K and $E = 3 \times 10^5$ V/cm. (b) Mobility activation Δ for PMPS, undoped and doped with 1 wt.% PPR.

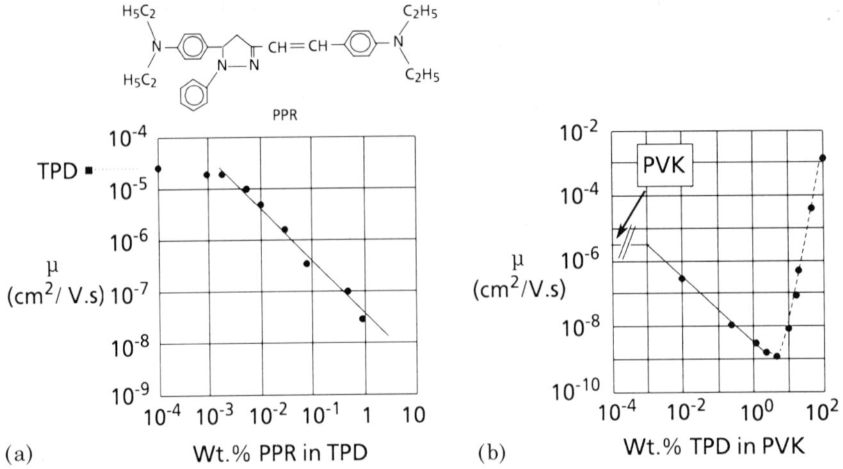

Fig. 5. The effect of dopants on the hole mobility at $E = 3 \times 10^5$ V/cm and 295 K. (a) PPR in 50% TPD/polycarbonate; (b) TPD in PVK.

DEH (*p*-diethylaminobenzaldehyde-diphenylhydrazone) ($\varepsilon^{ox}_{1/2} = 0.56$ V) also leads to the reduction in μ but to a lesser degree. On the other hand, doping with 1 wt.% of TPD ($\varepsilon^{0x}_{1/2} = 0.78$ V) does not affect the mobility in PMPS, which indicates that the transport states in PMPS are approximately at the same energy level as TPD.

The theory of single-trap-controlled hopping [9] predicts that at low concentration of dopants, the mobility decreases with the concentration of the dopant. This is clearly seen in Fig. 5(a) which shows the doping effect of the low $\varepsilon^{ox}_{1/2}$ PPR on hole transport in 50% TPD/polycarbonate [10].

Figure 5(a) also seems to suggest that further reduction in μ could be accomplished by increasing the dopant concentration further. However, as the concentration of the traps is increased, a critical concentration is reached at which the dopant-to-dopant average intersite distance is small enough that hopping among the dopant molecules becomes probable in the time scale of the experiment and thus a new transport channel is created (see Fig. 5(b) which shows how hole transport in PVK is modulated by low $\varepsilon^{ox}_{1/2}$ TPD) [11]. The function of the host system is then reduced to acting as an inert binder for the dopant which is now becoming a transport-active molecule. The extent to which the hole mobility is reduced thus depends, in the first approximation, on the relative difference of the oxidation potentials of the host system and the dopant. However, when the difference is too large, of the order of 0.6 V or greater, the release time for the traps can be so high that the dopant may in effect act as a deep trap. Deep traps are defined as traps with release time greater than the transit time of the carrier. This is, for example, observed when PMPS is doped with *N,N,N′,N′*-tetramethyl-*p*-phenylene diamine (TMPD) ($\varepsilon^{ox}_{1/2} = 0.10$ V). With 1% (by weight) of TMPD, no detectable

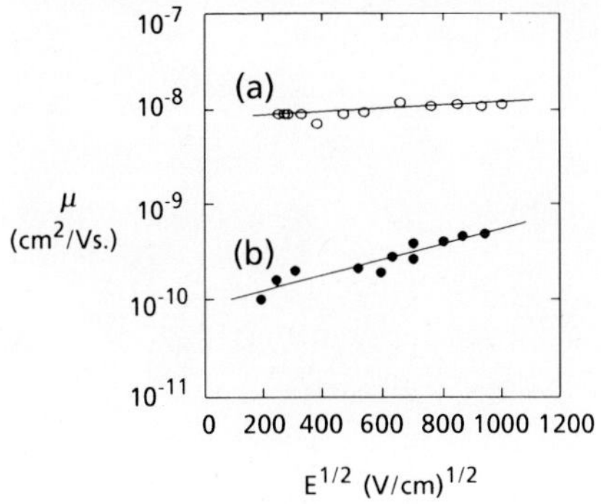

Fig. 6. The binder effect on μ in TPD: (a) polystyrene, (b) polycarbonate.

charge transport is seen since all photoinjected carriers are immobilized in TMPD sites.

The effect of the medium polarity

Recently published data suggest that the hole mobility depends to some extent on the polarity of the medium. For example, TPD, when molecularly dispersed in polystyrene, displays up to two orders of magnitude higher μ than in polycarbonate at the same volume concentration (the same intersite separation distance) [12] (Fig. 6). The difference is especially large at low concentrations of the transport-active molecule. Even the addition of polar transport-inactive molecules to the transporting polymer causes a reduction of the hole mobility [13].

Mobility enhancement

The previous paragraphs suggest that carrier drift mobilities can only be *reduced* by doping. However, as the recent results indicate, under certain circumstances, the hole mobility in the host polymer can actually be increased by heavy doping with molecules which themselves are transport-active. Figure 4 shows that hole mobility in neat (undoped) PMPS at $E = 2.5 \times 10^5$ V/cm and 295 K is 10^{-4} cm^2/V s and that light doping with TPD (1%) has no effect. However, heavy doping with TPD (see Fig. 7) resulted in hole mobilities which are not only greater than in neat undoped PMPS, but greater than TPD would show if it were molecularly dispersed in polycarbonate at the same concentrations (see Fig. 3). A similar synergistic effect was seen when PMPS was ‘doped’ with TPS and PDA (Fig. 8) [14]. The effect is even stronger with a structural variant of TPD, *N,N′*-bis(4-methylphenyl)-*N,N′*-bis(4-ethylphenyl)-[1,1′-(3,3′-dimethyl)biphenyl]-4-4′-diamine (ETPD) where heavy ‘doping’ resulted in hole mobility approaching 10^{-1} cm^2/V s at $E = 2.5 \times 10^5$ V/cm and 295 K which is the highest recorded value for disordered organic systems. In this case the hole mobility in PMPS was enhanced by almost three orders of magnitude.

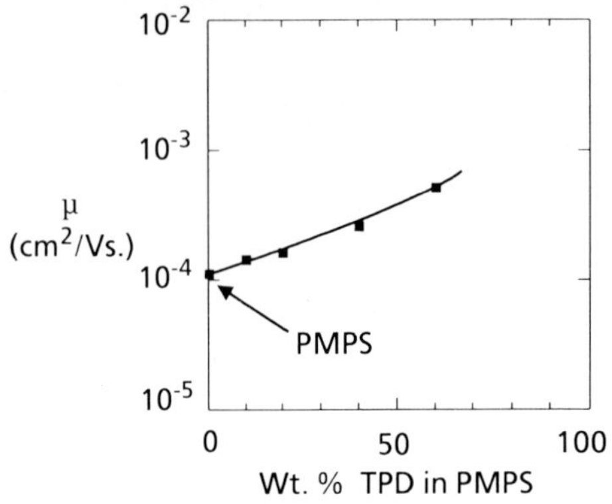

Fig. 7. Enhancement of μ in PMPS by heavy doping with TPD at $E = 2.5 \times 10^5$ V/cm and 295 K.

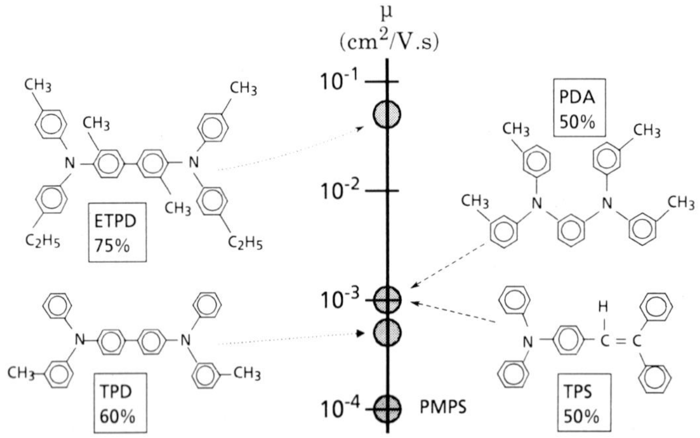

Fig. 8. Enhancement of μ in PMPS by heavy doping with charge transporting molecules: PDA [14], TPS [14], ETPD and TPD. The wt.% numbers indicate the highest practical concentration above which phase separation may occur.

In all observed cases, the reduction or the enhancement of hole drift mobility was accomplished without the concurrent change of the carrier range or conductivity.

Stability of SCLC

Space-charge-limited currents in materials like TPD/polycarbonate are extraordinary stable. Current equilibrium is established in a few transit times and remains stable for many days when the applied bias is left on. No signs of degradation have been detected [5].

Conclusions

As demonstrated by the examples of TPD/polycarbonate and PMPS, some organic charge transport polymers show trap-free non-dispersive charge transport, perfect ohmic contacts with carbon and metallic electrodes and provide the opportunity to tailor the drift mobility from values that are barely measurable ($<10^{-10}$ cm^2/V s) to values as high as 10^{-1} cm^2/V s *without limiting the carrier range* (without introducing deep traps) and *without increasing the conductivity*. These properties in combination with the known fact that these materials can be made conductive by doping with stong electron acceptors may open new possibilities for electronic device application in addition to the current use in electrophotography.

References

1 J. S. Facci, M. A. Abkowitz, W. W. Limburg, F. E. Knier, J. F. Yanus and D. S. Renfer, *J. Phys Chem.*, *95* (1991) 7908.
2 M. A. Abkowitz and M. Stolka, *Philos. Mag. Lett.*, *58* (1988) 239.
3 (a) A. Many and G. Rakavy, *Phys. Rev.*, *126* (1962) 1980; (b) P. Mark and W. Helfrich, *J. Appl. Phys.*, *33* (1962) 205.
4 H.-J. Yuh and M. Stolka, *Philos. Mag. B*, *58* (1988) 539.
5 M. A. Abkowitz and D. M. Pai, *Philos. Mag. B*, *53* (1986) 193.
6 M. Stolka, J. F. Yanus and D. M. Pai, *J. Phys. Chem.*, *88* (1984) 4707.
7 J. X. Mack. L. B. Schein and A. Peled, *Phys. Rev. B*, *39* (1989) 7500.
8 M. A. Abkowitz, M. Stolka, R. J. Weagley, K. M. McGrane and F. E. Knier, *Synth. Met.*, *28* (1989) C553.
9 F. Schmidlin, *Phys. Rev. B*, *16* (1977) 2362.
10 H.-J. Yuh, D. Abramsohn and M. Stolka, *Philos. Mag. Lett.*, *55* (1987) 277.
11 D. M. Pai, J. F. Yanus and M. Stolka, *J. Phys. Chem.*, *88* (1984) 4714.
12 H.-J. Yuh and D. M. Pai, *Mol. Cryst. Liq. Cryst.*, *183* (1990) 217.
13 A. V. Vannikov, A. Yu. Kryukov, A. G. Tyurin and T. S. Zhuravleva, *Phys. Status Solidi (a)*, *115* (1989) K47.
14 K. Yokoyama, S. Notsu and M. Yokoyama, *J. Chem. Soc., Chem. Commun.*, (1990) 805.

Electrical and optical characterization of poly(phenylene-vinylene) light emitting diodes

S. Karg, W. Riess, V. Dyakonov and M. Schwoerer
Physikalisches Institut und Bayreuther Institut für Makromolekülforschung, Universität Bayreuth, 8580 Bayreuth (Germany)

Abstract

We investigated the electrical and optical properties of poly(p-phenylene-vinylene) and light emitting devices based on this polymer. At room temperature these devices can be well described by a Schottky diode configuration. Dark $I-V$ characteristics reveal a good diode behaviour with a maximum rectification ratio of 10^6. White light illumination yields an open-circuit voltage of more than 1 V and a power conversion efficiency of about 0.1%, suggesting the possible application as solar cells. Complex impedance gives information about acceptor dopant concentration, diffusion voltage and depletion width. At low temperatures the $I-V$ characteristics become symmetrical and electroluminescence appears in both current directions, indicating that the Schottky junction is not necessary for the observation of electroluminescence in conjugated polymers.

Introduction

Conjugated polymers are currently considered as materials with a wide range of possible applications, mainly due to their interesting electrical and non-linear optical properties. Although semiconductor devices, for example diodes and MISFETs [1], have been successfully prepared from conjugated polymers, most attention was focused on the high electrical conductivity achievable by treating these materials with reducing or oxidizing agents. A new research interest in the semiconducting, 'undoped' properties of these substances has been stimulated by the observation of electroluminescence (EL) from light emitting diodes based on the conjugated polymer poly(p-phenylene-vinylene) (PPV) [2–4]. Technical application nevertheless implies the knowledge of the charge transport properties in both the polymer itself and the devices fabricated from it. In this contribution we report therefore on the electrical and optical properties of PPV films and PPV-based light emitting devices.

Synthesis and semiconductor properties of PPV

Our PPV was synthesized via a precoursor polymer route similar to that of Burroughes *et al.* [2]. Homogeneous PPV films with a thickness

0379-6779/93/$6.00

TABLE 1

Results of elementary analysis of three different PPV films

Sample[a]	*T* (°C)	C (%)	H (%)	O (%)	Cl (%)	S (%)
1	260	85.85	5.55	0.87	6.03	0.69
2	300	90.20	5.85	0.55	2.60	0.30
3	320	93.40	5.71	0.75	0.12	0.15

[a]Samples 1, 2 and 3 converted at 260, 300 and 320 °C respectively, under a pressure of 1 Pa for 15 h for 1 and 3 and 10 h for 2. The thickness of the freestanding films was between 5 and 10 μm.

between 0.1 and 100 μm and areas up to 20×15 cm^2 were prepared by coating a substrate with the PPV precursor polymer films using a doctor blade technique with subsequent thermal conversion. Essential for the purity of the PPV films and consequently for the electrical and optical properties are the conditions of the thermal elimination [5].

Table 1 shows the results of elementary analysis of three samples converted at different temperatures. Increasing the conversion temperature, decreases the contents of sulfur and chlorine drastically. The d.c. conductivity (Fig. 1) of two samples, taken from the same PPV film as samples 1 and 3 of the elementary analysis, is approximately thermally activated with an activation energy of about 0.6 eV. However, the conductivity of the purer sample is reduced by two orders of magnitude.

All the samples we investigated followed approximately Arrhenius law with activation energies between 0.5 and 0.8 eV. These values are about 0.4–0.7 eV smaller as expected from the optically measured band gap of 2.5 eV. This points to thermally activated hopping in band tails as the dominant mechanism of d.c. charge transport. Therefore, the reduced

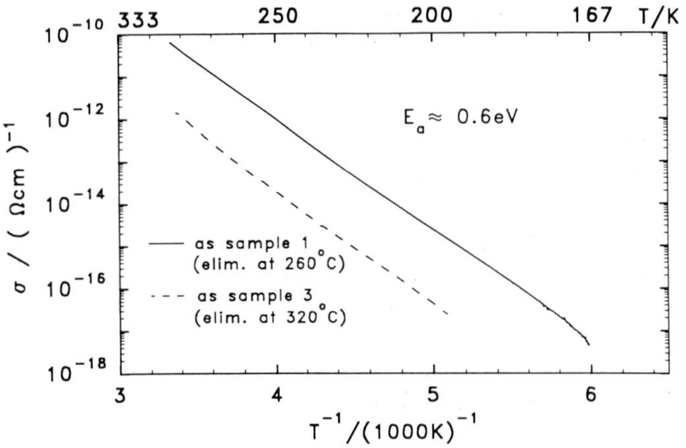

Fig. 1. Temperature dependence of the d.c. conductivity of two samples taken from the same PPV film as samples 1 and 3 of Table 1.

d.c. conductivity of better eliminated samples is, in this model, related to a decrease of defect states in band tails.

Characterization of electronic devices based on PPV

At room temperature the devices can be well described by a Schottky diode configuration with an ohmic contact (ITO, Au), the p-type semiconducting PPV layer (thickness 0.1–1 μm) and a vacuum evaporated rectifying electrode (Al, Mg, Ca). The good processability of PPV makes the fabrication of large emitting diodes possible with active areas of more than 2 cm^2.

I–V Characteristics in the dark and under illumination

Current versus applied voltage characteristics at room temperature in the dark and under illumination are presented in Fig. 2. In the dark the characteristics are asymmetric and we obtain rectification ratios between 10^4 and 10^6, values similar to those of other authors [3]. Usually from *I–V* characteristics conclusions can be drawn about the charge transport mechanism. However, none of the well-known theoretical models [6] yield a good description for our experimental data. We attribute these differences to the influence of a serial bulk resistance together with the existence of an interfacial oxide layer between PPV and the metal electrode.

Under white light illumination a photovoltaic effect is observed. The short-circuit current I_{sc} and the open-circuit voltage V_{oc} are given by the interceptions with the axes. We yield open-circuit voltages of more than

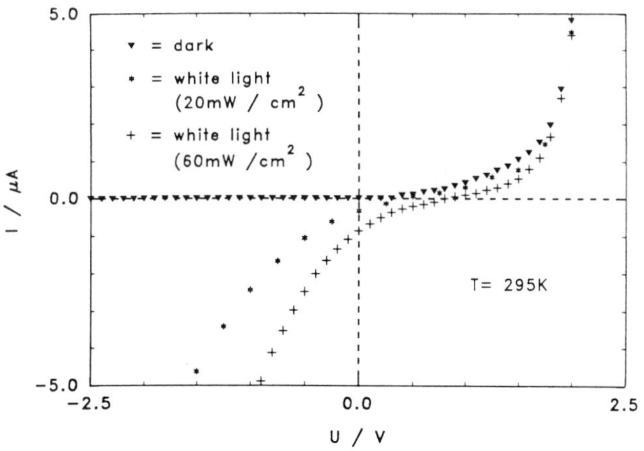

Fig. 2. *I–V* characteristics at room temperature in the dark and under white light illumination for an Al/PPV/ITO diode (thickness 0.5 μm).

1 V and under best conditions the short-circuit current density is 80 μA/cm^2 at a white light illumination of 100 mW/cm^2. Taking reflection losses into account, the power conversion efficiency of this solar cell is in the order of 0.1%. The low short-circuit current points to a significant influence of a serial resistance.

Impedance spectroscopy

Detailed information about the bulk resistance and the Schottky junction can be obtained by complex impedance measurements. Schottky barrier diodes are commonly analysed by using an equivalent circuit as shown in the insert of Fig. 3. C_j and R_j characterize the junction, whereas C_b and R_b represent the undepleted bulk. The complex plane representation of this equivalent circuit yields two perfect semicircles with diameters corresponding to the ohmic resistances. In our case all quantities are field, frequency and temperature dependent and therefore slight deviations from this ideal behaviour are expected. Figure 3 shows real and imaginary parts of the complex impedance for different applied bias voltages. The right semicircle falls off rapidly as the device is brought further into forward bias; it represents the junction. The resistive layer of the undepleted bulk is practically not affected and is still present under strong forward bias.

At low frequencies and high temperatures the bulk can be neglected and the equivalent circuit reduces to the junction part. Then under reverse bias the measured capacitance corresponds to the junction capacitance, which can be expressed according to the Schottky model as

$$\frac{C}{A} = \left[\frac{q\epsilon_0\epsilon_r N_a}{2} \times \frac{1}{V_d + V}\right]^{1/2} \tag{1}$$

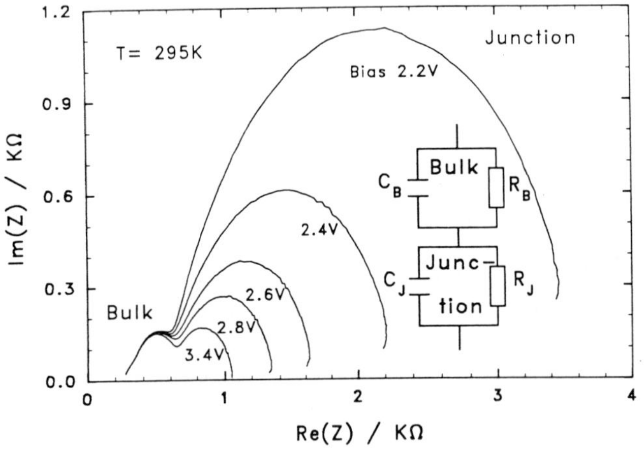

Fig. 3. Impedance plots under forward bias for an Al/PPV/ITO Schottky diode with an active area of 4 mm^2. The insert shows an equivalent circuit for the Schottky barrier.

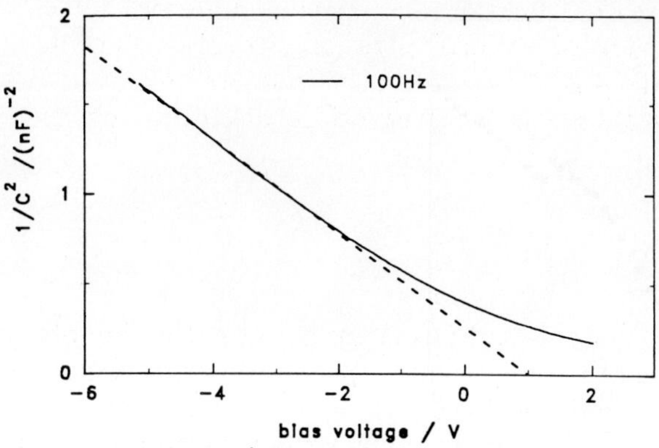

Fig. 4. $1/C^2$ vs. bias voltage of an Al/PPV/ITO device (area 4 mm^2).

giving a linear variation of $1/C^2$ with the bias voltage in reverse direction. In eqn. (1) A represents the device area, V_d the built-in potential or diffusion voltage and N_a the ionized acceptor dopant concentration. In Fig. 4 a plot of $1/C^2$ versus bias voltage measured at a frequency of 100 Hz is shown. The interception of the dashed line with the abscissa gives the diffusion voltage and the slope determines the acceptor dopant concentration. Inserting for the permittivity ϵ_r a measured value of 4.3, we find for this device values of $N_a \approx 10^{17}$ cm^{-3} and $V_d \approx 1$ V. Thus V_d corresponds to the maximum open-circuit voltage of the solar cells.

For an abrupt junction the Schottky barrier model yields for the depletion width

$$w = \left[\frac{2\epsilon_0 \epsilon_r (V + V_d)}{qN_a}\right]^{1/2} \tag{2}$$

Inserting the values for N_a and V_d we obtain a depletion width of $w \approx 0.1$ μm at zero bias, a reasonable value, compared to the film thickness of 0.5 μm.

Temperature dependent EL spectra and $I-V$ characteristics

Depending on the semiconducting properties of PPV, the film thickness and the rectifying electrode material (Al, Mg, Ca), visible luminescence appears at voltages exceeding a threshold typically 2–8 V in forward direction at room temperature. At 4 K the quantum yield of EL is about two orders of magnitude higher, but the threshold voltage increases significantly to about 30–60 V.

The EL spectra consist of several vibronic peaks moderately broadened by phonons and are essentially the same as those from photoluminescence, indicating that the same excited states are involved [2, 7].

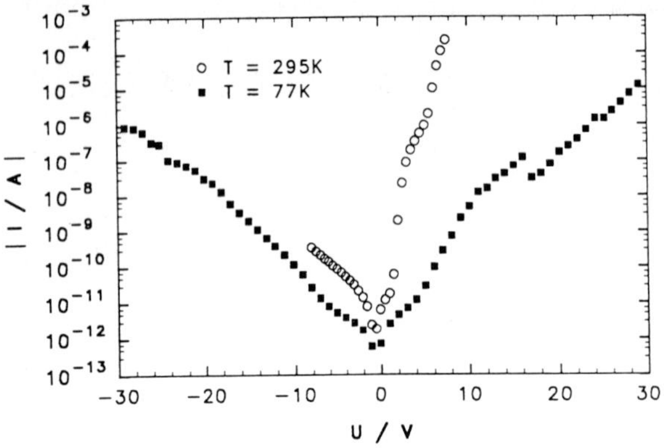

Fig. 5. *I–V* characteristics of an Al/PPV/ITO device at room temperature and at 77 K. The device area is about 4 mm^2.

I–V Characteristics of an Al/PPV/ITO device measured at room temperature and 77 K are presented in Fig. 5. While the characteristics at room temperature exhibit a diode behaviour with visible luminescence at about 4 V forward bias, the *I–V* characteristics at 77 K are almost symmetrical and consequently EL appears in both current directions, at this sample at about 35 V.

The approximately symmetrical behaviour of the device at low temperatures is confirmed by complex impedence measurements showing a voltage and frequency independent capacitance to be equal to the geometrical capacitance.

These results point out that the Schottky junction is no longer present and consequently not essential for EL at low temperatures. Hence the electrode material should not have a significant influence. We therefore investigated the *I–V* characteristics and EL of Au/PPV/Au and Al/PPV/Al devices at liquid nitrogen temperature. These devices showed symmetrical *I–V* characteristics combined with EL in both current directions above 30–60 V.

Conclusions

Depending on the temperature we observe two different mechanisms for EL in light emitting devices based on PPV. At room temperature a Schottky barrier exists in devices with the configuration Al/PPV/ITO. The radiative recombination of injected electrons from the Al electrode with holes of the p-type PPV is responsible for EL. With decreasing temperature the ionized acceptor dopant concentration gradually freezes out and the Schottky barrier vanishes. At high applied external electrical fields carriers can be injected from both metal electrodes into the PPV layer (independent from the polarity). This injection of carriers is responsible for the visible EL, which is now independent of the direction of the applied voltage and the electrode material.

A Schottky barrier is essential for light emission at low voltages at room temperature, however not fundamentally necessary for EL in conjugated polymers. Consequently, electroluminescence is a general feature of conjugated polymers and requires only a high electrical field. Detailed investigations of the influence of the purity of the basic PPV films and electrode materials on the quantum efficiency of the EL are in progress and will be reported later.

Acknowledgements

We thank J. Gmeiner and M. Meier for sample preparation and many helpful discussions. This work was supported by BASF/BMFT and Fonds der Chemischen Industrie.

References

1 J. H. Burroughes and R. H. Friend in J. J. Bredas and R. Silbey (eds.), *Conjugated Polymers*, Kluwer, Dordrecht, 1991, p. 555.
2 J. H. Burroughes, D. D. C. Bradley, A. R. Brown, R. N. Marks, K. Mackay, R. H. Friend, P. L. Burns and A. B. Holmes *Nature (London), 347* (1990) 539.
3 D. Braun and A. J. Heeger, *Appl. Phys. Lett., 58* (1991) 1982.
4 G. Grem, G. Leditzky, B. Ullrich and G. Leising, *Adv. Mater., 4* (1992) 36.
5 J. Gmeiner, S. Karg, M. Schwoerer and P. Strohriegl, in preparation.
6 S. M. Sze, *Physics of Semiconductor Devices*, Wiley, New York, 1981.
7 U. Rauscher, H. Bässler, D. D. C. Bradley and M. Hennecke, *Phys. Rev. B, 42* (1990) 9830.

Synthetic Metals, *54* (1993) 435–445

All-organic field-effect transistors made of π-conjugated oligomers and polymeric insulators

Gilles Horowitz*, Françoise Deloffre, Francis Garnier, Riadh Hajlaoui, Mohamed Hmyene and Abderrahim Yassar

Laboratoire des Matériaux Moléculaires, C.N.R.S., 2 rue Henry Dunant, F-94320 Thiais (France)

Abstract

All-organic field-effect transistors (FET) based on various oligothiophenes and insulators are reviewed. These devices generally work in the accumulation mode, but can also be operated in the depletion regime. The performances of organic FETs, as mirrored by their field-effect mobility μ_{FET}, are strongly dependent on the nature of both the semiconducting oligomer and the insulator. A general trend is the increase of μ_{FET} as the chain length increases. α-Substitution with alkyl chains also increases the mobility, whereas β-substitution leads to an almost complete disappearance of the field effect. The use of an organic insulator with a high dielectric constant, cyanoethylpullulane (CYEP), resulted in a considerable increase of μ_{FET}. All these behaviors are attributed to structural effects, and more specifically to changes in crystallinity. The study of the depletion regime gives indications on the concentration of dopants and free carriers. Free carriers are two to three orders of magnitude less numerous than dopants. Finally, measurements on freshly made and aged devices give information on the process of oxygen doping in thiophene oligomers.

Introduction

Considerable efforts have been devoted during the last decade towards the fabrication of organic-based electronic devices. A long term goal for organic materials in electronics is the realization of devices at the molecular scale. Turning to short term applications, organic field-effect transistors (FETs) have interest in devices where large areas are required, such as liquid-crystal displays (LCD). To date, the more widely used organic materials for making FETs have been the conjugated polymers [1–4]. More recently, π-conjugated oligomers have known a growing interest [5–9]. Advantages of oligomers reside mainly in their well-defined physical and chemical properties, as opposed to the often ill-defined conjugated polymers. Oligomers are also model molecules of their parent polymers [10]. Importantly, the influence of the chain length on the physical and chemical properties of the materials can be easily accessed with oligomers.

Figure 1 shows a schematic diagram of an organic FET. It is constructed according to the thin-film transistor (TFT) architecture. The

*Author to whom correspondence shoud be addressed.

0379-6779/93/$6.00

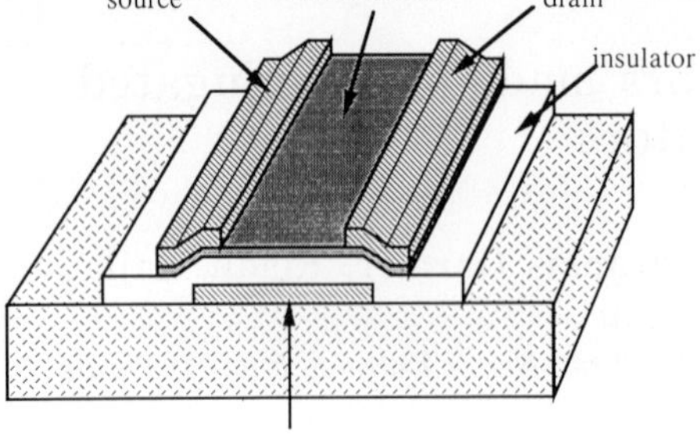

Fig. 1. View of a thin-film transistor (TFT).

semiconducting layer is constituted by a thin film of the organic compound. Contacts are taken through two metal electrodes, the source and drain, that form ohmic contacts to the organic semiconductor. The modulation of the source–drain current is operated through a third electrode, the gate, which is isolated from the semiconducting film by an insulating layer. In most reported organic FETs, the insulating material is thermally grown silicon oxide, but it can also be replaced by a polymeric organic material, thus realizing an all-organic FET [1, 7].

In this work, all-organic FETs made of various thiophene oligomers and polymeric insulators are described. The influence of the oligomer chain length, as well as various substitutions, is presented. The role of the insulator is reported. The results are discussed and related to structural effects.

Experimental

The synthesis of unsubstituted [10] and β-substituted [11] oligothiophenes has already been extensively described. It generally proceeds by homocoupling shorter oligomers. The organic FETs are made on glass substrates. The gate electrode (silver) is firstly sputtered as a 0.1 mm wide stripe. The insulating layer, which is then deposited, consists of either casted cyanoethylpullulane (CYEP) from an acetone solution, or spin-coated polymethylmethacrylate (PMMA) from toluene solution. CYEP was kindly donated by Shin-Etsu (Japan). PMMA was purchased from Aldrich. The third step consisted of vacuum evaporating the semiconducting oligothiophene. In the case of β-substituted oligomers, which are soluble, spin-coating from a chloroform solution has also been used. Finally, source and drain gold contacts were evaporated through a nickel mask that delimited two 50 μm wide stripes separated by 25 or 50 μm. The length of the stripes could be varied from 0.1 to 0.5 cm.

Current–voltage characteristics were measured with a Hewlett-Packard 4140B semiconductor d.c. parameter measurement system, monitored with an HP model 310 desktop computer. Capacitance measurements were carried out with an HP 4192A impedance analyzer.

Results and discussion

Determination of the parameters of an FET

The $I_d - V_d$ characteristic of a typical organic TFT is shown in Fig. 2. The device can operate in two different regimes. When the gate is biased negative to the source an accumulation layer forms at the semiconductor–insulator interface, owing to the fact that oligothiophenes, like the great majority of organic semiconductors, are p-type. The increase of the current can then be attributed to the increased concentration of charge carriers in the accumulation layer. At low drain voltages V_d, I_d varies linearly with V_d. As V_d approaches V_g, I_d tends to saturate due to the pinch off of the accumulation layer. Using the classical theory of field-effect transistors [12], the drain current is given by

$$I_d = \frac{W}{L} \mu_{FET} C_i (V_g - V_t) V_d \qquad (1)$$

in the linear regime and

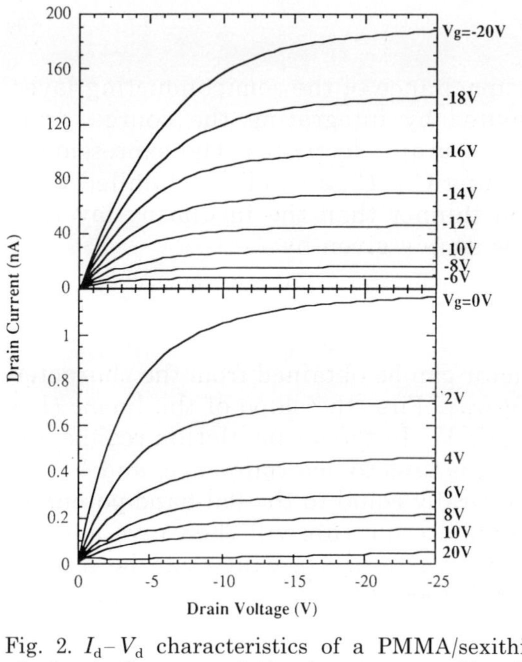

Fig. 2. $I_d - V_d$ characteristics of a PMMA/sexithiophene TFT. The device can operate whether in the accumulation (upper curves, $V_g < 0$) or the depletion (lower curves, $V_g > 0$) regime. 6T thickness: 25 nm; capacitance of the PMMA layer: 8 nF cm^{-2}; channel width and length: 25 μm and 2 mm. Note the different current scales in the two regimes.

$$I_{dsat} = \frac{W}{2L} \mu_{FET} \, C_i (V_g - V_t)^2 \tag{2}$$

in the saturation regime. Here, W and L are TFT width and length, respectively; μ_{FET} is the field-effect mobility, C_i the capacitance (per unit area) of the insulating layer, and V_t the threshold voltage. The important parameter that can be deduced from I_d–V_d measurements in the accumulation regime is then the mobility μ_{FET}. It can be estimated either from the slope of the I_d–V_d curve in the linear regime, or from the slope of a plot of $(I_{dsat})^{1/2}$ as a function of V_g.

When V_g is positive, the TFT operates in the depletion regime (lower curves in Fig. 2). In that case, I_d decreases as V_g increases. This decrease comes from the growing of an insulating depletion layer, the thickness of which increases as V_g increases (eqn. (3)).

$$W_{sc} = \frac{\varepsilon_s}{C_i} \left[\left(1 + \frac{2C_i^2 (V_g - V_{fb})}{qN\varepsilon_s} \right)^{1/2} - 1 \right] \tag{3}$$

Here ε_s is the dielectric constant of the semiconductor, V_{fb} the flat-band potential, q the absolute electron charge and N the dopant (acceptor or donor) concentration. As $V_g - V_{fb}$ increases, W_{sc} increases and I_d decreases, until W_{sc} reaches the total thickness d of the semiconductor layer. This occurs at the pinch-off voltage V_p, obtained by equating W_{sc} to d:

$$V_p = \frac{qNd^2}{2\varepsilon_s} \left[1 + 2\frac{C_s}{C_i} \right] \tag{4}$$

where $C_s = \varepsilon_s / d$ is the dielectric capacitance of the semiconducting layer. The drain current can be computed by integrating the source–drain conductance between the source and drain electrodes. The expression of I_d can be much simplified by assuming $C_i \ll C_s$, a condition fulfilled when the semiconducting layer is much thinner than the insulating layer. In that case, the current at low V_d is simply given by

$$I_d = \frac{WdqN\mu}{L} \left[1 - \frac{V_g - V_{fb}}{V_p} \right] V_d \tag{5}$$

The parameters of the semiconductor can be obtained from the characteristics of the TFT in the following way. First the slope of the linear (low V_d) regime is determined for various V_g. In the accumulation regime, the plot of this slope as a function of V_g is used to determine μ_{FET} and V_t (see eqn. (1)). This last parameter is generally equal to the flat-band potential. A similar plot in the depletion region (eqn. (5)) will then give another determination of the mobility, and a value of V_p, which can be used to estimate the dopant concentration N (eqn. (4)).

Role of the oligomer

Table 1 summarizes the characteristics of TFT made on thermally grown SiO_2 insulating layers. The semiconductors are unsubstituted and

TABLE 1

Conductivity and field-effect mobility of various unsubstituted and substituted oligothiophenes

Compound	Conductivity ($S\ cm^{-1}$)	Mobility ($cm^2\ V^{-1}\ s^{-1}$)
Unsubstituted *n*T	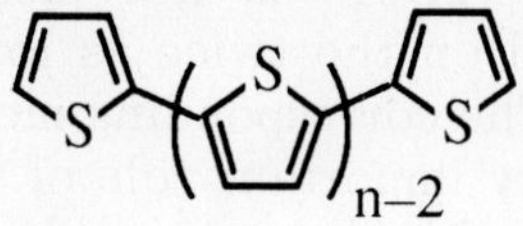	
3T	10^{-10}	not measurable
4T	10^{-9}	2×10^{-7}*
5T	10^{-8}	10^{-5}
6T	10^{-7}	2×10^{-3}
8T	10^{-7}	2×10^{-4}
Polythiophene	10^{-8}–10^{-6}	$10^{-5} - 10^{-4}$
'End-capped' *n*T	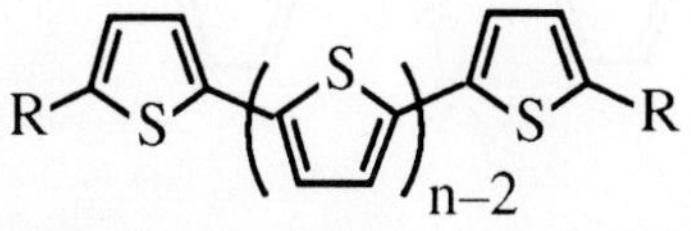	
Di-ethyl-3T	NA	1.9×10^{-7}*
Di-ethyl-4T	NA	5×10^{-5}*
Di-ethyl-5T	NA	9×10^{-4}*
Di-hexyl-6T	$\sigma_{\parallel}$: 10^{-5}; $\sigma_{\perp}$: 10^{-7}	1.5×10^{-2}
β-substituted *n*T	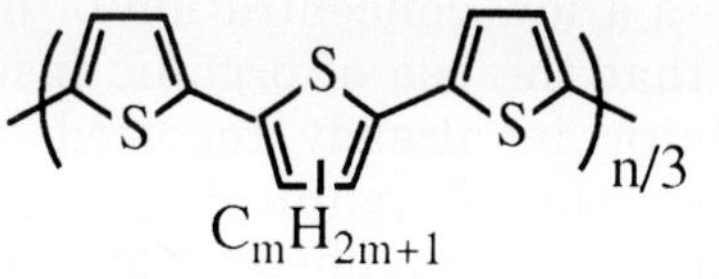	
Di-decyl-6T	10^{-13}	not measurable
Tetra-decyl-12T	10^{-9}	5×10^{-6}

Field-effect mobility was measured on FET with SiO_2 as the insulating material. Data with an asterisk are taken from ref. 9.

substituted oligothiophenes (*n*T, $n = 3, 4, 5, 6, 8, 12$). We have also mentioned in Table 1 results reported by Hotta and co-workers [9]. The effect of the length of the oligomer is observed on both the unsubstituted and substituted molecules. It is worth noting the increase of the field-effect mobility with the chain length is much faster than that of the conductivity.

It has been recently reported [9] that substitution in the α-position of alkyl groups ('end-capped' oligomers) considerably increases the field-effect mobility. We have synthesized α, α'-dihexyl-sexithiophene (DH6T) and can confirm this behaviour: a field-effect mobility of $2 \times 10^{-2}\ cm^2\ V^{-1}\ s^{-1}$ was obtained with this end-capped sexithiophene, a value ten times higher than that measured with unsubstituted 6T.

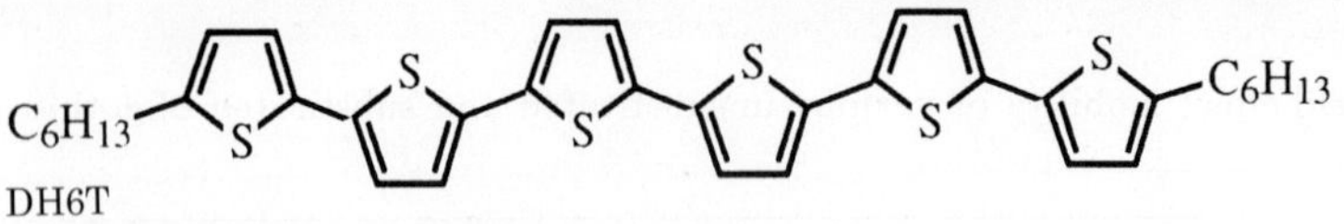

DH6T

The effect of β-substitution is the exact reverse: the field effect on β',β'''''-di-decyl-sexithiophene (2D6T) is not measurable. However, the field-effect could be detected on the twice as long oligomer, tetradecyl-duodecithiophene (4D12T). The corresponding mobility is lower than that of unsubstituted 6T. The very important role of α- and β-substitution on the charge transport in oligothiophene thin films is attributed to structural effects, as will be discussed further.

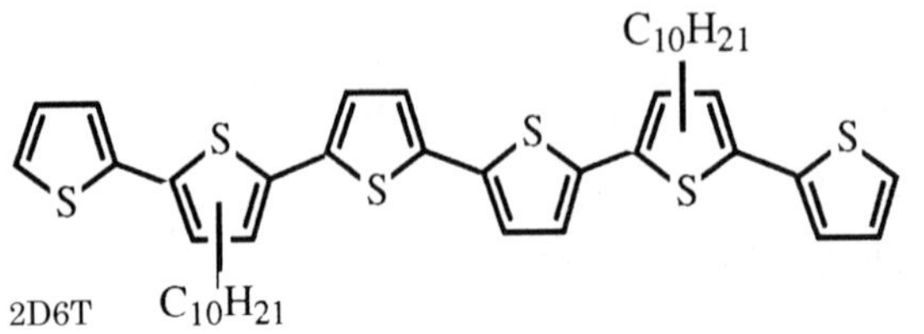

2D6T

Role of the insulator

The insulator plays an important role in a field-effect transistor. The properties required for a 'good' insulating material are (i) high dielectric constant, (ii) high resistivity and (iii) high dielectric strength. Furthermore, it should possess a low concentration of fixed and mobile charges.

We have shown that the use of organic insulators in place of SiO_2, the most used insulator in already reported organic FETs, generally increases the characteristics of organic devices [13]. As shown in Table 2, the field-effect mobility of sexithiophene is multiplied by a factor of about 2 when the insulator is polymethylmethacrylate (PPMA), and a factor of 50 with cyanoethylpullulane (CYEP), a cyanoethylated polysaccharide fabricated by Shin-Etsu, Japan. This last polymer presents a very high dielectric constant. Unfortunately, its other characteristics are rather poor, as can be seen in Table 2. The presence of mobile charges is also

TABLE 2

Electrical properties of various insulating materials used for making FET with sexithiophene as the semiconductor. The corresponding field-effect mobility is also indicated

Insulator	Conductivity ($S\ cm^{-1}$)	Dielectric strength ($V\ cm^{-1}$)	Dielectric constant	Field-effect mobility ($cm^2\ V^{-1}\ s^{-1}$)
SiO_2	10^{-14}–10^{-16}	10^7	3.9	2×10^{-3}
PMMA	10^{-14}–10^{-17}		2.5–4.5	4×10^{-3}
CYEP	2.5×10^{-13}	8×10^5	18.5	10^{-1}

PMMA: polymethylmethacrylate; CYEP: cyanoethylpullulane.

probable, as indicated by a large hysteresis in capacitance–voltage measurements. PMMA, which is already used in silicon-based devices, is much more suited for electronics. However, the very promising results obtained with CYEP are an indication that the search for other polymeric insulators must be continued.

Structural effects

The large effect of α- or β-substitution on the field-effect mobility, together with the influence of the nature of the insulating material, are clear indications of the main role of the structure of the oligomeric layer on its charge transport properties. There is today a convergent set of experimental evidences, mainly from X-ray measurements [14], for a high crystallinity in oligothiophene thin films. In a crystalline film, the oligothiophene molecules (up to the sexithiophene) appear as quasi-planar rigid rods oriented perpendicular to the surface. Unlike longer polythiophene chains, the most favorable direction for the charge transport is the one perpendicular to the rods, i.e. parallel to the surface. This is due to the proximity of the molecules in that direction. α-Substitution tends to enhance this anisotropy, and hence to improve the charge transport in the direction parallel to the surface. In contrast to α-substitution, β-substitution leads to a complete change of the crystalline organization of the film, as shown by the X-ray diffraction spectra of unsubstituted, α-substituted and β-substituted 6T films, reported in Fig. 3. The presence of long side chains tends to push the oligomer molecules apart, thus impeding the intermolecular charge transfer. Only with very long chains is the charge transport partly restored. However, duodecithiophene (12 thiophene rings) behaves more like a polymer than an oligomer. In particular, the rigidity and orientation of its chains (perpendicular, like in sexithiophene, or parallel to the insulator surface) have not yet been established.

Structural effects can also explain the decrease of the mobility from sexithiophene to octithiophene and polythiophenes. Structural defects, e.g. β-branching, are much more numerous in these insoluble compounds than in the shorter oligomers.

Operation in the depletion regime

As stated above, the measure of $I_d - V_d$ characteristics in the depletion regime may lead to the determination of the concentration of dopant. This is done by plotting the slope of the $I_d - V_d$ curve at low V_d as a function of the gate voltage, as shown in Fig. 4, where the effective conductivity σ_{eff} was computed from the slope of the $I_d - V_d$ curve at low V_d and the dimensions of the device. From eqns. (1) and (5), the plot should give a straight line, the slope of which is proportional to the mobility. Furthermore, this line should cross the x axis at V_t (with $V_t \approx V_{fb}$) in the accumulation regime ($V_g < 0$), and at $V_p + V_{fb}$ in the depletion regime ($V_g > 0$). The value of V_p can then be used to calculate the dopant concentration N.

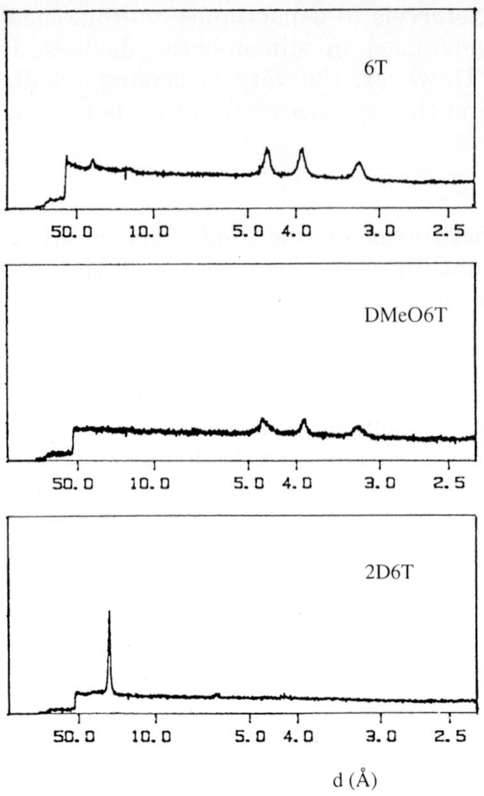

Fig. 3. X-ray diffraction spectra of unsubstituted (upper curve), α-substituted (dimethoxy, middle curve) and β-substituted (didecyl, lower curve) sexithiophene thin film.

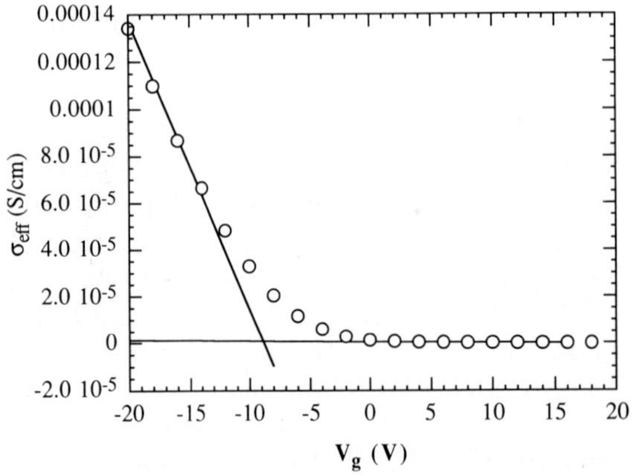

Fig. 4. Variation of the effective conductivity, calculated from the slope of the $I_d - V_d$ curves at low V_d, as a function of the gate voltage V_g.

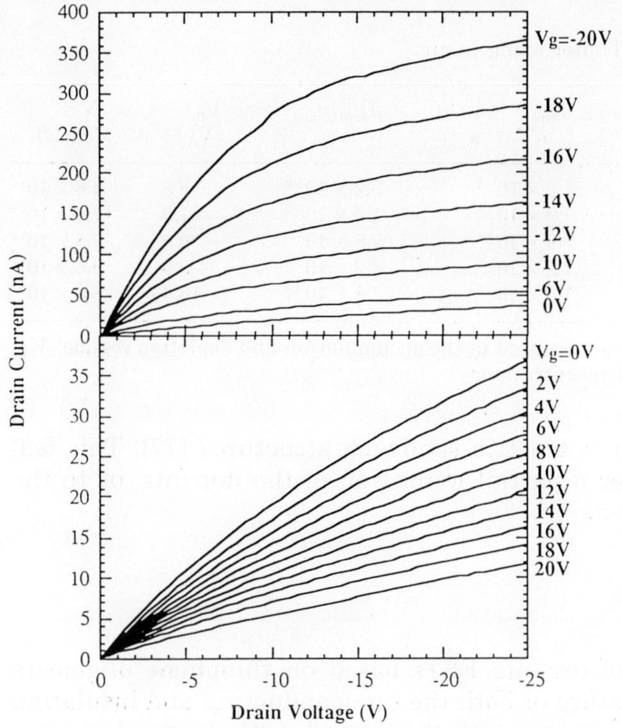

Fig. 5. $I_d - V_d$ characteristics of the same TFT as in Fig. 2 after 24 h in air.

Figure 5 shows the $I_d - V_d$ characteristic of the same TFT as in Fig. 1, after 24 h in air. The aged TFT is characterized by a residual current that adds to the saturation current in the accumulation regime, and cannot be completely suppressed in the depletion regime. We attribute this current to mobile charges in the 6T layer, probably coming from the well-documented oxygen doping of thiophene oligomers and polymers [15, 16]. The parameters deduced from the analysis of the characteristics of the freshly made and aged TFT are given in Table 3. Several interesting features of Table 3 can be discussed.

The oxygen doping clearly appears from the increase of the dopant concentration. Interestingly, N remains practically unchanged after the first exposition of the device to air.

The mobility calculated in the depletion regime is two to three orders of magnitude lower than the one obtained in the accumulation regime. This is due to the fact that we have assumed the concentration of free charge carriers to be equal to that of the dopants. The discrepancy can thus be raised by stating the ratio μ_{depl}/μ_{acc} corresponds to the ratio of the concentration of free carriers to the total carrier concentration. We have recently come to a similar conclusion from the analysis of current–

TABLE 3

Parameters of a PMMA/6T TFT after aging in air

Aging time (days)	μ_{acc} ($cm^2\ V^{-1}\ s^{-1}$)	μ_{depl} ($cm^2\ V^{-1}\ s^{-1}$)	μ_{depl}/μ_{acc}	V_{fb} (V)	N (cm^{-3})
0	3.8×10^{-3}	1.2×10^{-5}	3.2×10^{-3}	−8.8	4.8×10^{17}
1	5.4×10^{-3}	7.9×10^{-5}	1.5×10^{-2}	−7.9	7.0×10^{17}
12	5.0×10^{-3}	3.9×10^{-5}	7.8×10^{-3}	−10.0	7.8×10^{17}
60	3.6×10^{-3}	1.5×10^{-5}	4.2×10^{-3}	−9.9	9.6×10^{17}
81	3.3×10^{-3}	1.6×10^{-5}	4.8×10^{-3}	−10.4	8.8×10^{17}

μ_{acc} and μ_{depl}: field-effect mobility measured in the accumulation and depletion regime; V_{fb}: flat-band potential; N: dopant concentration.

voltage measurements on symmetric sandwich structures [17]. This feature may arise from either a partial ionization of the dopants, or to the presence of trapping defects.

Conclusions

The characteristics of organic FETs based on thiophene oligomers strongly depend on the nature of both the semiconducting and insulating materials. The mobility increases with the chain length. It also increases with α-substitution, whereas β-substitution kills the field effect. The mobility is improved by using organic insulators with high dielectric constants. The study of the depletion regime gives information on the concentration of both the doping centers and the free charge carriers. It appears that the latter is two to three orders of magnitude lower than the dopant concentration in unsubstituted sexithiophene.

Acknowledgements

This work was partly supported by the French MRT under contract No. 91 S 785. X-ray measurements were performed at Thomson L. C. R., Orsay, France.

References

1 F. Ebisawa, T. Kurokawa and S. Nara, *J. Appl. Phys.*, *54* (1983) 3255.
2 A. Tsumura, H. Koezuka and Y. Ando, *Synth. Met.*, *25* (1988) 11.
3 J. H. Burroughes, C. A. Jones and R. H. Friend, *Nature (London)*, *335* (1988) 137.
4 A. Assadi, C. Svensson, M. Willander and O. Inganäs, *Appl. Phys. Lett.*, *53* (1988) 195.
5 G. Horowitz, D. Fichou, X. Z. Peng, Z. G. Xu and F. Garnier, *Solid State Commun.*, *72* (1989) 381.

6 F. Garnier, G. Horowitz, X. Z. Peng and D. Fichou, *Adv. Mater.*, *2* (1990) 592.
7 G. Horowitz, X. Z. Peng, D. Fichou and F. Garnier, *J. Mol. Electron.*, *7* (1991) 85.
8 J. Paloheimo, P. Kuivalainen, H. Stubb, E. Vuorimaa and P. Yli-Lahti, *Appl. Phys. Lett.*, *56* (1990) 1157.
9 H. Akimichi, K. Waragai, S. Hotta, H. Kano and H. Sakati, *Appl. Phys. Lett.*, *58* (1991) 1500.
10 D. Fichou, G. Horowitz, B. Xu and F. Garnier, *Synth. Met.*, *39* (1990) 243.
11 D. Delabouglise, M. Hmyene, G. Horowitz, A. Yassar and F. Garnier, *Adv. Mater.*, *4* (1992) 107.
12 S. M. Sze, *Physics of Semiconductor Devices*, Wiley, New York, 1981.
13 X. Z. Peng, G. Horowitz, D. Fichou and F. Garnier, *Appl. Phys. Lett.*, *57* (1990) 2013.
14 S. Hotta and K. Waragai, *J. Mater. Chem.*, *1* (1991) 835.
15 G. Horowitz, X. Z. Peng, D. Fichou and F. Garnier, *J. Appl. Phys.*, *67* (1990) 528.
16 D. M. Taylor, H. L. Gomes, A. E. Underhill, S. Edge and P. I. Clemenson, *J. Phys. D*, *24* (1991) 2032.
17 G. Horowitz, D. Fichou, X. Z. Peng and P. Delannoy, *J. Phys. Fr.*, *51* (1990) 1489.

Synthetic Metals, 54 (1993) 447–452

Electrical characteristics of field-effect transistors formed with ordered α-sexithienyl

P. Ostoja*[,a], S. Guerri[a], S. Rossini[b], M. Servidori[a], C. Taliani[b] and R. Zamboni[b]
[a]*CNR, Ist. LAMEL,* [b]*CNR, Ist. ISM, via Castagnoli 1, 40126 Bologna (Italy)*

Abstract

X-ray observations on α-sexithienyl allowed us to reveal that this material, in the form of evaporated film, has an ordered crystalline structure, with an average dimension of the grains of 30 nm. Field-effect transistors, fabricated by using evaporated α-sexithienyl films as active element of the devices, showed good modulation characteristics, but low charge carrier mobilities. This fact is attributed to the small grain dimensions and it is suggested that better results could be obtained by using opportune evaporation substrates.

Introduction

Oligothiophenes are being investigated extensively as model compounds of the polythiophene. Moreover, their vacuum processability, together with their semiconducting properties, draws increasing attention because of their interesting electronic properties. In fact, the possibility of depositing these compounds in film form by vacuum evaporation permits the fabrication of prototype electronic devices, such as diodes [1] or field-effect transistors [2, 3].

A limitation in the performance of the devices made with organic compounds is their rather poor environmental stability [3–5] which prevents real applications, even if the almost innumerable possibilities of modification of these materials are likely to overcome this problem. On the other hand, it is possible to develop device fabrication procedures, in order to get more reliable devices.

A further limitation for electronic applications is the low mobility of the charge carriers which are injected into the active region of field-effect transistors. This low carrier mobility has been generally attributed to the low efficiency of the interchain charge hopping in conjugated polymers [6]. Mobility values which can be compared with those of hydrogenated amorphous silicon have been reported in the case of an all organic thin film transistor, fabricated by using α-sexithienyl as active layer [7]. This

*Author to whom correspondence should be addressed.

0379-6779/93/$6.00

result was attributed to the dielectric properties of the organic insulator used as gate insulating layer. On the other hand, it has been demonstrated that, in the case of alkyl-substituted oligothiophenes, simple methods, like solution casting and vacuum evaporation, allow ordered structures to be obtained [8]. Moreover, anisotropic conductivity measurements on crystals of these compounds showed that the carrier transport along the molecular long axis is the least advantageous, and electronic motion through the face-to-face molecular array is the dominant transport mechanism. This indicates that higher mobilities could be obtained by an opportune control of the structural order of the organic films. In fact, increased order is known to increase carrier mobilities: for example, the mobility of carriers in Durham polyacetylene is 1000 times larger in stretch-oriented films than in unoriented films [9].

Actually, it has been discovered that α-sexithienyl shows a thermotropic liquid crystalline phase with nematic order [10], and this offers a very important avenue of work to allow improved order and alignment in this material through liquid crystallinity and by the use of orienting substrates, this method being the one used for orienting this kind of material [11].

In this work we report on the structural characterization of evaporated α-sexithienyl films and on the electrical characterization of field-effect transistors, fabricated with these films.

Experimental

Details on the synthesis of α-sexithienyl and on the procedure of devices fabrication have been reported elsewhere [3]. In the present study, an accurate control of the temperature during evaporation allowed a good control of the evaporation rate and of the thicknesses of the evaporated films. Transistors with satisfactory characteristics were fabricated with film thickness varying between 5 and 100 nm, with different evaporation rates (from 1 up to 100 nm/min).

With the aim of preventing degradation phenomena [3–5], the organic film was protected with a cap of 100 nm of evaporated SiO_x. Before evaporation, the devices were kept in the same vacuum system at about 150 °C for 2 h.

X-ray analyses were performed with a Seifert MZ VI diffractometer, by using the conventional symmetric Bragg–Brentano geometry and an asymmetric, grazing incidence, configuration. For the latter, a thin film attachment, equipped with Soller slits (0.4° as acceptance angle) perpendicular to the scattering plane and a flat graphite monochromator (out-of-plane mosaicity of $0.27 \pm 0.02°$), was utilized in front of the detector. A fixed incidence angle of 0.5° between sample surface and incoming X-ray beam was chosen to make the signal from the crystalline film as high as possible. Cu Kα radiation was employed for both geometries. Sample rotation around the axis perpendicular to the sample surface was avoided, in order to ascertain the presence of possible preferred orientations in the grain distribution. The films for these structural observations were deposited using glass or fused silica as substrate. The evaporation procedure was the same as for the fabrication of the devices, but, in order to obtain significant X-ray spectra, the thickness was of the order of 100–200 nm.

Results and discussion

Figure 1(a) and (b) shows the diffraction patterns recorded with the asymmetric and symmetric geometries, respectively, on a film of α-sexithienyl, evaporated on a glass substrate up to a thickness of 125 nm. An anisotropic behaviour of the diffraction peak intensities is apparent from the Figure, which is directly related to an anisotropic distribution of the crystallites. In fact, while lattice planes inclined to the surface can diffract with an asymmetric configuration, only planes parallel to the film satisfy the Bragg condition in the symmetric geometry. If attention is paid to the first peak located at $2\theta = 3.84^\circ$, we see that its intensity increases by a factor of 30 on passing from Fig. 1(a) to (b). As the intensities of the 111 peak from a randomly oriented silicon powder (used as a control standard) were observed to have a ratio of about 8 under the same experimental conditions, we can conclude that lattice planes of 2.30 nm spacing are

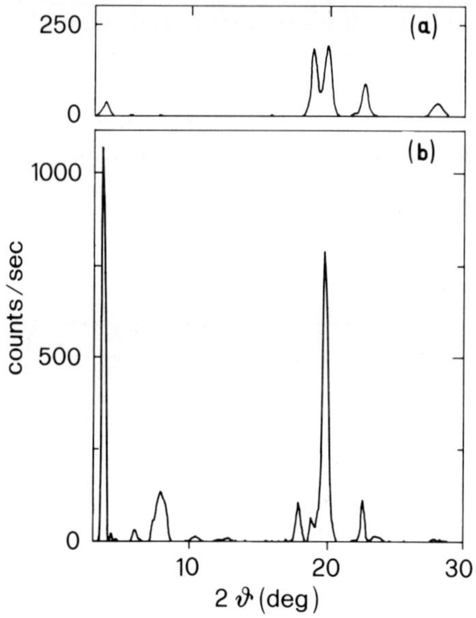

Fig. 1. X-ray diffraction patterns taken on an evaporated α-sexithienyl film at 0.5° grazing incidence (a) and symmetric Bragg–Brentano geometry (b). Note the anisotropy of the relative peak intensities in the two patterns.

mostly oriented parallel to the surface. The structure of α-sexithienyl was recently solved [12] as monoclinic with $a = 4.538$, $b = 0.7856$ and $c = 0.6027$ nm. Moreover, the molecular axis was found to make an angle of 32° with the long axis of the unit cell. This implies that the peak at $2\theta = 3.84°$ in Fig. 1 corresponds to the 200 reflection and that the a axis is perpendicular to the film surface. As to the crystallite size, the determination was made by evaluation of the full width at half maximum of the diffraction peaks. These values, after correction for the instrumental broadening and wavelength dispersion, gave a size of about 30 nm.

The patterns of Fig. 1 were drawn after background subtraction. The angular distribution of this diffuse intensity was the same as that produced by a bare glass substrate, which indicates that no appreciable amorphous phase is present in the films.

A regular layered structure, with high crystallinity, obtained by vacuum evaporation (or solution casting) has already been reported for dimethylquaterthiophene [8], and our observations confirm that even non-substituted oligothiophenes can be obtained in an ordered form even with simple evaporation methods. Films that we obtained at different evaporation rates (from 10 up to 100 nm/min) showed the same degree of structural order.

The electrical characteristics of field-effect transistors (FET), fabricated with an evaporation rate of about 10 nm/min, up to film thicknesses of 15 and 100 nm, are shown in Fig. 2(a) and (b), respectively. After the oligomer deposition, the devices were subjected to a heat treatment under vacuum for 2 h at about 150 °C. Thereafter a SiO_x layer was deposited by evaporation, up to a thickness of 100 nm. The devices protected with the SiO_x cap showed lower shifts in the electrical performance with respect to the unprotected ones. As can be deduced from Fig. 2, the lower film thickness determines better modulation conditions of the source–drain current I_{ds}, and lower values of I_{ds} for gate voltage $V_g = 0$. The electrical charge carrier mobilities, calculated from the transfer characteristics of the FETs, were

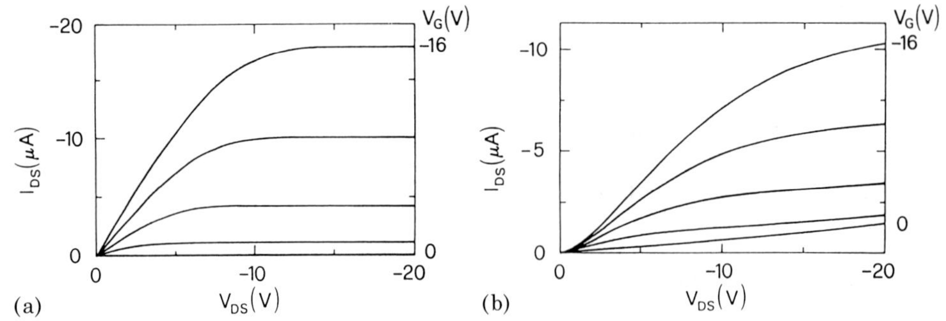

Fig. 2. Electrical characteristics of field-effect transistor with an α-sexithienyl film thickness of 15 nm (a) and 100 nm (b).

found to be relatively low (of the order of 10^{-5}–10^{-4} $cm^2/V\ s$). This indicates that, even in films presenting an ordered crystalline structure, the small (30 nm) dimension of the crystallites probably contributes to maintain the charge carrier mobilities at low values. Experiments performed in order to deposit α-sexithienyl films with larger crystallite dimensions by using orientation-inducing substrates, did not give satisfactory results until now. In fact, the use of thin films of polytetrafluoroethylene (PTFE), deposited mechanically on glass [13] did not seem to produce larger grains. Moreover, the anti-adhesive properties of PTFE cause a poor adhesion of the evaporated source–drain metal contacts. Clearly it will be necessary to investigate other materials, like, for instance, polyimides, which are widely used as orienting substrates in liquid-crystal displays, and which are compatible with the procedures of microelectronic technology.

Conclusions

In this paper we demonstrated that evaporated α-sexithienyl films present an ordered crystalline structure, with the molecular long axis perpendicular to the film surface. The small dimensions of the crystallites, however, do not allow the attainment of electrical mobility values of practical utility. In order to form α-sexithienyl films with larger grain sizes by using orientation-inducing substrates, further experimental work will be necessary.

Acknowledgements

The authors warmly thank Dr W. Porzio for permitting the use of his unpublished data on the α-sexithienyl cell structure. This work has been financed by P.F. Materiali Speciali per Tecnologie Avanzare.

References

1 G. Horowitz, D. Fichou, X. Z. Peng and P. Delannoy, *J. Phys. (Paris)*, *51* (1990) 1489.
2 G. Horowitz, X. Z. Peng, D. Fichou and G. Garnier, *J. Appl. Phys.*, *19* (1990) 2013.
3 P. Ostoja, S. Guerri, M. Impronta, P. Zabberoni, R. Danieli, S. Rossini, C. Taliani and R. Zamboni, *Adv. Mater. Opt. Electron.*, *1* (1992) 127.
4 D. M. Taylor, H. L. Gomes, A. E. Underhill, S. Edge and P. I. Clemenson, *J. Phys. D*, *24* (1991) 2032.
5 E. Punkka, H. Isotalo, M. Ahlskog and H. Stubb, *Mater. Res. Proc.*, *Boston MA*, *1991*, in press.
6 F. Garnier, G. Horowitz, X. Z. Peng and D. Fichou, *Synth. Met.*, *45* (1991) 163.
7 F. Garnier, G. Horowitz, X. Z. Peng and D. Fichou, *Adv. Mater.*, *2* (1990) 592.
8 S. Hotta and K. Waragai, *J. Mater. Chem.*, *1* (1991) 835.

9 A. S. Dhindas, M. R. Brice, J. P. Lloyd and M. C. Petty, *Thin Solid Films, 165* (1988) L97.
10 C. Taliani, R. Zamboni, G. Ruani, S. Rossini and R. Lazzaroni, *J. Mol. Electron., 6* (1990) 225.
11 R. Friend, *Nature (London), 352* (1991) 377.
12 W. Porzio, S. Destri, M. Mascherpa, S. Rossini and S. Brückner, *Proc. ICSM '92, Göteborg; Synth. Met., 55–57* (1993) in press.
13 J. C. Wittmann and P. Smith, *Nature (London), 352* (1991) 414.

The application of polypyrrole as counterelectrode in electrolytic capacitors

L. H. M. Krings*, E. E. Havinga and J. J. T. M. Donkers
Philips Research Laboratories, P.O. Box 80000, 5600 JA Eindhoven (Netherlands)

F. T. A. Vork
Philips Components, Central Development Passive Components, P.O. Box 218, 5600 MD Eindhoven (Netherlands)

Abstract

Solid electrolyte capacitors in which the counterelectrode MnO_2 is replaced by polypyrrole have been prepared and investigated. The dopant counterions of the polypyrrole electrode were varied and the influence on capacitor performance related to differences in material properties. This was verified both for zero-hour characteristics as for lifetime experiments at 125 °C in air. Capacitors with naphthalenesulfonate as counterion were very stable and showed an excellent high frequency performance, comparable with a ceramic multilayer capacitor.

Introduction

Since the discovery of highly conducting polypyrrole (PPY) films, made by electrochemical polymerization [1], a lot of work has been done in the field of conducting polymers on both materials research and applications. The problem of intractability of most conducting polymers has been tackled with success [2, 3]. The environmental stability of heavily doped polymers has been strongly improved and progress is still being made [4, 5]. Due to this success more and more conceived applications such as sensors, displays, memories, FETs, diodes, secondary batteries and anti-static coatings have been shown to be feasible.

Another very promising application of conducting polymers is their use as counterelectrode in solid electrolyte capacitors. Conventional solid electrolyte capacitors consist of a highly etched and anodized anode and a solid electrolyte manganese dioxide (MnO_2) as the counterelectrode. These capacitors have the advantage of a high capacitance per unit volume at low price. A disadvantage is that they have rather poor high frequency properties. A high series resistance (R_s) leads to high losses at frequencies above 10 kHz.

To improve the performance of these capacitors the MnO_2 ($\sigma = 0.1$ S/cm) should be replaced by a material with a higher specific conductivity.

*Author to whom correspondence should be addressed.

0379-6779/93/$6.00

Sanyo succeeded in developing a capacitor with TCNQ ($\sigma = 1$ S/cm) as counterelectrode [6]. The higher specific conductivity of TCNQ results in an improvement of the frequency characteristic. The application of polypyrrole ($\sigma = 10-100$ S/cm) will lead to an even better performance. Matshushita succeeded in developing a polypyrrole capacitor using a combination of MnO_2 and polypyrrole [7].

In this paper the results of measurements on capacitors with only polypyrrole as counterelectrode are presented. We have tried to relate the material properties to the capacitor characteristics. Conductivity and stability of polypyrrole films with different counterions are compared with the performance of capacitors made thereof.

Experimental

Polypyrrole films by electrochemical polymerization

Films of polypyrrole with various counterions were prepared and studied. The films were made potentiostatically on a platinum electrode. A two-electrode configuration was used without a reference electrode. During polymerization the cell potential was kept constant at 2.5 V. The composition of the electrolyte was typically 0.3 mol/l pyrrole and 0.1 mol/l supporting electrolyte. The solvent used for all the polypyrrole films was demineralized water. As supporting electrolyte polystyrenesulfonate sodium salt (PSS), tetraethylammonium-*p*-toluenesulfonate (TOS) and 2-naphthalenesulfonate sodium salt (NS) were used. All materials were used as received. The specific conductivity of the films was measured using the standard four probe method.

The polypyrrole films were also used to study the environmental stability. The films were placed in an oven at 125 °C under air. The conductivity of the films was measured as a function of time.

Preparation of the capacitors

The capacitor substrates (etched and anodized) were used as received. The capacitance after anodization (measured in 30 wt.% NH_4NO_3 at 100 Hz) was 6.8 μF and the working voltage was 10 V.

The impregnation of the capacitors was carried out in two steps [8]. Figure 1 shows a schematic representation of the procedure. In the first step a thin conducting layer of polypyrrole was deposited on the oxide surface by chemical polymerization. The capacitor substrates were immersed in a reactive solution containing both the monomer pyrrole (0.25 mol/l) and the oxidizing agent $(NH_4)_2S_2O_8$(0.25 mol/l) dissolved in a 1:1 (vol.%) mixture of water and ethyleneglycol. This step was carried out under reduced pressure. After about 20 min the substrates were removed from the reaction mixture, washed with water and dried. The whole process was repeated twice.

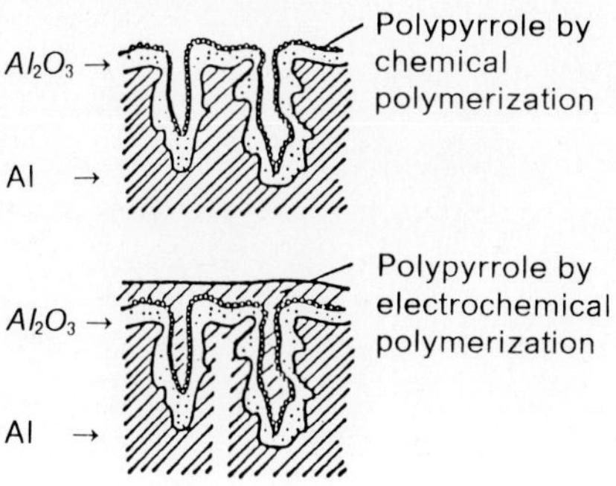

Fig. 1. Schematic representation of the impregnation process for the PPY capacitors.

The second step in the process was the electrochemical deposition of polypyrrole. The first layer served as the anode for the polymerization of pyrrole. The polymerization was carried out in a two-electrode configuration with a stainless steel cathode and no reference electrode. The pores were filled and a bulk layer of about 25 μm was formed on the outside.

The electrolyte during the electrochemical polymerization consisted of 0.3 mol/l pyrrole and 0.1 mol/l supporting electrolyte dissolved in demineralized water. The supporting electrolytes were the same as used for the preparation of the PPY films. The reaction was carried out potentiostatically with a cell potential of 2.5 V.

After impregnation the capacitors were provided with a layer of graphite and silver-epoxy. The anode and cathode wires were connected and finally the capacitor was encapsulated in an epoxy resin.

Results and discussion

Conductivity and stability of PPY films

Table 1 shows the specific conductivity values for the PPY films with various counterions. Only organic sulfonates were chosen because of their reported good environmental stability [9, 10]. TOS and NS show about the same value for specific conductivity while PSS is a factor of two lower.

TABLE 1

Thickness d, specific conductivity σ, degradation constant k and half-life $t_{1/2}$ for PPY films with various counterions

TOS	d (μm)	σ (S/cm)	k (h^{-1})	$t_{1/2}$(h)
PSS	16	12	0.0119	58.2
TOS	45	23	0.0146	47.5
NS	50	24	0.0059	117.5

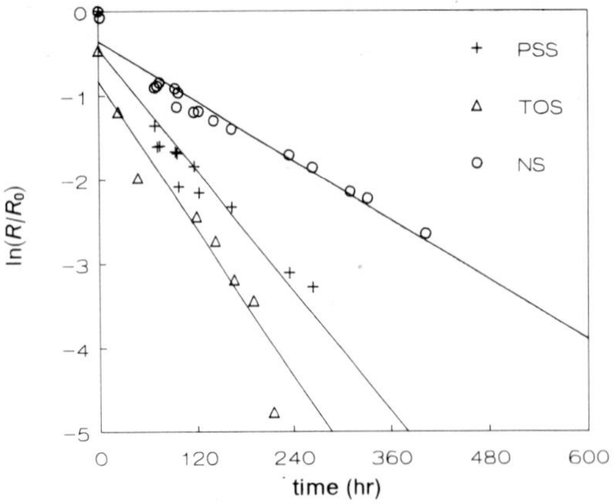

Fig. 2. The natural logarithm of the normalized resistance of the PPY films vs. time at 125 °C in air for three different counterions.

The films were used to study their environmental stability at 125 °C in air. In Fig. 2 the natural logarithm of the normalized resistance of the films R/R_0 is presented as a function of time. A straight line is obtained which implies that the resistance as a function of time can be described as a first order reaction

$$R = R_0 \times \exp(-k \times t)$$

where k is a constant depending on temperature [11]. In Table 1 the values of the degradation constant k and the half-life $t_{1/2} = (\ln 2)/k$ have been summarized. The value of $t_{1/2}$ for NS is about twice the value for PSS and TOS. Obviously the high molecular weight of PSS ($M_w = 70\,000$ g/mol) has no positive effect on the stability of PPY films compared to NS.

Characterization of the capacitors

Figure 3 shows the normalized capacitance C_s/C_{s0} (normalized at 100 Hz) as a function of frequency for the various counterions and for MnO_2. At low frequencies all three PPY capacitors follow the same slowly decreasing curve. Around 10^4 Hz PSS starts to deviate significantly resulting in a strong loss of capacitance. There is no significant difference between TOS and NS. On the other hand MnO_2 already deviates at low frequencies resulting in a loss of capacitance of more than 50%. This behaviour can be explained in terms of conductivity. The lower the specific conductivity the stronger the decrease in capacitance. Between 100 kHz and 1 MHz the capacitors reach the resonance frequency where they start to behave like an inductive element. In this region the capacitance is poorly defined.

In Fig. 4 the impedance Z is plotted as function of frequency for the NS capacitor. For comparison the characteristic of the MnO_2 capacitor and of a ceramic multilayer capacitor (CMC) are also plotted. The PPY

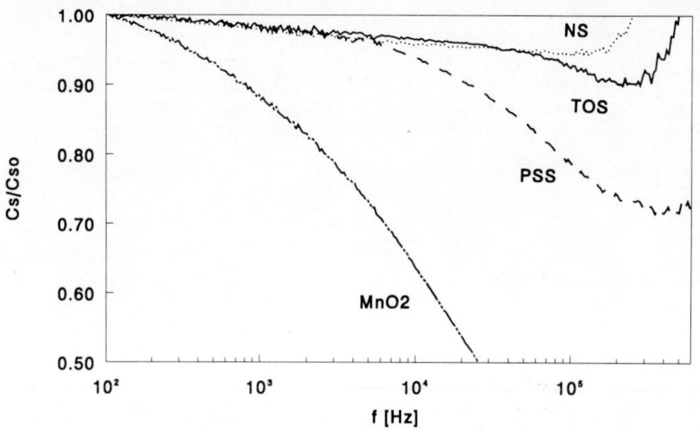

Fig. 3. The normalized capacitance C_s/C_{s0} vs. frequency for PPY capacitors with three different counterions.

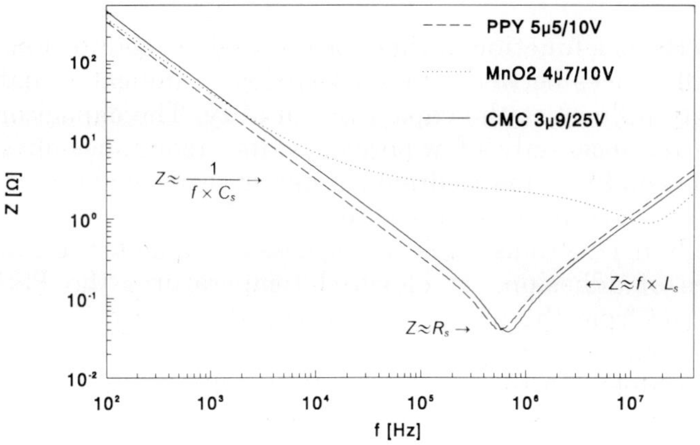

Fig. 4. Impedance Z vs. frequency for the PPY/NS capacitor compared to the MnO_2 and the ceramic multilayer capacitor (CMC).

capacitor behaves almost like an ideal capacitor, which means that the impedance Z is proportional to $1/fC_s$. The MnO_2 capacitor typically shows a non-ideal behaviour. Even at low frequencies the impedance is determined by both the capacitance and the series resistance. For the PPY capacitor the series resistance is only important around the minimum (resonance frequency). Above the resonance frequency the capacitor turns into an inductive element. The similarity in performance between the PPY capacitor and the CMC is very remarkable, especially considering the fact that CMCs are well known for their excellent high frequency characteristic.

Temperature stability of the capacitors

The capacitors were exposed to 125 °C in air for 600 h while a working voltage of 10 V was applied. Figure 5 shows the normalized capaci-

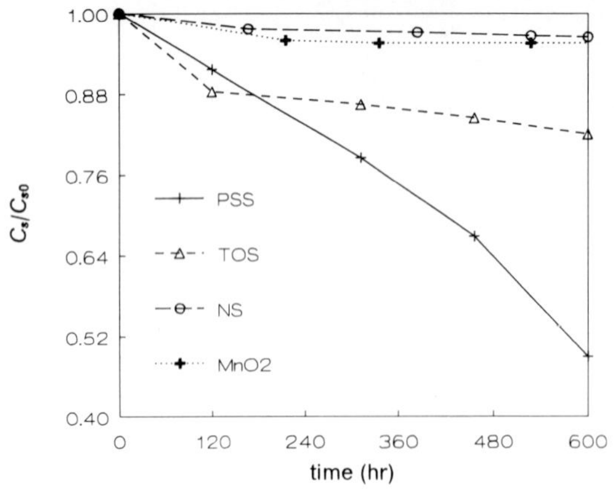

Fig. 5. The normalized capacitance C_s/C_{s0} as function of time for PPY capacitors, with different counterions, compared to the MnO_2 capacitor; $T = 125$ °C, in air, measured at 100 Hz.

tance measured at 100 Hz as a function of time for the various counterions and for MnO_2. The differences between the curves clearly indicates that the counterion strongly influences the capacitor stability. The capacitor with NS is very stable and loses only a few percent of its capacitance after 600 h. TOS is less stable and PSS shows a dramatic decrease of capacitance. This is an unexpected result considering the stability of the PPY films. PSS films are more stable than TOS films. This strong loss of capacitance can be explained in terms of adhesion. At elevated temperatures the PSS counterelectrode loosens from the oxide causing the strong decrease in capacitance. The same trend is found for the series resistance as a function of time (Fig. 6). It is interesting to see that there is no significant difference in stability between the PPY capacitor with NS and the MnO_2 capacitor.

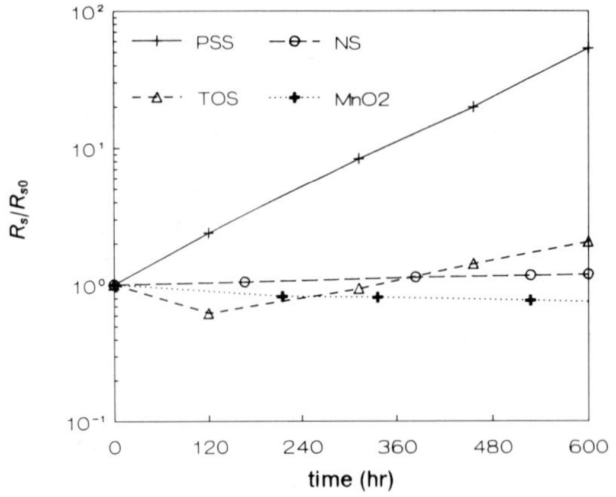

Fig. 6. The normalized resistance R_s/R_{s0} vs. frequency for PPY capacitors, with different counterions, compared to the MnO_2 capacitor; $T = 125$ °C, in air, measured at 100 Hz.

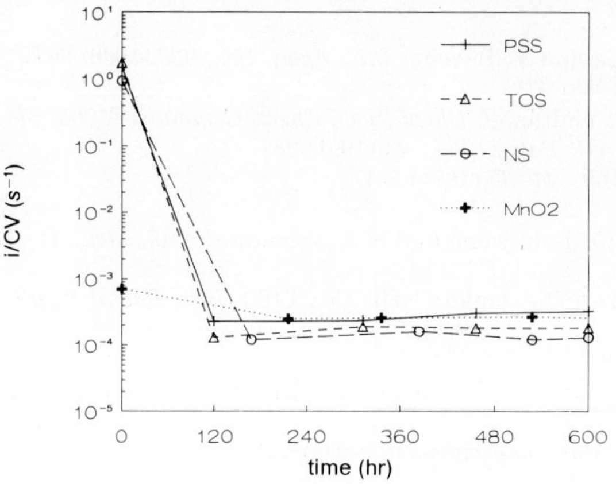

Fig. 7. The normalized leakage current i/CV vs. time for the PPY capacitors, with different counterions, compared to the MnO_2 capacitor; $T = 125$ °C, in air, measured at 100 Hz.

Figure 7 shows the behaviour of the leakage current during 600 h at 125 °C and 10 V working voltage. To compare the different types of capacitors the leakage current is normalized and plotted as i/CV (C = capacitance; V = working voltage). The PPY capacitors have a high initial leakage current. But after applying the working voltage the leakage current decreases dramatically by more than three orders of magnitude. The i/CV reaches a stable value equal to the MnO_2 capacitor independent of the type of counterion. This behaviour indicates that some kind of 'healing' (isolation or repair of weak spots in the oxide layer) occurs. The exact mechanism is not yet really understood. But the process of healing will prevent a catastrophic breakdown of the capacitor during lifetime. This is a very important requirement for reliability.

Conclusions

From the foregoing we can conclude that a substantial improvement of the high frequency performance of solid electrolyte capacitors can be realized by replacing MnO_2 by polypyrrole. The performance is even comparable with ceramic multilayer capacitors.

The characteristics strongly depend on the properties of the PPY counterelectrode and consequently on the inserted counterion. With NS the frequency performance and the stability at temperatures up to 125 °C in air are excellent.

There are indications that a 'healing mechanism' is active which prevents catastrophic breakdown of the capacitor. This is a basic requirement for a reliable capacitor performance.

References

1 A. Dall'Olio, G. Dascola, V. Varacca and V. Bocche, *C.R. Acad. Sci.*, *433* (1968) 267C.
2 G. L. Baker, *Adv. Chem. Ser.*, *28* (1988) 271.
3 N. Bates, M. Cross, R. Lines and D. Walton, *J. Chem. Soc.*, *Chem. Commun.*, (1985) 871.
4 T. R. Walton and R. F. Gratz, *J. Appl. Polym. Sci.*, *44* (1991) 387.
5 F. Jonas and L. Schrader, *Synth. Met.*, *41–43* (1991) 831.
6 S. Niwa, *Synth. Met.*, *18* (1987) 665.
7 Y. Kudoh, S. Tsuchiya, T. Kojima, M. Fukuyama and S. Yoshimura, *Synth. Met.*, *41–43* (1991) 1133.
8 M. Fukuda, H. Yamamoto and I. Isa (The Japan Carlit Co., Ltd.), *Eur. Patent Applic. No. 274 755* (1987).
9 H. Münstedt, *Polymer*, *27* (1986) 899.
10 T. Skotheim, M. Velasquez Rosenthal and C. H. Linkous, *J. Chem. Soc.*, *Chem Commun.*, (1985) 612.
11 L. A. Samuelson and M. A. Drury, *Macromolecules*, *19* (1986) 824.

Author Index

Subject Index